Methods of
Electronic Structure Theory

MODERN THEORETICAL CHEMISTRY

Editors: **William H. Miller,** *University of California, Berkeley*
Henry F. Schaefer III, *University of California, Berkeley*
Bruce J. Berne, *Columbia University, New York*
Gerald A. Segal, *University of Southern California, Los Angeles*

Methods of
Electronic Structure Theory

Edited by
Henry F. Schaefer III
University of California, Berkeley

PLENUM PRESS · NEW YORK AND LONDON

Library of Congress Cataloging in Publication Data

Main entry under title:

Methods of electronic structure theory.

(Modern theoretical chemistry; v. 3)
Includes bibliographical references and indexes.
1. Molecular theory. 2. Electrons. I. Schaefer, Henry F. II. Series.
QD461.M578   541'.28   76-46965
ISBN 0-306-33503-4 (v. 3)

© 1977 Plenum Press, New York
A Division of Plenum Publishing Corporation
227 West 17th Street, New York, N.Y. 10011

Printed in the United States of America

Contributors

C. F. Bender, Lawrence Livermore Laboratory, University of California, Livermore, California

Frank W. Bobrowicz, Arthur Amos Noyes Laboratory of Chemical Physics, California Institute of Technology, Pasadena, California and Batelle Memorial Institute, Columbus, Ohio

G. Das, Chemistry Division, Argonne National Laboratory, Argonne, Illinois

Thom. H. Dunning, Jr., Theoretical Division, Los Alamos Scientific Laboratory, University of California, Los Alamos, New Mexico

Arthur A. Frost, Department of Chemistry, Northwestern University, Evanston, Illinois

William A. Goddard III, Arthur Amos Noyes Laboratory of Chemical Physics, California Institute of Technology, Pasadena, California

R. F. Hausman, Jr., Lawrence Livermore Laboratory, University of California, Livermore, California

P. Jeffrey Hay, Theoretical Division, Los Alamos Scientific Laboratory, University of California, Los Alamos, New Mexico

Werner Kutzelnigg, Lehrstuhl für Theoretische Chemie, Ruhr-Universität, Bochum, Germany

Clyde W. McCurdy, Jr., Arthur Amos Noyes Laboratory of Chemical Physics, California Institute of Technology, Pasadena, California

Vincent McKoy, Arthur Amos Noyes Laboratory of Chemical Physics, California Institute of Technology, Pasadena, California

Wilfried Meyer, Institut für Physikalische Chemie, Johannes Gutenberg Universität, Mainz, Germany

Jules W. Moskowitz, Chemistry Department, New York University, New York, New York

Thomas N. Rescigno, Arthur Amos Noyes Laboratory of Chemical Physics, California Institute of Technology, Pasadena, California

Björn O. Roos, Institute of Theoretical Physics, University of Stockholm, Stockholm, Sweden

Isaiah Shavitt, Battelle Memorial Institute, Columbus, Ohio and Department of Chemistry, The Ohio State University, Columbus, Ohio

Per E. M. Siegbahn, Institute of Theoretical Physics, University of Stockholm, Stockholm, Sweden

Lawrence C. Snyder, Bell Laboratories, Murray Hill, New Jersey

Arnold C. Wahl, Chemistry Division, Argonne National Laboratory, Argonne, Illinois

Danny L. Yeager, Arthur Amos Noyes Laboratory of Chemical Physics, California Institute of Technology, Pasadena, California

Preface

These two volumes deal with the quantum theory of the electronic structure of molecules. Implicit in the term *ab initio* is the notion that approximate solutions of Schrödinger's equation are sought "from the beginning," i.e., without recourse to experimental data. From a more pragmatic viewpoint, the distinguishing feature of *ab initio* theory is usually the fact that no approximations are involved in the evaluation of the required molecular integrals. Consistent with current activity in the field, the first of these two volumes contains chapters dealing with methods *per se*, while the second concerns the application of these methods to problems of chemical interest. In a sense, the motivation for these volumes has been the spectacular recent success of *ab initio* theory in resolving important chemical questions. However, these applications have only become possible through the less visible but equally important efforts of those developing new theoretical and computational methods and models.

Henry F. Schaefer

Contents

Chapter 1.   Gaussian Basis Sets for Molecular Calculations

Thom. H. Dunning, Jr. and P. Jeffrey Hay

Chapter 4. The Self-Consistent Field Equations for Generalized Valence Bond and Open-Shell Hartree–Fock Wave Functions

Frank W. Bobrowicz and William A. Goddard III

Chapter 5. Pair Correlation Theories

Werner Kutzelnigg

Chapter 6. The Method of Configuration Interaction
Isaiah Shavitt

Chapter 7.   The Direct Configuration Interaction Method from Molecular Integrals
Björn O. Roos and Per E. M. Siegbahn

Chapter 8.   A New Method for Determining Configuration Interaction Wave Functions for the Electronic States of Atoms and Molecules: The Vector Method
R. F. Hausman, Jr. and C. F. Bender

Chapter 9.   The Equations of Motion Method: An
Approach to the Dynamical Properties of
Atoms and Molecules

Clyde W. McCurdy, Jr., Thomas N. Rescigno,
Danny L. Yeager, and Vincent McKoy

Chapter 10.   POLYATOM: A General Computer Program
for *Ab Initio* Calculations

Jules W. Moskowitz and Lawrence C. Snyder

Chapter 11.   Configuration Expansion by Means of Pseudonatural Orbitals

Wilfried Meyer

Contents of Volume 4

(continued)

Gaussian Basis Sets for Molecular Calculations

Thom. H. Dunning, Jr.
and
P. Jeffrey Hay

1. Introduction

In the following chapters the electronic structure of molecules will be discussed and the techniques of electronic structure calculations presented. Without exception the molecular electronic wave functions will be expanded in some convenient, but physically motivated, set of one-electron functions. Since the computational effort strongly depends on the number of expansion functions (see, e.g., the following chapters), the set of functions must be limited as far as possible without adversely affecting the accuracy of the wave functions. This chapter will discuss the choice of such functions for molecular calculations.

1.1. Slater Functions and the Hydrogen Molecule

To begin, let us consider the hydrogen molecule as a prototypical molecular system. In this it must be recognized that some features of the electronic structure of the hydrogen molecule are unique. Nonetheless, this simple molecule can be profitably exploited in order to develop many of the basic ideas about the use of one-electron functions to expand the electronic wave functions of molecules.

Thom. H. Dunning, Jr. and P. Jeffrey Hay • Theoretical Division, Los Alamos Scientific Laboratory, University of California, Los Alamos, New Mexico

At infinite internuclear separation the exact electronic wave function for the ground state of the hydrogen molecule is[1]

$$\Psi_{H_2}(1, 2) = \frac{1}{2^{1/2}} \mathscr{A} 1s_l(1)1s_r(2)[\alpha(1)\beta(2) - \beta(1)\alpha(2)] \tag{1}$$

(see also Chapter 4). In Eq. (1), $1s_l$ and $1s_r$ are hydrogen atom $1s$ orbitals,

$$1s = \left(\frac{1}{\pi}\right)^{1/2} e^{-r} \tag{2}$$

centered on the left (l) and right (r) nuclei. From Eq. (1) it is clear that we must first understand how to represent the wave functions of atoms before we will be able to understand how to represent the wave functions of molecules. Since the exact wave functions for the hydrogen atom are known, this is straightforward in the present case. For many-electron atoms this is not a trivial task and will be discussed in Section 2.

From Eq. (1) we see that for the hydrogen molecule, functions of the general form

$$r^{n-1} e^{-\zeta r} Y_{lm}(\theta, \phi) \tag{3}$$

are most appropriate for expanding the molecular wave function, at least at infinite internuclear separation. The functions (3) are referred to as *Slater functions*.[2-5]

At finite values of the internuclear distance the exact wave function for hydrogen is no longer of the form of Eq. (1). However, it has been found (see, e.g., Chapter 4), that Eq. (1) is an excellent approximation to the molecular wave function if the orbitals are adjusted to take into account the effects of molecular formation. To see what adjustments need to be made in the orbitals, we first note that the wave function (1) yields a dissociation energy (D_e) for the ground state of H_2 of 3.14 eV.* The experimental D_e is 4.72 eV, so that (1) results in an error of 33%. The most important change in the hydrogenic $1s$ orbital is a change in the size of the orbital as reflected in the orbital exponent ζ in (3). Minimizing the energy of H_2 with respect to the exponent leads to

$$\zeta_e = 1.17$$

$$D_e = 3.76 \text{ eV}$$

where ζ_e is the optimum exponent at the calculated equilibrium internuclear distance (R_e). Thus, optimizing the orbital exponent of the $1s$ orbital reduces the error in D_e by 39%.

---

*Since chemists are primarily interested in energy differences rather than total energies, we concentrate here on $D_e = E_{atoms} - E_{molecule}(R_e)$, where R_e is the equilibrium internuclear distance.

This same effect can be accounted for by using two functions on each center rather than just one, i.e.,

$$1s_i \rightarrow c_1 1s_i(\zeta_1) + c_2 1s_i(\zeta_2) \qquad (4)$$

The use of (4) replaces optimization of the nonlinear parameter ζ in (3) by optimization of the linear parameters (c_1, c_2), a far easier task as will be seen in later chapters. Taking $\zeta_1 = 1.0$ (the hydrogenic value) and $\zeta_2 = 1.5$ (to allow for contraction of the orbital), which are reasonable but certainly not optimum choices for the two exponents, we obtain

$$c_1 = 0.663 \qquad c_2 = 0.351$$

$$D_e = 3.74 \text{ eV}$$

in good agreement with the dissociation energy above.

Next in importance is polarization of the $1s$ orbital by inclusion of a $2p\sigma$ function, i.e.,

$$1s_i \rightarrow c_1 1s_i(\zeta_s) + c_2 2p\sigma_i(\zeta_p) \qquad (5)$$

The orbital (5) has more amplitude in the $+z$ direction than in the $-z$ direction, thus allowing a buildup of charge in the bonding region. Optimizing the parameters in (5) leads to[7]

$$\zeta_s = 1.19 \qquad \zeta_p = 2\zeta_s = 2.38$$

$$c_1 = 0.995 \qquad c_2 = 0.100$$

$$D_e = 4.02 \text{ eV}$$

Finally, it is possible (see Chapter 4) to allow the orbital to be a completely general function[8,9] and to determine the best form for this function. With this we obtain $D_e = 4.12$ eV, which is 87% of the experimental dissociation energy.

The effects illustrated above for H_2 are all important in describing the response of any atom to molecular formation. In general, however, we will be expanding the atomic orbitals in more than just a single function so that exponent optimization will be less important. The use of polarization functions remains of great importance for, as we shall see in later chapters, we cannot obtain semiquantitative accuracy in most molecules without the inclusion of polarization functions.

1.2. Gaussian Functions and the Hydrogen Atom

The above discussion presented one type of function often used in atomic and molecular calculations, the Slater function, and such functions have much to recommend them. Unfortunately, for molecular calculations the resulting many-center two-electron integrals are difficult, and therefore time consuming, to evaluate. To bypass the molecular integral problem, Boys[10] in 1950 suggested the use of functions of the form

$$x^l y^m z^n e^{-\zeta r^2} \qquad (6)$$

The functions (6) are referred to as *Gaussian functions*. For Gaussian functions all of the integrals needed for molecular calculations can be computed from simple formulas. As expected, Gaussian functions are less appropriate for expanding the electronic wave functions of atoms and molecules than Slater functions. This necessitates the use of a larger set of functions to obtain equivalent results.

As an example let us consider the use of Gaussian functions to describe the ground state of the hydrogen atom. Using one function and optimizing the orbital exponent, we obtain[11]

$$\zeta = 0.283$$

$$E = -0.42441 \text{ hartrees}$$

Since the energy of the hydrogen atom is -0.5 hartrees, the use of only one Gaussian function results in an error of 15%. This is inadequate for a system as simple as the hydrogen atom. However, using four functions and optimizing both the exponents and the expansion coefficients, we obtain[11]

$$E = -0.49928 \text{ hartrees}$$

which is in error by only 0.14%. Thus, it is possible to obtain an accurate description of the hydrogen atom and, as we shall see in the next section, of any atom if a sufficient number of Gaussian functions are used.

Gaussian lobe functions[12] can be used instead of the Cartesian functions (6). In the lobe method all functions, including p, d, etc., are expanded as linear combinations of $1s$ Gaussian functions, e.g., a $2pz$ function is represented by

$$N\{\exp\{-\zeta[x^2+y^2+(z-\delta)^2]\}-\exp\{-\zeta[x^2+y^2+(z+\delta)^2]\}\} \qquad (7)$$

In (7), N is a normalization factor and δ is the displacement of the $1s$ functions from the nucleus (at $x = y = z = 0$). The integrals over all functions can now be expressed in terms of the simple formulas for integrals over $1s$ functions. Lobe functions have been calculated for the hydrogen atom and are comparable in accuracy to Cartesian Gaussian functions.[13] The use of lobe expansions is, however, less effective for the representation of Cartesian functions for which $l+m+n \geq 2$ in (6), i.e., $3d$, $4f$, etc., functions. Lobe and Cartesian Gaussian functions have also been compared for the second row atoms.[16]

The use of the larger Gaussian basis sets increases the computer requirements for molecular calculations in two ways: (1) the numbers of (two-electron) integrals which must be computed increases rapidly with the number of functions (as N^4) and (2) the calculation of the electronic wave function is proportional to the number of integrals. (1) above presents no problem because integrals over Gaussian functions can be rapidly computed. Fortunately, there is also an easy solution to (2). For example, in a molecular calculation the four-term expansion of the hydrogen atom $1s$ orbital discussed above could be treated as only one function, the integrals being stored only for this *contracted* function. The procedure of contracting a large Gaussian basis set into a compact, but suitably flexible, set of expansion functions is the basis for the modern, efficient use of Gaussian functions in molecular calculations. As we shall see in later sections, however, care must be exercised when contracting atomic basis sets for use in molecular calculations.

2. Hartree–Fock Calculations on the First Row Atoms

As noted in Section 1, a convenient set of functions for molecular calculations is the set of Cartesian Gaussian functions

$$N_{lmn\zeta}x^l y^m z^n e^{-\zeta r^2} \tag{8}$$

In (6), $N_{lmn\zeta}$ is a normalization factor. In contrast to Slater functions, where $1s$, $2s$, $3s$, ... functions behave as $e^{-\zeta r}$, $re^{-\zeta r}$, $r^2 e^{-\zeta r}$, ..., all s Gaussian functions are taken to behave as $e^{-\zeta r^2}$, i.e., only $1s$ Gaussian functions are used. Similarly, all p_z functions are taken to be of the form $ze^{-\zeta r^2}$, i.e., as $2p_z$ functions; and all d_{xy} functions of the form $xye^{-\zeta r^2}$, i.e., $3d_{xy}$ functions. This choice has been found to be entirely adequate, even when representing higher atomic orbitals, e.g., $2s$, $3s$, etc., orbitals, and result[6] in substantial simplifications in the integral formulas. Note also that in (6) there are six $l+m+n=2$ Cartesian Gaussian functions. Five combinations of the functions (xy, xz, yz, x^2-y^2, and $3z^2-r^2$) are $3d$ functions and the sixth combination ($x^2+y^2+z^2$) is a $3s$ function.

One method of determining the exponents of a set of Gaussian functions for use in atomic calculations consists of fitting a linear combination of Gaussian functions to a Slater function.[17–21] This method will be discussed further in Chapter 1 of Volume 4 of this series. The *even tempered* Gaussian basis[22] has exponents chosen in a geometric progression,

$$\zeta_k = \alpha\beta^k \qquad k=1, 2, 3, \ldots, M$$

and α and β are optimized to give the lowest energy. This procedure has the merit of having only two parameters to be optimized per symmetry type (s, p, d, etc.), regardless of the number of expansion functions.

In the present chapter we will concentrate on energy-optimized Gaussian functions where each exponent has been variationally optimized in an atomic

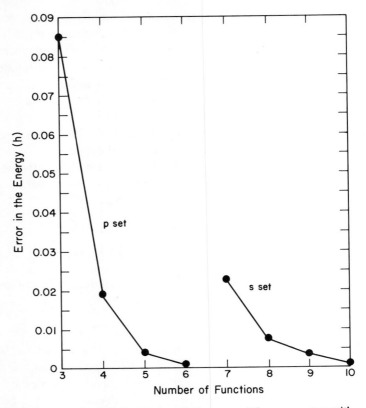

Fig. 1. Convergence of the Hartree–Fock energy of the oxygen atom with number of Gaussian functions.

calculation. In Figs. 1 and 2 we show the effect of increasing the number of Gaussian functions on the energy and properties of the oxygen atom (in the Hartree–Fock approximation). For the variation in the p set, the s set consisted of 11 functions; for the variation in the s set, the p set contained 7 functions. The $3p$ set is seen to yield unacceptably large errors; we shall, therefore, not consider this set further. The results of these, and other, calculations suggest that the $(9s5p)$ Gaussian set is a good compromise basis set for use in molecular calculations; it is capable of providing an accurate description of the atoms, while remaining of easily tractable size for many molecules. For the remainder of this section we shall concentrate on the use of the $(9s5p)$ set in atomic calculations.

2.1. Valence States of the First Row Atoms

The exponents for a given basis set are normally optimized only for the ground electronic state of the atom, e.g., the ^3P state of oxygen. In a molecule a

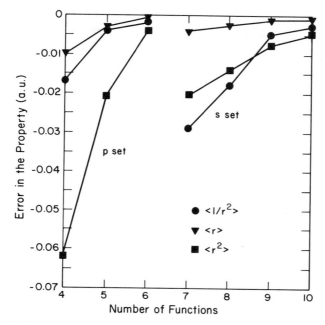

Fig. 2. Convergence of properties of the oxygen atom with number of Gaussian functions.

number of low-lying states of the atoms often contribute and so the atomic basis must also adequately represent such states. From calculations on the low-lying valence multiplets of the first row atoms, e.g., ^3P, ^1D, and ^1S for oxygen, we find that the $(9s\,5p)$ set gives excitation energies to within ± 0.05 eV of accurate Hartree–Fock results.

2.2. Rydberg States of the First Row Atoms

For the Rydberg states of the first row atoms, e.g., the $^5S(2s^2 2p^3 3s)$ state of the oxygen atom, the use of any of the preceding basis sets leads to errors of 5–25 eV (see Fig. 3). These *valence* basis sets must be augmented with diffuse functions appropriate for describing the Rydberg orbital.

Single s, p, and d exponents have been optimized to augment the $(9s\,5p)$ basis sets for the $n = 3$ and 4 Rydberg states of the first row atoms; these exponents are given in Table 1. The calculated excitation energies for the lowest $2p \to 3s$ transition *with* and *without* the diffuse functions are shown in Fig. 3. Although the errors are clearly unacceptable, 5–25 eV for the $(9s\,5p)$ basis alone, addition of the single Rydberg function gives excellent agreement (± 0.03 eV) with accurate Hartree–Fock calculations.

As noted in Section 1, a single function, without optimization, is often not sufficiently flexible to describe the distortions resulting from molecular

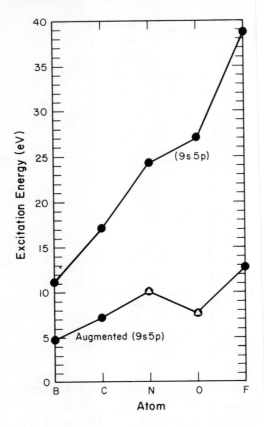

Fig. 3. Excitation energies for 3s Rydberg states of first row atoms. Triangles show results of accurate SCF calculations.

formation. As previously noted, this problem can be overcome by using more than one function. From calculations on Rydberg states with two diffuse functions we have derived splitting factors for converting the single exponents in Table 1 into two exponents. The splitting factors are

$$f_1 = 0.75$$
$$f_2 = 1.9$$

Table 1. Rydberg States: Exponents of the Functions to Augment the (9s5p) Basis Sets for the First Row Atoms

	Exponents				
Function	B	C	N	O	F
3s	0.019	0.023	0.028	0.032	0.036
3p	0.015	0.021	0.025	0.028	0.029
3d	0.015	0.015	0.015	0.015	0.015
4s	0.0047	0.0055	0.0066	0.0066	0.0066
4p	0.0041	0.0049	0.0051	0.0054	0.0054
4d	0.0032	0.0032	0.0032	0.0032	0.0032

and the two exponents (ζ_1, ζ_2) are related to the single exponent (ζ) by

$$\zeta_i = f_i \zeta, \qquad i = 1, 2$$

The splitting factors are, to a fair approximation, independent of the atom and of the state.

2.3. Positive Ions of the First Row Atoms

Ionization potentials calculated with the $(9s5p)$ basis agree quite well, within ± 0.05 eV, with accurate Hartree–Fock calculations.

2.4. Negative Ions of the First Row Atoms

Electron affinities obtained from calculations on the negative ions with the $(9s5p)$ set differ by as much as 0.5 eV from accurate Hartree–Fock calculations (see Fig. 4). As in the case of Rydberg excited states, it is necessary to augment the valence basis set with functions which can describe the expanded orbitals of the negative ion. Optimized $2p$ exponents to augment the $(9s5p)$ set are given in Table 2. (Addition of an s function has a negligible effect.) The exponents for these functions are seen to be intermediate between the valence and Rydberg exponents. Electron affinities calculated with the augmented basis sets are now in excellent agreement (± 0.04 eV) with accurate Hartree–Fock calculations (see Fig. 4).

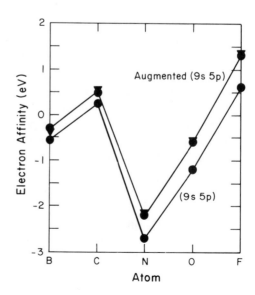

Fig. 4. Electron affinities for first row atoms. Triangles show results of accurate SCF calculations.

Table 2. *Negative Ions: Exponents of the 2p Functions to Augment the (9s5p) Basis Sets for the First Row Atoms*

Atom	Exponent
B	0.019
C	0.034
N	0.048
O	0.059
F	0.074

2.5. Contraction of Atomic Basis Sets

Although the basis sets discussed above provide accurate descriptions of the atoms, they are still not suitable for molecular calculations; even the modest $(9s5p)$ set would contain 24 functions per atom. These sets must first be *contracted* to a more convenient size. This is accomplished by taking the contracted basis functions to be fixed linear combinations of the Gaussian functions, now called the primitive functions or primitives, with the contraction coefficients taken from the atomic calculations.

One of the authors has given recommended contractions of the $(9s5p)$ primitive set to a *double zeta* $[4s2p]$ *set.*[24]*† In most instances the changes in the 1s orbital between the atom and the molecule are insignificant, so that a $(1s, 1s')$ representation of the core is unnecessary.‡ In Appendix 2 we give recommended contractions of the $(9s5p)$ sets to $[3s2p]$ sets where only the valence orbitals are described by two functions, i.e. $(1s, 2s, 2s', 2p, 2p')$. The total energies and properties obtained for the first row atoms with the $[3s2p]$ set are identical to those obtained with the $[4s2p]$ set.

Experience has shown that the contraction schemes most efficient for molecular calculations leave the outermost (smallest exponent) primitive functions uncontracted, making a single contracted function of the innermost (largest exponent) primitives. This permits the outer functions, having maximum amplitude in the interatomic regions, to respond to the changes which occur upon molecular formation. The inner functions describe regions which are largely atomic in character. The only difficulty arises when there is more than one atomic orbital of a given symmetry, e.g., the 1s and 2s orbitals of the first row atoms. In this case, some primitive functions may contribute to each of the orbitals in significant, but differing, ways. This function must then either be

---

*The term "double zeta" basis set refers to the use of two basis functions for each atomic orbital.
†Uncontracted sets are denoted by parentheses, while contracted sets are denoted by brackets.
‡An exception would be a positive ion with a 1s "hole."

left uncontracted, as in the $[4s2p]$ sets given earlier,[25] or included in more than one contracted function, as in the $[3s2p]$ sets given in Appendix 2.

In the $[3s2p]$ contracted sets given in Appendix 2, the four inner $2p$ primitive functions are combined into a single basis function $(2p)$ and the fifth $2p$ primitive is taken to be the second function $(2p')$. Similarly, the first eight $1s$ primitive functions make up the first two s basis functions $(1s, 2s)$ with the ninth primitive, the third s function $(2s')$, being left uncontracted. The seventh function contributes significantly to both the $1s$ and $2s$ atomic orbitals and so it has been included in both the $1s$ and $2s$ basis functions.

Evidence that the recommended contractions do not affect atomic excitation energies, ionization potentials, and electron affinities is given in Table 3. The results of the calculations on the oxygen atom with the $[3s2p]$ and augmented $[3s2p]$ sets compare well with accurate Hartree–Fock calculations.

The contraction schemes discussed above are *segmented* in the sense that, with minor exceptions, primitive functions occur in only one contracted basis function. If primitives are repeated, most integral programs recompute all of the integrals involved, see, e.g., Chapter 10. Recently a program has been developed which permits a *general* contraction in which the same primitive function may occur in numerous contracted functions at no expense in the integral computation time.[25] The use of the general contraction scheme effectively eliminates the problem of contracting atomic basis sets for use in molecular calculations. With this scheme all core orbitals can be represented by a single function, the atomic orbital. The valence basis set is then composed of one basis function which is a linear combination of all core primitive functions

Table 3. Calculations on the Oxygen Atom with the Contracted $(9s5p)/[3s2p]$ Gaussian Basis Set

Configuration	State	$(9s5p)/[3s2p]$		Accurate Hartree–Fock	
		E_{HF}, hartrees	ΔE, eV	E_{HF}, hartrees	ΔE, eV
Neutral states					
$2s^22p^4$	^3P	-74.79884	0.00	-74.80938^a	0.00
	^1D	-74.71845	2.19	-74.72921^a	2.18
	^1S	-74.59938	5.43	-74.61096^a	5.40
$2s^22p^33s$	^5S	-74.51734^b	7.66	-74.52980^c	7.61
$2s2p^5$	^3P	-74.17319	17.03	-74.18367^d	17.03
Ion states					
$2s^22p^5$	^2P	-74.77788^e	-0.57	-74.78948	-0.54
$2s^22p^3$	^4S	-74.36055	11.93	-74.37256	11.89

[a] H. F. Schaefer III, R. A. Klemm, and F. E. Harris, *J. Chem. Phys.* **51**, 4643–4650 (1969).
[b] Augmented with the $3s$ function in Table 1.
[c] P. S. Kelley, *Astrophys. J.* **140**, 1247–1268 (1964).
[d] C. C. J. Roothaan and P. S. Kelley, *Phys. Rev.* **131**, 1177–1182 (1963).
[e] Augmented with the $2p$ function of Table 2.

plus the innermost valence function, with the contraction coefficients taken from the valence atomic orbital. The outermost valence primitive functions are then left uncontracted. (Since the outermost valence functions are uncontracted, the core functions need contain no contribution from these functions.) In spirit, this scheme is analogous to the $[3s2p]$ sets given in Appendix 2.

3. *Hartree–Fock Calculations on Molecules Containing First Row Atoms*

In this section we illustrate the use of the basis sets derived in Section 2 in molecular calculations. Although the *atomic* basis sets give accurate descriptions of the atoms, to obtain high accuracy in molecular calculations these sets must be augmented with *polarization* functions (e.g., $2p$ for hydrogen and $3d$ for the first row atoms). Polarization functions are necessary to describe the distortions in the atomic orbitals resulting from molecular formation (see, e.g., Section 1). Specifically, we illustrate the use of the $[2s1p]$ hydrogen set and the $[3s2p1d]$ first row atom sets in molecular calculations. As noted previously, these sets represent a reasonable compromise between accuracy and computational convenience for calculations on small polyatomic systems. Further references for Gaussian basis sets are given in Appendix 1.

3.1. Convergence of the Atomic Basis Sets

In Table 4 the molecular properties of hydrogen fluoride are listed for selected atomic basis sets. The general contraction scheme as outlined in the last section has been used to contract the fluorine basis sets. The scale factors for the (4s)/[2s] and (5s)/[3s] hydrogen sets have been optimized, i.e., the exponents are given by

$$\zeta_i^{\text{molecule}} = \alpha \zeta_i^{\text{atom}}$$

where α, the scale factor, has been optimized.

The difference between the results obtained with the $(9s5p/4s)/[3s2p/2s]$ and $(11s7p/5s)/[4s5p/3s]$ sets are generally small, e.g., the atomization energies, ΔE_{HF}, differ by only 0.11 eV, or 2.5%.

3.2. Addition of Polarization Functions

From atomic calculations, such as those discussed in Section 2, the exponents and contraction coefficients for the s and p functions of the first row atoms can be obtained. Polarization functions (e.g., $2p$ for the hydrogen atom and $3d$ for the first row atoms) are, however, essential for an accurate

Table 4. Convergence of the Hartree–Fock Calculations on HF with Atomic Basis Sets[a]

Quantity	P^b: (8s4p/4s) C^b: [3s2p/2s]	(9s5p/4s)		(10s6p/5s)		(11s7p/5s)	
	[3s2p/2s]	[3s2p/2s]	[3s3p/2s]	[4s3p/3s]	[4s4p/3s]	[4s4p/3s]	[4s5p/3s]
E_{HF}	−99.98987	−100.02455	−100.02557	−100.03718	−100.03733	−100.04092	−100.04112
$\Delta E_{HF}[E_{HF}(\text{atoms})-E_{HF}(\text{molecules})]^c$	0.12527	0.12934	0.13036	0.13242	0.13257	0.13351	0.13371
Orbital energies: $\varepsilon_{1\sigma}$	−26.2762	−26.2833	−26.2861	−26.2930	−26.2928	−26.2946	−26.2939
$\varepsilon_{2\sigma}$	−1.5946	−1.6040	−1.6037	−1.6092	−1.6090	−1.6101	−1.6101
$\varepsilon_{3\sigma}$	−0.7400	−0.7531	−0.7530	−0.7590	−0.7590	−0.7604	−0.7606
$\varepsilon_{1\pi}$	−0.6312	−0.6432	−0.6429	−0.6484	−0.6484	−0.6496	−0.6498
Dipole moment, μ	−0.9177	−0.9424	−0.9406	−0.9397	−0.9388	−0.9419	−0.9410
Hellman–Feynman force, f_z	2.1311	2.0901	1.6279	1.7380	1.3283	1.4273	1.1714
Field gradients, q_{zz}(H)	−0.6260	−0.6342	−0.6225	−0.6282	−0.6257	−0.6144	−0.6106
q_{zz}(F)	−3.1098	−3.1114	−3.2439	−3.1629	−3.2356	−3.2710	−3.2061
Quadrupole moment,^{d} θ_{zz} (CM)	1.6102	1.6646	1.6728	1.7033	1.7000	1.6879	1.6857
Octapole moment,^{d} Ω_{zzz} (CM)	−2.1847	−2.2015	−2.2061	−2.1399	−2.1360	−2.0529	−2.0474

[a] All quantities are in atomic units.
[b] P = primitive set; C = contracted set.
[c] Fluorine atom energies: $E_{HF}(8s4p) = -99.36532$; $E_{HF}(9s5p) = -99.39593$; $E_{HF}(10s6p) = -99.40495$; $E_{HF}(11s7p) = -99.40760$. Hydrogen atom energies: $E_{HF}(4s) = -0.49928$; $E_{HF}(5s) = -0.49981$.
[d] Relative to the center of mass.

Table 5. Comparison of Polarization Functions and Bond Functions in Calculations on
O_2 and O_3

Molecule	State	Total energies[a]		[3s2p]+ bond functions	Ratio of energy lowering (bond/polarization)
		[3s2p]	[3s2p1d]		
O_2	$^3\Sigma_g^-$	−149.57114	−149.63319	−149.62634	0.89
O_3	$^1A_1$	−224.20709	−224.30638	−224.29741	0.91

[a] Energies are given in hartrees.

description of the electronic structure of molecules. These functions can only be determined from molecular calculations.

In Table 5 we illustrate the effect of polarization functions in calculations on the ground states of the oxygen and ozone molecules. A single set of $3d$ functions was added to a $[3s2p]$ set on each center [the $3s(x^2+y^2+z^2)$ function was not included]. The exponent of the $3d$ *function* ($\zeta = 0.97$) was obtained by minimizing the total energy of O_2. The addition of polarization functions lowered the energy of O_2 by 0.0621 hartree (1.69 eV) and of O_3 by 0.0993 hartree (2.70 eV) relative to the energy of the $[3s2p]$ set. Since $3d$ functions do not contribute to the atomic energies, the improvements of 1.5–3 eV in the energies of these molecules are purely molecular binding effects.

In Table 6 we illustrate the effect of polarization functions on the energy and properties of hydrogen fluoride. Comparing these results with those obtained with the analogous atomic basis sets in Table 4 we note significant

Table 6. Convergence of the Hartree–Fock Calculations on HF with Selected Basis Sets[a]

Quantity[b]	P^c: C^c:	(9s5p1d/4s1p) [3s2p1d/2s1p]	(9s5p2d/4s1p) [3s3p1d/2s1p]	(10s6p2d/5s1p) [4s3p2d/3s1p]	(11s7p2d/5s1p) [4s4p2d/3s1p]	Accurate[d]
E_{HF}		−100.05000	−100.05237	−100.06295	−100.06603	−100.0705
ΔE_{HF}		0.15479	0.15716	0.15819	0.15862	0.1612
$\varepsilon_{1\sigma}$		−26.2828	−26.2822	−26.2905	−26.2917	−26.2950
$\varepsilon_{2\sigma}$		−1.5933	−1.5925	−1.5989	−1.5999	−1.6013
$\varepsilon_{3\sigma}$		−0.7588	−0.7599	−0.7663	−0.7674	−0.7685
$\varepsilon_{1\pi}$		−0.6409	−0.6408	−0.6468	−0.6479	−0.6505
μ		−0.8137	−0.7706	−0.7791	−0.7826	−0.7609
f_z		0.9059	0.5923	0.6555	0.4557	−0.0506
$q_{zz}(H)$		−0.5735	−0.5647	−0.5654	−0.5588	−0.5226
$q_{zz}(F)$		−2.9554	−2.9452	−2.8780	−2.9315	−2.8733
$\Theta_{zz}(CM)$		1.7218	1.7806	1.7951	1.7835	1.732
$\Omega_{zzz}(CM)$		−2.5183	−2.5916	−2.5675	−2.5108	—

[a] All quantities are given in atomic units.
[b] See Table 4.
[c] P = primitive set; C = contracted set.
[d] A. D. McLean and M. Yoshimine, *J. Chem. Phys.* **47**, 3256 (1967).

changes in many of the properties. For example, the magnitude of the dipole moment is reduced by $\sim 15\%$. These calculations also point out that the $[3s2p1d/2s1p]$ set is indeed a reasonable compromise between accuracy and computational convenience.

For large molecular systems optimization of the exponents for the polarization functions is often not practical. Fortunately, the total energies are relatively insensitive to changes in the exponent. The following values represent reasonable choices for the $3d$ exponents for the first row atoms: B(0.70), C(0.75), N(0.80), O(0.85), and F(0.90). For the hydrogen $2p$ exponent 1.00 is a reasonable compromise.

One alternative to the use of higher angular momentum functions on the nuclei to describe polarization effects is the use of bond functions centered *between* the nuclei.[26,27] In Table 5 we also list the results of calculations on O_2 and O_3 using bond functions to complement the $[3s2p]$ set. In this case the set of bond functions consisted of a single s function and a single set of p functions located midway between the nuclei. The orbital exponents ($\zeta_s = 1.37$, $\zeta_p = 0.87$) were optimized for O_2. The bond functions account for 90% of the energy lowering obtained with the $3d$ set at a considerable savings in computation time (approximately a factor of 2). Bond functions would, however, appear to be less useful in calculations of potential energy surfaces for polyatomic molecules where ambiguities might arise in the location of the functions.

3.3 Calculations with the [3s2p1d] and [2s1p] Sets

As noted earlier, the $[3s2p1d]$ first row atom sets and the $[2s1p]$ hydrogen set are good sets for use in molecular calculations. In the following we compare the results of calculations with the $[3s2p1d]$ and $[2s1p]$ sets with accurate Hartree–Fock calculations for selected molecules.

3.3.1. The Valence States of O_2, CH_2, and H_2CO

Table 7 lists the total energies and excitation energies for the low-lying valence states of O_2, CH_2, and H_2CO. The discrepancies in the excitation energies never exceed 0.05 eV, even though the total energies may differ as much as 1.0 eV. In the case of H_2CO the inclusion of polarization functions is seen to be important, lowering the energy of the ground state ~ 0.5 eV relative to the two excited states.

3.3.2. The Lowest Rydberg State of NO

Calculations on the lowest Rydberg state of NO are reported in Table 8. In this case the $[3s2p1d]$ set has been augmented with a single $3s$ and $3p$ Rydberg

Table 7. *Calculated Hartree–Fock Total Energies and Excitation Energies for the Low-Lying Valence States of O_2, CH_2, and H_2CO*

Molecule	Configuration	State	Total energies, hartrees			Excitation energies, eV		
			$[3s2p]$	$[3s2p1d]$	Accurate	$[3s2p]$	$[3s2p1d]$	Accurate
O_2	$2\sigma_g^2 3\sigma_g^2 2\sigma_u^2 1\pi_u^4 1\pi_g^2$	$X^3\Sigma_g^-$	-149.57113	-149.63251	-149.6659^a	0.00	0.00	0.00
		$a^1\Delta_g$	-149.52332	-149.58403	-149.6172^a	1.30	1.32	1.33
		$b^1\Sigma_g^+$	-149.47574	-149.53536	-149.5683^a	2.60	2.64	2.66
CH_2	$2a_1^2 1b_2^2 3a_1 1b_1$	X^3B_1	—	-38.92407	-38.93078^b	–	0.00	0.00
	$2a_1^2 1b_2^2 3a_1^2$	$^1A_1$	—	-38.88270	-38.89090^b	–	1.13	1.09
H_2CO	$4a_1^2 1b_2 5a_1^2 1b_1^2 2b_2^2$	X^1A_1	-113.82706	-113.89108	-113.91494^c	0.00	0.00	0.00
	$4a_1^2 1b_2 5a_1^2 1b_1^2 2b_2 2b_1$	$\{\,^3A''$	-113.76644	-113.81297	-113.83469^c	1.65	2.13	2.18
		$\{\,^1A''$	-113.75589	-113.80216	-113.82476^c	1.94	2.45	2.45

[a] P. E. Cade and A. C. Wahl, *At. Data Nucl. Data Tables* **13**, 339 (1974).
[b] V. Staemmler, *Theor. Chim. Acta* **31**, 49 (1973).
[c] B. J. Garrison, H. F. Schaefer III, and W. A. Lester, Jr., *J. Chem. Phys.* **61**, 3039 (1974).

Table 8. *Calculated Hartree–Fock Total Energies and Excitation Energies*
for the Low-Lying Rydberg State of NO

Basis	Total energies, hartrees		Excitation energy, eV
	$X^2\Pi$	$A^2\Sigma^+$	
$[3s2p1d]+[1s1p]^a$	-129.26806	-129.06349	5.57
Accurate^b	-129.29531	-129.10365	5.22

^{a} Augmented with the $3s$ and $3p$ Rydberg functions from Table 1.
^{b} S. Green, *Chem. Phys. Lett.* **13**, 552 (1972).

function (see Table 1). For NO there is a larger discrepancy in the excitation energy, 0.35 eV. Since the calculated ionization potentials of the Rydberg state are essentially identical, 3.53 eV for the augmented $[3s2p1d]$ set and 3.54 eV for the accurate set, the difference must be attributed to differences in the valence orbitals, a somewhat puzzling finding in light of the results given in Sections 3.3.1 and 3.3.3.

3.3.3. The Positive Ions of N_2 and H_2O

Ionization potentials calculated for the three lowest ion states of N_2 and H_2O (see Table 9) with the $[3s2p1d]$ and $[2s1p]$ sets agree well with accurate Hartree–Fock calculations. The differences are less than 0.1 eV.

3.3.4. The Negative Ions of C_2 and NH

For the negative ions C_2^- and NH^- the calculations were carried out *with* and *without* the addition of diffuse functions from Table 2. With the augmented basis the agreement with accurate calculations is again satisfactory, in general, within 0.1 eV (see Table 10). Use of the $[3s2p1d]$ and $[2s1p]$ sets alone lead to similar discrepancies (0.2–0.6 eV) as was found in the atomic calculations in Section 2.

4. Basis Sets for the Second and Third Row Atoms

Basis sets for the atoms in the second and third rows of the periodic chart for use in molecular calculations are obtained in the same manner as for the first row atoms. References for basis sets for the second and third row atoms are given in Appendix 1; the contracted sets recommended in these papers should, however, be viewed with some caution.

Table 9. *Calculated Hartree–Fock Total Energies and Ionization Potentials for the Low-Lying States of N_2^+ and H_2O^+*

Molecule	Configuration	State	Total energies, hartrees			Ionization potentials, eV		
			[3s2p]	[3s2p1d]	Accurate	[3s2p]	[3s2p1d]	Accurate
N_2	$2\sigma_g^2 3\sigma_g^2 2\sigma_u^2 1\pi_u^4$	$X^1\Sigma_g^+$	-108.87805	-108.95794	-108.9928[a]	0.00	0.00	0.00
N_2^+	$2\sigma_g^2 3\sigma_g 2\sigma_u^2 1\pi_u^4$	$X^2\Sigma_g^+$	-108.29860	-108.37322	-108.4037[a]	15.77	15.91	16.03
	$2\sigma_g^2 3\sigma_g^2 2\sigma_u^2 1\pi_u^3$	$A^2\Pi_u$	-108.30953	-108.39720	-108.4270[a]	15.47	15.26	15.40
	$2\sigma_g^2 3\sigma_g^2 2\sigma_u 1\pi_u^4$	$B^2\Sigma_u^+$	-108.14350	-108.22310	-108.2596[a]	19.99	20.00	19.95
H_2O	$2a_1^2 3a_1^2 1b_2^2 1b_1^2$	X^1A_1	-76.00084	-76.04081	-76.06213[b]	0.00	0.00	0.00
H_2O^+	$2a_1^2 3a_1^2 1b_2^2 1b_1$	X^2B_1	-75.59167	-75.63546	-75.65488[b]	11.13	11.03	11.08
	$2a_1^2 3a_1 1b_2^2 1b_1^2$	$^2A_1$	-75.52088	-75.55347	-75.57184[b]	13.06	13.26	13.34
	$2a_1^2 3a_1^2 1b_2 1b_1^2$	$^2B_2$	-75.34267	-75.39512	-75.41512[b]	17.91	17.57	17.61

[a]P. E. Cade and A. C. Wahl, *At. Data Nucl. Data Tables* **13**, 339 (1974).
[b]T. H. Dunning, Jr., R. M. Pitzer, and S. Aung, *J. Chem. Phys.* **57**, 5044 (1972).

Table 10. *Calculated Hartree–Fock Total Energies and Electron Affinities of the Low-Lying States of C_2^- and NH^-*

Molecule	Configuration	State	Total energies					Electron affinities				
			$[3s2p]$	$[3s2p]$ $+[1p]^a$	$[3s2p1d]$	$[3s2p1d]$ $+[1p]^a$	Accurate	$[3s2p]$	$[3s2p]$ $+[1p]^a$	$[3s2p1d]$	$[3s2p1d]$ $+[1p]^a$	Accurate
C_2	$2\sigma_g^2 2\sigma_u^2 1\pi_u^4$	$X^1\Sigma_g^+$	-75.35645	-75.35699	-75.38940	-75.38977	-75.40620^b	0.00	0.00	0.00	0.00	0.00
C_2^-	$2\sigma_g^2 3\sigma_g 2\sigma_u^2 1\pi_u^3$	$^2\Pi_u$	-75.51734	-75.52165	-75.55125	-75.55566	-75.56327^b	-4.38	-4.48	-4.40	-4.51	-4.27
	$2\sigma_g^2 3\sigma_g 2\sigma_u^2 1\pi_u^4$	$^2\Sigma_g^+$	-75.50702	-75.51230	-75.53919	-75.54450	-75.55670^b	-4.10	-4.23	-4.08	-4.21	-4.10
	$2\sigma_g^2 3\sigma_g 2\sigma_u 1\pi_u^4$	$^2\Sigma_u^+$	-75.38685	-75.39251	-75.41520	-75.42126	-75.43396^b	-0.83	-0.97	-0.70	-0.86	-0.76
NH	$2\sigma^2 3\sigma^2 1\pi^2$	$X^3\Sigma^-$	-54.94838	-54.94940	-54.96547	-54.96606	-54.97838^c	0.00	0.00	0.00	0.00	0.00
NH^-	$2\sigma^2 3\sigma^2 1\pi^3$	$^2\Pi$	-54.86892	-54.89256	-54.88450	-54.90853	-54.92138^c	-2.16	-1.54	-2.20	-1.57	-1.55

[a] Augmented with the 2p negative ion functions from Table 2.
[b] P. E. Cade and A. C. Wahl, At. Data Nucl. Data Tables **13**, 339 (1974).
[c] P. E. Cade, Proc. Phys. Soc., London **91**, 842 (1967).

Table 11. Comparison of Segmented and General Contractions
for the SO_2 Molecule

	Total energies	
Basis	Segmented contraction[a]	General contraction
$(11s7p1d/9s5p)$	-547.15092	-547.15175
$(11s7p1d/9s5p1d)$	-547.20763	-547.20797
$(12s9p1d/10s5p1d)^b$	-547.2089	—

[a] See Appendix 3 for the sulfur basis set and Ref. 24 for the oxygen basis set.
[b] S. Rothenberg and H. F. Schaefer III, *J. Chem. Phys.* **53**, 3014 (1971).

Segmented basis sets for the second row atoms, aluminum through chlorine, are given in Appendix 3, Table A.3. The $(11s7p)$ primitive sets were taken from Huzinaga.[28] The resulting $[6s4p]$ contracted sets are comparable to the $[4s2p]$ sets given earlier for the first row atoms. The atomic energies obtained with the contracted and uncontracted sets are given in Table A.4. Exponents for diffuse functions appropriate for calculations on Rydberg states and negative ions are reported in Tables A.5 and A.6.

For atoms beyond the first row obtaining segmented basis sets, with a minimum of duplicated functions and only modest sacrifices in the energy, becomes increasingly difficult. This is to be contrasted with the nearly effortless manner with which primitive sets can be contracted using the general contraction scheme.[25] In addition, with the general contraction method only one function need be used per core orbital, and for atoms beyond the first row this leads to a significant reduction in the time required to compute wave functions.

In Table 11 the segmented and general contractions are compared for the SO_2 molecule. The segmented set consists of the $[6s4p]$ sulfur set given in Appendix 3 and the $[4s2p]$ oxygen set given earlier. In the general set the same primitive sets are contracted to $[4s3p]$ and $[3s2p]$ for the respective atoms. The $1s$, $2s$, and $3s$ basis functions for sulfur are merely the $1s$, $2s$, and $3s$ atomic orbitals expressed in terms of the first ten primitive functions. The $3s'$ function is the eleventh primitive function. Similar considerations apply to the sulfur p set and the oxygen s and p sets. Note that the general set contains seven fewer functions than the segmented set; the segmented set could not be so contracted without excessive duplication of primitive functions or increases in the energy. Both contracted sets were augmented with $3d$ functions on the sulfur ($\zeta = 0.6$) and oxygen ($\zeta = 0.8$).

The energy of SO_2 obtained with the generally contracted set is slightly lower than that obtained with the segmented contracted set and is in good agreement with published calculations which used larger primitive sets.

Appendix 1. Bibliography of Gaussian Basis Sets for Ab Initio Calculations on the Electronic Structure of Molecules

A.1. Atomic Gaussian Basis Sets

A.1.1. H–Ne

S. Huzinaga, Gaussian-type functions for polyatomic systems. I., *J. Chem. Phys.* **42**, 1293–1302 (1965); $(1s)$–$(10s)$ for H and He; $(9s5p)$ and $(10s6p)$ for Li to Ne.

S. Huzinaga and Y. Saki, Gaussian-type functions for polyatomic systems. II., *J. Chem. Phys.* **50**, 1371–1381 (1969); $(11s7p)$.

S. Huzinaga, Approximate Atomic Wave Functions. I., Department of Chemistry Report, University of Alberta, Edmonton, Alberta, Canada 1971; $(6s3p)$ through $(11s7p)$.

D. R. Whitman and C. J. Hornback, Optimized Gaussian basis SCF wavefunctions for first-row atoms, *J. Chem. Phys.* **51**, 398–402 (1969); $(2s1p)$ through $(8s4p)$.

B. Roos and P. Siegbahn, Gaussian basis for the first and second row atoms, *Theor. Chim. Acta* **17**, 209–215 (1970); $(7s3p)$.

F. B. van Duijneveldt, IBM Technical Research Report No. RJ945 (1971); $(4s2p)$ through $(13s9p)$.

A.1.2. Na–Ar

A. Veillard, Gaussian basis set for molecular calculations containing second-row atoms, *Theor. Chim. Acta* **12**, 405–411 (1968); $(12s9p)$.

S. Huzinaga, Approximate Atomic Wavefunctions. II., Department of Chemistry Report, University of Alberta, Edmonton, Alberta, Canada 1971; $(9s5p)$ through $(12s9p)$.

B. Roos and P. Siegbahn, Gaussian basis sets for the first and second row atoms, *Theor. Chim. Acta* **17**, 209–215 (1970); $(10s6p)$.

A.1.3. K–Zn

B. Roos, A. Veillard, and G. Vinot, Gaussian basis sets for molecular wavefunctions containing third-row atoms, *Theor. Chim. Acta* **20**, 1–11 (1971); $(12s6p4d)$.

A. J. H. Wachters, Gaussian basis set for molecular wavefunctions containing third-row atoms, *J. Chem. Phys.* **52**, 1033–1036 (1970).

A.1.4. Miscellaneous

T. H. Dunning, Jr., Gaussian basis sets for the atoms gallium through krypton, *J. Chem. Phys.* (to be published); $(13s9p5d)$ and $(14s11p5d)$.

T. H. Dunning, Jr., Gaussian basis sets for the atoms tin through xenon (unpublished).

A.2. Segmented Contraction of Gaussian Basis Sets

A.2.1. B–F

H. Basch, M. B. Robin, and N. A. Kuebler, Electronic states of the amide group, *J. Chem. Phys.* **47**, 1201–1210 (1967); **49**, 5007–5018 (1968); $(10s5p)/[4s2p]$.

T. H. Dunning, Jr., Gaussian basis functions for use in molecular calculations. I. Contraction of $(9s5p)$ atomic basis sets for the first-row atoms, *J. Chem. Phys.* **53**, 2823–2833 (1970); $(9s5p)/[4s2p]$, $[4s3p]$, and $[5s3p]$.

T. H. Dunning, Jr., Gaussian basis functions for use in molecular calculations. III. Contraction of $(10s6p)$ atomic basis sets for the first-row atoms, *J. Chem. Phys.* **55**, 716–723 (1971); $(10s6p)/[5s3p]$ and $[5s4p]$.

A.2.2. Si–Cl

S. Rothenberg, R. H. Young, and H. F. Schaefer III, Ground state self-consistent-field wave functions and molecular properties for the isoelectronic series SiH_4, PH_3, H_2S and HCl, *J. Am. Chem. Soc.* **92**, 3243–3250 (1970); $(12s9p)/[6s4p]$.

T. H. Dunning, Jr., Gaussian basis functions for use in molecular calculations. Contraction of $(12s9p)$ atomic basis sets for the second-row atoms, *Chem. Phys. Lett.* **7**, 423–427 (1970); Cl $(12s9p)/[6s4p]$ and $[6s5p]$.

A.2.3. Sc–Cu

H. Basch, C. J. Hornback, and J. W. Moskowitz, Gaussian-orbital basis sets for the first-row transition metal atoms, *J. Chem. Phys.* **51**, 1311–1318 (1969); $(15s8p5d)/[4s2p1d]$.

See, in addition, the papers listed in Section A.1 above.

A.3. General Contraction of Gaussian Basis Sets

R. C. Raffenetti, General contraction of Gaussian atomic orbitals: Core, valence, polarization and diffuse basis sets; molecular integral evaluation, *J. Chem. Phys.* **58**, 4452–4458 (1973).

A.4. Polarization Functions

A.4.1. Nuclear-Centered Functions

B. Roos and P. Siegbahn, Polarization functions for the first and second row atoms in Gaussian type MO-SCF calculations, *Theor. Chim. Acta* **17**, 199–208 (1970).

T. H. Dunning, Jr., Gaussian basis functions for use in molecular calculations. IV. The representation of polarization functions for the first-row atoms and hydrogen, *J. Chem. Phys.* **55**, 3958–3966 (1971).

A.4.2. Bond-Centered Functions

S. Rothenberg and H. F. Schaefer III, Methane as a numerical experiment for polarization basis function selection, *J. Chem. Phys.* **54**, 2765–2766 (1971).

T. Vladimiroff, Comparison of the use of $3d$ polarization functions and bond functions in Gaussian Hartree–Fock calculations, *J. Phys. Chem.* **77**, 1983–1985 (1973).

Appendix 2. The [3s2p] Contracted Gaussian Basis Sets for the First Row Atoms, Including a [2s] Hydrogen Set

Table A.1. Contracted Segmented Gaussian Basis Sets for the First Row Atoms, Lithium through Neon, Obtained from the (9s5p) Primitive Sets of Huzinaga[a]

Lithium		Beryllium		Boron		Carbon	
ζ_s	[3s]	ζ_s	[3s]	ζ_s	[3s]	ζ_s	[3s]
921.3	0.001367	1741.	0.001305	2788.	0.001288	4233.	0.001220
138.7	0.010425	262.1	0.009955	419.	0.009835	634.9	0.009342
31.94	0.049859	60.33	0.048031	96.47	0.047648	146.1	0.045452
9.353	0.160701	17.62	0.158577	28.07	0.160069	42.50	0.154657
3.158	0.344604	5.933	0.351325	9.376	0.362894	14.19	0.358866
1.157	0.425197	2.185	0.427006	3.406	0.433582	5.148	0.438632
0.4446	0.169468	0.8590	0.160490	1.306	0.140082	1.967	0.145918
0.4446	−0.222311	2.185	−0.185294	3.406	−0.179330	5.148	−0.168367
0.07666	1.116477	0.1806	1.057014	0.3245	1.062594	4.4962	1.060091
0.02864	1.000000	0.05835	1.000000	0.1022	1.000000	0.1533	1.000000
ζ_p^b	$[2p]^b$	ζ_p^c	$[2p]^c$	ζ_p	[2p]	ζ_p	[2p]
1.488	0.038770	6.710	0.016378	11.34	0.017988	18.16	0.018539
0.2667	0.236257	1.442	0.091553	2.436	0.110343	3.986	0.115436
0.07201	0.830448	0.4103	0.341469	0.6836	0.383072	1.143	0.386188
0.02370	1.000000	0.1397	0.685428	0.2134	0.647895	0.3594	0.640114
		0.04922	1.000000	0.07011	1.000000	0.1146	1.000000

Nitrogen		Oxygen		Fluorine		Neon	
ζ_s	[3s]	ζ_s	[3s]	ζ_s	[3s]	ζ_s	[3s]
5909.	0.001190	7817.	0.001176	9995.	0.001166	12100.	0.001200
887.5	0.009099	1176.	0.008968	1506.	0.008876	1821.	0.009092
204.7	0.044145	273.2	0.042868	350.3	0.042380	432.8	0.041305
59.84	0.150464	81.17	0.143930	104.1	0.142929	132.5	0.137867
20.00	0.356741	27.18	0.355630	34.84	0.355372	43.77	0.362433
7.193	0.446533	9.532	0.461248	12.22	0.462085	14.91	0.472247
2.686	0.145603	3.414	0.140206	4.369	0.140848	5.127	0.130035
7.193	−0.160405	9.532	−0.154153	12.22	−0.148452	14.91	−0.140810
0.7000	1.058215	0.9398	1.056914	1.208	1.055270	1.491	1.053327
0.2133	1.000000	0.2846	1.000000	0.3634	1.000000	0.4468	1.000000
ζ_p	[2p]	ζ_p	[2p]	ζ_p	[2p]	ζ_p	[2p]
26.79	0.018254	35.18	0.019580	44.36	0.020876	56.45	0.020875
5.956	0.116461	7.904	0.124200	10.08	0.130107	12.92	0.130032
1.707	0.390178	2.305	0.394714	2.996	0.396166	3.865	0.395679
0.5314	0.637102	0.7171	0.627376	0.9383	0.620404	1.203	0.621450
0.1654	1.000000	0.2137	1.000000	0.2733	1.000000	0.3444	1.000000

[a] The contracted basis functions are separated by the short horizontal lines.
[b] Obtained from calculations on the $1s^2 2p(^2p)$ state of the lithium atom.
[c] Obtained from calculations on the $1s^2 2s 2p(^3P)$ state of the beryllium atom.

Table A.2. Contracted Segmented Gaussian Basis Sets for the Hydrogen Atom Obtained from the (4s) Primitive Set of Huzinaga

ζ_s	[2s]
13.36	0.032828
2.013	0.231204
0.4538	0.817226
0.1233	1.000000

Appendix 3.   The [6s4p] Contracted Gaussian Basis Sets for the Second Row Atoms, Including Diffuse Functions for Rydberg and Negative Ion States

Table A.3. *Contracted Segmented Gaussian Basis Sets for the Second Row Atoms, Aluminum through Chlorine, Obtained from the (11s7p) Primitive Sets of Huzinaga* [a]

Aluminum		Silicon		Phosphorus	
[6s]	ζ_s	ζ_s	[6s]	ζ_s	[6s]
23490.	0.002509	26740.	0.002583	30630.	0.002619
3548.	0.018986	4076.	0.019237	4684.	0.019479
823.5	0.092914	953.3	0.093843	1094.	0.095207
237.7	0.335935	274.6	0.341235	315.3	0.345742
78.6	0.647391	90.68	0.641675	104.1	0.636288
78.6	0.111937	90.68	0.121439	104.1	0.130706
29.05	0.655976	33.53	0.653143	38.42	0.650274
11.62	0.283349	13.46	0.277624	15.45	0.272308
3.465	1.000000	4.051	1.000000	4.656	1.000000
1.233	1.000000	1.484	1.000000	1.759	1.000000
0.2018	1.000000	0.2704	1.000000	0.3409	1.000000
0.07805	1.000000	0.09932	1.000000	0.1238	1.000000
ζ_p	[4p]	ζ_p	[4p]	ζ_p	[4p]
141.5	0.017882	163.7	0.011498	187.7	0.013158
33.22	0.120375	38.35	0.077726	43.63	0.090494
10.39	0.411580	12.02	0.263595	13.60	0.305054
3.593	0.595353	4.185	0.758269	4.766	0.713579
3.593	0.211758	4.185	-1.173045	4.766	-0.792573
1.242	0.837795	1.483	1.438335	1.743	1.429987
0.3040	1.000000	0.3350	1.000000	0.4192	1.000000
0.07629	1.000000	0.09699	1.000000	0.1245	1.000000

Sulfur		Chlorine	
ζ_s	[6s]	ζ_s	[6s]
35710.	0.002565	40850.	0.002532
5397.	0.019405	6179.	0.019207
1250.	0.095595	1425.	0.095257
359.9	0.345793	409.2	0.345589
119.2	0.635794	135.5	0.636401
119.2	0.130096	135.5	0.120956
43.98	0.651301	50.13	0.648511
17.63	0.271955	20.21	0.275487
5.42	1.000000	6.283	1.000000
2.074	1.000000	2.460	1.000000
0.4246	1.000000	0.5271	1.000000
0.1519	1.000000	0.1884	1.000000
ζ_p	[4p]	ζ_p	[4p]
212.9	0.014091	240.8	0.014595
49.60	0.096685	56.56	0.099047
15.52	0.323874	17.85	0.330562
5.476	0.691756	6.350	0.682874
5.476	-0.626737	6.350	-0.561785
2.044	1.377051	2.403	1.351901
0.5218	1.000000	0.6410	1.000000
0.1506	1.000000	0.1838	1.000000

[a] The contracted basis functions are separated by the short horizontal lines.

Table A.4. Energies for the (11s7p) and [6s4p] Gaussian Basis Sets for the Valence States of the Second Row Atoms, Aluminum through Chlorine

Atom	State	Hartree–Fock energy	
		(11s7p)	[6s4p]
Al	^2P	−241.85555	−241.85498
Si	^3P	−288.83004	−288.82948
	^1D		−288.78931
	^1S		−288.72991
P	^4S	−340.68948	−340.68882
	^2D		−340.61786
	^2P		−340.57105
S	^3P	−397.46860	−397.46777
	^1D		−397.41466
	^1S		−397.33560
Cl	^2P	−459.43839	−459.43736

Table A.5. Rydberg States: Exponents of the Functions to Augment the (11s7p) Basis Sets for the Second Row Atoms

Function	Exponents				
	Al	Si	P	S	Cl
$4s$	0.014	0.017	0.020	0.023	0.025
$4p$	0.011	0.014	0.017	0.020	0.020
$3d$	0.015	0.015	0.015	0.015	0.015

Table A.6. Negative Ions: Exponents of the 3p Functions to Augment the (11s7p) Basis Sets for the Second Row Atoms

Atom	Exponent
Al	0.017
Si	0.027
P	0.035
S	0.041
Cl	0.049

References*

1. W. Heitler and F. London, Wechselwirkung neutraler Atome und homopolar Bindung nach der Quantenmechanik, *Z. Phys.* **44**, 455–472 (1927).
2. E. Clementi and D. L. Raimondi, Atomic screening constants from SCF functions, *J. Chem. Phys.* **38**, 2686–2689 (1963); He–Kr.
3. E. Clementi, D. L. Raimondi and W. P. Reinhardt, Atomic screening constants from SCF functions. II. Atoms with 37 to 86 electrons, *J. Chem. Phys.* **47**, 1300–1307 (1967); Rb–Rn.
4. S. Huzinaga and C. Arnau, Simple basis set for molecular wavefunctions containing first- and second-row atoms, *J. Chem. Phys.* **53**, 451–452 (1970); He–Ar.
5. E. Clementi, R. Matcha, and A. Veillard, Simple basis sets for molecular wavefunctions containing third-row atoms, *J. Chem. Phys.* **47**, 1865–1866 (1967).
6. S. C. Wang, The problem of the normal hydrogen molecule in the new quantum mechanics, *Phys. Rev.* **31**, 579–586 (1928).
7. N. Rosen, The normal state of the hydrogen molecule, *Phys. Rev.* **38**, 2099–2114 (1931).
8. C. A. Coulson and I. Fischer, Notes on the molecular orbital treatment of the hydrogen molecule, *Philos. Mag.* **40**, 386–393 (1949).
9. W. A. Goddard III, Improved quantum theory of many-electron systems. II. The basic method, *Phys. Rev.* **157**, 81–93 (1967).
10. S. F. Boys, Electronic wavefunctions. I. A general method of calculation for the stationary states of any molecular system, *Proc. R. Soc. London Ser. A*, **200**, 542–554 (1950).
11. S. Huzinaga, Gaussian-type functions for polyatomic systems I, *J. Chem. Phys.* **42**, 1293–1302 (1965).
12. H. Preuss, Bemerkungen zum self-consistent-field-verfahren und zur Methode der Konfigurationenwechselwirkung in der Quantenchemie, *Z. Naturforsch.* **11**, 823 (1956).
13. J. L. Whitten, Gaussian expansions of hydrogen atom wavefunctions, *J. Chem. Phys.* **39**, 349 (1963).
14. J. L. Whitten, Gaussian lobe function expansions of Hartree–Fock solutions for the first row atoms and ethylene, *J. Chem. Phys.* **44**, 359 (1966).
15. J. D. Petke, J. L. Whitten, and A. W. Douglas, Gaussian lobe function expansions of Hartree–Fock solutions for the second row atoms, *J. Chem. Phys.* **51**, 256–262 (1969).
16. S. Shih, R. J. Buenker, S. D. Peyerimhoff, and B. Wirsan, Comparison of Cartesian and lobe function Gaussian basis sets, *Theor. Chim. Acta* **18**, 277–289 (1970).
17. W. J. Hehre, R. F. Stewart, and J. A. Pople, Self-consistent molecular-orbital methods. I. Use of Gaussian expansions of Slater-type atomic orbitals, *J. Chem. Phys.* **51**, 2657–2664 (1969); H, Li–F.
18. W. J. Hehre, R. Ditchfield, R. F. Stewart, and J. A. Pople, Self-consistent molecular-orbital methods. IV. Use of Gaussian expansions of Slater-type orbitals. Extension to second-row molecules, *J. Chem. Phys.* **52**, 2769–2773 (1970); Na–Ar.
19. R. Ditchfield, W. J. Hehre, and J. A. Pople, Self-consistent molecular-orbital methods. IX. An extended Gaussian-type basis for molecular orbital studies of organic molecules, *J. Chem. Phys.* **54**, 724–728 (1971); H, C–F.
20. W. J. Hehre and J. A. Pople, Self-consistent molecular-orbital methods. XIII. An extended Gaussian-type basis for boron, *J. Chem. Phys.* **56**, 4233–4234 (1972).
21. R. F. Stewart, Small Gaussian expansions of Slater-type orbitals, *J. Chem. Phys.* **52**, 431–438 (1970).
22. R. C. Raffenetti, Optimal even-tempered Gaussian atomic orbital bases: First row atoms, *Int. J. Quant. Chem.* **9S**, 289–295 (1975).
23. R. D. Bardo and K. Ruedenberg, Even tempered atomic orbitals III. Economic deployment of Gaussian primitives in expanding atomic SCF orbitals, *J. Chem. Phys.* **59**, 5956–5965 (1973).
24. T. H. Dunning, Jr., Gaussian basis functions for use in molecular calculations. I. Contraction of (9s5p) atomic basis sets for the first row atoms, *J. Chem. Phys.* **53**, 2823–2833 (1970).

*References 2, 3, 17, and 18 are for minimum basis sets, References 4, 5, 19, and 20 are for double zeta basis sets, and Reference 21 gives the general expansions of $1s$ to $5g$ Slater functions.

25. R. C. Raffenetti, General contraction of Gaussian atomic orbitals: core, valence, polarization and diffuse basis sets; molecular integral evaluation, *J. Chem. Phys.* **58**, 4452–4458 (1973).
26. S. Rothenberg and H. F. Schaefer III, Methane as a numerical experiment for polarization basis function selection, *J. Chem. Phys.* **54**, 2765–2766 (1971).
27. T. Vladimiroff, Comparison of the use of $3d$ polarization functions and bond functions in Gaussian Hartree–Fock calculations, *J. Phys. Chem.* **77**, 1983–1985 (1973).
28. S. Huzinaga, Approximate atomic wavefunctions. II., Department of Chemistry Technical Report, University of Alberta, Edmonton, Alberta, Canada 1971.

2

The Floating Spherical Gaussian Orbital Method

Arthur A. Frost

1. Desiderata for a Simple Ab Initio Method

One important object of molecular quantum mechanics is the calculation of the stationary state energies of molecular systems as functions of the geometrical configuration of their nuclei. By finding the energy minima one can predict the equilibrium geometric form of molecules, including internuclear distances and bond angles. The procedure implies the use of the Born-Oppenheimer approximation which is known to be excellent for the ground electronic state of most molecules.

The term *ab initio* is considered by some to imply a highly accurate calculation, but it is used here only to mean that for a given assumed wave function the expectation value of the energy is evaluated using the exact many-electron electrostatic Hamiltonian, and that any parameters in the wave function are determined by the variation principle rather than being set by semiempirical methods.

Ab initio calculations of hundreds of molecules using various degrees of sophistication in their basis sets have been reported and summaries presented.[1-4] Typically, a self-consistent field iteration procedure [5,6] is used. More accurate results, however, are obtained by configuration interaction or perturbation methods.[7] In general, the more sophisticated and accurate the method the more time-consuming and expensive the calculation, so the result may be limited to just one geometry, obtained from experiment.

Arthur A. Frost • Department of Chemistry, Northwestern University, Evanston, Illinois

Therefore, it may be desirable to forego high accuracy in energy in order to predict geometric structure if energy differences are still reasonably good. The method to be described accomplishes such a purpose.

1.1. A Single Slater Determinant

The simplest many-electron wave function with the necessary antisymmetry is conveniently represented by a single Slater determinant. Moreover, since most molecules have an even number of electrons and have singlet ground states, the Slater determinant for such cases can be formed from doubly occupied orbitals. If then the forms of the individual orbitals are varied in all possible ways, so as to minimize the electronic energy for a given nuclear configuration, one has as a result a restricted Hartree–Fock calculation.[8]

The energy formula for a system of $2n$ electrons in a single Slater determinant is remarkably simple despite the fact that a $2n$th order determinant expands to $(2n)!$ terms resulting in $(2n)!^2$ integrals over the many-electron Hamiltonian. For a Slater determinantal wave function formed from closed shells of orthonormal orbitals, χ_i, the electronic energy formula is well known[9]:

$$E_{el} = 2\sum_i [i|i] + \sum_{i,j}(2[ii|jj] - [ij|ji])$$

(1)

Summations are from 1 to n unless otherwise indicated;

$$[i|i] = \int \chi_i^* h \chi_i \, dv$$

(2)

are one-electron integrals involving the one-electron Hamiltonian operator

$$h = -\tfrac{1}{2}\nabla^2 - \sum_\nu^N (Z_\nu/r_\nu)$$

(3)

Z_ν being the atomic number of nucleus ν r_ν the distance of the electron from that nucleus; N is the number of nuclei;

$$[ij|kl] = \int \chi_i^*(1)\chi_j(1)\chi_k^*(2)\chi_1(2)(1/r_{12}) \, dv_1 \, dv_2$$

(4)

are the two-electron repulsion integrals.

Equation (1) is obtained from

$$E_{el} = \int \psi^* H \psi \, d\tau_1 \cdots d\tau_{2n}$$

(5)

with the total Hamiltonian

$$H = \sum_{i=1}^{2n} h_i + \sum_{i<j}^{2n} r_{ij}^{-1}$$

(6)

1.2. Easy Integrals—Economical Computation

For a simple method one needs to avoid the difficult many-center integrals over Slater-type orbitals. Fortunately, Boys[10,11] showed the simplification of using Gaussian orbitals, with the spherical Gaussians simplest of all. In principle, any orbital can be approximated by a linear combination of spherical Gaussians.[12,13] In some situations it might be that a localized orbital could be sufficiently well represented by a single such Gaussian. In any case the use of Gaussians to simulate Slater-type orbitals is well developed.[14] Gaussian orbitals do result in a more economical calculation as compared with Slater-type orbitals for the same final accuracy in the energy. [15]

1.3. Relation to Chemical Concepts

It would be desirable to have a simple *ab initio* method that would correlate with chemical concepts such as bonds, lone pairs, and inner shells. The usual molecular orbital theory does this only indirectly, in that a given molecular orbital may be bonding, antibonding, or nonbonding in different parts of a molecule. Valence bond theory does accomplish this but requires the complexity of a linear combination of Slater determinants.

Fortunately a theorem of Fock[16] enables one to transform the molecular orbitals into localized orbitals which may be expected to have the desired property. The construction of localized molecular orbitals out of canonical molecular orbitals has been particularly emphasized by Edmiston and Ruedenberg.[17]

Conversely, it is possible to form a Slater determinant from a set of initial local orbitals assumed on the basis of chemical considerations[18–20] and obtain molecular orbitals as an afterthought if desired.[21,22]

2. The Simple FSGO Method Defined

2.1. Gaussians as Local Orbitals

The floating spherical Gaussian orbital (FSGO) method in its simplest form applies only to systems with an even number of electrons in their lowest singlet energy states. Following the desiderata discussed in Section 1 it is assumed that each pair of electrons occupies a single spherical Gaussian orbital centered at an arbitrary position in the molecule. The many-electron wave function is taken to be a single Slater determinant formed from double occupancy of these orbitals. The orbitals are allowed to float and to vary in size so that the energy is minimized. Each orbital may be identified as bonding if it is located between nuclei, inner shell if it is associated with a nucleus and compressed in size, and lone pair if it is near a nucleus and larger than an inner shell orbital.

A normalized spherical Gaussian orbital is defined as

$$\varphi_i = (2a_i/\pi)^{\frac{3}{4}} \exp[-a_i(\mathbf{r} - \mathbf{R}_i)^2] \tag{7}$$

The orbital exponent a_i and the components of the orbital center vector $\mathbf{R}_i$ are variation parameters. For the present model it is useful to replace the parameter a_i by

$$a_i = 1/\rho_i^2 \tag{8}$$

so that

$$\varphi_i = (2/\pi\rho_i^2)^{\frac{3}{4}} \exp[-(\mathbf{r} - \mathbf{R}_i)^2/\rho_i^2) \tag{9}$$

The new parameter ρ_i has dimensions of length and may be called an "orbital radius." Through integration it is found that about 74% of the electron density of a given FSGO is within a sphere of radius ρ_i. The value of ρ_i is therefore a useful measure of the size of the orbital.

These orbitals are *local* and directly represent the electron pairs of traditional Lewis[23] valence theory. As compared with *localized* orbitals obtained from molecular orbitals[17] it should be realized that the FSGOs are nonorthogonal. Although the picture of electron pairs occupying spherical orbitals is simple to contemplate, it does have the disadvantage that certain properties such as electron density are easier to understand in terms of orthogonal orbitals. However, a Löwdin[9] symmetric orthogonalization of the FSGOs does generate orthogonal orbitals with both positive and negative lobes similar to typical localized orbitals.

The concept of using FSGOs for electron pairs in molecules originated with Kimball and Neumark,[24,25] who published calculations on the He atom and the H_2 molecule. The simple FSGO model is also a special case of the "KGO–Verfahren" of Preuss.[26] With the advent of high-speed digital computers it was a simple matter to carry out calculations by the FSGO method on numerous molecules.[27-44]

2.2. Absolute Minimal Basis Set

The simplest typical molecular orbital calculation involves what is called a *minimal basis set* of atomic orbitals. Such a set includes $1s$ orbitals on H atoms and all inner shell and valence orbitals on heavier atoms. For example, for a water molecule there would be a $1s$ orbital for each of the two H atoms and $1s$, $2s$, $2p_x$, $2p_y$, and $2p_z$ orbitals for the O atom making seven orbitals altogether. On the other hand the water molecule, with its ten electrons, requires only five doubly occupied orbitals which in the FSGO model are just five spherical

Gaussians as basis functions. Therefore, the latter may be referred to as an *absolute minimal basis set*. Owing to its relation to Lewis valence theory, it has also been called a Lewis basis set.

In a typical system with a minimal basis set there are more orbitals than there are electron pairs. This results in the need to carry out SCF iterations leading to a set of occupied molecular orbitals and some extra unoccupied or virtual orbitals. With an absolute minimal basis set no such iterations are required and the energy is a straightforward calculation as given in the next section.

As compared with a minimal basis set the FSGO method is more economical of computer time but it does have the limitation that molecules with equivalent resonance structures, such as benzene, cannot be treated unless the method is expanded to a larger basis.

2.3. Mathematical Derivation

For n electron pairs the $2n$-electron wave function is taken to be the normalized Slater determinant

$$\psi = N|\varphi_1(1)\bar{\varphi}_1(2)\varphi_2(3)\bar{\varphi}_2(4)\cdots\varphi_n(2n-1)\bar{\varphi}_n(2n)| \tag{10}$$

with the normalizing factor

$$N = 1/[(2n)!]^{\frac{1}{2}} \det S \tag{11}$$

The bars over alternate orbitals indicate β spin as opposed to α for the others, and det S is the determinant of the orbital overlap matrix S with elements

$$S_{ij} = \int \varphi_1^* \varphi_j \, dv \tag{12}$$

Since the FSGOs φ_i are nonorthogonal the energy formula given in Eq. (1) does not apply. However, a linear transformation of the φ_i, to a set of orthogonal orbitals χ_i which leaves the wave function ψ invariant[16] may be used to derive the desired formula.[9,28]

Letting the χ_i be the Löwdin symmetric orthogonal orbitals related to φ_i

$$\chi_i = \sum_j \varphi_j (T^{\frac{1}{2}})_{ji} \tag{13}$$

where $T = S^{-1}$ is the inverse overlap matrix. All wave functions are real and the matrices are real and symmetrical. The integral sums of Eq. (1) become

$$\sum_i [i|i] = \sum_i \sum_j \sum_k (j|k)(T^{\frac{1}{2}})_{ij}(T^{\frac{1}{2}})_{ik}$$

$$= \sum_j \sum_k (j|k) \, T_{jk} \tag{14}$$

where

$$(j|k) = \int \varphi_j^* h \varphi_k \, dv \tag{15}$$

and

$$\sum_{ij} [ii|jj] = \sum_{klpq} (kl|pq) \, T_{kl} T_{pq} \tag{16}$$

$$\sum_{ij} [ij|ji] = \sum_{klpq} (kl|pq) T_{kq} T_{lp} \tag{17}$$

where

$$(kl|pq) = \int \varphi_k(1)^* \varphi_1(1) \varphi_p(2)^* \varphi_q(2) r_{12}^{-1} \, dv_1 \, dv_2 \tag{18}$$

Note that the parentheses refer to integrals over nonorthogonal orbitals while the brackets involve orthogonal orbitals.

Substituting in Eq. (1) yields

$$E_{el} = 2 \sum_{jk} (j|k) T_{jk} + \sum_{klpq} (kl|pq)[2 T_{kl} T_{pq} - T_{kq} T_{lp}] \tag{19}$$

The total energy is then

$$E = E_{el} + \sum_{\mu < \nu}^{N} \frac{Z_\mu Z_\nu}{r_{\mu\nu}} \tag{20}$$

after adding the internuclear repulsions among the N nuclei.

2.4. Computer Programming and Computations

Formulas for the integrals over the Gaussian orbitals as in Eqs. (12), (15), and (18) were first given by Boys[10] and then elaborated by Shavitt.[11] The function

$$F_0(t) = \int_0^1 \exp(-tu^2) \, du \tag{21}$$

enters into several integrals. An efficient program for this has been given by Shipman and Christoffersen.[45]

A typical program[32] for the overall calculation requires an initial set of coordinates of the nuclei and orbital centers, together with orbital radii for input. The computation proceeds by evaluating the overlap integrals and the elements of the inverse overlap matrix. These are stored for later use in evaluating the energy integrals. Although there can be a multitude of two-electron repulsion integrals [Eq.(18)] depending upon four indices, these do not need to be stored since they are summed in Eq. (19) as they are produced.

After, an evaluation of the energy, one or more of the parameters is successively varied according to a pattern search method until a minimum is reached to the desired precision.

As an alternative to using a special program as described above, it is possible to make a calculation on the FSGO model by using a standard SCF-LCAO-MO program such as POLYATOM II.[46] A floating orbital not centered on a nucleus can be incorporated by defining a fictitious nucleus with zero nuclear charge at the given location.

3. FSGO Results

3.1. Calculation of LiH as a Simple Example

The LiH diatomic molecule[28] would be expected to have two FSGOs, one small one near the Li nucleus representing the inner shell and one larger one for bonding. The nuclei and the orbital centers are expected to be collinear. Placing the Li nucleus at the origin and the other centers along the x axis there are five parameters to be varied, three coordinates and the two orbital radii. In one particular calculation with the initial values shown in parentheses the final results obtained were :

H nucleus coordinate	3.2260 (3.2000)	bohr
Inner shell coordinate	−0.0076 (−0.0050)	
Inner shell radius	0.7074 (0.7000)	
Bonding orbital coordinate	2.8829 (2.9000)	
Bonding orbital radius	2.4351 (2.4000)	

With other initial values the same results are obtained, requiring more or less computer time depending upon how much optimization is needed.

Figure 1 illustrates the results for LiH as well as for the isoelectronic systems HeH⁻ and BeH⁺. The circles are drawn with a radius equal to the orbital radius. Similar diagrams have been used by Robb *et al.*[47,48] to represent more conventional localized orbitals.

The bond length is predicted to be 3.226 bohr, or 1.707 Å, which is about 7% higher than the experimental value. The final energy is −6.5727 hartrees or about 82% in magnitude as compared with a Hartree–Fock energy of −7.9873.[49] As will be seen below, the error in the bond length is much greater than usual by this method. However, an 82%–85% error in the energy as compared with Hartree–Fock is almost universal. The energy error is primarily due to the lack of a cusp in the inner shell orbital; such orbitals provide the greatest contribution to the total energy. It is evident that energy differences for different nuclear configurations are more accurate than the total

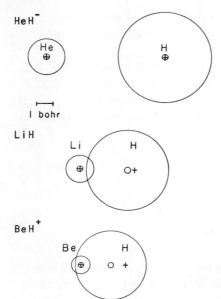

Fig. 1. FSGO results for LiH and isoelectronic systems. Each circle represents a spherical Gaussian orbital and is drawn with a radius proportional to the orbital radius.

energies. That the bonding orbital in LiH is centered near the hydrogen nucleus of course corresponds to the highly ionic character of the molecule.

Since the wave fuction is fully floating and has scale factor variations, the energy-minimized function satisfies both the Hellmann–Feynman theorem and the virial theorem. In connection with the former it is interesting to note the importance of the slight floating of the inner shell orbital off the Li nucleus and away from the H. It has been shown[28] that this slight effect is of great importance in satisfying the force balance on the Li nucleus.

3.2. Ethylene as a Further Example

The question of how to handle multiple bonds arises in the calculations for ethylene.[31] In a closed-shell single determinant calculation such as that considered here, it is as a consequence of the invariance of the many-electron wave function to linear transformations that the σ–π interpretation of the double bond is equivalent to the "banana bond" or localized orbital picture. Therefore, since the FSGOs are nodeless and cannot individually represent a π bond, it is only natural to use the localized orbital picture and place a pair of FSGOs (each doubly occupied and centered) above and below the plane of the molecule.

Figures 2 and 3 illustrate the complete FSGO structure after optimizing orbital parameters and nuclear coordinates. In order to simplify the calculation the molecule was assumed to have D_{2h} symmetry. Furthermore, it was found

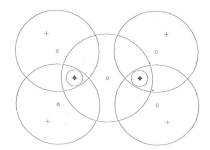

Fig. 2. FSGO structure of ethylene projected in the plane of the molecule. Large circles represent bonding orbitals. Smaller circles represent C atom inner shells. Very small circles indicate centers of orbitals.

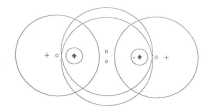

Fig. 3. FSGO structure of ethylene projected perpendicular to the molecular plane. Large overlapping circles in the center represent the double bond.

that, in general, the inner shell orbital on first row atoms such as carbon could be assumed to be centered on the nucleus and have a fixed orbital radius not too different from that in the corresponding two-electron ion.[29] These assumptions have a negligible effect on the resulting geometry and energy.

However, a difficulty arises with respect to the optimization of the positions of the double-bond orbitals. If the parameter defining the distances of the centers of these orbitals above and below the molecular plane is allowed to vary automatically in the computer program execution, the calculation "blows up" since the energy minimum occurs as the orbitals tend to coalesce resulting in complete overlap of the two orbitals and a singular determinant of the overlap matrix S. Although round-off errors prevent the numerical calculations from reaching the limit of zero interorbital distance the situation can be treated analytically,[30] showing that the coalescence limit is equivalent to having exact σ and π Gaussian orbitals. A similar result is found for three FSGOs in forming a triple bond or a triple set of lone pairs in which the limit is equivalent to an exact σ and a pair of π orbitals. In practice it has been found adequate to keep a pair of FSGOs representing a double bond (or two lone pairs as in H_2O) separated by a fixed amount such as 0.1 to 0.2 bohrs between centers.

The geometric results for ethylene and some other hydrocarbons are given in Table 1. It is seen that bond lengths and bond angles for these compounds reproduce the experimental results to within one or two percent. In particular the well-known decrease in length of the C–C bond in going from single to double to triple bond is accurately given as well as the corresponding small changes in the C–H bond length.

Table 1. Bond Lengths and Bond Angles in Small Hydrocarbons

Molecule	Calculated		Experimental[c]
	FSGO[a]	SCF[b]	
Methane			
C—H	1.115 Å	1.083 Å	1.085 Å
Ethane			
C—C	1.501	1.538	1.531
C—H	1.120	1.086	1.096
∠ H—C—H	108.2°	108.2°	107.8°
Ethylene			
C=C	1.351	1.306	1.330
C—H	1.101	1.082	1.076
∠ H—C—H	118.7°	115.6°	116.6°
Acetylene			
C≡C	1.214	1.168	1.203
C—H	1.079	1.065	1.061

[a] Reference 31.
[b] SCF results by Lathan *et al.*, Ref. 71.
[c] Experiment results quoted in Ref. 71.

3.3. General Results

3.3.1. Bond Lengths

Perhaps the greatest computational achievement of the simple FSGO method is the ability to calculate bond lengths in molecules to within an accuracy of a few percent without using any semiempirical parameters. Specific examples have been given in Sections 3.1 and 3.2. A summary of results on numerous diatomic and polyatomic molecules[27–36] is shown in Fig. 4., where the concentration of points along the 45° diagonal indicate the quality of the agreement between theory and experiment.

It can be seen that deviations tend to occur at the longer bond lengths (e.g., LiH) where too high values are predicted and at smaller bond lengths (e.g., HF) where too low values are obtained. This systematic deviation from experiment may possibly be due to the rapid falloff of the radial dependence of FSGOs as compared with hydrogen-like or Slater-type orbitals.

Since bond lengths for hydrocarbons are in the intermediate range, it is not surprising that such molecules are so successfully treated by this method. Several higher hydrocarbons both chain and ring compounds have been calculated.[38,42]

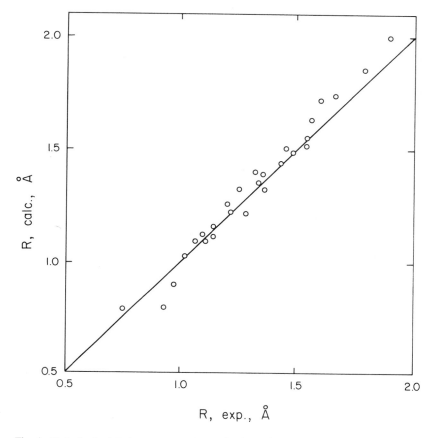

Fig. 4. Plot of calculated versus experimental values of bond lengths in diatomic and polyatomic molecules.

Included in Fig. 4 are results for second row elements[35,36] for which the L inner shell octet was represented by a set of four FSGOs arranged tetrahedrally about the nucleus.

3.3.2. Bond Angles

Although bond angles in hydrocarbons and also in some borohydrides[34,43] have been successfully predicted it must be admitted that poor results are obtained for such simple molecules as H_2O and NH_3. In these molecules the angles are low by about 17%, the calculated value for H_2O being 88.4° as compared with the experimental value of 104.5° and for NH_2, 87.6° calculated compared with 106.6° experimental.[30]

Although it is disconcerting to have this large discrepancy, it should be realized that because of the weak bending forces as compared with those for bond stretching it would be expected that the position of an energy minimum on the potential energy hypersurface would be less well defined for the former.

The fact that the bond angles in H_2O and NH_3 are too small may again be due to the too rapid falloff of the FSGOs. Using linear combinations of two FSGOs for each occupied orbital[33] does modify this falloff and does improve the angles. An attempt was made[40] to understand the bond angle in H_2O as due to interactions of pairs of FSGOs but the situation is greatly complicated by the presence in the energy formula of such a large number of two-electron integrals between nonorthogonal orbitals, as in Eq. (19).

3.3.3. Energy

It has previously been stated that the total energy in a typical simple FSGO calculation is only 82%-85% in magnitude as compared with the Hartree-Fock calculations. Of course the correlation energy is also missing, but is of much smaller magnitude. The importance of the total energy in these calculations is only as a device for determining best values of the parameters in the variation method.

On the other hand, the one-electron energies may be of interest. These energies are not obtained directly as they would be in a typical self-consistent-field calculation. But after the total energy is minimized and the orbital and nuclear parameters optimized the elements of the Fock matrix F can be calculated and the canonical molecular orbitals and one-electron energies obtained from the eigenvalue problem

$$FC = SCE \tag{22}$$

as in typical LCAO–MO theory.[5] No iterations are required since there are no unoccupied MOs. Calculations of such energies have been made for a number of hydrocarbons[39,50] and the results compared with photoelectron and ESCA spectra. Even though the total energies of molecules are poorly reproduced by the simple FSGO method it is possible to obtain semiquantitative results for the heats of reaction involving hydrocarbons.[51]

3.3.4. Electron Density, Dipole Moments, and Electrostatic Potentials

Once the wave function is fixed the electron density may be obtained by integration. For a Slater determinantal function formed from doubly occupied orthonormal orbitals χ_i, the electron density is

$$\gamma = 2 \sum_i \chi_i^* \chi_i \tag{23}$$

or the sum of the densities in the separate orbitals. For the nonorthogonal FSGOs the corresponding result, obtained by a linear transformation such as in Eq. (13), is

$$\gamma = 2 \sum_{j,k} \varphi_j^* \varphi_k T_{jk} \qquad (24)$$

The general result is that the electron density is highest in the neighborhood of the FSGO centers, as might be expected. Of course, the density is obviously deficient in not having cusps at the nuclei.

An interesting special situation arises in the case of coalescing orbitals, as in the double bond in ethylene discussed above. As two FSGOs approach each other the cross term involving them in Eq. (24) becomes large and negative. The result is that in the limit there will be two maxima separated by a finite distance.

Another similar effect occurs in the water molecule. It was found[30] that the FSGOs representing lone pairs, upon optimization, took up positions on a line perpendicular to the molecular plane and passing through it just inside the oxygen nucleus. This result seemed to violate the usual concept of lone pairs extending in the opposite direction. However, the cross terms in Eq. (24) are such that the symmetrically orthogonalized orbitals do have their densities in the expected direction.[40]

The electron density can be compared with experiment only through the resulting electrostatic potential or the corresponding multipole moments. In particular dipole moments have been calculated for a number of molecules.[27,28,51] Because of the spherical symmetry of the FSGOs and the fact that the overlap of two FSGOs is also an FSGO at an intermediate position, it turns out that the dipole moment can be simplified to a point charge calculation.[28] This has been emphasized by Hall and Tait,[52,53] who also showed the possibility of approximating the electrostatic potential in the neighborhood of the molecule by such means. If there are n FSGOs there will be $n(n+1)/2$ such point charges for the electronic distribution so that for a large molecule the number of point charges can be excessive. Shipman[54] has shown that this number can be reduced to n and still give the exact charge and dipole moment without serious error in the electrostatic potential near the molecule.

3.3.5. Energy Barriers and Molecular Conformations

In principle, the simple FSGO method should be able to predict molecular conformations as well as barriers to internal rotation. Conformations of several hydrocarbons have been investigated and turn out well. For example, results on the puckered rings in cyclobutane[38] and cyclopentane[42] have been published and preliminary results on cyclohexane[55] are as expected.

Rotational barriers in ethane,[31,56] propane,[38,51] and the butanes[38] have been calculated. Although the simple FSGO method gives results that are almost twice the expected values in energy, there is qualitative agreement in comparisons between molecules.

Another energy barrier of interest is the inversion barrier of the ammonia molecule. The simple FSGO model cannot adequately represent the lone pair in the planar form of the molecule, and it is therefore not surprising that an absurdly high energy is obtained.

3.3.6. Force Constants

Since the present method obtains the equilibrium geometry by searching for the minimum on the nuclear potential energy hypersurface, it can also calculate force constants as the curvature at the minimum. Hartree–Fock calculations generally give rather poor high values for force constants and it is not surprising that the present method likewise yields poor results.[57,58]

That force constants are high is no doubt related to the high values for rotational barriers and is a result of limiting the wave function to a single Slater determinant and thus omitting correlation energy.

3.3.7. Simple FSGO Results by Other Workers

Linnett and co-workers have carried out a considerable number of calculations by the simple method and various extensions of it. In addition to references given above, simple FSGO calculations have been applied to the ion clusters H_n^+, LiH_n^+.[59,60] Crystalline LiH both in the bulk[61] and surface properties[62] has been treated. Ethyl and propyl carbonium ions were studied in detail.[63] Orthogonal energy-localized orbitals, similar to those of Edmiston and Ruedenberg,[17] were produced from FSGOs in the fluoride ion.[64]

Ray[65] has calculated lithium and lithium hydride ions as well as protonated ethane.[66] Other proton affinities have been handled by Tamássy-Lentei and Szaniszló[67] as well as the small atom ions Li^+–He and Li_2^{++}.[68] Tinland and Decoret[69] studied the conformation of 1, 4-dioxadiene.

4. Comparison with Standard SCF Methods

In view of the very approximate energy values obtained in the FSGO method it can only be compared with the simplest of SCF methods using a minimal basis set.

Pople and Beveridge[70] have presented criteria for a suitable approximate method and favor a minimal basis set SCF-LCAO-MO calculation, inasmuch as no preconceived notions about electron pair bonds, for example, should be permitted. However, by assuming electron pairs in bonds the FSGO method can use an absolute minimal basis set and avoid SCF iterations, thus economizing on computer time.

Although SCF-LCAO-MO calculations are often made only for the experimentally known molecular geometry, Lathan *et al.*[71] have optimized the geometry of numerous small hydrocarbons. Some of these results for the STO-3G minimal basis set are shown in Table 1 along with the FSGO results. It is evident that both methods give reasonably good agreement with experiment.

5. Extensions of the Simple Method

5.1. Ellipsoidal Gaussians

Several workers have concerned themselves with the attempt to improve on the simple FSGO model by replacing the spherical Gaussians by ellipsoidal Gaussians (FEGOs or EGTOs). In a single bond one might expect to have a prolate spheroidal orbital with long axis along the internuclear axis. Lone-pair orbitals might possibly be either prolate or oblate. In any case an increase in flexibility of the wave function can lead to lower energy and improved wave function.

Katriel and Adam[72] and Simons and Schwartz[73] treated small diatomic and triatomic molecules and ions. van Duijnen and Cook[74] considered H_2O, NH_3, and several hydrocarbons. Vescelius[44] made calculations on similar molecules, as well as others. Results on the same molecules by different workers were in good agreement and as generally expected for the preferred shape of bonding orbitals. The lone pair of the ammonia molecule interestingly turns out to be oblate with its symmetry axis as the axis of the molecule.

Total energies obtained by this method are slightly improved over those from the simple FSGO. However, the calculations required numerical integrations and so were much more involved than for FSGO. It is doubtful that it is worthwhile to press FEGOs further.

5.2. Group Orbitals

By "group orbital" is meant a Lewis basis set orbital for a bond, lone pair, or inner shell that is a linear combination of two or more FSGOs. One such case is the use of double concentric FSGOs. This provides two improvements, one being the approach to a cusp on the inner shell, the other being improvement in the falloff of the orbital at larger distances. Rouse[73] studied the first row hydrides and additional hydrocarbons and found that the total energies were improved from 84% up to 96% of the Hartree–Fock values. Also bond angles in H_2O and NH_3 were improved somewhat.

Another aspect of the group orbital concept is to represent a bonding orbital as a linear combination of two or more FSGOs positioned along the bond and possibly including FSGOs on the nuclei as well. It was shown by Rothenberg and Schaefer,[75] in an SCF methane calculation, that to include in the basis set "bond functions," or FSGOs, placed along the C–H bonds resulted in an improved energy value more efficiently than using conventional polarization p or d orbitals on the atoms.

Barthelat and Durand,[76] using the Lewis or absolute minimal basis sets, treated methane with an inner shell group orbital of up to four FSGOs and C–H bond orbitals formed from one or two floating FSGOs. Nelson[41] made extensive calculations on ethane in which up to four FSGOs for each C–H and C–C bond were used, some or all of which were floating. The rotational barrier was greatly improved in the process at the expense of having to perform extensive optimizations on the many parameters. Blustin and Linnett[56] have also treated ethane using both one and two FSGOs for the C–C bond, and have paid particular attention to the various contributions to the energy as they affect the rotational barrier.

Another aspect of group orbitals is the possibility of improving the form of lone-pair orbitals. Tan and Linnett[77] studied the effect on the bond angle in NH_3 by taking the N lone-pair orbital to be a linear combination of as many as five FSGOs, one of which was taken to be a negative lobe on the axis inside the molecule. As the orbital was successively improved the bond angle increased from the simple FSGO value to 106.3°, surprisingly close to the experimental value of 106.7°. Similar calculations were performed by K. Müller.[78]

5.3. SCF Calculations with FSGOs

In the preceding section group orbitals for bonding, etc., were taken to be linear combinations of two or more FSGOs which could float to optimize the function. Having then more FSGOs than electron pairs a more general wave function can be obtained by forming molecular orbitals as linear combinations of the whole set of FSGOs as basis functions, followed by the SCF iteration procedure to optimize the coefficients. Such a method is called SCF-MO-P(LCGO) by Preuss.[12,79] An example is the application to H_2O by Schmittinger.[80] Ford et al.[81] made similar calculations on H_2O as well as CH_2O and hydrocarbons.

Tropis and Durand[82] did an SCF study of ethylene using several floating orbitals. Archibald et al.[83] have used a rather large floating orbital basis to calculate several properties of NH_3.

SCF methods have an advantage over the simple FSGO method in treating open shells. An exception is the work of Blustin and Linnett[84] on small odd-electron systems.

5.4. Molecular Fragment Method

Christoffersen and co-workers[85–88] have made extensive SCF calculations with FSGOs by building up large molecules out of small fragments. The small fragments, such as CH_4 as a fragment for saturated hydrocarbons, and a planar CH_3 radical, as a fragment for unsaturated hydrocarbons, are first optimized wtih respect to nonlinear parameters. The combination of fragments in forming a large molecule provides a set of basis orbitals numbering in excess of the number of electron pairs. An SCF iteration procedure is then carried out to find canonical molecular orbitals in linear combinations of the FSGOs.

For example, in forming ethane from two methane fragments an H atom is removed from each methane. The two vacant C—H bonding FSGOs are then taken as a pair of orbitals on the C—C bond axis of ethane. There is then one extra orbital in the basis and one virtual molecular orbital is produced along with the nine occupied MOs.

Rather large molecules have been studied by this procedure including several of biochemical significance, such as peptides.[87]

6. Conclusions

The FSGO method began with the idea, from a pedagogical standpoint, of representing molecular electronic and geometric structure in the simplest possible *ab initio* manner. It gives a representation of the molecule which is closely related to old familiar chemical concepts such as the electron pairs of G. N. Lewis. By a linear transformation of the local FSGO orbitals it also relates to modern molecular orbital theory.

Although because of the crudity of the basis orbitals it could not have been expected to produce good numerical values of molecular properties, nevertheless certain properties, especially bond lengths, are given to surprising accuracy. Conformations of molecules also are well represented.

Extensions of the theory using an extended set of floating orbitals can give more precise results but if, and until, a standard procedure for incorporating more orbitals in the basis is devised it is doubtful that FSGO methods can compete with well-developed SCF-LCAO-MO methods. The exception might be in the case of very large molecules where accurate calculations would be exceedingly expensive in computer time. In such a case a simplified method could be useful in producing preliminary results. Further simplifications of the simple FSGO method could even be useful such as using a Mulliken approximation to simplify or eliminate integrals[89] or the use of pseudopotentials to represent inner shells of heavier atoms.[90]

ACKNOWLEDGMENT

The National Science Foundation has provided a research grant to Northwestern University to aid this research.

Addendum

A recent paper by Amos and Yaffe[91] adds to the results in Section 3.3.4 above. They have used perturbation theory to calculate electric polarizabilities of several small molecules, H_2, LiH, H_2O, NH_3, CH_4, and other hydrocarbons up to C_3H_6, using both FSGO and FEGO where the latter calculations exist.

Talaty, Schwartz, and Simons[92] have made a careful comparison of the FSGO method with conventional *ab initio* methods as well as with the INDO semiempirical procedure as applied to several molecules: N_2H_2, Li_2O, C_3H_4, and O_3. The results illustrate the strengths and weaknesses of the FSGO model.

References

1. H. F. Schaefer III, *The Electronic Structure of Atoms and Molecules*, Addison-Wesley Publishing Co., Reading, Mass. (1972).
2. L. C. Snyder and H. Basch, *Molecular Wave Functions and Properties*, John Wiley and Sons, New York (1972).
3. W. G. Richards, T. E. H. Walker, and R. K. Hinkley, *A Bibliography of Ab Initio Molecular Wave Function*, Oxford University Press, London (1972).
4. F. E. Harris, Quantum chemistry, *Annu. Rev. Phys. Chem.* **23**, 415–438 (1972); also previous reviews referenced therein.
5. C. C. J. Roothaan, New developments in molecular orbital theory, *Rev. Mod. Phys.* **23**, 69 (1951).
6. G. G. Hall, The molecular orbital theory of chemical valency. VIII. A method of calculating ionization potentials, *Proc. R. Soc. London, Ser. A* **205**, 541 (1951).
7. E. Clementi and D. R. Davis, Electronic structure of large molecular systems, *J. Comput. Phys.* **2**, 223 (1967).
8. J. C. Slater, *Quantum Theory of Molecules and Solids*, Vol. 1, McGraw-Hill Book Co., New York (1963).
9. P.-O. Löwdin, On the non-orthogonality problem connected with the use of atomic wave functions in the theory of molecules and crystals, *J. Chem. Phys.* **18**, 365 (1950).
10. S. F. Boys, Electronic wave functions. I. A general method of calculation for the stationary states of any molecular system, *Proc. R. Soc. London, Ser. A* **200**, 542 (1950).
11. I. Shavitt, The gaussian function in calculations in statistical mechanics and quantum mechanics, *Methods Comput. Phys.* **2**, 1 (1963).
12. H. Preuss, Bemerkungen zum Self-consistent-field-Verfahren und zur Methode der Konfigurationenwechselwirkung in der Quantenchemie, *Z. Naturforsch.* **11a**, 823 (1956).
13. J. L. Whitten, Gaussian expansion of hydrogen-atom wave functions, *J. Chem. Phys.* **39**, 349 1963.
14. W. J. Hehre, R. F. Stewart, and J. A. Pople, Self-consistent molecular-orbital methods. I. Use of Gaussian expansions of Slater-type atomic orbitals, *J. Chem. Phys.* **51**, 2657 (1969).

15. R. P. Hosteny, R. R. Gilman, T. H. Dunning, A. Pipano, and I. Shavitt, Comparison of Slater and contracted Gaussian basis sets in SCF and CI calculations on H_2O, *Chem Phys. Lett.* **7**, 325 (1970).

16. V. Fock, Näherungsmethode zur Lösung des quantenmechanischen Mehrkörperproblems, *Z. Phys.* **61**, 126 (1930); see also Ref. 5.

17. C. Edmiston and K. Ruedenberg, Localized atomic and molecular orbitals, *Rev. Mod. Phys.* **35**, 457 (1963).

18. D. Peters, Computation of the SCF molecular orbital wavefunction in terms of localized orbitals, *J. Chem. Phys.* **46**, 4427 (1967).

19. J. R. Hoyland, *Ab initio* bond-orbital calculations. I. Application to methane, ethane, propane and propylene, *J. Am. Chem. Soc.* **90**, 2227 (1968).

20. J. H. Letcher and T. H. Dunning, Localized orbitals. I. σ bonds *J. Chem. Phys.* **48**, 4538 (1968).

21. R. D. Brown, A quantum-mechanical treatment of aliphatic compounds. Part I. Paraffins, *J Chem. Soc.* **1953**, 2615 (1953).

22. W. C. Herndon, Unparameterized localized orbital calculations for saturated hydrocarbons, *Chem. Phys. Lett.* **10**, 460 (1971).

23. G. N. Lewis, The atom and the molecule, *J. Am. Chem. Soc.* **38**, 762 (1916); *Valence and the Structure of Atoms and Molecules*, Chemical Catalog Co., New York (1923).

24. G. E. Kimball and G. F. Neumark, Use of Gaussian wave functions in molecular calculations, *J. Chem. Phys.* **26**, 1285 (1957).

25. G. F. Neumark, Free Cloud Approximation to Molecular Orbital Calculation, Ph.D. dissertation, Columbia University (1951); University Microfilms Publication No. 2845.

26. H. Preuss, Ein neues Programm zur quantentheoretischen Berechnung von Molekülen und Atomsystem. I. Grundlagen, *Mol. Phys.* **8**, 157 (1964).

27. A. A. Frost, B. H. Prentice III, and R. A. Rouse, A simple floating localized orbital model of molecular structure, *J. Am. Chem. Soc.* **89**, 3064 (1967).

28. A. A. Frost, Floating spherical Gaussian orbital model of molecular structure. I. Computational procedure. LiH as an example, *J. Chem. Phys.* **47**, 3707 (1967).

29. A. A. Frost, Floating spherical Gaussian orbital model of molecular structure. II. One- and two-electron-pair systems, *J. Chem. Phys.* **47**, 3714 (1967).

30. A. A. Frost, A floating spherical Gaussian orbital model of molecular structure. III. First-row atom hydrides, *J. Phys. Chem.* **72**, 1289 (1968).

31. A. A. Frost and R. A. Rouse, A floating spherical Gaussian orbital model of molecular structure. IV. Hydrocarbons, *J. Am. Chem. Soc.* **90**, 1965 (1968).

32. A. A. Frost, R. A. Rouse, and L. Vescelius, A floating spherical Gaussian orbital model of molecular structure. V. Computer programs, *Int. J. Quantum Chem.* **IIS**, 43 (1968).

33. R. A. Rouse and A. A. Frost, Floating spherical Gaussian orbital model of molecular structure. VI. Double-Gaussian modification, *J. Chem. Phys.* **50**, 1705 (1969).

34. A. A. Frost, A floating spherical Gaussian orbital model of molecular structure VII. Borazane and diborane, *Theor. Chim. Acta* **18**, 156 (1970).

35. S. Y. Chu and A. A. Frost, Floating spherical Gaussian orbital model of molecular structure. VIII. Second-row atom hydrides, *J. Chem. Phys.* **54**, 760 (1971).

36. S. Y. Chu and A. A. Frost, Floating spherical Gaussian orbital model of molecular structure. IX. Diatomic molecules of first-row and second-row atoms, *J. Chem. Phys.* **54**, 764 (1971).

37. A. A. Frost, The potential energy surface of the H_3 system using floating Gaussian orbitals, *Adv. Chem. Phys.* **21**, 65 (1971).

38. J. L. Nelson and A. A. Frost, A floating spherical Gaussian orbital model of molecular structure. X. C_3 and C_4 saturated hydrocarbons and cyclobutane, *J. Am. Chem. Soc.* **94**, 3727 (1972).

39. J. L. Nelson, and A. A. Frost, A floating spherical Gaussian orbital model of molecular structure. ESCA chemical shifts for inner shell electrons for small hydrocarbons, *Chem. Phys. Lett.* **13**, 610 (1972).

40. M. Afzal and A. A. Frost, A floating spherical Gaussian orbital model of molecular structure. XII. Analysis of energy terms affecting the geometry of the water molecule, *Int. J. Quantum Chem.* **VII**, 51 (1973).

41. J. L. Nelson and A. A. Frost, Local orbitals for bonding in ethane, *Theor. Chim. Acta* **29**, 75 (1973).

42. J. L. Nelson, C. C. Cobb, and A. A. Frost, FSGO investigation of several conformers of cyclopentane, *J. Chem. Phys.* **60**, 712 (1974).

43. J. Bicerano and A. A. Frost, FSGO calculations of octahydrotriborate anion and tetraborane, *Theor. Chim. Acta* **35**, 71 (1974).

44. L. Vescelius and A. A. Frost, Ellipsoidal Gaussian molecular calculations, *J. Chem. Phys.* **61**, 2983 (1974).

45. L. Shipman and R. E. Christoffersen, High speed evaluation of $F_0(x)$, *Comput. Phys. Commun.* **2**, 201 (1971).

46. D. B. Neumann, H. Basch, R. L. Kornegay, L. C. Snyder, J. W. Moskowitz, C. Hornback, and S. P. Liebmann, POLYATOM (Version 2) System of Programs for Quantitative Theoretical Chemistry, Program 199, Quantum Chemistry Program Exchange, Indiana University, Bloomington, Indiana, 1972.

47. M. A. Robb and I. G. Csizmadia, The generalized separated electron pair model. III. An application to three localization schemes for CO, *Int. J. Quantum Chem.* **6**, 367 (1972).

48. M. A. Robb, W. J. Haines, and I. G. Csizmadia, A theoretical definition of the "size" of electron pairs and its stereochemical implications, *J. Am. Chem. Soc.* **95**, 42 (1973).

49. P. E. Cade and W. M. Huo, Electronic structure of diatomic molecules. VI. A. Hartree-Fock wavefunctions and energy quantities for the ground states of the first-row hydrides, AH, *J. Chem. Phys.* **47**, 614 (1967).

50. M. Jungen, Das "floating Gaussian orbital"-Modell als Hilfsmittel zur interpretation von Photoelektronen-Spektren, *Theor. Chim. Acta* **22**, 255 (1971).

51. P. H. Blustin and J. W. Linnett, Applications of a simple molecular wavefunction. Part 1. Floating spherical Gaussian orbital calculations for propylene and propane, *J. Chem. Soc., Faraday Trans.* 2, **70**, 274 (1974).

52. G. G. Hall, Point charge models for molecular properties, *Chem. Phys. Lett.* **20**, 501 (1973).

53. A. D. Tait and G. G. Hall, Point charge models for LiH, CH_4 and H_2O, *Theor. Chim. Acta* **31**, 311 (1973).

54. L. L. Shipman, Derivation of a total charge and dipole moment-preserving population analysis for FSGO wavefunctions, *Chem. Phys. Lett.* **31**, 361 (1975).

55. T. D. Davis and A. A. Frost, FSGO investigation of several conformers of cyclohexane, *J. Am. Chem. Soc.* **97**, 7410 (1975).

56. P. H. Blustin and J. W. Linnett, Applications of a simple molecular wavefunction. Part 2. Torsional barrier in ethane, *J. Chem. Soc., Faraday Trans.* 2 **70**, 290 (1974).

57. J. R. Easterfield and J. W. Linnett, Applications of a simple molecular wavefunction, Part 4. The force fields of BH_4^-, CH_4 and NH_4^+, *J. Chem. Soc., Faraday Trans.* 2 **70**, 317 (1974).

58. S. L. Schulman, *Ab initio* Calculation of Spectroscopic Constants Using the FSGO Model, Ph.D. dissertation, Northwestern University, Evanston, Illinois (1973).

59. J. Easterfield and J. W. Linnett, The ions H_n^+ and the possibility of LiH_n^+ and BeH_n^+, *Chem. Commun.* **1970**, 64.

60. J. Easterfield and J. W. Linnett, Theoretical calculations on the ion cluster $Li(H_2)_n^+$ and $BeH(H_2)_n^+$, *Nature* **226**, 141 (1970).

61. W. D. Erickson and J. W. Linnett, Gaussian orbital calculations of solids. Crystalline lithium hydride, *J. Chem. Soc., Faraday Trans.* 2 **68**, 693 (1972).

62. W. D. Erickson and J. W. Linnett, An *ab initio* Gaussian orbital calculation of the (100) surface of crystalline lithium hydride, *Proc. R. Soc. London, Ser. A* **331**, 347 (1972).

63. P. H. Blustin and J. W. Linnett, Applicatons of a simple molecular wavefunction. Part 3. Ethyl and propyl carbonium ions, *J. Chem. Soc., Faraday Trans.* 2 **70**, 297 (1974).

64. R. A. Suthers and J. W. Linnett, On the localized nature of FSGO wavefunctions, *Chem. Phys. Lett.* **25**, 84 (1974).

65. N. K. Ray, Floating spherical Gaussian orbital model calculation for LiH_2^+, Li_2H^+, and Li_3^+, *J. Chem. Phys.* **52**, 463 (1970).

66. N. K. Ray, Theoretical study of the structure of protonated ethane $(C_2H_7^+)$, *Theoret. Chim. Acta* **23**, 111 (1971).

67. I. Tamássy-Lentei and J. Szaniszló, Calculation of the proton affinity of several small molecules by the FSGO method, *Acta Phys. Acad. Sci. Hung.* **35**, 201 (1974).

68. I. Tamássy-Lentei and J. Szaniszló, Interaction of atoms and ions with two electrons, *Acta Phys. Chim. Debrecina.* **18**, 61 (1972).

69.. B. Tinland and C. Decoret, An *ab initio* study of the conformation of 1,4-dioxadience, *J. Mol. Struct.* **9**, 205 (1971).

70. J. A. Pople and D. L. Beveridge, *Approximate Molecular Orbital Theory*, McGraw-Hill Book Co., New York (1970), p. 58.

71. W. A. Lathan, W. J. Hehre, and J. A. Pople, Molecular orbital theory of the electronic structure of organic compounds. VI. Geometries and energies of small hydrocarbons, *J. Am. Chem. Soc.* **93**, 808 (1971).

72. J. Katriel and G. Adam, Comparative study of ellipsoidal Gaussians: H_2 and He_2^{++}, *J. Chem. Phys.* **53**, 302 (1970).

73. G. Simons and A. K. Schwartz, Floating ellipsoidal Gaussian orbital computations on small molecules, *J. Chem. Phys.* **60**, 2272 (1974).

74. P. T. van Duijnen and D. B. Cook, *Ab initio* calculations with small ellipsoidal gaussian basis sets. I and II, *Mol. Phys.* **21**, 475 (1971); **22**, 637 (1971).

75. S. Rothenberg and H. F. Schaefer III, Methane as a numerical experiment for polarization basis function selection, *J. Chem. Phys.* **54**, 2764 (1971).

76. J. C. Barthelat and P. Durand, Orbitales moléculaires localisées: structure électronique de la molecule de méthane, *Theoret. Chim. Acta* **27**, 109 (1972).

77. L. P. Tan and J. W. Linnett, The lone-pair orbital in NH_3 and the calculation of the HNH angle, *J. Chem. Soc. D.* **1973**, 736.

78. K. Müller, private communication, 1972.

79. H. Preuss, Das SCF-MO-P(LCGO)-Verfahren und seine Varianten, *Int. J. Quantum Chem.* **II**, 651 (1968).

80. P. Schmittinger, Das Wassermolekül nach dem SCF-MO-P(LCGO)-Verfahren, *Z. Natur-forsch.* **26a**, 1411 (1971).

81. B. Ford, G. G. Hall, and J. C. Packer, Molecular modelling with spherical Gaussians, *Int. J. Quantum Chem.* **IV**, 533 (1970).

82. M. Tropis and P. Durand, Description des orbitales σ et π de la molécule d'éthylene par une base réduite de fonctions Gaussionnes sphériques, *C. R. Acad. Sci. Paris* **C276**, 1775 (1973).

83. R. M. Archibald, D. R. Armstrong, and P. G. Perkins, Molecular calculations using spherical Gaussian orbitals. Part 1.—Optimisation of the atomic parameters for first-row atoms, *J. Chem. Soc., Faraday Trans. 2* **70**, 1557 (1974).

84. P. H. Blustin and J. W. Linnett, Applications of a simple molecular wavefunction. Part 5 —Floating spherical Gaussian orbital open-shell calculations: Introduction, *J. Chem. Soc., Faraday Trans. 2* **70**, 327 (1974).

85. R. E. Christoffersen, D. W. Genson, and G. M. Maggiora, *Ab initio* calculations on large molecules using molecular fragments. Hydrocarbon characterizations, *J. Chem. Phys.* **54**, 239 (1971).

86. R. E. Christoffersen, *Ab initio* calculations on large molecules, *Adv. Quantum Chem.* **6**, 333 (1972).

87. L. L. Shipman and R. E. Christoffersen, *Ab initio* calculations on large molecules using molecular fragments. Model peptide studies. Polypeptides of glycine, *J. Am. Chem. Soc.* **95**, 1408, 4733 (1973).

88. T. D. Davis, G. M. Maggiora, and R. E. Christoffersen, *Ab initio* calculations on large molecules using molecular fragments. Unrestricted Hartree–Fock calculations on the low-lying states of formaldehyde and its radical ions, *J. Am. Chem. Soc.* **96**, 7878 (1974).

89. G. Nicolas and P. Durand, Orbitales moléculaires localisées ou fonctions de Loge et approximation de Mulliken, *C. R. Acad. Sci. Paris* **C272**, 1482 (1971).

90. J. C. Barthelet and P. Durand, Pseudopotentials and localized molecular orbitals. Application to the methane molecule, *Chem. Phys. Lett.* **16**, 63 (1972).

91. A. T. Amos and J. A. Yaffe, The Frost model and perturbation theory, *Chem. Phys. Lett.* **31**, 57 (1975).

92. E. R. Talaty, A. K. Schwartz, and G. Simons, Simple *ab initio* studies of the isomers of N_2H_2, Li_2O, C_3H_4 and O_3, *J. Am. Chem. Soc.* **97**, 972 (1975).

3

The Multiconfiguration Self-Consistent Field Method

Arnold C. Wahl
and
G. Das

1. Introduction

The multiconfiguration self-consistent field (MCSCF method), which consists of optimizing both the mixing of several configurations as well as the orbitals of which they are composed, would appear to be a natural extension of the one-configuration SCF method. However, owing to a number of factors the method has not received a great deal of attention until rather recently. First, as recently as a decade ago, SCF wave functions were believed to be "sufficiently" accurate for most of the interesting properties of a chemical system. Second, the successful implementation of the single configuration SCF method was difficult and the implementation of the more complex MCSCF framework proved to be significantly more arduous.

While for one-center problems exact numerical solutions to the Hartree–Fock equations had become a reality in the fifties, thanks mainly to the efforts of Hartree and Hartree,[1,2] and have since been perfected and extended to incorporate relativistic effects in the hands of several authors,[3,4] the progress on the molecular front has not been quite as complete. SCF calculations on chemically interesting systems have so far been carried out only in Roothaan's LCAO formalism. The choice of basis set, although fairly well standardized for electronic ground states is still a serious problem for excited states. The

Arnold C. Wahl and G. Das • Chemistry Division, Argonne National Laboratory, Argonne, Illinois

iterative use of large supermatrices required by the formalism makes a heavy demand on both time and machine-accessible storage of the computer. In particular, the open-shell systems can still be a source of frustration even in the SCF approximation, mainly because of convergence problems.

The belief that the SCF can yield satisfactory accuracy, though justified with respect to some properties in some systems, can be very wrong with respect to others. That MOSCF (molecular orbital SCF) can lead to a meaningless chemical prediction was dramatically established by calculations[5] on the F_2 molecule. These calculations, using a carefully optimized "accurate" basis set, yielded a negative binding energy for F_2 as large in magnitude as the actual value of 1.6 eV. It therefore follows that, at least with an MO description, one can have qualitatively incorrect results with the SCF wave function.

Quantum chemistry apart, configuration-mixing or configuration interaction (CI) as it is popularly called can be very important in atomic spectroscopy, electron-atom scattering, and other atomic phenomena. van der Waals interaction, i.e., the attractive long-range interaction of two systems, cannot be explained without going beyond SCF. Generally speaking, the SCF wave function can only be regarded as a starting point for more accurate calculation and should be used very cautiously in making predictions on the chemical behavior of a system.

Of course it must always have been clear that an exact solution of the Schrödinger equation can be represented by a linear superposition of an infinitely large number of configurations. Even something akin to the MCSCF concept was attempted by Frenkel[6] as early as the 1930s. Presumably, better "absolute accuracy" in the wave function can be achieved by considering just a few additional configurations. But it is by no means clear whether, for example, the relative accuracy between the many-configuration wave functions of the separated atoms and the corresponding one when they combine to form a molecule (measured usually in terms of the interaction potential thus obtained) is also improved, unless one considers a rather large set of such configurations.

The need for using more than one configuration in the wave function is usually discussed in terms of electron correlation, defined to be the total difference of the exact nonrelativistic Hartree–Fock energy and the exact nonrelativistic energy of a given system. There is, however, much more physical content in the configuration description than in the concept of electron correlation, as we will see later. It will also be seen that provided the right configurations are chosen, the MCSCF procedure in general leads to a better description of the interaction phenomena between atomic or molecular systems than a single SCF configuration.

There are, of course, other methods for obtaining electron correlation. Closely parallel is what is known in the literature as the iterative natural orbital (INO) method of Bender and Davidson.[7] We shall later describe this method in some detail. The higher-order perturbation theory[8] has been used in various systems with varying degrees of success. The "pair-theory" used by Nesbet,[9] Sinanoğlu, and others,[10] which consists in evaluating correlation of all pairs or orbitals occupied in the Hartree–Fock configuration, has been successful in calculations on many systems, although the lack of a variational principle makes some view the results with suspicion. The Kelly–Bethe–Goldstone diagram techniques have also been highly successful in various contexts. The equations-of-motion method of McKoy *et al.*[12] and the many-body approach of Schneider *et al.*[13] are particularly successful in evaluating and predicting excitation energies, ionization potentials, reactive-scattering cross section, and other dynamic properties of many atoms and molecules. the preference for using the MCSCF technique, of course, is the fact that for many steady-state phenomena, and particularly to evaluate accurate and reliable wave functions and interaction potentials, the MCSCF technique has proven itself to be quite effective.

2. CI-Matrix Elements and Solution of the Secular Equation

We assume that the total wave function is a linear superposition of configurations given by

$$\Psi^{(n)} = \sum_a A_a^{(n)} \Phi_a \tag{1}$$

where the superscript (n) corresponds to the nth root of the secular equation

$$\det|E_{ab} - E\,\delta_{ab}| = 0 \tag{2}$$

It is assumed above that the configurations $|\Phi_a|$ are orthonormal and E_{ab} is the "matrix element" of the Hamiltonian:

$$E_{ab} = \langle \Phi_a | H | \Phi_b \rangle \tag{3}$$

Before we discuss how to construct E_{ab}, we have first to discuss Φ_a. The first step is, of course, to select a starting set of "spin-orbitals" $\{\Phi_i^{so}\}$. The SCF calculations can obviously furnish the orbitals that are occupied in the SCF configuration. However, the extra orbitals needed to construct the configurations Φ_a are often very different from the "virtual" orbitals obtained from the SCF calculations. We shall later discuss various ways in which the starting non-SCF orbitals can be chosen.

We can write Φ_a formally as

$$\Phi_a = \sum u_{at} \Phi_t \tag{4}$$

where the Φ_t are Slater determinants written as

$$\Phi_t = |n_1^t, n_2^t, \ldots, n_m^t\rangle \tag{5}$$

the n_i^t being occupation numbers (either 0 or 1) of the spin-orbitals i, and the u_{at} are coupling coefficients defined by the symmetry of the state represented by $\Psi^{(n)}$.

We shall now specialize to the "restricted" form of the spin-orbitals by way of grouping the spin-orbitals in what have been customarily called "shells," characterized by species index λ and an ordering index i. For example, "closed" shells are composed of the fully occupied spin-orbitals differing by subspecies indices and spin-projections but constrained to have the same spatial form referred to as the orbital functions or simply orbitals. For "open" shells the spin-orbitals may not be fully occupied but they occur with the same spatial form in all the determinants of Eq. (4), while assuming various values of the subspecies indices and spin-projections. Such a restricted picture, of course, is not an approximation in an MCSCF formalism since the deficiencies caused by such restriction can be immediately removed by including appropriate addition configurations. It is, however, well-established that the deficiencies are usually negligible and no special efforts are necessary to remove them.

Thus the spin-orbitals in various shells will have the form

$$\phi_{i\lambda\alpha\xi} \equiv \phi_{i\lambda}f_{\lambda\alpha}\xi \tag{6}$$

where $\phi_{i\lambda}$ defines the spatial form of the shells, λ the symmetry species, and $f_{\lambda\alpha}$ and ξ are, respectively, the orbital and spin functions defining the subspecies and the spin-projection.

We shall now express the matrix elements (3) in terms of $\phi_{i\lambda}$. We shall drop the symmetry index λ, for brevity. There are various cases we will have to distinguish.

1. Diagonal Matrix Elements: $a = b$

$$H_{aa}^{(1)} = \langle \varphi_a | H | \Phi_a \rangle = \sum_i n_{ia} \langle i | h + \sum_\nu \sum_j n_{ja} t_{ija}^\nu I_{jj}^\nu | i \rangle \tag{7}$$

where h is the one-electron bare-nuclei Hamiltonian, the n_{ia} are the occupancies of the shells i in the configuration $\Phi_{a'}$ and the t_{ija}^ν are the vector coupling coefficients multiplying the two-electron integrals I_{jj}^ν, considered in what follows.

2. Off-Diagonal Elements: Φ_a and Φ_b differing by a "single excitation." Let the single excitation $\Phi_a \to \Phi_b$ correspond to an electron jumping from the shell i to shell j. we shall designate such an excitation as $a \to b \equiv i \to j$. The matrix element is given by

$$H_{ab}^{(2)} = \langle \Phi_a | H | \Phi_b \rangle \equiv \langle i | F_{ab,ij} | j \rangle \tag{8}$$

where

$$F_{ab,ij} = n_{ab,ij} h + \sum_k t_{abijk}^\nu I_{kk}^\nu$$

n_{ab}, $t_{ab,k}^\nu$ being one-electron and two-electron vector-coupling constants, respectively.

3. *Off-Diagonal Elements*: Φ_a and Φ_b differ by a "diagonal double excitation," where two electrons jump from shell i to shell j $(a \to b \equiv ii \to jj)$. The matrix element is given by

$$H^{(3)}_{ab} \equiv \langle \Phi_a | H | \Phi_b \rangle = \sum_{\nu} d^{\nu}_{ab} \langle i | I^{\nu}_{jj} | i \rangle \tag{9}$$

4. *Off-Diagonal Elements*: Φ_a and Φ_b differ by a "split-shell double excitation," where two electrons from the shells i and j jump to the shells k and l $(a \to b \equiv ij \to kl)$. The matrix element is given by

$$H^{(4)}_{ab} \equiv \langle \Phi_a | H | \Phi_b \rangle = \sum_{\nu} t^{\nu}_{ab} \langle i | I^{\nu}_{kl} | j \rangle \tag{10}$$

5. *Off-Diagonal Elements*: Φ_a and Φ_b have identical occupation numbers differing only in the couplings of the various open-shells. The form of the matrix element is

$$H^{(5)}_{ab} \equiv \langle \Phi_a | H | \Phi_b \rangle = \sum_{i \neq j, O_{ab}} t^{\nu}_{ab,ij} \langle i | I^{\nu}_{jj} | i \rangle \tag{11}$$

where O_{ab} is the set of all open shells occupied in Φ_a and Φ_b.

We shall write down the two-electron integrals I^{ν}_{ij} assuming the orbitals to have the symmetry-species of a diatomic molecule, i.e., those of the symmetry group $C_{\infty v}$. For this symmetry group the subspecies part $f_{\lambda \alpha}$ in Eq. (6) is simply $(2\pi)^{-1/2} \exp[im_{\lambda} \alpha \phi]$, where α can assume the values ± 1 and m_{λ} is a positive number characterizing the symmetry species λ and corresponds to the quantum number of the angular momentum projection along the molecular axis. We further restrict ourselves to those cases of configuration mixing for which none of the integrals involve more than two different symmetry species. Only six types of integrals arise in that case. They are [with the orbitals (i, j) belonging to the same species λ, say, and the orbitals (k, l) to μ]:

(i) *Coulomb Integrals of the First Kind*

$$J^{(1)}_{ijkl} = \langle i | I^{(1)}_{kl} | j \rangle = \int \int \phi_i(1) \phi_j(1) (1/r_{12}) \phi_k(2) \phi_l(2) \, d\mathbf{r}_1 \, d\mathbf{r}_2 \tag{12}$$

(ii) *Coulomb Integrals of the Second Kind*

$$J^{(2)}_{ijkl} = \langle i | I^{(2)}_{kl} | j \rangle = \int \int \phi_i(1) \phi_j(1) \frac{\exp[2im(\phi_1 - \phi_2)]}{r_{12}} \phi_k(2) \phi_l(2) \, d\mathbf{r}_1 \, d\mathbf{r}_2 \tag{13}$$

where $m_i = m_j = m_k = m_l = m$. Obviously $J^{(2)}$ integrals for $m = 0$ are identical with the Coulomb integrals of the first kind. Since at most two symmetry species are involved the Coulomb integrals of the second kind are to be considered only if all the shells i, j, k, l occurring in the integral belong to one of the non-σ species.

(iii) Symmetric Exchange Integrals of the First Kind

$$K_{ijkl}^{(+)s} \equiv \langle i|I_{kl}^{(3)}|j\rangle$$

$$= \frac{1}{2} \int\int [\phi_i(1)\phi_j(2) + \phi_j(1)\phi_i(2)]\frac{\exp[i(m_\lambda - m_\mu)(\phi_1 - \phi_2)]}{r_{12}}$$

$$\times \phi_k(1)\phi_l(2)\, d\mathbf{r}_1\, d\mathbf{r}_2 \qquad (14)$$

(iv) Symmetric Exchange Integrals of the Second Kind

$$K_{ijkl}^{(-)s} \equiv \langle i|I_{kl}^{(4)}|j\rangle$$

$$= \frac{1}{2} \int\int [\phi_i(1)\phi_j(2) + \phi_j(1)\phi_i(2)]\frac{\exp[i(m_\lambda + m_\mu)(\phi_1 - \phi_2)]}{r_{12}}$$

$$\times \phi_k(1)\phi_l(2)\, d\mathbf{r}_1\, d\mathbf{r}_2 \qquad (15)$$

(v) Antisymmetric Exchange Integrals of the First Kind

$$K_{ijkl}^{(+)a} \equiv \langle i|I_{kl}^{(5)}|j\rangle$$

$$= \frac{1}{2} \int\int [\phi_i(1)\phi_j(2) - \phi_j(1)\phi_i(2)]\frac{\exp[i(m_\lambda - m_\mu)(\phi_1 - \phi_2)]}{r_{12}}$$

$$\times \phi_k(1)\phi_l(2)\, d\mathbf{r}_1\, d\mathbf{r}_2 \qquad (16)$$

(vi) Antisymmetric Exchange Integrals of the Second Kind

$$K_{ijkl}^{(-)a} \equiv \langle i|I_{kl}^{(6)}|j\rangle$$

$$= \frac{1}{2} \int\int [\phi_i(1)\phi_j(2) - \phi_j(1)\phi_i(2)]\frac{\exp[i(m_\lambda + m_\mu)(\phi_1 - \phi_2]}{r_{12}}$$

$$\times \phi_k(1)\phi_l(2)\, d\mathbf{r}_1\, d\mathbf{r}_2 \qquad (17)$$

On forming the matrix elements we, either by straightforward diagonalization or by a Shavitt[14]-type procedure, obtain the eigenvalue of interest and the corresponding "mixing coefficients" of Eq. (2). It can be shown[15] that the nth eigenvalue of the secular matrix is an upper bound of the nth eigenvalue of the Hamiltonian operator itself. Construction of the matrix elements and the diagonalization of the Hamiltonian matrix are the two steps that constitute what is conventionally called CI (configuration interaction). If we have a very large number of configurations corresponding to all possible interaction and interplays of the valence electrons and an SCF-optimized core, the CI process is very often able to yield accurate wave functions.

The use of CI to go beyond SCF is by now well explored and well established and many successful calculations have been reported based on CI.

In the applications of CI, one does not ask whether the method can yield accurate results, since it is obvious that in the limit of a complete set of configurations, CI wave function is the exact wave function. The question is whether the basis set used is large enough or if the configuration set is large enough. With a large basis, however, the possible number of configurations also becomes large such that solution of the secular equation very quickly becomes quite a big task even with various existing ingenious schemes to handle large matrices. There exists a by-now fairly well-established means of reducing the number of configurations without seriously affecting the accuracy of the CI wave function. The theory of natural orbitals (NO), on which this method is based, was first proposed by Löwdin.[16] Consider the density function defined by

$$\rho(\mathbf{r}, \mathbf{r}') = N \int \int \int \Psi^{(n)}(\mathbf{r}, \mathbf{r}_2, \mathbf{r}_3, \ldots, \mathbf{r}_N) \Psi^{(n)}(\mathbf{r}', \mathbf{r}_2, \mathbf{r}_3, \ldots, \mathbf{r}_N) \, d\mathbf{r}_2 \, d\mathbf{r}_3 \cdots d\mathbf{r}_N$$

(18)

where $\Psi^{(n)}$ is defined as in Eq. (1). Then obviously in terms of the orbitals $\{\phi_i\}$,

$$\rho(r, r') = \sum_{ij} n_{ij} \phi_i(\mathbf{r}) \phi_j(\mathbf{r})$$

(19)

where

$$n_{ij} = \sum_{ab} n_{ab,ij} A_a^{(n)} A_b^{(n)}$$

The NOs are defined to be linear combinations of $\{\phi_i\}$ such that the density matrix n_{ij} is diagonal, i.e.,

$$\rho(r, r') = \sum_i n_i \phi_{NO,i}^*(\mathbf{r}) \phi_{NO,i}(\mathbf{r}')$$

(20)

The importance of NOs derives from the fact that they remove or "weaken" most of the "cross" terms from the matrix representation of the Hamiltonian. In other words, number of important configurations is drastically reduced. The method of iterative natural orbital (INO), first proposed and applied by Bender and Davidson,[7] starts by constructing a relatively small CI wave function, obtains NOs from this function, and uses them as the new orbitals for the next cycle until the method attains near self-consistency. This, as we will see later, can also be used as an effective means of solving the MCSCF problem.

3. MCSCF Theory and Orbital Optimization

3.1. Formulation

The basic assumption of the MCSCF theory is that if one knows how to pick the right description for the orbitals, most of the important chemical and physical aspects of interaction in a given system are represented by a wave function consisting of a few terms. These terms, the configurations, not only improve the total energy, but also the description of the physical and chemical processes that are associated with wave function changes.

The criteria for choosing the right description of the orbitals is of course the minimization of the total energy. Let us write the expression for the total energy as

$$E_{tot} = \sum_{a,b} A_a^{(n)} A_b^{(n)} E_{ab} \tag{21}$$

Consider the change in E_{tot} as the orbitals $\{\phi_i\}$ are varied by $\{\delta\phi_i\}$ in the form

$$\delta E_{tot} = 2 \sum_i \delta\phi_i F_i \phi_i \tag{22}$$

where the F_i terms are the so-called Fock operators.

The constraints under which the minimization has to be carried out are those of the orthonormality of the orbitals, i.e., for a minimum in E_{tot}, we must set the variation

$$\delta E_{tot} - \sum \varepsilon_{ij}(\delta\phi_i^+ \phi_j + \delta\phi_j^+ \phi_i) = 0 \tag{23}$$

where the ε_{ij} terms are the Lagrangian multipliers. This immediately yields the so-called Fock equations,

$$F_i \phi_i = \sum_j \varepsilon_{ij} \phi_j \tag{24}$$

with the additional "Hermiticity" conditions on the Lagrangian multipliers,

$$\varepsilon_{ij} = \varepsilon_{ji} \tag{25}$$

From Eq. (24), we can write

$$\varepsilon_{ij} = \phi_j^+ F_i \phi_i$$

such that Eq. (25) can be rewritten as

$$\phi_j^+(F_i - F_j)\phi_i = 0 \tag{26}$$

Before we discuss the solution of Eqs. (24) and (26), we consider first the actual forms of F_i and how they differ from the monoconfiguration SCF equations.

$$\delta E_{tot} = 2 \sum_a \delta A_a^{(n)} \sum_b A_b^{(n)} E_{ab} + \sum_{a,b} A_a^{(n)} A_b^{(n)} \delta E_{ab} \tag{27}$$

The first term on the right-hand side is zero since

$$\sum_b A_b^{(n)} E_{ab} = E^{(n)} A_a^{(n)} \tag{28}$$

and

$$\sum A_a^{(n)} \, \delta A_n^{(n)} = 0 \qquad (29)$$

by virtue of the normalization condition.

Thus it is necessary to consider only the variations involving the Fock operators F_i. We shall now write down explicitly the contributions to these operators from the five types of matrix elements considered in the last section in the form:

$$F_i = \sum_{n=1}^{5} F_i^{(n)}$$

We have

$$F_i^{(1)} = \sum_a A_a^2 n_{ia} \left(h + \sum_\nu \sum_j n_{ja} t_{ija}^\nu I_{jj}^\nu \right) \qquad (30)$$

$$F_i^{(2)} = \sum_{ab} \left[\sum_{k,l \neq i} \theta_{ab,kl} \sum_\nu t_{ab,i}^\nu I_{kl}^\nu + \sum_{j \neq i} \theta_{ab,ij} \left(f_{abj} \phi_i^+ + \phi_i f_{abj}^+ \right) \right] \qquad (31)$$

$$\theta_{ab,kl} = \begin{cases} 1 & \text{if } a \to b \equiv k \to 1 \\ 0 & \text{otherwise} \end{cases}$$

$$\qquad (32)$$

$$f_{abj} = \tfrac{1}{2} [n_{ab} h + \sum_\nu \sum_k t_{ab,k}^\nu I_k^\nu] \phi_j$$

$$F_i^{(3)} = \sum_{ab} \theta'_{ab,ii,jj} \sum_\nu t_{ab}^\nu I_{jj}^\nu$$

$$\theta'_{ab,ij,kl} = \begin{cases} 1 & \text{if } a \to b \equiv (ij) \to (kl) \\ 0 & \text{otherwise} \end{cases}$$

$$\qquad (33)$$

$$F_i^{(4)} = \sum_{ab,jkl} \theta'_{ab,ijkl} \left(f'_{ab,jkl} \phi_i^+ + \phi_i f'^+_{ab,jkl} \right) \qquad (34)$$

$$f'_{ab,jkl} = \tfrac{1}{2} \sum_\nu t_{ab}^\nu I_{kl}^\nu \phi_j \qquad (35)$$

$$F_i^{(5)} = \sum_{\nu ab,j} \theta''_{ab,ij} \sum_\nu t_{ab}^\nu I_{jj}^\nu$$

$$\qquad (36)$$

$$\theta''_{ab,ij} = \begin{cases} 1 & \text{if } i,j \subset O_{ab} \\ 0 & \text{otherwise} \end{cases}$$

It can be shown that with the Fock operators defined above, the total energy (17) can be re-expressed as

$$E_{\text{tot}} = \tfrac{1}{2} \sum_i \langle i | n_i h + F_i | i \rangle \tag{37}$$

where

$$n_i = \sum_a n_{ia} A_a^2 \tag{38}$$

3.2. Solution of the Fock Equations

It is obvious that the MCSCF equations are vastly more complex and numerous than the SCF equations, although the formal analogies do exist. While the orbitals in the SCF are usually only weakly coupled, the nature of the excited orbitals in the MCSCF theory is strongly dependent upon the orbitals from which they receive the excited electrons. This, therefore, makes it necessary to adopt a different approach to the solution of the MCSCF equations than that conventionally used for SCF.

The solution of the Fock equations for molecules has so far been successfully carried out only in the Roothaan's[17,18] LCAO framework. This consists of selecting a basis set adequate for orbital representation. The most commonly used basis functions are the Slater-type (STF) and Gaussian-type (FTD) functions. We then express all the orbitals to be used as a linear combination of these basis functions:

$$\phi_i = \sum c_{ip} \chi_p \tag{39}$$

Both the energy and the Fock equations are re-expressed in terms of $\{c_i\}$. The orthogonality constraints of the orbitals become

$$\sum_{pq} c_{ip} S_{pq} c_{jq} = \delta_{ij} \tag{40}$$

where $\{S_{pq}\}$ is the so-called "overlap" matrix:

$$S_{pq} = \langle \chi_p | \chi_q \rangle \tag{41}$$

Obviously the Fock equations now are matrix equations with Fock operators replaced by Fock matrices.

Thus, for example, Eq. (30) becomes

$$\mathbf{F}_i^{(1)} = \sum_a n_{ia} A_a^2 \left(\mathbf{h} + \sum_\nu \sum_j n_{ja} t_{ija}^\nu \mathcal{J}^\nu D_j^\nu \right) \tag{42}$$

where $\mathbf{h}$, given by

$$h_{pq} = \langle \chi_p | h | \chi_q \rangle \tag{43a}$$

is the one-electron matrix, $\mathscr{I}^\nu$ is given by

$$\mathscr{I}^\nu_{pq,rs} = \langle \chi_p | I^\nu_{rs} | \chi_q \rangle \tag{43b}$$

which is the νth "super-matrix," and $\mathbf{D}_j$, the density matrix, is defined as

$$D_{jpq} = c_{jp} c_{jq} \tag{43c}$$

The Fock equations take the general form

$$\mathbf{F}_i \mathbf{c}_i = \sum \varepsilon_{ij} \mathbf{S}' \mathbf{c}_j \tag{44}$$

with the matching conditions

$$\mathbf{c}_i^+ (\mathbf{F}_i - \mathbf{F}_j) \mathbf{c}_j = 0 \tag{45}$$

Using the orthogonality conditions and defining the matrices $\{\mathbf{R}_i\}$ as

$$\mathbf{R}_i = \sum_{j \neq i} [\mathbf{S} \mathbf{c}_j (\mathbf{F}_i \mathbf{c}_j)^+ + (\mathbf{F}_i \mathbf{c}_j) \mathbf{S} \mathbf{c}_j^+] \tag{46}$$

the Fock equations, Eq. (44), are reduced to a simple eingenvalue problem:

$$[\mathbf{F}_i - \mathbf{R}_i] \mathbf{c}_i = \varepsilon_i \mathbf{S} \mathbf{c}_i \tag{47}$$

In Eq. (46) the second term within the square brackets, which is identically zero when operating on $\mathbf{c}_i$, is inserted in order to have a Hermitian $\mathbf{R}_i$.

Let $\mathbf{c}_i^0$ be an approximate starting set of vectors and let

$$\mathbf{c}_i = \mathbf{c}_i^0 + \delta \mathbf{c}_i \equiv \mathbf{c}_i^0 + \delta \mathbf{c}_i^\nu + \delta \mathbf{c}_i^0 \tag{48}$$

where the $\mathbf{c}_i$ are the correct solutions to the Fock equations and $\delta \mathbf{c}_i$ has been broken up into a part $\delta \mathbf{c}_i^\nu$ that is orthogonal to all $\{\mathbf{c}_j^0\}$ and lies in the "virtual" space, and a part $\delta \mathbf{c}_i^0$ which is orthgonal to $\mathbf{c}_i^0$ but is really a linear combination of all $\{\mathbf{c}_j^0\}, j \neq i$. We notice that if $\mathbf{F}_j^0$ and $\mathbf{R}_i^0$ correspond to the vector $\mathbf{c}_i^0$, we have

$$\mathbf{c}_j^+ (\mathbf{F}_i^0 - \mathbf{R}_i^0) \mathbf{c}_i^0 = 0 \tag{49}$$

We now solve for $\delta \mathbf{c}_i^\nu$ and $\delta \mathbf{c}_i^0$ separately. Using Eq. (48) in Eq. (47) we have

$$[\mathbf{F}_i^{(1)} - \mathbf{R}_i^0 - \varepsilon_i^0 \mathbf{S}] \delta \mathbf{c}_i^\nu = -[\mathbf{F}_i^0 - \mathbf{R}_i^0 - \varepsilon_i^0 \mathbf{S}] \mathbf{c}_i^0 \tag{50}$$

where

$$\mathbf{F}_i^{(1)} \equiv \mathbf{F}_i^0 - (\mathbf{F}_i^0 \mathbf{c}_i^0)(\mathbf{S} \mathbf{c}_i^0)^+ - (\mathbf{S} \mathbf{c}_i)(\mathbf{F}_i^0 \mathbf{c}_i^0)^+ \tag{51}$$

and

$$\varepsilon_i^0 = \mathbf{c}_i^{0+} \mathbf{F}_i \mathbf{c}_i^0 \tag{52}$$

We observe that owing to Eq. (49), the $\delta \mathbf{c}_i^\nu$ as obtained from Eq. (50) continue to be outside the "occupied" space. In deriving Eq. (50), we have disregarded the variations of the two-electron terms in $\mathbf{F}_i$. Had we considered those terms, Eq. (47) would be replaced by a set of simultaneous inhomogeneous matrix equations in δc_i^ν. Solving such a set of equations would, however, be a great deal harder than Eq. (50), whose solution is obtained by simple matrix inversion. For three reasons it is unnecessary to consider variations of the F_i'. First, they are usually small compared with the terms on the left-hand side of Eq. (50). Second, Eq. (50) can be looked upon as defining the optimization of the vector c_i, when the rest of the vectors are kept frozen. Obviously such a process, when converged, i.e., when at the end of an nth iteration all $\{\delta c_i^\nu\}$ are zero, leads to the correct solution. Let $a_j, j = n+1, \ldots, m$ along with $\{c_i^0\}$ be the eigenvectors of the matrix $\mathbf{F}^{(1)}$ defined by Eq. (51) and $\varepsilon_j^{(1)} - \varepsilon_i^0$ be the corresponding eigenvalues. Then one can write

$$\delta c_i^\nu = - \sum_{j=n+1}^{m} \frac{\mathbf{a}_j^+ \mathbf{F}_i \mathbf{c}_i^0}{\varepsilon_j^{(1)} - \varepsilon_i^0} \mathbf{a}_j \tag{53}$$

This is to be compared to the first-order perturbative expression given by

$$\delta c_i'^\nu = - \sum_{j=n+1}^{m} \frac{\mathbf{a}_j^+ F_i \mathbf{c}_i^0}{E(i \to j) - E_0} \mathbf{a}_j \tag{54}$$

where E_0 is the energy corresponding to $\{c_i^0\}$ and $E(i \to j)$ is the energy when the orbital $\{c_i^0\}$ is singly excited to the virtual state $\{a_j\}$. The denominators in Eqs. (53) and (54) can be shown to be the same except for small "exchange" or "Coulomb-hybrid" terms such as $\langle i|I_{ij}^{(1)}|j\rangle$ or $\langle i|I_{ij}^{(1)}|i\rangle|\delta c_i'^\nu|$. It can be expected that in most cases δc_i^ν is correct, at least with respect to the sign of the coefficients multiplying a_j, which is an important criterion for the convergence of the essentially Newton–Raphson process adopted above for the solution of the Fock equations. We shall later discuss Eq. (54) in connection with alternative methods of solution of the MCSCF equations.

Solving for δc_i^0 is, however, quite a different story. Equation (45) is used for this purpose. We assume

$$\delta c_i^0 = \sum_{k \neq i} u_{ik} c_k^0 \tag{55}$$

We shall satisfy Eq. (45) in the first order in u_{ik}s and neglecting all interaction terms involving the occupied and virtual space. Replacing c_i and c_j by $c_i^0 + \delta c_i^0$ and $c_j^0 + \delta c_j^0$, where the δc's are defined by Eq. (55), we observe that

$$u_{ik} = -u_{ki} \tag{56}$$

Equation (45) leads to

$$u_{ij}(f_{ij,jj} - f_{ij,ii} + g_{ij}) + \sum_{k \neq i,j} [u_{jk} f_{ij,ik} + u_{ik} f_{ij,jk}] = -\Delta \varepsilon_{ij} \tag{57}$$

where

$$f_{ij,kl} = \mathbf{c}_k^{0+}(\mathbf{F}_i^0 - \mathbf{F}_j^0)\mathbf{c}_l^0$$

$$\Delta\varepsilon_{ij} = \mathbf{c}_i^{0+}(\mathbf{F}_i^0 - \mathbf{F}_j^0)\mathbf{c}_j^0 \qquad (58)$$

$$g_{ij} = \mathbf{c}_i^{0+}\frac{\delta(F_i^0 - F_j^0)}{\delta u_{ij}}\mathbf{c}_j$$

with $\delta M/\delta u_{ij}$ defined as

$$\frac{\delta M}{\delta u_{ij}} = \lim_{u_{ij}\to 0} \frac{M(\mathbf{c}_1, \mathbf{c}_2, \ldots, \mathbf{c}_i + u_{ij}\mathbf{c}_j, \ldots, \mathbf{c}_j + u_{ji}\mathbf{c}_i, \ldots) - M(\mathbf{c}_1, \mathbf{c}_2, \ldots, \mathbf{c}_n)}{u_{ij}} \qquad (59)$$

Equations (57) are simultaneous linear algebraic equations with $n(n-1)/2$ unknown variables. If $n(n-1)/2$ is large (≥ 20), which very often is the case, it is advantageous to solve these equations by an iterative process rather than a process of matrix inversion. We use the fact that the $k \neq i, j$ terms are, in general, small compared with the u_{ij} terms. A value for u_{ij} is first obtained by setting to zero all u_{ki}, u_{kj}, $k \neq i, j$. These are then used in the next step to evaluate $k \neq i, j$ terms and new u_{ij}s are obtained. The process is iterated until convergence is obtained.

3.3. MCSCF Theory of the Excited States[19,20]

Although our formulation did not specialize to the ground state, it is necessary to modify the theory somewhat for dealing with excited states which are not the lowest of a symmetry species. Henceforth, by excited states we shall refer only to this class of excited states, since those that are the lowest in a given symmetry do not require any new formulation.

The first important observation to be made is that the nth root of the secular equation, as already mentioned, is a strict upper bound to the correct nth excited state of the Hamiltonian. Thus if we perform MCSCF calculations, and the process converges, the resulting energy is the true MCSCF energy. For an excited state characterized by a predominant configuration that differs from all lower states either by a double or a higher-order excitation, it is found that straightforward application of the above MCSCF formalism on the corresponding higher root of the secular equation can be successfully carried out. This is what we would expect for the following reason. We can look upon every MCSCF iteration as a process in which the starting wave function $\Psi_0^{(n)}$ is mixed with a many-electron wave function Ψ_1 differing from $\Psi_0^{(n)}$ by "single excitation" of the occupied orbitals of Ψ_0 into "virtuals" or into each other. It can be shown that Ψ_1 thus constructed is orthogonal to $\Psi_0^{(n)}$. When $\Psi_0^{(n)}$ is predominantly a double excitation with respect to $\Psi_0^{(n-1)}$, say, Ψ_1 is obviously also orthogonal to $\Psi_0^{(n-1)}$. Thus unless Ψ_1 has an energy that lies between E_0^n and E_0^{n-1}, both $\Psi_0^{(n)}$ and $\Psi_0^{(n-1)}$ will be simultaneously improved by the mixing of Ψ_1. This implies that the solutions of the secular equation and the Fock equations will not be at the expense of one another. For a situation in which $\Psi_0^{(n)}$ is predominantly a single excitation from $\Psi_0^{(n-1)}$, this is obviously no longer the case, since Ψ_1 is no longer orthogonal to $\Psi_0^{(n-1)}$, although it certainly is orthogonal to $\Psi_0^{(n)}$. It is therefore very likely that by way of orbital optimization an improper mixing of Ψ_1 with $\Psi_0^{(n)}$ will take place leading to a divergent or at best an oscillatory iterative process.

As we have described above the solution of the Fock equation is done in two steps: (i) obtain increments of the vectors outside the space spanned by the vectors and (ii) perform rotational transformations among the occupied vectors to match the Lagrangian multipliers. As evidenced by Eq. (54), the first step involves excitation to virtual states, such that the part of Ψ_1 coming from this step will be orthogonal to both $\Psi^{(n)}$ and $\Psi^{(n-1)}$. It is in the second part, that is, the rotation of the vectors among themselves, that one has to give explicit consideration to the lower states. Let us consider a pairwise rotation of the vectors c_1 and c_2 into each other,

$$c_1' = c_1 + u c_2$$
$$c_2' = c_2 - u c_1 \tag{60}$$

and study the corresponding effects on the total energy of the first excited state. Extension to higher-order excited states will be obvious. Let the wave functions for the lowest and the first excited states change to $\Psi^{(0)\prime}$ and $\Psi^{(1)\prime}$, as given by

$$\Psi^{(0)\prime} = \Psi^{(0)} + u \Psi_1^{(0)}$$
$$\Psi^{(1)\prime} = \Psi^{(1)} + u \Psi_1^{(1)} \tag{61}$$

Since $\Psi^{(0)\prime}$ and $\Psi^{(1)\prime}$ are orthogonal to the first order:

$$\langle \Psi^{(0)} | \Psi_1^{(1)} \rangle = - \langle \Psi^{(1)} | \Psi_1^{(0)} \rangle \tag{62}$$

Consider now the secular equation with respect to $\Psi^{(0)\prime}$ and $\Psi^{(1)\prime}$. It can be shown that, in general, up to second order in u

$$E^{(n)\prime} = \langle \Psi^{(n)\prime} | H | \Psi^{(n)\prime} \rangle$$
$$= E^{(n)} + 2u(\varepsilon_{12}^{(n)} - \varepsilon_{21}^{(n)}) + u^2(f_{22,12}^{(n)} - f_{11,12}^{(n)} + g_{12}^{(n)}) \tag{63}$$

using the notations of Eqs. (58), the superscript denoting that the parameters are to be calculated with respect to the wave function $\Psi_0^{(n)}$. We therefore have

$$\begin{vmatrix} \langle E^{(0)\prime} - \lambda \rangle & \langle \Psi^{(0)\prime} | H | \Psi^{(1)\prime} \rangle \\ \langle \Psi^{(0)\prime} | H | \Psi^{(1)\prime} \rangle & E^{(1)\prime} - \lambda \end{vmatrix} = 0 \tag{64}$$

If $|E^{(0)} - E^{(1)}| \gg \langle \Psi^{(0)\prime} | H | \Psi^{(1)\prime} \rangle$ we can write

$$\lambda^{(1)} = E^{(1)\prime} + \frac{\langle \Psi^{(0)\prime} | H | \Psi^{(1)\prime} \rangle^2}{E^{(1)} - E^{(0)}}$$
$$= E^{(1)} + 2u(\varepsilon_{12}^{(1)} - \varepsilon_{21}^{(1)}) + u^2(f_{22,12}^{(1)} - f_{11,12}^{(1)} + g_{12}^{(1)\prime}) \tag{65}$$

where

$$g'_{12} = g_{12} + \frac{1}{u^2} \frac{\langle \Psi^{(0)\prime}|H|\Psi^{(1)\prime\prime}\rangle^2}{E^{(1)} - E^{(0)}}$$

$$= g_{12} + \frac{[\langle \Psi_1^{(0)}|H|\Psi_1\rangle + \langle \Psi^{(0)}|H|\Psi_1^{(1)}\rangle]^2}{E^{(1)} - E^{(0)}}$$

(66)

Thus for excited states, we have only to modify the parameters g_{ij} to g'_{ij} as given by Eq. (66). Calculation of the numerator in Eq. (66) can be carried out as follows. We have

$$\Psi^{(n)\prime} = \sum A_a^{(n)}\Phi_a + \sum_a \delta A_a^{(n)}\Phi_a + \sum A_a^{(n)}\delta\Phi_a$$

(67)

such that

$$\langle \Psi^{(0)\prime}|H|\Psi^{(1)\prime}\rangle = \sum_{a,b}(A_a^{(0)}\delta A_b^{(1)} + A_b^{(0)}\delta A_a)H_{ab}$$

$$+ \tfrac{1}{2}\sum_{a,b}(A_a^{(0)}A_b^{(1)} + A_a^{(1)}A_b^{(0)})$$

$$\times[\langle\Phi_a|H|\delta\Phi_b\rangle + \langle\Phi_b|H|\delta\Phi_a\rangle]$$

(68)

The first term on the right-hand side is zero by virtue of the A's being solutions of the secular equations. The expression within the square brackets is simply the contribution to the Fock operators for the vectors c_1 and c_2 from the matrix elements $\langle\Phi_a|H|\Phi_b\rangle$.

4. Alternative Methods for Attaining MCSCF Wave Functions

Various other methods have been reported in the literature for obtaining MCSCF wave functions. We shall describe in some detail two of these methods, (i) that of Hinze and co-workers[20,21] based upon a reduction of the Fock equations to a "two-by-two" process and (ii) the method of generalized Brillouin's theorem first proposed by Levy and Berthier[22] and later extended by Ruedenberg and Rafenetti.[23]

4.1. Hinze's Method of Solution of the MCSCF Equations

If $\{\phi_i\}$, $i = 1, 2, \ldots, n$ are the occupied orbitals and $\{\phi_i\}$, $i = n+1, \ldots, m$ form a set of "virtual orbitals" in a "basic-function" space of m basis functions, then we can generalize Eqs. (26) by using the Fock equation to include also these virtuals by defining the corresponding Fock operators as null operators; i.e., $F_R = 0$, $k = n+1, \ldots, m$. In this framework Eqs. (26) are the necessary and sufficient conditions to be satisfied by the optimum orbitals. Using the formulas contained in Eqs. (57), one can satisfy these conditions by rotating into each other a pair of orbitals out of the total set $\{\phi_i\}$, $i = 1, \ldots, m$ and iterating all pairs.

Hinze also reports an "averaging" method[21] of solving MCSCF equations for excited states. The method consists of obtaining the Fock operators for the excited states on the basis of an averaged-out energy expression given by

$$\bar{E}^{(n)} = \sum_{k=0}^{n} a_k E_k \tag{69}$$

where the E_k are the eigenvalues of the secular equation. The orbitals optimized in this "averaged field" obviously are not optimized for either the lower states or for the excited states. This deficiency can, however, be largely removed subsequent to the MCSCF step by adding single excitations to the MCSCF set of configurations and performing a larger CI.

4.2. Generalized Brillouin's Theorem

In Hartree–Fock theory the Brillouin's theorem states that the matrix element between a fully optimized closed-shell Hartree–Fock configuration and a configuration that represents any single excitation from the Hartree–Fock vanishes. For open-shell cases and MCSCF one can derive a similar theorem. Assume that $\Psi^{(n)}$ of Eq. (1) is fully optimized in the MCSCF sense. Then we have

$$\langle \delta \Psi^{(n)} | H | \Psi^{(n)} \rangle = 0 \tag{70}$$

and since $\sum_{a,b} \delta A_a^{(n)} A_b \langle \Phi_a | H | \Phi_b \rangle = 0$ by virtue of the A's being the solution of the secular equation, we have

$$\sum_a A_a^{(n)} \langle \delta \Phi_a | H | \Psi^{(n)} \rangle = 0 \tag{71}$$

The variation Φ_a is, of course, to be constructed by varying the orbitals. In order that orthonormality be maintained to the first order, the orbital variations must be of the form

$$\delta \phi_i = \sum_{j \neq i} u_{ij} \phi_j$$

where $\tag{72}$

$$u_{ij} + u_{ji} = 0$$

In Eq. (72), summation over j goes over all the occupied and virtual orbitals. Let us set $u_{ij} = 0$, except for $i = k$ and $j = l$. Let $\Phi_a(k \to l)$ and $\Phi_a(l \to k)$ denote single excitations from Φ_a. In case they are not possible (i.e., ϕ_k or ϕ_l is not occupied in Φ_a) or are prohibited by Pauli principle, we formally set them to zero. Then we arrive at the following generalized form of the Brillouin's theorem:

$$\langle \sum A_n^{(a)}[\Phi_a(k \to l) - \Phi_a(l \to k)]|H|\Psi^{(n)}\rangle = 0 \tag{73}$$

4.3. Solution of the MCSCF Problem by Using Generalized Brillouin's Theorem

The following iterative procedure will lead to the same rigorously optimized MCSCF wave function as the Fock equation approach.

1. Starting with the set of configurations $\{\Phi_a\}$, formed out of the orbitals $\{\phi_i\}$, the CI secular equation is solved to obtain $\{A_a^{(n)}\}$.

2. The singly excited functions $\Psi(i \to j)$ are formed and the new CI wave function defined:

$$\Psi_{CI} = \Psi + \sum_{i<j} u_{ij}\Psi(i \to j) \tag{74}$$

The linear coefficients u_{ij} are then obtained by solving a second "super-CI" secular equation.

3. A new set of orbitals $\{\phi_i'\}$ is then formed according to

$$\phi_i' = \phi_i + \sum_{j \neq i} u_{ij}\phi_j$$

with

$$u_{ij} = -u_{ji} \tag{75}$$

4. The orbitals ϕ_i' are reorthonormalized and then with $\{\phi_i\}$ replaced by ϕ_i' steps 1 and 2 are repeated until convergence.

The convergence of the above iterative procedure, of course, depends crucially on the smallness of the coefficients u_{ij}. For those cases where u_{ij} may not be small, Ruedenberg[20] uses the following modification: After step 1 has been carried out, a natural orbital analysis is carried out yielding the natural orbitals $\phi_i^{(nat)}$ for the smaller CI consisting of the configuration selected for the MCSCF. Let

$$\phi_i = \sum_j N_{ij}\phi_j^{(nat)} \tag{76}$$

Step 3 is now replaced by a natural orbital analysis of the super-CI wave function, yielding a new set of natural orbitals $\{\phi_i'^{(nat)}\}$. The new orbitals $\{\phi_i'\}$ are then constructed as

$$\phi_i' = \sum_j N_{ij}\phi_j'^{(nat)} \tag{77}$$

where N_{ij} is obtained from Eq. (76).

5. Applications

A successful MCSCF study can be carried out only if the proper basis set and proper configurations have been selected. In fact, since the convergence of an MCSCF computation is so heavily dependent on the input conditions, the use of wrong configurations or a wrong basis can both lead to a divergent situation. A great deal of experience and data have been accumulated over the past several years in this connection, leading to some standardizations and guiding principles. We shall, in what follows, deal with these aspects of MCSCF computation and present examples covering a wide range of problems, both to elucidate as well as illustrate the quality of the results one can expect from the MCSCF method.

5.1. Choice of Configurations: Optimized Valence Configurations

There are various concepts involved in selecting configurations in addition to the Hartree–Fock wave function. The traditional distinction of three types of correlation, viz., in–out, left–right, and angular, works best for two-valence electron problems. The in–out type represents the fact that if owing to correlation one of the electrons is found in the interior of the system the other electron will more likely be found in the outer region. The left–right correlation corresponds to the electrons tending to stay near different nuclei, while the angular correlation relates to the electrons preferring to separate out by larger angle, while approximately staying the same region around the nuclei. Consider the H_2 correlation problem. The above three types of correlation are represented by the following configurations: (1) $2\sigma_g^2$, (2) $1\sigma_u^2$, and (3) $1\pi_u^2$, respectively, the Hartree–Fock configuration being of course $1\sigma_g^2$.

In many-body theory the language of excitation has been used as an effective guide to the construction of higher and higher order terms in building up a wave function. One can adopt the same language for the construction of configurations for the MCSCF theory. In doing so it is very important to distinguish between two types of excitations—the *intraatomic* and the *interatomic* type. Those excitations which improve the wave functions of isolated atoms beyond the corresponding Hartree–Fock functions are the intraatomic type and those which lead to improved description of the total wave function when the atoms approach each other are the interatomic type. The MCSCF with only the latter type of excitations, along with the Hartree–Fock lead to what we have called in numerous published papers the optimized valence configuration (OVC) wave functions. Such wave functions have been found to predict accurate interaction curves for diatomic and polyatomic molecules. What makes OVC an attractive method is the fact that it usually consists of a small number of configurations. The excitations these configurations represent are also easily interpretable and fall into four broad categories:

(i) the "dissociative" ones that represent charge transfers between centers required to insure proper dissociation of the molecule into Hartree–Fock constituents; (ii) the "redistribution" type that represent redistribution of electrons in asymptotically degenerate atomic states, prohibited by symmetry in the atoms but accessible in the molecule; (iii) the "charge transfer excitation" or split shell type that features one electron exchanged between the open-shells of the atoms followed by a "local" excitation at either of the centers; and (iv) the "dispersive" type, where two electrons from two different centers undergo single excitations into "instantaneously" polarized states of the atoms, accounting for the van der Waals attraction. Following are several examples of typical MCSCF calculations along with references.

Example 1. The Li_2 Molecule

This is one of the simplest and earliest cases treated by MCSCF methods. The Hartree–Fock configuration for Li_2 is $1\sigma_g^2 1\sigma_u^2 2\sigma_g^2$. The dissociative configuration is $1\sigma_g^2 1\sigma_u^2 2\sigma_u^2$. That is, if these two configurations are simultaneously optimized in MCSCF framework, the asymptotic wave function will consist of two Hartree–Fock Li atoms in their ground state.

The redistribution-type configurations cannot be constructed since the atoms are nondegenerate. The charge transfer configurations for Li_2 will be of the type $1\sigma_g^2 3\sigma_u 2\sigma_u 2\sigma_g$ or $1\sigma_g 3\sigma_g 1\sigma_u^2 2\sigma_u^2 \sigma_g$. These yield small effects since the configurations represent high-energy states. The dispersion-type configuration is $1\sigma_g^2 1\sigma_u^2 1\pi_u^2$. This configuration represents two electrons at large distance going from the $2s$ to the $2p$ shells of the atoms. Since $2s$ and $2p$ are near degenerate, this configuration contributes very strongly to the interaction curve. Results obtained from such calculation on the alkalis Na_2, $NaLi$, and Li_2 are given in Table 1.

Example 2. The O_2 Molecule

The ground state has the symmetry $^3\Sigma_g^-$. We make the following correspondence:

$$1\sigma_g \rightarrow 1s_A + 1s_B$$

$$1\sigma_u \rightarrow 1s_A - 1s_B$$

$$2\sigma_g \rightarrow 2s_A + 2s_B$$

$$2\sigma_u \rightarrow 2s_A - 2s_B$$

$$3\sigma_g \rightarrow 2p\sigma_A + 2p\sigma_B$$

$$3\sigma_u \rightarrow 2p\sigma_A - 2p\sigma_B$$

$$1\pi_u \rightarrow 2p\pi_A + 2p\pi_B$$

$$1\pi_g \rightarrow 2p\pi_A - 2p\pi_B$$

Table 1. Typical MCSCF Results for Diatomic Molecules Utilizing Optimized Valence Configuration Selection Rules[a]

System		R_e, bohr		ω_e, cm^{-1}		D_e, eV	
		OVC	Experiment	OVC	Experiment	OVC	Experiment
H_2[b]	$(X\,^1\Sigma)$	1.40	1.40	4398	4400	4.63	4.75
NaLi[c]	$(X\,^1\Sigma)$	5.48	5.45	248.5	251	0.85	unknown
Li_2[d]	$(X\,^1\Sigma)$	5.089	5.051	345.3	351.4	0.99	1.03
Na_2[e]	$(X\,^1\Sigma)$	5.9313	5.818	155.7	159.2	0.719	0.73
CH[f]	$(X\,^2\Pi)$	2.086	—	—	—	3.43	3.65
NH[g]	$(X\,^3\Sigma)$	—	—	—	—	3.37	3.40
OH[h]	$(X\,^2\Pi)$	1.840	1.834	3723	3735	4.53	4.63
FH[i]	$(X\,^1\Sigma^+)$	1.7328[n]	—	—	—	6.18	6.12
F_2[j]	$(X\,^1\Sigma^+)$	—	—	—	—	1.67	1.68
O_2^-[k]	$(X\,^2\Pi)$	—	—	—	—	4.14	4.16
CO[l]	$(X\,^1\Sigma^+)$	2.132[n]	—	—	—	11.33	11.38
O_2[m]	$(^3\Sigma_g^-)$	2.31	2.28	1481	1580	3.4[n]	5.08

[a] All of these calculated values are based on a consistent simple model designed to evaluate only changes occurring in the correlation energy with molecular formation. See G. Das and A. C. Wahl, *J. Chem. Phys.* **56**, 1769 (1972).
[b] G. Das and A. C. Wahl, *J. Chem. Phys.* **44**, 87 (1966).
[c] P. J. Bertoncini, G. Das, and A. C. Wahl, *J. Chem. Phys.* **52**, 5112 (1970).
[d] G. Das and A. C. Wahl, *J. Chem. Phys.* **44**, 87 (1966); G. Das, *J. Chem. Phys.* **46**, 1568 (1967).
[e] P. J. Bertoncini and A. C. Wahl, *J. Chem. Phys.*, in press.
[f] D. Neumann and M. Krauss, in press.
[g] W. J. Stevens, *J. Chem. Phys.* **58**, 1264 (1973).
[h] W. J. Stevens, G. Das, A. C. Wahl, D. Neumann, and M. Krauss, *J. Chem. Phys.* **61**, 3686 (1974).
[i] D. Neumann and M. Krauss, *Mol. Phys.* **27**, 917 (1974).
[j] G. Das and A. C. Wahl, *J. Chem. Phys.* **56**, 3532 (1973).
[k] M. Krauss, D. Neumann, A. C. Wahl, G. Das, and W. Zemke, *Phys. Rev. A* **7**, 69 (1973); W. Zemke, G. Das, and A. C. Wahl, *Chem. Phys. Lett.* **14**, 310 (1972).
[l] F. P. Billingsley and M. Krauss, *J. Chem. Phys.* **60**, 4130 (1974).
[m] G. Das and A. C. Wahl, unpublished results. Only configurations for proper dissociation included, thus low value of D_e.
[n] Energy calculated at one point, the experimental R_e.

The Hartree–Fock configuration is $1\sigma_g^2 1\sigma_u^2 2\sigma_g^2 2\sigma_u^2 3\sigma_g^2 1\pi_u^2 1\pi_g^2$. Then it can be shown that the dissociative configurations are, apart from the constant factor $[\text{core}] \equiv [1\sigma_g^2 1\sigma_u^2 2\sigma_g^2 2\sigma_u^2)$,

$$3\sigma_g^2 1\pi_u^4 1\pi_g^2$$

$$3\sigma_u^2 1\pi_u^4 1\pi_g^2$$

$$3\sigma_g^2 1\pi_u^2 1\pi_g^4$$

$$3\sigma_u^2 1\pi_u^2 1\pi_g^4$$

$$3\sigma_g 3\sigma_u 1\pi_u^3 1\pi_g^3 \quad (2)$$

where in the last entry the number in the parentheses indicates that there exist two spin couplings with the same occupancy, giving rise to the state $^3\Sigma_g^-$.

The distributive configurations are $3\sigma_g^2 3\sigma_u^2 1\pi_u^2 1\pi_g^2$ (2). The important charge transfer excitations are $3\sigma_g 3\sigma_u 1\pi_u^4 1\pi_g 2\pi_g$ (2). The dispersion excitations cannot be important for the $^3\Sigma_g^-$ state since they will use higher atomic shells in the atoms, and effects from such excitations are negligible in comparison with the other type of excitations. In Tables 1 and 2 results of calculations on a number of systems of similar complexity are presented. In Fig. 1 a typical MCSCF potential curve is given for OH.

Example 3. van der Waals Systems

Obviously, the dispersion excitations are the most predominant type of excitations for rare gas systems, because dissociative configurations do not exist, the Hartree–Fock already properly dissociates, the redistributive configuration cannot be constructed since the atomic shells are all closed, and the

Table 2. *Diatomic Dipole Moments Computed from SCF and OVC Wave Functions*[a]

Molecule[b]		SCF	OVC[c]	Experiment
C^-H^+	$(^2\Pi)$	1.570^d	1.53^i	1.46 ± 0.06^n unknown
N^-H^+	$(^3\Sigma)$	1.627^d	1.537^j	
O^-H^+	$(^2\Pi)$	1.780^d	1.655^k	1.660 ± 0.010^o
F^-H^+	$(^1\Sigma)$	1.942^d	1.805^i	1.797^p
C^+N^-	$(^2\Sigma^+)$	2.301^e	1.481^l	1.45 ± 0.08^q
C^-O^+	$(^1\Sigma^+)$	-0.274^f	0.156^m	0.112 ± 0.005^r
Na^+Li^-	$(^1\Sigma)$	0.679^g	1.141^g	0.47 ± 0.03^s
C^+O^-	$(^3\Pi)$	2.47^h	1.67^h	1.37^t

[a] All values in debyes.
[b] The dipole moments are given as positive for the indicated polarity.
[c] These calculations were all done at or near the experimental equilibrium. Vibrational averaging has not been taken into account except where noted in the footnotes.
[d] P. E. Cade and W. M. Huo, *J. Chem. Phys.* **45**, 1063 (1966).
[e] S. Green, *J. Chem. Phys.* **57**, 4694 (1972).
[f] W. M. Huo, *J. Chem. Phys.* **43**, 624 (1965).
[g] W. J. Stevens and A. C. Wahl, unpublished work. These results were obtained in an effort to improve previously published values by greatly expanding the STO basis set. The previous values were 0.95 D and 1.24 D for the SCF and OVC dipole moments, respectively. See P. Bertoncini, G. Das, and A. C. Wahl, *J. Chem. Phys.* **52**, 5112 (1970).
[h] D. Neumann and M. Krauss, *Mol. Phys.*, in press.
[i] D. Neumann and M. Krauss, *Mol. Phys.* **27**, 917 (1974).
[j] W. J. Stevens, *J. Chem. Phys.* **58**, 1264 (1973).
[k] W. J. Stevens, G. Das, A. C. Wahl, D. Neumann, and M. Krauss, *J. Chem. Phys.* **61**, 3686 (1974).
[l] G. Das, T. Janis, and A. C. Wahl, *J. Chem. Phys.* **61**, 1274 (1974).
[m] F. P. Billingsley and M. Krauss, *J. Chem. Phys.* **60**, 4130 (1974). This value has been vibrationally averaged.
[n] D. H. Phelps and F. W. Dalby, *Phys. Rev. Lett.* **16**, 3 (1966).
[o] F. X. Powell and D. R. Lide, *J. Chem. Phys.* **42**, 4201 (1965).
[p] J. S. Muenter and W. Klemperer, *J. Chem. Phys.* **52**, 6033 (1970).
[q] R. Thomson and F. W. Dalby, *Can. J. Phys.* **46**, 2815 (1968).
[r] R. A. Toth, R. H. Hunt, and E. K. Plyler, *J. Mol. Spectros.* **32**, 74 (1969).
[s] P. J. Dagdigian, J. Graff, and L. Wharton, *J. Chem. Phys.* **55**, 4980 (1971).
[t] W. Klemperer, *J. Chem. Phys.* **56**, 5758 (1972).

charge transfer excitations imply both high-energy states and an exponential-type dependence on the internuclear separation. The dispersion excitation for He_2 is

$$1s_A 2p\sigma_A 1s_B 2p\sigma_B \quad (2)$$

$$1s_A 2p\pi_A 1s_B 2p\pi_B \quad (2)$$

For higher-order accuracy ($\leq 10\%$) in the molecular interactions (as well as some unusual cases), however, we have to consider both the interatomic and the intraatomic excitations (Table 3). The latter are usually numerous; although, unless a good total energy (rather than just the interaction) is also desired, one need only consider the dominant ones. We shall demonstrate why this is so. Consider the excitation $1s^2 \to nl^2$ in He. The strength of this excitation is determined by the coupling team:

$$E_{12} = \int [\phi_{nl}^{(1)} \phi_{nl}^{(2)} \phi_{1s}^{(1)} \phi_{1s}^{(2)} | r_{12}] d\mathbf{r}_1 \, d\mathbf{r}_2 \tag{78}$$

Maximization of E_{12} (which is almost equivalent to the minimization of the total MCSCF energy consisting of $1s^2$ and nl^2, since the change in the diagonal

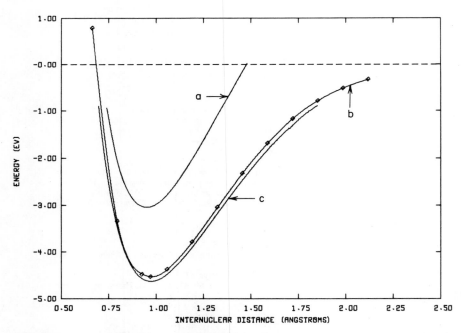

Fig. 1. Comparison of Hartree–Fock and MCSCF potential curves with experiment for OH. (a) Hartree–Fock; (b) OVC XIV; (c) RKR. [W. Stevens, G. Das, A. C. Wahl, M. Krauss, and D. Neumann, *J. Chem. Phys.* **61**, 3689 (1974).]

Table 3. Comparison of Experimental and
Theoretical Well Depths for He$_2$

Procedure	°K
Hartree–Fock	Repulsive
Interatomic terms only included[a]	12
Interatomic and intraatomic computed separately and added[b]	10.8
Interatomic and intraatomic computed simultaneously[c]	~10.8
Experiment[d]	10.3–11.2

[a] D. R. McLaughlin and H. F. Schaefer III, *Chem. Phys. Lett.* **12**, 244 (1972); P. Bertoncini and A. C. Wahl, *Phys. Rev. Lett.* **25**, 991 (1970).
[b] P. J. Bertoncini and A. C. Wahl, *J. Chem. Phys.* **58**, 1259 (1973).
[c] W. Meyer, private communication; B. Liu and A. D. McLean, to be published; B. Liu and A. D. McLean, *J. Chem. Phys.* **59**, 4557 (1973).
[d] P. E. Siska, J. M. Parson, T. P. Schafer, and Y. T. Lee, *J. Chem. Phys.* **55**, 5762 (1971); J. M. Farrar and Y. T. Lee, *J. Chem. Phys.* **56**, 5801 (1972); H. G. Bennewitz, H. Busse, H. D. Dohmann, D. E. Oates, and W. Schrader, *Z. Phys.* **253**, 435 (1972); H. G. Bennewitz, H. Busse, H. D. Dohmann, D. E. Oates, and W. Schrader, *Phys. Rev. Lett.* **29**, 533 (1972).

terms is relatively mild compared with that of E_{12}) yields the integral equation:

$$\int [\phi_{1s}^{(2)}\phi_{nl}^{(2)}|r_{12}\phi_{1s}^{(1)}] = \varepsilon\phi_{nl}^{(1)} + \varepsilon'\phi_{1s}^{(1)}$$

where ε' is the Lagrangian multiplier arising from the orthogonality requirement $\langle\phi_{1s}|\phi_{nl}\rangle = 0$. This is equivalent to the differential equation

$$\nabla^2 f = -n\phi_{1s}^2 f, \qquad \eta = 4\pi/\varepsilon$$

where

$$f = \phi_{nl}/\phi_{1s}$$

For s-type excitation and $\phi_{1s} \sim e^{-\zeta r}$, an analytical solution is readily obtained:

$$\phi = [\exp(-3\zeta r)/r] \sum_{k=0}^{\infty} a_k \exp(-2\zeta kr) + e^{-\zeta r} \sum_{k=0}^{\infty} b_k \exp(-2\zeta kr)$$

where

$$a_k = \left(-\frac{\eta}{4\zeta^2}\right)^{k+1} \frac{b_0}{[(k+1)!]^2} \frac{s(k+1)}{\zeta}, \qquad b_k = \left(-\frac{\eta}{4\zeta^2}\right)^k \left[\frac{b_0}{(k!)^2}\right]$$

$$s(k) = \sum_{r=1}^{k} r^{-1}$$

Table 4. Comparison of Experimental and Theoretical Determinations of the Position r_m and Depth ε of the van der Waals Well in Rare Gas Hydrides

Type	r_m, Å		ε, meV, °K	
	Experimental	Theoretical	Experimental	Theoretical
HeH	3.72^a	3.59^e	$(0.46, 5.3)^a$	$(0.57, 6.6)^e$
NeH	3.50^b	3.67^e	$(1.35, 15.6)^b$	$(1.34, 15.5)^e$
ArH	3.61^b	3.57^c	$(4.16, 48.2)^b$	$(4.16, 48.2)^c$
KrH	3.57^b	in progress^e	$(5.90, 68.4)^b$	in progress^e
XeH	3.82^d	in progress^e	$(7.08, 82.1)^d$	in progress^e

[a] J. P. Toennies, W. Welz, and G. Wolf, *Chem. Phys. Lett.* **44**, 5 (1976).
[b] G. Wolf, Ph.D. thesis, Max Planck Institut für Stromungsforschung, 1976.
[c] Albert F. Wagner, G. Das, and Arnold C. Wahl, *J. Chem. Phys.* **60**, 1885 (1974).
[d] W. Welz, Ph.D. thesis, Max Planck Institut für Stromungsforschung, 1976.
[e] Albert F. Wagner, G. Das, and Arnold C. Wahl, to be published.

The eigenvalues ε form the discrete spectrum given by

$$\sum_{k=0}^{\infty} s(k+1)\frac{(-\eta/4\zeta^2)^k}{[(k+1)!]^2} = 0$$

From the expression for a_n and b_n it is obvious that the functions that contribute less (contribution being measured by the magnitude of ε) are also those that are more localized.

Thus, for studying molecular interactions, only the dominant intraatomic excitations are to be included. Calculations of this sort are summarized in Tables 3 and 4 and recent potential curves for ArH given in Fig. 2.

There are some unusual cases where some of the intraatomic correlation terms defined in the conventional way can be as important as the Hartree–Fock. Consider the Be atom. The Hartree–Fock for this atom is $1s^2 2s^2$. However $1s^2 2p^2$ is almost degenerate and the matrix element $\langle 1s^2 2s^2 | H | 1s^2 2p^2 \rangle$ is large. Thus for systems involving Be, the intraatomic term $1s^2 2p^2$ plays an important role in the molecular interaction and should be considered in all physically valid approximations of the wave function.

Example 4. Triatomic Molecules

Drawing upon the experience gained with diatomic molecules the MCSCF technique, utilizing OVC configuration selection principles has been applied to a number of atmospheric molecules and to the calculation of energy surfaces for LiH_2 and Li_2H. Energy surfaces for LiH_2 and curves for LiH_2 are given in Figs. 3 and 4.

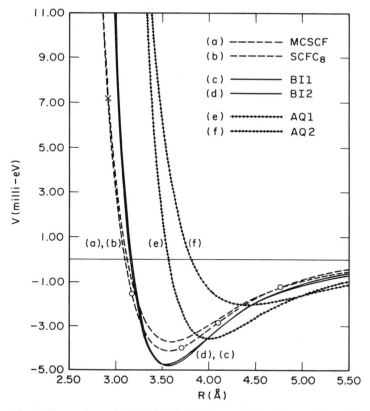

Fig. 2. Comparison of MCSCF and SCFC$_8$ potentials with experimentally derived potentials SCFC$_8$ corresponds to addition of SCF repulsion and C$_6$, C$_8$ attractive potentials. [See A. F. Wagner, G. Das, and A. C. Wahl, *J. Chem. Phys.* **60**, 1885 (1974).] BI1 and BI2 are from experimental work of R. W. Buches, Jr., B. Lantzsch, J. P. Toennies, and K. Walaschewski, *Faraday Disc. Chem. Soc.* **55**, 167 (1973). AQ1 and AQ2 are from experimental work of V. Aquilanti, G. Linti, F. Vecchio-Cattivi, and G. G. Volpi, *Chem. Phys. Lett.* **15**, 305 (1972).

5.2. Choice of Basis Sets

The selection of basis functions to represent excited orbitals follows the same intuitive and iterative course as with the choice for the Hartree–Fock orbitals prior to the basis set standardizations, such as those of Clementi[17] or Bagus–Gilbert's[18] for STOs and those of Pople, Dunning, and others[16] for GTOs. Fortunately, in studying molecule interactions one is usually concerned with valence excitations only, for which one rarely needs any extra functions than that for the Hartree–Fock occupied orbitals. Charge transfer excitations and even the predominant atomic excitations to be used for accurate interaction curves are well represented by the so-called "accurate" basis sets. The only new functions that have to be chosen are those to represent the dispersion excitations. The simple prescription for determining these functions is as follows.

Let us consider the dispersion states in which the electrons localized in the states ϕ_{0A} and ϕ_{0B} are excited into the polarized states ϕ'_{0A} and ϕ'_{0B}, respectively. The matrix element for such excitation is given by

$$\langle\phi_{0A}\phi_{0B}|H|\phi'_{0A}\phi'_{0B}\rangle \equiv \int \frac{\phi_{0A(1)}\phi'_{0A'(1)}\phi_{0B(2)}\phi'_{0B'(2)}}{r_{12}}\,dT_1\,dT_2$$

$$\approx \sum_{lm} C_{lm} \frac{\int r^{l+2}Y_{lm}\phi_{0A}\phi'_{0A}\,d\mathbf{r}\int r^{l+2}\phi_{0B}\phi'_{0B}Y_{lm}\,d\mathbf{r}}{R^{l+1}} \quad (85)$$

for large enough R (usually within the van der Waals range), with C_{lm} being algebraic constants.

We require that each term on the right-hand side be individually maximized. This gives us a set of functions ϕ_{lA} and ϕ_{lB} for ϕ'_{0A} and ϕ'_{0B} given by

$$\phi_l = \phi_0 r^l Y_{lm}(\theta, \phi) \quad (86)$$

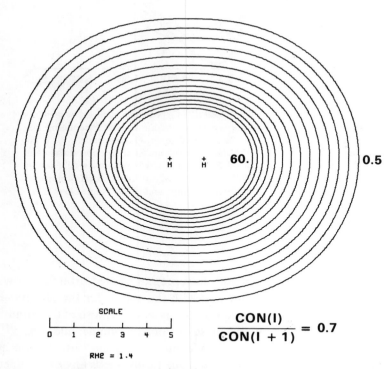

Fig. 3. MCSCF energy surface for LiH$_2$ with H$_2$ at its equilibrium geometry. Outermost contour value is 0.5 kcal/mole rising to the innermost contour of value 60 kcal/mole with a ratio of 0.7 between adjacent contours. [A. Karo, A. Wagner, and A. C. Wahl, in preparation.]

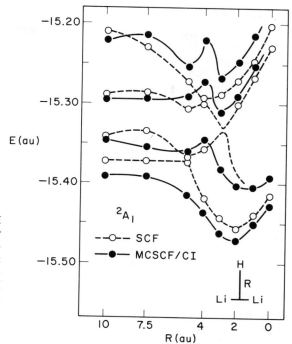

Fig. 4. SCF and MCSCF/CI adiabatic energy curves for four $^2A_1$ states of Li_2H. In this calculation the Li_2 distance was fixed at 5.051 bohrs, the equilibrium distance of the Li_2 molecule, as the H was brought toward the Li_2 axis in C_{2v} symmetry. [W. B. England, N. H. Sabelli, and A. C. Wahl, *J. Chem. Phys.* **63**, 4596 (1975)].

In an actual MCSCF calculation of VDW wave functions one very nearly maximizes the magnitude of the coupling integral $\langle \phi_{0A}\phi_{0B}|H|\phi'_{0A}\phi'_{0B}\rangle$ since the diagonal terms are not quite as sensitive with respect to the selection of the orbitals as the off-diagonal ones, so long as they occupy space similar to the functions ϕ_{0A}. Thus, the basis functions for the dispersive excitations to be selected are those functions generated by multiplying by $r^l Y_{lm}(\theta, \phi)$ the functions predominantly used by the orbitals $\phi'_{0A}\phi_{0B}$.

6. Summary and Conclusion

The MCSCF procedure, alone for simple systems and in conjunction with straightforward configuration interaction, has been shown to be a valuable technique for obtaining accurate wave functions for a wide variety of systems.

As recently shown, the MCSCF procedure, in much the same way as the SCF method, can be extended to heavy-atomic systems by using pseudopotentials. Extension to large polyatomic systems, under certain approximations, is an interesting subject for future study.

Thus, it appears clear that the MCSCF method and its variants will emerge as some of the very potent theoretical tools in the study of molecular systems and reactions.

References

1. D. R. Hartree, *Calculation of Atomic Structure*, John Wiley and Sons, New York (1957).
2. D. R. Hartree and W. Hartree, *Proc. R. Soc. London, Ser. A* **154**, 588 (1936).
3. C. Froese-Fischer, *Comput. Phys. Commun.* **1**, 152 (1970).
4. J. P. Desclaux and Y. K. Kim, *J. Phys. B* **8**, 1177 (1975).
5. A. C. Wahl, *J. Chem. Phys.* **41**, 2600 (1964).
6. J. Frenkel, *Wave Mechanics, Advanced General Theory*, Clarendon Press, Oxford (1934).
7. C. F. Bender and E. R. Davidson, *Phys. Rev.* **183**, 23 (1969).
8. H. P. Kelly, *Adv. Chem. Phys.* **14**, 129 (1969).
9. R. K. Nesbet, *Adv. Chem. Phys.* **9**, (1965); **14**, 7 (1969).
10. O. Sinanoğlu, *Adv. Chem. Phys.* **14**, 237 (1969).
11. H. J. Silverstone and O. Sinanoğlu, *J. Chem. Phys.* **44**, 1899 (1966).
12. T. Shibuya and V. McKoy, *Phys. Rev. A* **2**, 2208 (1970).
13. B. S. Yarlagadda, H. S. Taylor, R. Yaris, and B. Schneider, *Chem. Phys. Lett.* **22**, 381 (1973).
14. I. Shavitt, C. F. Bender, A. Pipano, and R. P. Hosteny, *J. Comput. Phys.* **11**, 90 (1973).
15. J. K. L. MacDonald, *Phys. Rev.* **43**, 830 (1933).
16. P.-O. Löwdin, *Adv. Chem. Phys.* **2**, 207 (1959).
17. C. C. J. Roothaan, *Rev. Mod. Phys.* **23**, 69 (1951).
18. C. C. J. Roothaan and P. S. Bagus, *Methods Comput. Phys.* **2**, 47 (1963).
19. G. Das, *J. Chem. Phys.* **58**, 5104 (1973).
20. G. Das, T. Janis, and A. C. Wahl, *J. Chem. Phys.* **61**, 1274 (1974).
21. J. Hinze, *J. Chem. Phys.* **59**, 6424 (1973).
22. B. Levy and G. Berthier, *Int. J. Quantum Chem.* **2**, 307 (1968); **3**, 247 (1969).
23. R. C. Rafenetti and K. Ruedenberg, *Int. J. Quantum Chem.* **34**, 625 (1970).

4

The Self-Consistent Field Equations for Generalized Valence Bond and Open-Shell Hartree–Fock Wave Functions

Frank W. Bobrowicz
and
William A. Goddard III

1. Introduction

The basic starting point for calculating *ab initio* wave functions of molecules is generally the Hartree–Fock (HF) wave function, which in the simplest case involves two electrons (one with each spin) in each orbital ϕ_i with the total wave function antisymmetrized in order to satisfy the Pauli principle

$$a[(\phi_1\alpha)(\phi_1\beta)(\phi_2\alpha)(\phi_2\beta)\cdots(\phi_n\alpha)(\phi_n\beta)]$$
$$= a(\phi_1\phi_1\phi_2\phi_2\cdots\phi_n\phi_n\alpha\beta\alpha\beta\cdots\alpha\beta) \quad (1)$$

Here a is the antisymmetrizer or determinant operator* and α and β are the usual spin functions. In Eq. (1) as elsewhere, we arrange products of spatial functions and spin functions in order of increasing electron numbers.

*$a \equiv \sum_{\nu} \zeta_{\tau}\tau$ where the sum is over all permutations τ and where $\zeta_{\tau} = \pm 1$ is the parity of τ.

Frank W. Bobrowicz • Arthur Amos Noyes Laboratory of Chemical Physics, California Institute of Technology, Pasadena, California, and Batelle Memorial Institute, Columbus, Ohio. *William A. Goddard III* • Arthur Amos Noyes Laboratory of Chemical Physics, California Institute of Technology, Pasadena, California

As is well known, this type of wave function generally does not lead to an adequate description of bond dissociation, nor does it provide a consistent level of treatment for excited states. A simple generalization of Eq. (1) is used to allow selected electron pairs to be described in terms of overlapping singly occupied orbitals,

$$\phi_i(1)\phi_i(2) \rightarrow [\phi_{ia}(1)\phi_{ib}(2) + \phi_{ib}(1)\phi_{ia}(2)] \tag{2}$$

as in the simple valence bond wave function. If the various orbitals of such a wave function are solved for self consistency, the resulting wave function is referred to as a *generalized valence bond* (GVB) wave function.[1-3] In this chapter we will develop the variational equations which are solved to obtain GVB wave functions, and we will describe the various steps for solving the GVB wave functions of general systems. This discussion necessarily includes open-shell Hartree–Fock wave functions as a special case and the procedures developed are also appropriate for more general wave functions.

A common procedure for solving SCF wave functions[4] has been to consider all occupied orbitals as fixed except for one, say ϕ_k, and to require that the first-order change in the energy be zero upon variation of this orbital, consistent with maintaining orthogonality between all occupied orbitals. A typical approach is to write the variational equations as pseudoeigenvalue equations and to solve these equations iteratively. Although often successful, this procedure is not reliable for complicated wave functions.

Our approach is to start with the expression for the total energy and to expand it (through second order) in terms of simultaneous corrections in all orbitals.[5,6] Because of the orthogonality conditions the variations in different orbitals are coupled. Including these couplings, we derive the equations for the optimal simultaneous *correction* in each orbital. Solution of these equations leads to quadratic convergence to the optimum orbitals. Various simplifications are also discussed.

2. The Energy Expression

Given the usual nonrelativistic Hamiltonian

$$\mathcal{H} = \sum_{p=1}^{N} h(p) + \sum_{p>q=1}^{N} \frac{1}{r_{pq}} \tag{3}$$

where N is the number of electrons and the operator $h(p)$ contains all one-electron terms involving electron p, the total energy of any electronic wave function can be written as[7]

$$E = \sum_{i,j}^{n} D_j^i h_{ij} + \sum_{\substack{i,j \\ k,l}}^{n} D_{kl}^{ij}(ik|jl) \tag{4}$$

where

$$h_{ij} = \langle \phi_i | h | \phi_j \rangle$$

$$(ik | jl) \equiv \langle \phi_i(1)\phi_j(2) | (1/r_{12}) | \phi_k(1)\phi_l(2) \rangle$$

and $\{\phi_i\}$ is a set of n orbital basis functions in terms of which the wave function is described. Here D_j^i and D_{kl}^{ij} are appropriate density matrix elements.

For a wide class of useful wave functions (including open-shell HF and important cases of GVB), once appropriately expressed in terms of orthonormal orbitals

$$\langle \phi_i | \phi_j \rangle = \delta_{ij} \tag{5}$$

the one- and two-electron density matrices in (4) can be brought into diagonal form so that the only nonzero elements are

$$D_i^i \equiv 2f_i, \qquad D_{ij}^{ij} \equiv a_{ij}, \qquad D_{ji}^{ij} \equiv b_{ij}$$

where the energy coefficients $\{f_i, a_{ij} = a_{ji}, b_{ij} = b_{ji}\}$ are independent of the orbitals $\{\phi_i\}$. For such wave functions the energy expression takes on the greatly simplified form,

$$E = 2 \sum_{i=1}^{n} f_i h_{ii} + \sum_{i,j=1}^{n} (a_{ij}J_{ij} + b_{ij}K_{ij}) \tag{6}$$

where

$$J_{ij} \equiv (ii|jj) = \langle \phi_i | J_j | \phi_i \rangle = \langle \phi_j | J_i | \phi_j \rangle$$

$$K_{ij} \equiv (ij|ij) = \langle \phi_i | K_j | \phi_i \rangle = \langle \phi_j | K_i | \phi_j \rangle$$

indicate the usual Coulomb and exchange *energies*, respectively. The Coulomb and exchange *operators* J_i and K_i are defined here for later convenience. In this chapter we will be concerned mainly with the problem of solving for the self-consistent solutions of wave functions whose energy expressions conform to Eq. (6).

3. Wave Functions

While Eq. (6) does not apply to all electronic wave functions, this energy expression *is* applicable to many of the most important wave functions useful in dealing with chemical problems at the conceptual level. In this section we will examine such wave functions, paying particular attention to both the simplifying aspects of their functional forms and their inherent limitations.

3.1. Hartree–Fock Wave Functions

The HF wave function is of singular importance in our basic understanding of atomic and molecular electronic structure. For mainly historical reasons, however, this wave function defies a neat unambiguous definition, since a fairly wide variety of wave functions of differing forms are classified under this appellation. Therefore, we will discuss HF wave functions by way of representative examples, without attempting an exhaustive treatment.

3.1.1. Closed Shell

Consider the four-electron singlet wave function involving two doubly occupied orbitals,

$$\mathcal{a}[\phi_1\alpha\phi_1\beta\phi_2\alpha\phi_2\beta] = \mathcal{a}[\phi_1\phi_1\phi_2\phi_2\alpha\beta\alpha\beta] \tag{7}$$

Allowing orbitals ϕ_1 and ϕ_2 to be nonorthogonal,

$$\langle\phi_1|\phi_2\rangle = S_{12}, \qquad \langle\phi_1|\phi_1\rangle = \langle\phi_2|\phi_2\rangle = 1 \tag{8}$$

the energy of (7) is

$$E = [2(1-S_{12}^2)(h_{11}+h_{22}-2S_{12}h_{12})+J_{11}+J_{22}+2(2-S_{12}^2)J_{12}-2(1-3S_{12}^2)K_{12}$$
$$-4S_{12}\langle\phi_1|(J_1+J_2)|\phi_2\rangle]/(1-S_{12}^2)^2 \tag{9}$$

However, writing ϕ_2 as

$$\phi_2 = \phi_2' + \lambda\phi_1$$

where

$$\langle\phi_1|\phi_2'\rangle = 0 \tag{10}$$

(7) becomes

$$\mathcal{a}[\phi_1\phi_1\phi_2\phi_2\alpha\beta\alpha\beta] = \mathcal{a}[\phi_1\phi_1\phi_2'\phi_2'\alpha\beta\alpha\beta] + \lambda\mathcal{a}[\phi_1\phi_1\phi_1\phi_2\alpha\beta\alpha\beta]$$
$$+ \lambda\mathcal{a}[\phi_1\phi_1\phi_2\phi_1\alpha\beta\alpha\beta] + \lambda^2\mathcal{a}[\phi_1\phi_1\phi_1\phi_1\alpha\beta\alpha\beta] \tag{11}$$

But, because of the Pauli principle (as embodied in the antisymmetrizer $\mathcal{a}$), we find that

$$\mathcal{a}[\cdots\phi_i\alpha\cdots\phi_i\alpha\cdots] = \mathcal{a}[\cdots\phi_i\beta\cdots\phi_i\beta\cdots] = 0 \tag{12}$$

Therefore, (11) reduces to simply

$$\mathcal{a}[\phi_1\phi_1\phi_2\phi_2\alpha\beta\alpha\beta] = \mathcal{a}[\phi_1\phi_1\phi_2'\phi_2'\alpha\beta\alpha\beta] \tag{13}$$

That is, the antisymmetrizer projects away any nonorthogonalities, and hence it is no restriction to take these orbitals as orthogonal in the first place.* In this case the energy expression (9) becomes

$$E_c = 2h_{11} + 2h_{22} + J_{11} + J_{22} + 4J_{12} - 2K_{12} \tag{14}$$

Since

$$J_{ii} = K_{ii} \tag{15}$$

we can add and subtract these *self-terms* to obtain the more symmetric expression

$$E_c = 2h_{11} + 2h_{22} + (2J_{11} - K_{11}) + (2J_{22} - K_{22}) + (4J_{12} - 2K_{12}) \tag{16}$$

which is of the form of (6) with

$$f_1 = f_2 = 1$$
$$a_{11} = a_{22} = a_{12} = 2$$
$$b_{11} = b_{22} = b_{12} = -1$$

Since ϕ_1 and ϕ_2 are maximally occupied ($f_i = 1$), we refer to them as *closed-shell* orbitals. Orbitals leading to $f_i \neq 1$ are referred to collectively as *open-shell* orbitals. A closed-shell orbital can be (and always is) taken to be orthogonal to all other orbitals, without restriction. A wave function composed entirely of closed-shell orbitals

$$\mathcal{a}[\phi_1\alpha\phi_1\beta\phi_2\alpha\phi_2\beta \cdots \phi_n\alpha\phi_n\beta] \tag{17}$$

is called a closed-shell wave function and has the general energy expression†

$$E_c = \sum_i^n 2h_{ii} + \sum_{i,j}^n (2J_{ij} - K_{ij}) \tag{18}$$

3.1.2. Closed-Shell/Open-Shell Energy Partitioning

Generally, wave functions involving open-shell orbitals also contain several closed-shell orbitals as well. Therefore, it is convenient to develop a

---

*Also because of the antisymmetrizer we find that starting with two orbitals ϕ_1 and ϕ_2 and transforming to a new set of orbitals $\bar{\phi}_1$ and $\bar{\phi}_2$,

$$\bar{\phi}_i = \sum_j \phi_j U_{ji}$$

leads to

$$\mathcal{a}[\bar{\phi}_1\bar{\phi}_1\bar{\phi}_2\bar{\phi}_2\alpha\beta\alpha\beta] = (\det U)^2 \mathcal{a}[\phi_1\phi_1\phi_2\phi_2\alpha\beta\alpha\beta]$$

Thus the closed-shell wave function is *invariant* under orthogonal transformations.
†We have seen that the form of (18) is invariant under orthogonal transformations of the orbitals.

scheme whereby energy expressions may be derived without explicitly considering the closed-shell orbitals. Consider the doublet wave function

$$\mathscr{A}[\phi_1\phi_1\phi_2\phi_2\phi_3\alpha\beta\alpha\beta\alpha] \tag{19}$$

Since all the orbitals can be taken as orthonormal, the energy of (19) is

$$E = (2h_{11}+2h_{22})+h_{33}+(2J_{11}-K_{11}+2J_{22}-K_{22}+4J_{12}-2K_{12})$$
$$+(2J_{13}-K_{13}+2J_{23}-K_{23}) \tag{20}$$

Hence

$$f_1 = f_2 = 2f_3 = 1$$
$$a_{11} = a_{22} = a_{12} = 2a_{13} = 2a_{23} = 2$$
$$b_{11} = b_{22} = b_{12} = 2b_{13} = 2b_{23} = -1$$
$$a_{33} = b_{33} = 0$$

Comparing with (16) we see that (20) can be written as

$$E = E_c + h_{33} + 2J_{13} - K_{13} + 2J_{23} - K_{23}$$
$$= E_c + h^c_{33} \tag{21}$$

where

$$h^c \equiv h + 2J_1 - K_1 + 2J_2 - K_2$$

Thus, insofar as the open-shell orbital (ϕ_3) is concerned all effects due to the closed-shell orbitals are included in the new "one-electron" operator h^c.

In general, if orbital i is a closed-shell orbital and orbital j is any other orbital, the only nonvanishing density matrix elements in (4) involving both of these orbitals are

$$D^{ij}_{ij} = 2D^i_j, \qquad D^{ij}_{ji} = -D^i_j$$

As a consequence, the energy for any wave function involving both closed-shell and open-shell orbitals has the form

$$E = E_c + \sum_{i,j}^{\text{open}} D^i_j h^c_{ij} + \sum_{\substack{i,j \\ k,l}}^{\text{open}} D^{ij}_{kl}(ik|jl) \tag{22}$$

where

$$h^c \equiv h + \sum_p^{\text{closed}} (2J_p - K_p)$$

$$\tag{23}$$

$$E_c \equiv 2\sum_p^{\text{closed}} h_{pp} + \sum_{p,q}^{\text{closed}} (2J_{pq} - K_{pq})$$

If the open-shell energy can be simplified to the form of (6), then (22) reduces to

$$E = E_c + \sum_i^{\text{open}} 2f_i h_{ii}^c + \sum_{i,j}^{\text{open}} (a_{ij} J_{ij} + b_{ij} K_{ij}) \qquad (24)$$

Hence, letting {core} denote the presence of an arbitrary number of closed-shell orbitals,

$$\{\text{core}\} = \phi_{c_1} \alpha \phi_{c_1} \beta \phi_{c_2} \alpha \phi_{c_2} \beta \cdots \phi_{c_m} \alpha \phi_{c_m} \beta \qquad (25)$$

the energy for the doublet wave function

$$\mathcal{A}[\{\text{core}\}\phi_n \alpha] \qquad (26)$$

becomes simply

$$E = E_c + h_{nn}^c \qquad (27)$$

3.1.3. High-Spin Open Shell

For the wave function

$$\mathcal{A}[\{\text{core}\}\phi_1 \phi_2 \alpha \alpha] \qquad (28)$$

where the electrons in orbitals ϕ_1 and ϕ_2 are coupled into a triplet state, we see from the analysis in (10)–(13) that it is no restriction to take these two orbitals as orthogonal. Therefore, the energy of (28) is

$$E = E_c + h_{11}^c + h_{22}^c + J_{12} - K_{12} \qquad (29)$$

Adding and subtracting self-terms leads to the more symmetric expression

$$E = E_c + h_{11}^c + h_{22}^c + \tfrac{1}{2}(J_{11} - K_{11}) + \tfrac{1}{2}(J_{22} - K_{22}) + J_{12} - K_{12} \qquad (30)$$

so that

$$f_1 = f_2 = a_{11} = a_{22} = a_{12} = -b_{11} = -b_{22} = -b_{12} = \tfrac{1}{2}$$

For the general high-spin open-shell wave function,

$$\mathcal{A}[\{\text{core}\}\phi_1 \alpha \phi_2 \alpha \cdots \phi_n \alpha] \qquad (31)$$

we obtain the energy expression*

$$E = E_c + \sum_i h_{ii}^c + \tfrac{1}{2}\sum_{i,j} (J_{ij} - K_{ij}) \qquad (32)$$

We refer to wave functions such as (31) as *multiplet* wave functions and to the high-spin coupled open-shell orbitals involved as the *multiplet-shell* orbitals.

---

*Under a transformation $\mathbf{U}$ of the open-shell orbitals of a high-spin open-shell wave function, the total wave function changes as (det U). Thus at most a sign change occurs under orthogonal transformations. As a result, the form of (32) is invariant under orthogonal transformations among the high-spin orbitals.

3.1.4. Orthogonality Restrictions

If in the open-shell singlet wave function,

$$\mathcal{A}[\{core\}\phi_1\phi_2(\alpha\beta-\beta\alpha)]=\mathcal{A}[\{core\}(\phi_1\phi_2+\phi_2\phi_1)\alpha\beta] \qquad (33a)$$

(where the electrons in ϕ_1 and ϕ_2 are coupled into a singlet pair), we take ϕ_1 and ϕ_2 as orthogonal,

$$\langle\phi_1|\phi_2\rangle=0 \qquad (33b)$$

the energy expression is

$$E=E_c+h_{11}^c+h_{22}^c+J_{12}+K_{12} \qquad (34)$$

so that

$$f_1=f_2=a_{12}=b_{12}=\tfrac{1}{2}$$
$$a_{11}=a_{22}=b_{11}=b_{22}=0$$

In this case, the energy expression cannot be put into a more symmetric form by adding and subtracting self-terms.

For all the HF wave functions considered in Sections 3.1.1–3.1.3, we found that because of the functional form of these wave functions the antisymmetrizer projected away any orbital nonorthogonalities. Thus, the sole effect of constraining the orbitals to be orthogonal was to permit us to express energies in the simplified form of (6), rather than (4). However, if in (33a) we allow ϕ_1 and ϕ_2 to be nonorthogonal,

$$\langle\phi_1|\phi_2\rangle=S_{12}$$

and rewrite ϕ_2 as

$$\phi_2=\phi_2'+\lambda\phi_1$$

where

$$\langle\phi_1|\phi_2'\rangle=0$$

then expansion of (33a) leads to

$$\mathcal{A}[\{core\}\phi_1\phi_2'(\alpha\beta-\beta\alpha)]+2\lambda\mathcal{A}[\{core\}\phi_1\phi_1\alpha\beta] \qquad (35)$$

The second term in (35) is nonvanishing, and hence in this case imposing orthogonality between ϕ_1 and ϕ_2 does, in general, represent a restriction. Thus, in this instance (33b) is *not* just a convenient computational tool.*

The restrictive nature of the orbital orthogonality constraint, which is *always* imposed upon HF wave functions, is not confined solely to the two

---

*The exception would be the case where orbitals ϕ_1 and ϕ_2 necessarily belong to different symmetry classes and hence lead to (33b) without restriction.

orbitals comprising an open-shell singlet pair. It applies in general with respect to all open-shell orbitals which are not multiplet (high-spin) coupled to one another. For example, in the wave function

$$a[\phi_1\phi_1\phi_2\phi_3\phi_4\phi_5\alpha\beta(\alpha\beta-\beta\alpha)\alpha\alpha] \tag{36}$$

while the conditions,

$$\langle\phi_1|\phi_i\rangle=0, \qquad i=2,3,4,5$$
$$\langle\phi_4|\phi_5\rangle=0 \tag{37}$$

are completely unrestrictive, the constraints,

$$\langle\phi_k|\phi_l\rangle=0, \qquad k=2,3 \text{ and } l=4,5$$
$$\langle\phi_2|\phi_3\rangle=0 \tag{38}$$

are actual restrictions. Whether or not (38) is imposed upon (36) can have a significant effect upon the energy (and potential usefulness) of this wave function.

3.2. GVB Wave Functions

3.2.1. General Considerations

The Hartree–Fock wave function forms the conceptual basis for understanding the electronic structures of broad classes of molecules. However, upon applying HF wave functions to studies of chemical reactions, a serious deficiency becomes apparent. These wave functions are generally incapable of describing bond dissociation.[7] To illustrate, consider the case of the hydrogen molecule. Near the equilibrium internuclear distance R_e the ground state molecule is adequately described by the closed-shell HF wave function

$$a[\phi\phi\alpha\beta] \tag{39}$$

However, in the dissociative limit the system consists of two isolated hydrogen atoms and is accurately described by the open-shell singlet HF wave function

$$a[\phi_1\phi_2(\alpha\beta-\beta\alpha)] \qquad \langle\phi_1|\phi_2\rangle=0 \tag{40}$$

Since (39) consists of a single doubly occupied orbital, it can never describe the separated-atom limit. Conversely, the *required* orthogonality between ϕ_1 and ϕ_2 prevents (40) from describing the molecular bond. Hence, while each limit is described by an HF wave function, the inherent restrictions of the HF formalism preclude the possibility of smoothly going from one limit to the other.

The obvious solution of this dilemma is to relax the orthogonality restriction in (40), thereby giving rise to the GVB wave function

$$\mathcal{A}[\phi_1'\phi_2'(\alpha\beta - \beta\alpha)] = \mathcal{A}[(\phi_1'\phi_2' + \phi_2'\phi_1')\alpha\beta] \tag{41}$$

where

$$\langle\phi_1'|\phi_2'\rangle = S_{12} \neq 0$$

Since in (41) no restrictions are placed upon either ϕ_1' or ϕ_2', this wave function can behave properly at all internuclear separations with S_{12} (for the optimum orbitals) varying from zero at infinite separation $[(41) = (40)]$ to ~ 0.8 at R_e [that is, $(41) \approx (39)$].[1] Furthermore, the added functional freedom associated with this wave function allows for the incorporation of optimal ionic and covalent character in the wave function, thereby leading to a significantly stronger chemical bond than does (39). That is, wave function (41) allows the electrons to better correlate their motions, thus reducing the energy of the system.

For workers conditioned to thinking of chemical systems in terms of traditional molecular HF wave functions it is useful to regard (41) as being a generalization of (39), in which we replace the conventional closed-shell orbital description of a singlet electron pair by a *GVB pair* consisting of two nonorthogonal orbitals coupled into a singlet,

$$\phi_a\phi_a\alpha\beta \rightarrow \phi_{1a}\phi_{2a}(\alpha\beta - \beta\alpha) \tag{42a}$$

or equivalently

$$\phi_a\phi_a\alpha\beta \rightarrow (\phi_{1a}\phi_{2a} + \phi_{2a}\phi_{1a})\alpha\beta \tag{42b}$$

Using this pair correlation prescription (42), the HF wave function

$$\mathcal{A}[\phi_1\phi_1\alpha\beta \cdots \phi_m\phi_m\alpha\beta\phi_{m+1}\alpha \cdots \phi_n\alpha] \tag{42c}$$

would be extended to the wave function of GVB form,

$$\mathcal{A}[\phi_{11}'\phi_{21}'(\alpha\beta - \beta\alpha) \cdots \phi_{1m}'\phi_{2m}'(\alpha\beta - \beta\alpha)\phi_{m+1}\alpha \cdots \phi_n\alpha] \tag{43}$$

where

$$\langle\phi_{1i}'|\phi_{2i}'\rangle = S_i$$

In (43) each singlet electron pair is allowed to correlate. However, for most chemical problems many such correlations can be ignored. Since only selected electron pairs need be represented by GVB pairs (e.g., those describing chemical bonds), we normally generalize (42c) to a wave function of the form

$$\mathcal{A}[\{core\}\phi_{11}'\phi_{21}'(\alpha\beta - \beta\alpha) \cdots \phi_{1m}'\phi_{2m}'(\alpha\beta - \beta\alpha)\phi_{m+1}\alpha \cdots \phi_n\alpha] \tag{44}$$

For example, the excitation energies between the lower electronic states of O_2 are essentially unaffected by correlation effects involving the $1s$ core orbitals of each oxygen. Indeed, for the lower few states it is not necessary to correlate even the $2s$ core orbitals of each oxygen.[8]

In wave functions such as (43) or (44) it is computationally useful to require that all orbitals other than the two within a given singlet pair be orthogonal. This *strong orthogonality* constraint[9,10] is of course a restriction. However, since it applies between orbitals that are normally not expected to overlap significantly, strong orthogonality is usually not a serious restriction.*

The wave function developed here singlet couples as many orbital pairs as is possible for a given eigenstate of spin. Thus for a four-electron singlet state we have

$$a[\phi_{11}\phi_{21}\phi_{12}\phi_{22}(\alpha\beta - \beta\alpha)(\alpha\beta - \beta\alpha)] \tag{45}$$

However this *perfect pairing* is not the only coupling scheme possible. For example,

$$a\{\phi_{11}\phi_{21}\phi_{12}\phi_{22}[2\alpha\alpha\beta\beta + 2\beta\beta\alpha\alpha - (\alpha\beta + \beta\alpha)(\alpha\beta + \beta\alpha)]\}$$

represents another linearly independent way of coupling these orbitals into a singlet state. The unrestricted GVB wave function[1] allows for these other possibilities as well, having the form

$$a[\phi_1\phi_2\cdots\phi_n\theta(1\cdots n)] \tag{46}$$

where the spin function θ is allowed full functional freedom within the confines of being an eigenstate of spin

$$\hat{S}^2\theta = S(S+1)\theta$$

Furthermore, in the GVB wave function no orbital orthogonality conditions are imposed. The perfect-pairing spin function is especially inappropriate for systems such as the allyl radical[11] or benzene,[12] in which two or more bonding schemes are of comparable importance, and for describing the dissociation of molecules leading to triplet or higher spin states on the separated fragments (e.g., O_2).[13] However, for most cases the orbitals obtained from GVB–PP are quite adequate.

The GVB wave function, in which the perfect pairing and strong orthogonality restrictions are both imposed,[3] is referred to as the GVB–PP (perfect pairing) wave function to distinguish it from the general GVB wave function. In addition, we have developed an intermediate level of calculation in which the spin function is optimized while retaining the strong orthogonality

---

*These orthogonality conditions have been relaxed in several cases[1,2,5] where it has been found that the orbitals in different singlet pairs of perfect pairing wave functions such as (43) and (44) generally have small overlaps (this arises basically from the Pauli principle).

restrictions.[5] This has been quite useful, for example, for considering allylic systems. These more general GVB wave functions lead to an energy expression not conforming to (6), and hence in this chapter we will restrict ourselves to the GVB–PP approximation.

3.2.2. The GVB–PP Energy Expression (The Natural Orbital Representation)

The energy of the simplest GVB wave function (41) is

$$E = (1+S_{12}^2)^{-1}[h_{11}+h_{22}+2S_{12}h_{12}+J_{12}+K_{12}] \tag{47}$$

an expression not conforming to (6). However, letting

$$\phi_1' = (\sigma_1^{1/2}\phi_1+\sigma_2^{1/2}\phi_2)/(\sigma_1+\sigma_2)^{1/2}$$
$$\phi_2' = (\sigma_1^{1/2}\phi_1-\sigma_2^{1/2}\phi_2)/(\sigma_1+\sigma_2)^{1/2} \tag{48}$$

where $\sigma_1>0$, $\sigma_2>0$, and $\langle\phi_1|\phi_2\rangle=0$ allows (41) to be rewritten as

$$a[(\sigma_1\phi_1\phi_1-\sigma_2\phi_2\phi_2)\alpha\beta] = \sigma_1 a[\phi_1\phi_1\alpha\beta] - \sigma_2 a[\phi_2\phi_2\alpha\beta] \tag{49}$$

giving rise to the energy expression

$$E = \frac{\sigma_1^2}{\sigma_1^2+\sigma_2^2}(2h_{11}+J_{11}) + \frac{\sigma_2^2}{\sigma_1^2+\sigma_2^2}(2h_{22}+J_{22}) - \frac{2\sigma_1\sigma_2}{\sigma_1^2+\sigma_2^2}K_{12} \tag{50}$$

Equation (50) does have the desired form of (6), differing from a typical HF expression only in that the energy coefficients (f, a, b) are now functions of the pair coefficients σ_1 and σ_2, rather than being fixed.

Thus, while it is conceptually the nonorthogonal GVB orbitals that are the most useful, computationally it is much more convenient to replace the GVB pair,

$$\phi_{1i}'\phi_{2i}'(\alpha\beta-\beta\alpha), \qquad \langle\phi_{1i}'|\phi_{2i}'\rangle = S_i \tag{51}$$

with its *natural orbital*[14] representation,

$$(\sigma_{1i}\phi_{1i}\phi_{1i}-\sigma_{2i}\phi_{2i}\phi_{2i})\alpha\beta, \qquad \langle\phi_{1i}|\phi_{2i}\rangle = 0 \tag{52}$$

where

$$\phi_{1i}' = (\sigma_{1i}^{1/2}\phi_{1i}+\sigma_{2i}^{1/2}\phi_{2i})/(\sigma_{1i}+\sigma_{2i})^{1/2}$$
$$\phi_{2i}' = (\sigma_{1i}^{1/2}\phi_{1i}-\sigma_{2i}^{1/2}\phi_{2i})/(\sigma_{1i}+\sigma_{2i})^{1/2} \tag{53}$$
$$\langle\phi_{1i}|\phi_{2i}\rangle = 0, \qquad \sigma_{ji}>0$$

and where for future convenience we take

$$\sigma_{1i}^2+\sigma_{2i}^2 = 1 \tag{54}$$

Once written in terms of these natural orbital pairs, the energy of any GVB–PP wave function always has the form of (6). For example, casting the two-pair wave function,

$$\mathcal{A}[\{core\}\phi_1'\phi_2'\phi_3'\phi_4'(\alpha\beta - \beta\alpha)(\alpha\beta - \beta\alpha)] \tag{55}$$

in its natural orbital form

$$\mathcal{A}[\{core\}(\sigma_1\phi_1\phi_1 - \sigma_2\phi_2\phi_2)(\sigma_3\phi_3\phi_3 - \sigma_4\phi_4\phi_4)\alpha\beta\alpha\beta] \tag{56}$$

leads to

$$E = E_c + \sum_{i=1}^{4} \sigma_i^2(2h_{ii}^c + J_{ii}) + 2 \sum_{i=1}^{2}\sum_{j=3}^{4} \sigma_i^2\sigma_j^2(2J_{ij} - K_{ij})$$
$$- 2\sigma_1\sigma_2 K_{12} - 2\sigma_3\sigma_4 K_{34} \tag{57}$$

Thus, the GVB–PP wave function (in natural orbital form) having the general form

$$\mathcal{A}[\{core\}\{pair\}\{open\}] \tag{58}$$

where*

$$\{core\} = \phi_{c_1}\phi_{c_1}\alpha\beta\phi_{c_2}\phi_{c_2}\alpha\beta\cdots$$
$$\{pair\} = (\sigma_{p_1}\phi_{p_1}\phi_{p_1} - \sigma_{p_2}\phi_{p_2}\phi_{p_2})\alpha\beta(\sigma_{p_3}\phi_{p_3}\phi_{p_3} - \sigma_{p_4}\phi_{p_4}\phi_{p_4})\alpha\beta\cdots$$
$$\{open\} = \phi_i\alpha\phi_{i+1}\alpha\cdots$$

leads to an energy expression in the desired form (6), where

$$f_i = 1 \qquad \text{if } \phi_i \text{ is a core orbital}$$
$$f_i = \sigma_i^2 \qquad \text{if } \phi_i \text{ is a pair orbital}$$
$$f_i = \tfrac{1}{2} \qquad \text{if } \phi_i \text{ is an open orbital} \tag{59}$$
$$a_{ij} = 2f_if_j$$
$$b_{ij} = -f_if_j$$

except that

$$b_{ij} = -\tfrac{1}{2} \qquad \text{if } \phi_i \text{ and } \phi_j \text{ are both open orbitals}$$

$$\left.\begin{array}{l} a_{ii} = f_i \\ b_{ii} = 0 \end{array}\right\} \qquad \text{if } \phi_i \text{ is a pair orbital}$$

$$\left.\begin{array}{l} a_{ij} = 0 \\ b_{ij} = -\sigma_i\sigma_j \end{array}\right\} \qquad \text{if } \phi_i \text{ and } \phi_j \text{ are in the same pair}$$

*The {open} wave function can also be taken of the form in (33) or as a product of (33) with a multiplet open shell. The modifications of the coefficients (59) are straightforward.

Upon solving for the GVB natural orbitals ϕ_{1i} and ϕ_{2i} and the pair coefficients σ_{1i} and σ_{2i}, the GVB orbitals ϕ'_{1i} and ϕ'_{2i} are obtained from (48), where the overlap S_i is given by

$$S_i = \frac{\sigma_{1i} - \sigma_{2i}}{\sigma_{1i} + \sigma_{2i}} \tag{60}$$

3.3. Separated-Pair Wave Functions

Written in its natural orbital form, we see that another prescription for obtaining a GVB–PP wave function from the usual HF wave function is

$$\phi_p \phi_p \alpha\beta \rightarrow (\sigma_{1p}\phi_{1p}\phi_{1p} - \sigma_{2p}\phi_{2p}\phi_{2p})\alpha\beta, \qquad \langle\phi_{1p}|\phi_{2p}\rangle = 0 \tag{60a}$$

That is, an electron pair normally described by a closed-shell orbital in the HF wave function is instead described by a geminal expansion consisting of two orthogonal doubly occupied orbitals. In a *separated-pair wave function*[15,16] this pair correlation prescription is extended to the general form,*

$$\phi_p \phi_p \alpha\beta \rightarrow \sum_{i=1}^{n_p} \sigma_{ip}\phi_{ip}\phi_{ip}\alpha\beta, \qquad \langle\phi_{ip}|\phi_{jp}\rangle = \delta_{ij} \tag{60b}$$

so that the pair functional consists of an expansion of arbitrary length. That these orbitals can be taken as orthogonal without restriction becomes obvious upon realizing that the general singlet functional,

$$\sum_{i,j} d_{ij}\phi'_{ip}\phi'_{jp}\alpha\beta$$

where

$$\langle\phi'_{ip}|\phi'_{ji}\rangle = \delta_{ij}, \qquad d_{ij} = d_{ji}$$

is directly related to (60b) through a general orthogonal transformation of the form

$$\phi'_{ip} = (\phi_{ip} + \lambda_{ij}\phi_{jp})/(1 + \lambda_{ij}^2)^{1/2}, \qquad \lambda_{ij} = -\lambda_{ji} \qquad \text{for all } i \text{ and } j$$

Thus, a separated-pair functional can, in principle, be used to introduce any amount of correlation between the two singlet coupled electrons by taking n_p to be arbitrarily large. In practice, all significant correlation effects can normally be taken into account using a relatively small expansion through an optimal choice of the orbitals involved.[15,16]

---

*Note that for two-term wave functions we define a minus sign with σ_2 so that $\sigma_1, \sigma_2 > 0$. For separated-pair wave functions we let all terms enter with $+\sigma_i$ coefficient so that some $\sigma_i < 0$.

The ability of (60b) to correlate the two electrons within a given singlet pair to an arbitrary degree can at times be quite important and can obviously lead to results that are quantitatively superior to those possible using the GVB pair functional (60a). However, for most chemical problems the dominant correlations (both qualitative and quantitative) can usually be accomodated by the simpler GVB pair functional. Additional effects, which could be incorporated through use of (60b) with $n_p > 2$, are generally of less importance, especially when compared with other types of correlations possible in a many-electron system.

The GVB wave function (46) obtained by relaxing the perfect-pairing and strong orthogonality constraints associated with a GVB–PP wave function can incorporate many of these additional kinds of correlations as well, and we regard the GVB–PP wave function as a restricted GVB wave function, *not* as a restricted separated-pair wave function. This distinction, which on the basis of comparing (60a) and (60b) is rather obscure, crystallizes upon recalling that the GVB pair functional in (60a) is simply a computationally convenient representation for a pair of nonorthogonal orbitals coupled into a singlet. Therefore, just as is the case for HF and GVB wave functions, a GVB–PP wave function is readily susceptible to physical interpretation through the fundamental *conceptualization* of associating each electron with a special orbital. On the other hand, for a separated-pair wave function ($n_p > 2$) this is not possible since each correlated electron pair is described in terms of more than two orbitals.

On a computational level, however, there really is no formal distinction. That is, using the conventional normalization criterion

$$\sum_{i=1}^{n_p} \sigma_{ip}^2 = 1$$

the energy expression and actual energy coefficients as given in (59) are the same for both GVB–PP and separated-pair wave functions and hybrids thereof. The only difference is that now a "pair" can be composed of more than two orbitals. This being the case, we find it convenient to refer to such wave functions collectively as *correlated-pair* wave functions and the orbitals involved in such pairs are then simply called *correlated-pair* orbitals.

4. The Basic Variational Equations

Having demonstrated that energy expression (6) is applicable to a wide range of useful wave functions, we will now concern ourselves with the task of solving for the optimal (self-consistent) orbitals of (6); generally those leading to the lowest energy. Since we will be dealing solely with (6), where the energy coefficients $\{f_i, a_{ij}, b_{ij}\}$ are regarded as arbitrary parameters, the results obtained are quite general in that they apply to any wave function (actual or contrived) involving orthonormal orbitals whose energy conforms to (6). Moreover, our underlying approach is not rooted in the simple homogeneous form of (6) and can therefore be readily extended to the more general case.

In solving this problem, we will use the variational principle as our guide. That is, we will seek those orbitals $\{\phi_i\}$ for which the energy is stationary (at a minimum) with respect to any arbitrary allowed variations in any of the orbitals. For convenience of notation we will take all orbitals as real. Extension to complex orbitals is straightforward.

4.1. The Variational Equation and Fock Operators

Starting with the optimal orbitals $\{\phi_i\}$ for (6) having energy E, consider a *slightly* modified set of orbitals $\{\phi_i'\}$ having energy

$$E' = \sum_i 2f_i h_{\phi_i'\phi_i'} + \sum_{i,j} (a_{ij}J_{\phi_i'\phi_j'} + b_{ij}K_{\phi_i'\phi_j'}) \tag{61}$$

Since the form of (61) is applicable only for orthonormal orbitals, we *must* require that this new set satisfy

$$\langle \phi_i' | \phi_j' \rangle = \delta_{ij} \tag{62}$$

Recognizing that the only meaningful change in an orbital is one which is orthogonal to that orbital, we can write the modified orbital ϕ_i' in terms of the optimal orbital ϕ_i as

$$\phi_i' = (\phi_i + \delta\phi_i)/(1 + \langle \delta\phi_i | \delta\phi_i \rangle)^{1/2} \tag{63a}$$

where

$$\langle \phi_i | \delta\phi_i \rangle = 0 \tag{63b}$$

and $\delta\phi_i$ is the variation in this orbital. Expanding E' in terms of $\{\phi_i\}$ and $\{\delta\phi_i\}$ yields

$$E' = E + \delta^{(1)}E + \delta^{(2)}E + \cdots \tag{64}$$

where the first-order variation in the energy is

$$\delta^{(1)}E = 4 \sum_i \left\langle \delta\phi_i \left| \left\{ f_i h + \sum_j (a_{ij}J_j + b_{ij}K_j) \right\} \right| \phi_i \right\rangle \tag{65}$$

Defining the *generalized Fock operator*,

$$F_i = f_i h + \sum_j (a_{ij}J_j + b_{ij}K_j) \tag{66}$$

allows (65) to be written as

$$\delta^{(1)}E = 4 \sum_i \langle \delta\phi_i | F_i | \phi_i \rangle$$

Since the energy E is at a minimum with respect to changes in the orbitals, it must then be stationary through first-order variations in these orbitals. That is,

$$\delta^{(1)}E = 0 \tag{67}$$

if the $\{\phi_i\}$ are indeed optimal. Hence, a necessary condition for $\{\phi_i\}$ to be optimal is that

$$\sum_i \langle \delta\phi_i | F_i | \phi_i \rangle = 0 \tag{68}$$

for *all* orbital variations consistent with (62) and (63).

It is important to keep in mind that this result (68) with F_i in the form of (66) is valid only when the orthogonality condition (62) is maintained (at least through first order).

Having defined the Fock operators (66), we find that the energy can be written as

$$E = \sum_i [f_i h_{ii} + \langle \phi_i | F_i | \phi_i \rangle] \tag{69a}$$

From this expression we see that these operators are somewhat ambiguous in that,

$$F'_i = F_i + g_i(J_i - K_i) \tag{69b}$$

(where g_i is a completely arbitrary parameter) is an equally valid Fock operator for orbital i. This result is of course simply a reflection of (15).

Fock Operator Examples (Shells)

Using (66) we find from (18) that the Fock operator for the orbitals of *closed-shell* wave function (17) is

$$F_c = h + \sum_j (2J_j - K_j) \tag{70a}$$

for all orbitals. That is, we obtain the same operator for each closed-shell orbital. Attainment of this result was the main reason for adding and subtracting self-terms in the energy expression for such wave functions. Had we neglected to do so, (66) would have led to a different operator for each orbital,

$$F_i = h + J_i + \sum_{j \neq i} (2J_j - K_j) \tag{70b}$$

However, we could also arrive at (70a) by using (69b) and choosing $g_i = 1$. Extending to wave functions involving open-shell orbitals as well, the closed-shell Fock operator becomes

$$F_c = h + \sum_j f_j(2J_j - K_j) \tag{71}$$

For high-spin (*multiplied*) *coupled open-shell* orbitals, as in wave function (31), we obtain

$$2F_m = h^c + \sum_j (J_j - K_j) \tag{72}$$

Thus, all the Fock operators for multiplet coupled orbitals can be the same. This remains true even if additional open-shell orbitals are present such as in (36) and (55). Note, however, that

$$F_c \neq F_m \tag{73}$$

so that even for the simplest wave functions involving both closed-shell and open-shell orbitals there is no single Fock operator applicable to all orbitals.

In most other instances each orbital has its own distinct Fock operator. For example, for the *open-shell singlet* wave function (33) we obtain

$$2F_1 = h^c + J_2 + K_2$$
$$2F_2 = h^c + J_1 + K_1 \tag{74}$$

and there is no way of rewriting the energy expression [or choosing g_1 and g_2 in (69b)] to obtain the same operator for each orbital. Likewise, in a *correlated-pair* wave function, the Fock operator for the kth orbital of the lth correlated pair,

$$F_{kl} = \sigma_{kl}^2 \left[h^c + J_{kl} + \sum_{j \neq 1l, 2l \ldots} f_j (2J_j - K_j) \right] + \sum_{i \neq k}^{n_l} \sigma_{kl} \sigma_{il} K_{il} \tag{75}$$

is unique to this orbital.

By convention, orbitals which can have the same Fock operator are said to belong to the same *shell*. From the preceding examples we see that the closed and multiplet shells may contain several orbitals. On the other hand, the orbitals in a correlated-pair or HF open-shell singlet pair each belong to a unique shell. If a shell is capable of multiple occupancy, the distinct Fock operator applicable to all of the orbitals in the shell is referred to as the Fock operator for that shell.* For a shell containing only one orbital, the Fock operator for that shell is generally taken as the one containing no self-exchange term ($b_{\nu\nu} = 0$).

4.2. The Variational Conditions

Since Eq. (68) must be satisfied for all allowed orbital variations, we can explore its implications by focusing attention on all possible changes in a particular orbital, say ϕ_ν. In solving for the wave functions of molecules we will

---

*Since the F operator for such a shell is symmetric with respect to the orbitals of that shell, it is invariant to any orthogonal (unitary) transformation among these orbitals. Therefore, from (69a) it follows that the energy is also invariant to such transformations between these orbitals.

exclusively consider the *basis set expansion* approach. That is, all orbitals and their variations will be expanded in terms of a set of P linearly independent basis functions, thus defining the P-dimensional Hilbert space in terms of which the system is to be described. In the following we will make a distinction between the occupied orbitals, that is the n orbitals involved in (6) (at any stage in the iterative process), and the $P-n$ *virtual orbitals*, also constructable in terms of the P-dimension basis being employed.

Allowing *only* orbital ϕ_ν to vary, (68) reduces to

$$\langle \delta\phi_\nu | F_\nu | \phi_\nu \rangle = 0 \tag{76}$$

However, from (62) and (63), for $j \neq \nu$, we have

$$0 = \langle \phi_j | \phi_\nu' \rangle$$

$$= \frac{\langle \phi_j | \phi_\nu + \delta\phi_\nu \rangle}{(1 + \langle \delta\phi_\nu | \delta\phi_\nu \rangle)^{1/2}} \tag{77}$$

where

$$\langle \phi_i | \phi_j \rangle = \delta_{ij}, \qquad \langle \delta\phi_\nu | \phi_\nu \rangle = 0$$

Thus

$$\langle \phi_j | \delta\phi_\nu \rangle = 0 \qquad \text{for all } j \tag{78}$$

That is, only those variations which are orthogonal to all orbitals may be considered. Therefore (76) provides us with a necessary condition for ϕ_ν to be optimum with respect to any variations orthogonal to all occupied orbitals.

Since (76) applies only to variations orthogonal to all occupied orbitals, we must still consider the case where orbital ν is allowed to vary within the space of other occupied orbitals. For example, consider the case where ϕ_ν' is allowed to overlap with ϕ_μ,

$$\phi_\nu' = (\phi_\nu + \lambda\phi_\mu)/(1 + \lambda^2)^{1/2} \tag{79}$$

so that

$$\delta\phi_\nu = \lambda\phi_\mu \tag{80}$$

Since orbital orthogonality must be preserved, orbital μ *must* change simultaneously and must satisfy

$$\delta\phi_\mu = -\lambda\phi_\nu \tag{81}$$

That is, ϕ_ν' as given in (79) dictates that ϕ_μ' must have the form

$$\phi_\mu' = (\phi_\mu - \lambda\phi_\nu)/(1 + \lambda^2)^{1/2} \tag{82}$$

Allowing no additional variations, Eq. (68) reduces to

$$0 = \langle \delta\phi_\nu | F_\nu | \phi_\nu \rangle + \langle \delta\phi_\mu | F_\mu | \phi_\mu \rangle$$

$$= \lambda \langle \phi_\nu | (F_\nu - F_\mu) | \phi_\mu \rangle \tag{83}$$

from which we obtain

$$\langle \phi_\nu | (F_\nu - F_\mu) | \phi_\mu \rangle = 0 \tag{84}$$

as a necessary condition for orbitals ϕ_ν and ϕ_μ to be optimum with respect to each other.*

Condition (76) ensures that $\delta^{(1)}E = 0$ for any variation $\delta\phi_\nu$ (in orbital ϕ_ν) that is *orthogonal* to all occupied orbitals, and condition (84) (for all μ) ensures that $\delta^{(1)}E = 0$ for any variation $\delta\phi_\nu$ *overlapping* any occupied orbital. Thus conditions (76), for all ν, and (84), for all μ and ν, are sufficient (and necessary) to ensure that $\delta^{(1)}E = 0$ for all variations of all orbitals. Together these equations are referred to as the *variational conditions*.

4.3. The Orbital Correction Equation

Although the variational conditions determine whether a given set of orbitals is optimal, they do not tell us how to improve upon a nonoptimal set. That is, they do *not* tell us how to *solve* for an optimal set of orbitals. We will now consider that we have some initial guess $\{\phi_i^0\}$ for the orbitals and that we want to find the corrections $\{\Delta\phi_i\}$ leading to the optimum orbitals $\{\phi_i\}$. In order to do this we return to the energy expression and re-expand, retaining all terms through second order in $\{\Delta\phi_i\}$. Applying the variational principle to the resulting expression then leads to equations that are valid through first order[†] in $\{\Delta\phi_i\}$ and that can be solved to obtain the optimal corrections.

Since orbital orthonormality must always be preserved, these corrections cannot be fully independent of each other. Expanding each orbital as

$$\phi_i = (\phi_i^0 + \Delta\phi_i)/(1 + \langle \Delta\phi_i | \Delta\phi_i \rangle)^{1/2} \tag{85}$$

where

$$\langle \phi_i^0 | \Delta\phi_i \rangle = 0$$

ensures that

$$\langle \phi_i | \phi_i \rangle = 1$$

---

*Note that if orbitals μ and ν are in the same shell ($F_\mu = F_\nu$), then (84) is always satisfied.
†We will also retain some higher-order terms (*vide infra*).

Since $\{\phi_i^0\}$ and $\{\phi_i\}$ each constitute an orthogonal set, we obtain

$$0 = \langle \phi_i | \phi_j \rangle$$

$$= \frac{(\langle \Delta\phi_i | \phi_j^0 \rangle + \langle \phi_i^0 | \Delta\phi_j \rangle + \langle \Delta\phi_i | \Delta\phi_j \rangle)}{(1 + \langle \Delta\phi_i | \Delta\phi_i \rangle)^{1/2}(1 + \langle \Delta\phi_j | \Delta\phi_j \rangle)^{1/2}} \qquad i \neq j \qquad (86)$$

Thus, $\{\Delta\phi_i\}$ must be constrained to satisfy

$$\langle \Delta\phi_i | \phi_j^0 \rangle + \langle \phi_i^0 | \Delta\phi_j \rangle + \langle \Delta\phi_i | \Delta\phi_j \rangle = 0 \qquad i \neq j \qquad (87)$$

A logical starting point for considering a variational determination of $\{\Delta\phi_i\}$ is the expression for the change in energy due to these corrections,

$$\Delta E = E - E^0$$

$$= 2 \sum_i f_i (h_{ii} - h_{i_0 i_0}) + \sum_{i,j} [a_{ij}(J_{ij} - J_{i_0 j_0}) + b_{ij}(K_{ij} - K_{i_0 j_0})] \qquad (88)$$

Substituting (85) into this expression and collecting terms yields

$$\tfrac{1}{2}\Delta E = \sum_i (1 + \langle \Delta i | \Delta i \rangle)^{-1} \{ f_i [2\langle \Delta i | h | i_0 \rangle + \langle \Delta i | h | \Delta i \rangle - \langle \Delta i | \Delta i \rangle h_{i_0 i_0}]$$

$$+ \sum_j (1 + \langle \Delta j | \Delta j \rangle)^{-1} [a_{ij}(2\langle \Delta i | J_{j_0} | i_0 \rangle + \langle \Delta i | J_{j_0} | \Delta i \rangle - \langle \Delta i | \Delta i \rangle J_{i_0 j_0}$$

$$+ 2(i_0 \Delta i | j_0 \Delta j) + 2\langle \Delta i | J_{\Delta j} | i_0 \rangle + \tfrac{1}{2} J_{\Delta i \Delta j} - \tfrac{1}{2}\langle \Delta i | \Delta i \rangle \langle \Delta j | \Delta j \rangle J_{i_0 j_0})$$

$$+ b_{ij}(2\langle \Delta i | K_{j_0} | i_0 \rangle + \langle \Delta i | K_{j_0} | \Delta i \rangle - \langle \Delta i | \Delta i \rangle K_{i_0 j_0} + (\Delta i \Delta j | i_0 j_0)$$

$$+ (i_0 \Delta j | j_0 \Delta i) + 2\langle \Delta i | K_{\Delta j} | i_0 \rangle + \tfrac{1}{2} K_{\Delta i \Delta j} - \tfrac{1}{2}\langle \Delta i | \Delta i \rangle \langle \Delta j | \Delta j \rangle K_{i_0 j_0})]\} \qquad (89)$$

where we have used the shorthand notation

$$i = \phi_i, \qquad i_0 = \phi_i^0, \qquad \Delta i = \Delta\phi_i$$

Bear in mind here that (89) is subject to the constraint of (87).

In principle, $\{\Delta\phi_i\}$ could be variationally obtained from (89) and (87) exactly. However, in practice this would eventually require us to solve for the (real) solutions for an extremely complicated set of coupled high-order equations. The alternative is to consider an iterative procedure for obtaining $\{\phi_i\}$. In this way, during each iteration we need only seek corrections leading to *better* orbitals for use in the next iteration. Since only approximate solutions for $\{\Delta\phi_i\}$ will now be needed, we can base such a procedure on an approximation to (89) in which troublesome high-order terms are neglected. We will now consider such an approximation.

Rewriting (89) in terms of the zero-order Fock operators

$$F_{i_0} = f_i h + \sum_j (a_{ij} J_{j_0} + b_{ij} K_{j_0}) \qquad (90)$$

we obtain

$$\Delta E = \sum_i (1+\langle \Delta i|\Delta i\rangle)^{-1}\{2\langle \Delta i|F_{i_0}|i_0\rangle + \langle \Delta i|F_{i_0}|\Delta i\rangle - \langle \Delta i|\Delta i\rangle\langle i_0|F_{i_0}|i_0\rangle$$

$$+ \sum_i [2a_{ij}(i_0\Delta i|j_0\Delta j) + b_{ij}(\Delta i\Delta j|i_0 j_0) + b_{ij}(i_0\Delta j|j_0\Delta i)]\} + O(\Delta^3) \quad (91)$$

where $O(\Delta^3)$ contains only terms third order and higher in $\{\Delta\phi_i\}$ and is of the form

$$O(\Delta^3) = \sum_{i,j} (1+\langle \Delta i|\Delta i\rangle)^{-1}(1+\langle \Delta j|\Delta j\rangle)^{-1}O_{ij}(\Delta^3)$$

Defining $\Delta F_{i_0}(j_0 \to \Delta j)$ as the first-order correction to F_{i_0} due to the change $\Delta\phi_j$ in orbital j, the total first-order correction is

$$\Delta F_{i_0} = \sum_j \Delta F_{i_0}(j_0 \to \Delta j) \quad (92)$$

and we find that

$$\langle \Delta i|\Delta F_{i_0}|i_0\rangle = \sum_j \langle \Delta i|\Delta F_{i_0}(j \to \Delta j)|i_0\rangle$$

$$= \sum_j [2a_{ij}(i_0\Delta i|j_0\Delta j) + b_{ij}(\Delta i\Delta j|i_0 j_0) + b_{ij}(i_0\Delta j|j_0\Delta i)] \quad (93)$$

Therefore, defining $\Delta\varepsilon$ as the part of $\frac{1}{2}\Delta E$ containing terms up through order Δ^2, we obtain the result

$$\Delta\varepsilon = \sum_i (1+\langle \Delta i|\Delta i\rangle)^{-1}\{2\langle \Delta i|F_{i_0}|i_0\rangle + \langle \Delta i|F_{i_0}|\Delta i\rangle$$

$$+ \langle \Delta i|\Delta F_{i_0}|i_0\rangle - \langle \Delta i|\Delta i\rangle\langle i_0|F_{i_0}|i_0\rangle\} \quad (94a)$$

Alternatively, we can define a total ε as

$$\varepsilon = \Delta\varepsilon + \sum_i \langle i_0|F_{i_0}|i_0\rangle$$

$$\approx E^0 + \sum_{i,j} (a_{ij}J_{i_0 j_0} + b_{ij}K_{i_0 j_0})$$

leading to

$$\varepsilon = \sum_i (1+\langle \Delta i|\Delta i\rangle)^{-1}\{\langle(i_0+\Delta i)|F_{i_0}|(i_0+\Delta i)\rangle + \langle \Delta i|\Delta F_{i_0}|i_0\rangle\} \quad (94b)$$

In deriving (94a, b) we have purposely retained individual orbital renormalization terms. In this way we impart to $\Delta\varepsilon$ and ε the essential functional dependence upon $\{\Delta\phi_i\}$ consistent wih preservation of orbital normalization. However, we have made no attempt to incorporate orbital orthogonality constraints. Consequently, in variationally determining $\{\Delta\phi_i\}$ using (94), we must still explicitly require that $\{\Delta\phi_i\}$ satisfy

$$\langle i_0|\Delta\phi_j\rangle + \langle j_0|\Delta\phi_i\rangle + \langle \Delta\phi_i|\Delta\phi_j\rangle = 0 \qquad i \neq j$$
$$\langle i_0|\Delta\phi_i\rangle = 0 \tag{95}$$

Equations (94) and (95) constitute the basic orbital correction equations which we must solve to obtain corrections leading to new and better orbitals,

$$\phi_i^1 = (\phi_i^0 + \Delta\phi_i)/(1 + \langle \Delta\phi_i|\Delta\phi_i\rangle)^{1/2} \tag{96}$$

These orbitals $\{\phi_i^1\}$ will then serve as the initial guesses for another iteration until self-consistency is achieved. Since all terms through second order are retained in (94), these correction terms will be valid through first order and hence the iterative solution for the optimum orbitals $\{\phi_i\}$ should converge quadratically. Substituting (95) into (94) and requiring that $\partial\Delta\varepsilon/\partial\Delta i = 0$ for all i leads to a set of coupled equations [see (127) below] that must be solved to obtain the optimal corrections $\{\Delta\phi_i\}$. For a system with P basis functions and n occupied orbitals, there are nP such equations determining the optimal corrections, the formation and solution of which is currently impractical for cases of most interest. As a result, in the following section we will partition the problem into several more restricted variations, each of which will be sequentially applied.

5. The Iterative Self-Consistent Field Equations

Rigorously applied, Eqs. (94) and (95) permit simultaneous solution for all $\Delta\phi_i$. This straightforward approach ultimately requires (for the case of n orbitals and P basis functions) simultaneous solution of at least nP coupled nonlinear equations. However, by appropriately decoupling these equations the computational problem can be greatly simplified while retaining adequate convergence. Thus, we partition each iterative cycle into two separate steps.

a. Optimization of each orbital with respect to the space orthogonal to all orbitals, hereafter referred to as the *virtual space*. This step utilizes an *o*rthogonality *c*onstrained *b*asis *s*et *e*xpansion (OCBSE) and is referred to as the OCBSE step.

b. Optimization of the orbitals with respect to each other. This step is referred to as *orbital mixing*.

5.1. OCBSE Optimization

Let us first consider optimization of just one orbital (ϕ_ν) while keeping all other orbitals fixed. Thus, (94b) reduces to

$$\varepsilon_\nu = (1+\langle\Delta_\nu|\Delta_\nu\rangle)^{-1}[\langle\nu_0+\Delta_\nu|F_{\nu 0}|\nu_0+\Delta_\nu\rangle \tag{97}$$
$$+\langle\Delta_\nu|\Delta F_{\nu 0}(\nu_0\to\Delta_\nu)\nu_0\rangle]$$

where from (95) the change $\Delta\phi_\nu$ in orbital ν must satisfy

$$\langle i_0|\Delta\phi_\nu\rangle = 0 \qquad \text{for all } i \tag{98a}$$

Hence, $\Delta\phi_\nu$ can only involve the virtual space, that is the space orthogonal to all occupied orbitals. Evaluating the Fock correction term in (97), we obtain

$$\varepsilon_\nu = (1+\langle\Delta_\nu|\Delta_\nu\rangle)^{-1}$$
$$\times[\langle\nu_0+\Delta_\nu|F_{\nu 0}|\nu_0+\Delta_\mu\rangle+(2a_{\nu\nu}+b_{\nu\nu})K_{\nu 0\Delta\nu}+b_{\nu\nu}J_{\nu 0\Delta\nu}] \tag{98b}$$

However, since the operator

$$F'_\nu = F_\nu + b_{\nu\nu}(J_\nu - K_\nu)$$
$$= f_\nu h + \sum_{j\neq\nu}(a_{j\nu}J_j + b_{j\nu}K_j) + (a_{\nu\nu}+b_{\nu\nu})J_\nu \tag{98c}$$

is equally valid for this orbital, we can work with the simpler yet equivalent expression

$$\varepsilon_\nu = (1+\langle\Delta_\nu|\Delta_\nu\rangle)^{-1}[\langle\nu_0+\Delta_\nu|F'_{\nu 0}|\nu_0+\Delta_\nu\rangle+2a'_{\nu\nu}K_{\nu 0\Delta\nu}] \tag{98d}$$

where

$$a'_{\nu\nu} \equiv a_{\nu\nu} + b_{\nu\nu}$$

As usual, we solve for the orbitals in terms of a finite basis set,

$$\phi_j = \sum_{i=1}^{P} d_{ij}\chi'_i \tag{99}$$

The primitive basis functions $\{\chi'_i\}$ used in such expansions are generally of a type which facilitate evaluation of molecular integrals (e.g., atomic-like functions of Slater or Gaussian form). However, during the iterative SCF procedure it is convenient to redefine the basis (in terms of these primitives) as an orthonormal set consisting of the n current orbitals $\{\phi_i^0\}$ plus $P-n$ *virtuals*, $\{\chi_i\}$. In this way we can automatically satisfy (98) by explicitly excluding the other occupied orbitals $\{\phi_i^0, i \neq \nu\}$ from the expansion for ϕ_ν

$$\phi_\nu = \left(\phi_\nu^0 + \sum_{i=1}^{P-n} d_{i\nu}\chi_i\right) \bigg/ \left(1 + \sum_i d_{i\nu}\right)^{1/2} \tag{100}$$

so that

$$\Delta\phi_\nu = \sum_{i=1}^{P-n} d_{i\nu}\chi_i \tag{101}$$

This is referred to as the orthogonality constrained basis set expansion, or OCBSE.

Substituting (101) in (98d) we obtain

$$\varepsilon_\nu = (1 + \sum_i d_{i\nu}^2)^{-1}[\langle\nu_0|F_{\nu_0}'|\nu_0\rangle + 2\sum_i d_{i\nu}\langle\chi_i|F_{\nu_0}'|\nu_0\rangle$$

$$+ \sum_{i,j} d_{i\nu}d_{j\nu}\langle\chi_i|(F_{\nu_0}' + 2a_{\nu\nu}'K_{\nu_0})|\chi_j\rangle] \tag{102}$$

If we define

$$\chi_0 \equiv \phi_\nu^0$$

and let

$$C_{i\nu} = d_{i\nu}d_{0\nu} \tag{103}$$

then (102) becomes

$$\sum_{i,j=0}^{P-n} C_{i\nu}C_{j\nu}(\bar{F}_{\nu_0})_{ij} = \varepsilon_\nu \sum_{i=0}^{P-n} C_{i\nu}^2 \tag{104}$$

where in matrix notation

$$\bar{F}_{\nu_0}' = F_{\nu_0}' + X_{\nu_0}' \tag{105a}$$

and

$$(X_{\nu_0}')_{ij} = 2a_{\nu\nu}'(1 - \delta_{i,0})(1 - \delta_{j,0})\langle\chi_i|K_{\nu_0}|\chi_j\rangle \tag{106a}$$

The variation equation from (104) is

$$\bar{F}_{\nu_0}C_\nu = C_\nu\varepsilon_\nu \tag{105b}$$

Thus, with OCBSE the optimum orbital ϕ_ν is obtained by merely diagonalizing the $\bar{F}_{\nu_0}$ matrix and selecting the appropriate root. Usually we want the orbital leading to the lowest total energy and, therefore, we select the root having the most negative eigenvalue. We find that even when there are large changes in the orbital (105b) generally has a sufficiently wide radius of convergence to accommodate such changes. In certain circumstances we can deliberately force convergence to self-consistent wave functions describing excited states by choosing appropriate higher roots [e.g., by selecting from the roots of (105b) the one that has the highest overlap with ϕ_ν^0].

A common approach to variational equations has been to take the *first-order* variational equation[17]

$$F_\nu' C_\nu = \varepsilon_\nu C_\nu$$

that must be satisfied by the optimum orbital ϕ_ν, replace the operator F_ν' with F_{ν_0}', and solve

$$\mathbf{F}_{\nu_0}' \mathbf{C}_\nu = \boldsymbol{\varepsilon}_\nu \mathbf{C}_\nu \tag{106b}$$

to obtain a new orbital. Comparing with (105b) we see that this procedure is correct *if* there are no self-terms, that is, is $a_{\nu\nu} = b_{\nu\nu} = 0$ in (98c). Otherwise (106b) should include the terms (106a) and as a result (106b) is expected to have a much smaller radius of convergence than (105b).

By virtue of both the $\mathbf{F}'$ and $\mathbf{X}'$ matrices, the $\bar{\mathbf{F}}$ matrix in (105a) is different for each orbital of a shell. However, letting orbital ν belong to shell s we can rewrite (105a) in terms of the Fock operator $\mathbf{F}_{s_0}$ for the shell s to give

$$(\mathbf{F}_{s_0} + \mathbf{X}_{\nu_0}^S)\mathbf{C}_\nu = \mathbf{C}_\nu \boldsymbol{\varepsilon}_\nu \tag{107a}$$

where

$$(X_{\nu_0}^S)_{ij} = (1 - \delta_{i,0})(1 - \delta_{j,0})\langle\chi_i|[(2a_{SS} + b_{SS})K_{\nu_0} + b_{SS}J_{\nu_0}]|\chi_j\rangle \tag{107a}$$

In this form, the $\bar{\mathbf{F}}$ matrices for each orbital in a given shell differ only by virtue of the $\mathbf{X}$ matrices. If orbital ν is self-consistent,* this fact is signaled by

$$\langle\chi_i|F_{\dot{s}}|\phi_{\nu_0}\rangle = 0 \qquad i = \text{all virtuals} \tag{107c}$$

That is, the matrix $\mathbf{X}_{\nu_0}^S$ is not needed for verification of having attained self-consistency for orbital ν. The sole role of this matrix is to provide the rigorous single iteration redefinition of an as yet unconverged orbital. In order to attain the computation advantages of solving for a whole shell of orbitals with one operator, we can neglect this *self-correction* matrix and take our OCBSE optimization equation to be simply

$$\mathbf{F}_{S_0}\mathbf{C}_\nu = \mathbf{C}_\nu \boldsymbol{\varepsilon}_\nu \tag{108a}$$

This approximation generally leads to sufficiently rapid convergence and is usually used in place of (107a) without adverse effect. If orbital ν belongs to its own unique shell, there is little to be gained by this approximation and we should use the more correct expression (105a).

---

*Within this context the "self-consistency" of orbital ν is with respect to the fixed field due to $\{\phi_i^0; i \neq \nu\}$.

The major advantage in using (108a) occurs in cases where several orbitals belong to the same shell. For example, for wave functions involving any number of closed-shell orbitals (108a) leads to

$$\mathbf{F}_{co}\mathbf{C}_i = \mathbf{C}_i\boldsymbol{\varepsilon}_i \qquad i = \text{all closed-shell orbitals} \tag{108b}$$

Therefore, since each orbital corresponds to a different (orthogonal) solution of the same eigenvalue equation, we can simultaneously obtain new guesses for *all* these orbitals by simply constructing and diagonalizing a *single* $\mathbf{F}_{co}$ matrix (constructed over the OCBSE basis excluding any current open-shell orbitals). This contrasts with the case of the more rigorous equations (105a) or (107a) which lead to

$$\bar{\mathbf{F}}_i \neq \bar{\mathbf{F}}_j$$

and hence to different eigenvalue problems for each closed-shell orbital. In such instances the significant computational simplification obtained using (108) is expected to usually outweigh the advantages of a somewhat increased convergence which could result from (107a).

5.2. Orbital Mixing

Thus far we have considered optimization only with respect to virtual (unoccupied) space, having determined that this can be accomplished via diagonalization of the appropriate $\mathbf{F}_i$ or $\bar{\mathbf{F}}_i$ matrices (OCBSE). We must now turn our attention to optimization with respect to occupied space, that is, we must deal with the problem of optimizing the orbitals with respect to one another.

5.2.1. Independent Pairwise Optimization

For simplicity we will first allow for mixing only between two orbitals at a time, so that

$$\phi_i = (\phi_i^0 + \lambda_{ij}\phi_j^0)/(1+\lambda_{ij}^2)^{\frac{1}{2}}$$
$$\phi_j = (\phi_j^0 + \lambda_{ji}\phi_i^0)/(1+\lambda_{ji}^2)^{\frac{1}{2}} \tag{109}$$

However, from (90) we have

$$0 = \langle\phi_i^0|\Delta\phi_j\rangle + \langle\phi_j^0|\Delta\phi_i\rangle + \langle\Delta\phi_j|\Delta\phi_i\rangle$$
$$= \lambda_{ji} + \lambda_{ij} \tag{110}$$

Thus, preservation of orbital orthogonality requires that

$$\lambda_{ji} = -\lambda_{ij}$$

so that

$$\Delta\phi_i = \lambda_{ij}\phi_j^0$$
$$\Delta\phi_j = -\lambda_{ij}\phi_i^0 \tag{111}$$

Therefore, from (94a) we obtain

$$\Delta\varepsilon_{ij}(1+\lambda_{ij}^2) = 2\lambda_{ij}\langle i_0|(F_{i_0}-F_{j_0})|j_0\rangle$$
$$+ \lambda_{ij}^2[\langle j_0|(F_{i_0}-F_{j_0})|j_0\rangle\rangle + \langle i_0|(F_{j_0}-F_{i_0})|i_0\rangle + \gamma_{ij}] \tag{112}$$

where

$$\lambda_{ij}\gamma_{ij} = \langle i_0|(\Delta F_{i_0}-\Delta F_{j_0})|j_0\rangle$$

(This makes γ_{ij} independent of λ_{ij}.)

If we can take $F_{i_0} = F_{j_0}$, then (112) reduces to

$$\Delta\varepsilon_{ij} = 0$$

which correctly reflects the fact that the energy is invariant with respect to mixing between orbitals in the same shell. Therefore, in cases where all orbitals belong to the same shell (e.g., an HF closed-shell singlet wave function), the iterative cycle consists solely of OCBSE optimization.

Assuming that orbitals i and j are in different shells, λ_{ij} is obtained by requiring that $\Delta\varepsilon_{ij}$ be stationary with respect to λ_{ij},

$$\frac{d(\Delta\varepsilon_{ij})}{d\lambda_{ij}} = 0$$

to give

$$0 = (1-\lambda_{ij}^2)\langle i_0|F_{j_0}-F_{i_0}|j_0\rangle - \lambda_{ij}[\langle i_0|F_{j_0}-F_{i_0}|i_0\rangle - \langle j_0|F_{j_0}-F_{i_0}|j_0\rangle + \gamma_{ij}] \tag{113}$$

where from (91) and (92)

$$\gamma_{ij} = 2(a_{ii}+a_{jj}-2a_{ij})K_{i_0j_0} + (b_{ii}+b_{jj}-2b_{ij})(J_{i_0j_0}+K_{i_0j_0}) \tag{114}$$

In keeping with variational condition (84) if

$$\langle i_0|(F_{i_0}-F_{j_0})|j_0\rangle = 0$$

then $\lambda_{ij} = 0$, since these orbitals are already optimal (self-consistent) with respect to each other. Assuming this is not the case, solving (113) for λ_{ij} gives

$$\lambda_{ij} = \frac{-1}{2\lambda_{ij}^0} \pm \left[\left(\frac{1}{2\lambda_{ij}^0}\right)^2 + 1\right]^{1/2} \tag{115a}$$

where

$$\lambda_{ij}^0 \equiv \frac{\langle i_0|F_{jo}-F_{io}|j_0\rangle}{\langle i_0|F_{jo}-F_{io}|i_0\rangle - \langle j_0|F_{jo}-F_{io}|j_0\rangle + \gamma_{ij}} \tag{115b}$$

Usually we want the solution leading to the lowest energy and therefore we choose the root which minimizes $\Delta\varepsilon_{ij}$. However, under certain circumstances, such as in excited state calculations, a "least change" root selection criterion may be more appropriate.

Ignoring the λ_{ij}^2 in (113) [which is equivalent to neglecting the renormalization term $(1+\lambda_{ij}^2)$ in (112)] leads to

$$\lambda_{ij} = \lambda_{ij}^0 \tag{115c}$$

as the solution. This approximation can only be valid for $\lambda_{ij}^2 \ll 1$. However, it is important to realize that (115c) may well result in $\lambda_{ij}^2 > 1$. On the other hand, the "least change" root of (115a) [which reduces to (115c) in the limit of $\lambda_{ij}^{0\,2} \ll 1$] will supply a $\lambda_{ij}^2 < 1$ result even when $(\lambda_{ij}^0)^2 > 1$. Moreover, as we shall see (112) can be sufficiently well behaved over a wide enough range for the $\lambda_{ij}^2 > 1$ root of (115a) to be meaningful as well. Therefore, restriction of (115a) to the approximation (115c) is inappropriate since it robs us of important control over the iterative process and unnecessarily restricts the radius of convergence.

Being derived from (94), Eq. (113) is only an approximation to the exact equation for λ_{ij}. A rigorous derivation of the expression for the change in energy under orthogonal mixing between orbitals i and j leads to the following (quartic) variational equation for λ_{ij},

$$0 = \langle i_0|F_{jo}-F_{io}|j_0\rangle + \lambda_{ij}[\langle i_0|F_{jo}-F_{io}|i_0\rangle - \langle j_0|F_{jo}-F_{io}|j_0\rangle + \gamma_{ij}]$$
$$+ 3\lambda_{ij}^2\langle i_0|J_{io}-J_{jo}|j_0\rangle\alpha_{ij}$$
$$+ \lambda_{ij}^3[\langle i_0|F_{jo}-F_{io}|i_0\rangle - \langle j_0|F_{jo}-F_{io}|j_0\rangle + \gamma_{ij} + \alpha_{ij}(J_{jojo}+J_{iojo}-2J_{ioio}-4K_{iojo})]$$
$$- \lambda_{ij}^4[\langle i_o|F_{jo}-F_{io}|j_0\rangle + \alpha_{ij}\langle i_0|J_{io}-J_{jo}|j_0\rangle] \tag{115d}$$

where

$$\alpha_{ij} = a_{ii} + b_{ii} + a_{jj} + b_{jj} - 2a_{ij} - 2b_{ij}$$

Since we are only interested in the real roots for this equation, if the terms

$$\alpha_{ij}\langle i_0|(J_{io}-J_{jo})|j_0\rangle \tag{115e}$$

and

$$\alpha_{ij}(J_{ioio}+J_{jojo}-2J_{ioio}-4K_{iojo})$$

were to vanish, then (115d) would reduce to (113). In several important instances this is in fact the case. For example, if one orbital is a simple singly occupied orbital ($f_j = \frac{1}{2}$, $a_{jj} + b_{jj} = 0$) and the other is a closed-shell or correlated pair orbital, then (113) is quite rigorous, since $\alpha_{ij} = 0$. Even when these terms (115e) do not vanish, we generally find that the solutions to (115a) are sufficiently close to the real roots of (115d) so that the considerable added computational effort required to solve (115d) exactly is usually not justified. Bear in mind that we are considering here only the rotation* of two occupied orbitals and hence ignore the changes in the optimum λ_{ij} due to simultaneous rotations of other orbitals.

If the Fock correction term γ_{ij} were to vanish, then (115a) would be equivalent to diagonalizing the matrix

$$(\mathbf{F}_{io} - \mathbf{F}_{jo})$$

over the space $\{\phi_i^0, \phi_j^0\}$. Unfortunately this term can be quite important and should be retained since its neglect can reduce convergence and may even lead to oscillatory or divergent results.

Ideally, a pairwise optimization scheme would involve intermediate orbital redefinition. That is, once λ_{ij} has been determined orbitals i and j would be redefined before proceeding to the next pair. However, the computational ramifications of such a procedure are prohibitive due to the necessity of reprocessing the two-electron integrals. Therefore, we must use the same set of initial orbitals to calculate each mixing coefficient. Once this is done, the new orbitals are constructed from the results. There are many plausible ways of doing this. Of the various possibilities, we generally favor a sequential redefinition procedure based upon increasing coefficient magnitude.

Since each mixing coefficient is determined in a completely *independent* manner, it is not surprising to find that this method tends toward somewhat oscillatory behavior. However, this very simple procedure generally converges successfully.

5.2.2. Simultaneous Orbital Optimization

Pairwise orbital mixing, no matter how rigorously applied, represents maximum decoupling between all possible orbital changes within the space of $\{\phi_i^0\}$. That is, a pairwise approach allows us to deal with only *one* degree of freedom at a time. Therefore, even though each pair mixing coefficient may be determined in a quadratically convergent manner, such a procedure can only be expected to lead to overall linear convergence. Furthermore, since we have variational control over only one variable at a time, we can expect to converge only to solutions directly accessible through individual pairwise changes. To overcome these limitations, it is necessary to allow orbital rotations to occur

---

*We will often use the vector space analogy of a Hilbert space, referring to a mixing of orbitals as a rotation.

simultaneously so that coupling between them can be taken into account. That is, we must consider the problem of optimizing all orbitals with respect to one another at the same time.

We will first reexpress the orbital expansions in terms of a set of linearly independent expansion coefficients. Ignoring (for ease of notation) same-shell invariances and possible orbital symmetry partitions, each orbital can be written in unnormalized form as

$$\phi_\nu = \sigma_\nu^0 + \sum_{i \neq \nu} d_{i\nu} \phi_i^0 \tag{116}$$

where we have ignored normalization since it drops out of the orthogonality conditions. However, preservation of orbital orthogonality requires that

$$0 = \langle \phi_i | \phi_j \rangle = d_{ij} + d_{ji} + \sum_{k \neq i,j} d_{ki} d_{kj} \qquad i \neq j \tag{117}$$

Hence, owing to the first-order term $(d_{ij} + d_{ji})$ appearing in (117), the variables $\{d_{ij}\}$ do not constitute a linearly independent set. However, since first-order dependencies can be resolved by requiring that

$$d_{ij} = -d_{ji} \qquad i \neq j$$

we are naturally led to an expression of the form

$$\phi_\nu' = \phi_\nu' + \sum_{i \neq \nu} \zeta_{i\nu} \lambda_{(i\nu)} \phi_i^0 \tag{118}$$

where

$$\zeta_{ij} = -\zeta_{ji} = 1 \qquad \text{for } i > j$$

and (ij) is a pair index ranging from 1 to $n(n-1)/2$,

$$(ij) = [\max(i,j) - 1][\max(i,j) - 2]/2 + \min(i,j)$$

While (118) does provide us with an expansion in terms of the linearly independent variables $\{\lambda_{(ij)}\}$, orthogonality is preserved only through first order. However, higher-order nonorthogonalities can easily be removed by orthogonalizing $\{\phi_i'\}$ using the Gram–Schmidt procedure,

$$\phi_\nu = \phi_\nu' - \sum_{i < \nu} \frac{\langle \phi_\nu' | \phi_i \rangle}{\langle \phi_i | \phi_i \rangle} \phi_i \tag{119}$$

Since explicit expansion of (119) through second order gives

$$\phi_\nu = \phi_\nu^0 + \sum_{i \neq \nu} \zeta_{i\nu} \lambda_{(i\nu)} \phi_i^0 + \sum_{i < \nu} \sum_{j \neq i,\nu} \zeta_{ij} \zeta_{j\nu} \lambda_{(ij)} \lambda_{(j\nu)} \phi_i^0 + O(\lambda^3) \tag{120}$$

we obtain the sought-after correction expansion,

$$\Delta\phi_\nu = \sum_{i\neq\nu} \zeta_{i\nu}\lambda_{(i\nu)}\phi_i^0 + \sum_{i<\nu}\sum_{j\neq i,\nu} \zeta_{ij}\zeta_{j\nu}\lambda_{(ij)}\lambda_{(j\nu)}\phi_i^0 + O(\lambda^3) \tag{121}$$

Having derived a correction expansion in terms of independent variables explicitly satisfying orthogonality condition (95) through second order, we can now apply (94) to obtain an optimization equation for $\{\lambda_{(ij)}\}$. Substituting (121) into the various terms of (94a) and expanding each through second order in $\{\lambda_{(ij)}\}$, we obtain

$$\Delta\varepsilon = \sum_i [1 + \sum_{j\neq i}\lambda_{(ij)}^2]^{-1}\{2\sum_{j\neq i}\zeta_{ji}\lambda_{(ij)}\langle i_0|F_{i_0}|j_0\rangle$$

$$+ \sum_{j\neq i}\lambda_{(ij)}^2[\langle j_0|F_{i_0}|j_0\rangle - \langle i_0|F_{i_0}|i_0\rangle]$$

$$+ 2\sum_{j<i}\sum_{k\neq i,j}\zeta_{jk}\zeta_{ik}\lambda_{(jk)}\langle i_0|F_{i_0}|j_0\rangle$$

$$+ \sum_{j\neq i}\sum_{k\neq i,j}\zeta_{ji}\zeta_{ki}\lambda_{(ij)}\lambda_{(ik)}\langle j_0|F_{i_0}|k_0\rangle$$

$$+ \sum_{j\neq i}\sum_l\sum_{k\neq l}\zeta_{ji}\zeta_{kl}\lambda_{(ij)}\lambda_{(kl)}\langle j_0|\Delta F_{i_0}(l_0\to k_0)|i_0\rangle\} \tag{122}$$

Unfortunately, despite the simplifying approximations already made the variational determination of $\{\lambda_{(ij)}\}$ using (122) is still too complicated and we must make further approximations. The major difficulty with this equation is the disparity between the various orbital renormalization terms,

$$[1 + \sum_{j\neq i}\lambda_{(ij)}^2]^{-1} \tag{123a}$$

Since we have already neglected many third-order and higher terms in arriving at (122), we could of course simply choose to ignore these terms and still retain second-order validity. However, by completely neglecting all reciprocal dependence upon $\{\lambda_{(ij)}\}$ we would be forcing ourselves into a solution over which we could have no control and which would be valid only for a $\{\lambda_{ij}^2\ll 1\}$ result. [See the discussion of (115a) and (115c).] This is a rather drastic step and one which we prefer to avoid. An alternate approach to this problem is to consider a formulation which, while making the necessary approximations to these renormalization terms, still preserves a fundamental reciprocal dependence with respect to $\{\lambda_{(ij)}\}$. This can be accomplished by replacing these terms (123a) by the single term

$$[1 + \sum_{k>l}\lambda_{(kl)}^2]^{-1} \tag{123b}$$

That is, instead of simply setting each of these denominators to a common fixed value of unity, we replace them with a common term exhibiting the correct functional dependence with respect to each individual mixing coefficient. In this way, we introduce additional errors only with respect to third-order (and higher) coupling between *different* mixing coefficients.

After making this common denominator approximation, evaluating Fock corrections, and collecting terms, Eq. (122) becomes

$$
\Delta\varepsilon\left[1+\sum_{i>j}\lambda_{(ij)}^2\right]=\sum_{i>j}\{2\lambda_{(ij)}\langle i_0|F_{j_0}-F_{i_0}|j_0\rangle
$$

$$
+\lambda_{(ij)}^2[\langle i_0|F_{j_0}-F_{i_0}|i_0\rangle-\langle j_0|F_{j_0}-F_{i_0}|j_0\rangle+\gamma_{ij}^{ij}]\}
$$

$$
+2\sum_{i>j>k}\{\lambda_{(ij)}\lambda_{(ik)}[\langle j_0|F_{i_0}-F_{j_0}|k_0\rangle+\gamma_{ik}^{ij}]
$$

$$
-\lambda_{(ij)}\lambda_{(jk)}[\langle i_0|F_{j_0}-F_{i_0}|k_0\rangle+\gamma_{jk}^{ii}]
$$

$$
+\lambda_{(ik)}\lambda_{(jk)}[\langle i_0|F_{k_0}-F_{i_0}|j_0\rangle+\gamma_{kj}^{ki}]\}
$$

$$
+2\sum_{\substack{(ij)>(kl)\\ i>j,k>l\\ i\neq j\neq k\neq l}}\lambda_{(ij)}\lambda_{(kl)}\gamma_{kl}^{ij} \tag{124}
$$

where

$$
\gamma_{kl}^{ij}\equiv 2[a_{ik}+a_{jl}-a_{il}-a_{jk}](i_0j_0|k_0l_0)
$$

$$
+[b_{ik}+b_{ji}-b_{il}-b_{jk}][(i_0k_0|j_0l_0)+(i_0l_0|j_0k_0)] \tag{125}
$$

If we define the symmetric matrix **B** by

$$
B_{0,0}=0
$$

$$
B_{0,(ij)}=\langle i_0|(F_{j_0}-F_{i_0})|j_0\rangle, \qquad i>j
$$

$$
B_{(ij),(ij)}=[\langle i_0|(F_{j_0}-F_{i_0})|i_0\rangle-\langle j_0|F_{j_0}-F_{i_0})|j_0\rangle+\gamma_{ij}^{ij}], \qquad i>j \tag{126}
$$

$$
B_{(ij),(ik)}=\zeta_{ij}\zeta_{ik}[\langle j_0|(F_{i_0}-F_{\max(j_0,k_0)})|k_0\rangle+\gamma_{ik}^{ij}], \qquad i\neq j\neq k
$$

$$
B_{(ij),(kl)}=\gamma_{kl}^{ij}, \qquad i>j, k>l, i\neq j\neq k\neq l
$$

and let

$$
C_{(ij)}=\lambda_{(ij)}C_0
$$

then (124) can be written as

$$
\Delta\varepsilon\sum_{i=0}^{nn}C_i^2=\sum_{i,j=0}^{nn}C_iC_jB_{i,j} \tag{127a}
$$

where $nn \equiv n(n-1)/2$ (and where, for example, $i \neq j \neq k$ is understood as i,j,k, all different). Thus, the rotation expansion coefficients $\{C_i\}$, and hence the corrected orbitals $\{\phi_i\}$ for the current iteration, can be obtained by simply diagonalizing the **B** matrix and selecting the appropriate root,

$$\sum_j B_{ij}C_j = C_i \Delta\varepsilon \qquad (127b)$$

Equation (127b) allows for simultaneous orbital rearrangement within the space of $\{\phi_i^0\}$ including the effects of each rotation $C_{(ij)}$ due to simultaneous application of all other rotations $C_{(kl)}$. Since (127b) is valid through second order for all mixing coefficients $\{\lambda_{(ij)}\}$ and since this expression couples the changes in all orbitals to one another, it is capable of leading to a controlled rapidly convergent solution. The coupling between individual mixing coefficients included here can be of crucial importance in permitting the essential simultaneous readjustments necessary to obtain the desired solution (some solutions are not readily accessible from a given set of trial functions $\{\phi_i^0\}$ through individual pairwise considerations alone).

5.2.3. Individual Pairwise Optimization

Mixing all orbitals with respect to each other simultaneously is clearly the most unrestrictive and desirable approach to optimizing orbitals within the minimal basis space of $\{\phi_i^0\}$. However, even for the relatively simple approximation to this problem presented above, the computational effort involved in constructing (and diagonalizing) the **B** matrix required to properly couple the various mixing coefficients is prohibitive for general cases. Therefore, we usually assume that our starting guesses are reasonable enough so that convergence to a desired solution can be achieved through pairwise considerations alone. That is, while being fully aware of the limitations involved, we assume that coupling between individual pairwise changes can be ignored. Our problem thus reduces to finding the most efficient way of accomplishing this much simpler task.

We have already discussed the simplest possible approach to this problem in Section 5.2.1., whereby each mixing coefficient is obtained from $\{\phi_i^0\}$ in a completely independent fashion using (115a) and an appropriate root selection criterion. While variationally determining $\{\lambda_{(ij)}\}$ in this manner usually does lead to successful results, this method has a tendency to exhibit somewhat oscillatory behavior and can at times converge relatively slowly. An alternate prescription that generally leads to superior results can be obtained from (127) by neglecting those elements of the **B** matrix that serve only to couple different mixing coefficients. That is, of the B_{ij} with $i,j \neq 0$ we include only the diagonal element. Within this approximation, our optimization equation reduces to

$$\Delta\varepsilon\Big[1 + \sum_{i>j} \lambda_{(ij)}^2\Big] = \sum_{i>j} \big[2\lambda_{(ij)}B_{0,(ij)} + \lambda_{(ij)}^2 B_{(ij),(ij)}\big] \qquad (128)$$

Since we need only evaluate the matrix elements $\{B_{0,(ij)}; B_{(ij),(ij)}\}$ (no other off-diagonal elements), this approximation greatly simplifies matters relative to (127). In fact, since (115b) can be written as

$$\lambda^{0}_{ij} = -B_{0,(ij)}/B_{(ij),(ij)}$$

we see that (128) requires no more information than is needed for an independent pairwise mixing procedure. The resulting matrix **B** is quite sparse (two nonzero elements per row) and hence this matrix can be diagonalized rapidly.[18]

While (128) does not involve any direct coupling between different mixing coefficients (and is therefore still only valid through first order, overall), in solving this equation for $\{\lambda_{(ij)}\}$ through diagonalization of the resultant approximate **B** matrix, each coefficient is nonetheless linked to all others in such a way that each is determined as "weighted" with respect to its relative contribution to $\Delta\varepsilon$. Since this contribution is at least a gross reflection of the relative energy contribution for each pairwise change, this weighting is a desirable property. In evaluating each coefficient independently as in (115a), however, even this most minimal correlation is not possible. As a consequence, it is not surprising that an independent evaluation scheme (115a) tends to overestimate total orbital changes* and can therefore lead to oscillatory behavior. In solving (128) instead, those changes being of greatest importance are weighted more heavily than all others, thereby minimizing this effect in a quite natural manner.

5.2.4. Summary of Orbital Mixing Considerations

Inclusion of all coupling terms in the energy expression through second order leads to the variational Eqs. (127) which include the effect on each rotation $C_{(ij)}$ of the simultaneous rotations of all other orbital pairs $\{C_{(kl)}\}$. The resulting Eqs. (127) is thus the proper one for iteratively solving for the optimum orbitals. Indeed we could also include in (127) the rotations between occupied and virtual orbitals and hence obtain quadratic convergence. Unfortunately at present (127) seems too costly, despite the rapid convergence to be expected.

At the other extreme we considered in (115a) the rotation of one pair of orbitals at a time. To sequentially solve rigorously for each rotation is impractical at the present time since the two-electron integrals must be reprocessed after each rotation. Thus when using (115a) we generally take a set of starting orbitals and solve for all $n(n-1)/2$ rotation coefficients $\lambda_{(ij)}$, each of which assumes the same set of starting coefficients. These rotations are not all consistent with each other and some procedure of sequentially applying them

---

*This effect can sometimes be used to our advantage in providing us with fortuitous extrapolatory results that can lead to an increased rate of convergence.

must be selected. This approach has been generally successful but is often to blame for the slow convergence that sometimes occurs.

Our compromise between these two extremes is (128), which determines all the rotations simultaneously but utilizes only quantities already required for (115a). In addition, (128) leads to no special storage or matrix diagonalization difficulties. Although determining the rotations simulataneously, this approach neglects the special effects upon the energy of simultaneous rotation of the orbitals. As a result, although rapidly convergent it cannot be expected to converge quadratically.

Because of its good overall convergence capabilities and computational simplicity, the orbital mixing optimization approximation (128) developed here has been found to be of excellent general utility. This procedure is also quite compatible with the OCBSE optimization method described previously in that both steps have similar computational requirements and handle their respective tasks to within almost the same degree of overall validity. Therefore, in the union of these two complementary steps we obtain a practical and yet highly efficient iterative prescription for obtaining SCF solutions for the orbitals in wave functions of the type we have been dealing with.

5.3. Correlated Pair Coefficient Optimization

For wave functions whose energy expressions involve fixed energy parameters $\{f_i, a_{ij}, b_{ij}\}$, self-consistency is achieved solely through orbital optimization, as described in Sections 5.1 and 5.2. Such is the case for HF wave functions. However, for wave functions containing correlated pairs, such as the GVB wave function in (58), these parameters are functions of the variable pair coefficients $\{\sigma_{ij}\}$. Therefore, in order to obtain fully self-consistent solutions for such wave functions, these coefficients must also be variationally determined.

5.3.1. The General Case

This task is most easily accomplished by adding yet a third step to the iterative cycle. In this step we optimize $\{\sigma_{ij}\}$ while keeping the orbitals fixed. To see how this is done let us focus attention on a single such pair. Abstracting from the total energy only those terms that depend on the νth *correlated pair*, we obtain

$$E_\nu = \sum_{i=1}^{n_\nu} [2f_{i\nu}h^c_{i\nu,i\nu} + (a_{i\nu,i\nu} + b_{i\nu,i\nu})J_{i\nu,i\nu}$$

$$+ \sum_{j \neq 1\nu,2\nu} (a_{j,i\nu}J_{j,i\nu} + b_{j,i\nu}K_{j,i\nu})] + \sum_{i \neq j}^{n\nu} (a_{i\nu,j\nu}J_{i\nu,j\nu} + b_{i\nu,j\nu}K_{i\nu,j\nu}) \qquad (129)$$

where we have allowed n_ν natural orbitals in the correlated pair ($n_\nu = 2$ for GVB). Using (59) to write (129) in terms of $\{\sigma_{i\nu}\}$ leads to

$$\tfrac{1}{2}E_\nu = \sum_{i,j}\sigma_{i\nu}\sigma_{i\nu}Y^\nu_{i,j} \tag{130a}$$

where

$$Y^\nu_{i,i} = h^c_{i\nu,i\nu} + \tfrac{1}{2}J_{i\nu,i\nu} + \sum_{j \neq 1\nu,2\nu\cdots} f_j(2J_{j,i\nu} - K_{j,i\nu})$$

$$Y^\nu_{i,j} = \tfrac{1}{2}K_{i\nu,j\nu} \quad \text{for } i \neq j$$

and

$$\sum_i^{n_i}\sigma^2_{i\nu} = 1$$

Therefore, $\{\sigma_{i\nu}\}$ can be obtained by simply diagonalizing the $\mathbf{Y}^\nu$ matrix,

$$\mathbf{Y}^\nu\boldsymbol{\sigma}_\nu = \varepsilon_\nu\boldsymbol{\sigma}_\nu \tag{130b}$$

and normally choosing the lowest energy root.

Since the diagonal elements of the $\mathbf{Y}^\nu$ matrix depend upon $\{f_j, j \neq 1\nu, 2\nu, \ldots\}$, the $\{\sigma_{ij}\}$ must be iteratively determined for wave functions involving more than one correlated pair. This can be most rapidly accomplished through a sequential procedure involving intermediate redefinition of these matrices.

5.3.2. Simplifications for GVB Pairs

If we are dealing with a GVB pair, then there are only two orbitals involved and the pair coefficients $\sigma_{1\nu}$ and $\sigma_{2\nu}$ can be determined in a very simple manner since the pair energy contribution is simply

$$2E_\nu = (\sigma^2_{1\nu}Y^\nu_{1,1} + \sigma^2_{2\nu}Y^\nu_{2,2} + \sigma_{1\nu}\sigma_{2\nu}K_{1\nu,2\nu})/(\sigma^2_{1\nu} + \sigma^2_{2\nu}) \tag{131}$$

[Note that the third term of (50) differs from that of (131) because we had built a minus sign into (49).] Therefore, letting

$$\alpha_\nu = \sigma_{2\nu}/\sigma_{1\nu}$$

and deleting from (131) the constant $Y^\nu_{1,1}$, we obtain

$$\mathscr{E}_\nu = (\alpha^2_\nu Y_\nu - \alpha_\nu K_{1\nu,2\nu})/(1 + \alpha^2_\nu) \tag{132}$$

where

$$\mathscr{E}_\nu = 2E_\nu - Y^\nu_{1,1}$$

and

$$Y_\nu = Y^\nu_{2,2} - Y^\nu_{1,1}$$

Requiring the energy to be stationary with respect to α_ν, $d\mathscr{E}_\nu/d\alpha_\nu = 0$, and assuming that $K_{1\nu,2\nu} \neq 0$ ($K_{1\nu,2\nu} = 0$ results in collapse of the pair), leads to a quadratic equation having the solutions

$$\alpha_\nu = \frac{Y_\nu}{K_{1\nu,2\nu}} \pm \left[1 + \left(\frac{Y_\nu}{K_{1\nu,2\nu}}\right)^2\right]^{1/2} \tag{133}$$

However, since it is always the case that $K_{1\nu,2\nu} \geq 0$, the root leading to the most negative energy must correspond to $\alpha_\nu \leq 0$. Therefore, we obtian the simple final result

$$\sigma_{1\nu} = 1/(1+\alpha_\nu^2)^{1/2}$$
$$\sigma_{2\nu} = \alpha_\nu/(1+\alpha_\nu^2)^{1/2}$$

(134)

where

$$\alpha_\nu = \frac{Y_\nu}{K_{1\nu,2\nu}} - \left[1 + \left(\frac{Y_\nu}{K_{1\nu,\,2\nu}}\right)^2\right]^{1/2}$$

(135)

5.4. The Iterative Sequence

The computer programs implementing the variational methods described in the previous sections are continually evolving and hence likely to change significantly over the next few years. Even so, it is appropriate to outline the steps involved in the current programs (referred to as GVBTWO*) in order to provide the reader with a clearer picture of how to put together the various steps.

5.4.1. Initialization

a. *Energy Coefficients.* The energy coefficients (f_i, a_{ij}, b_{ij}) for the common cases such as described in Section 3 are built into the program so that only the spin state and identity of open-shell orbitals need be supplied. For correlated-pair wave functions, one provides the number of pairs, the number of orbitals in each pair and initial guesses for the pair coefficients $\{\sigma_{ji}\}$ from which the program calculates the energy coefficients.

b. *Transformation Matrix.* Each occupied orbital ϕ_i is initially specified in terms of the atomic basis (AO) employed $\{\chi_i\}$ over which the one- and two-electron integrals have been evaluated,

$$\phi_i = \sum_j \chi_j T_{ji}$$

The program constructs orthonormal virtual orbitals (if they are not provided). The full set of orthonormal orbitals (occupied and virtual) is referred to as the MO basis. This basis is partitioned into symmetry types since the program is designed to take full advantage of the fact that orbital symmetries will be

---

*Developed by F. W. Bobrowicz, W. R. Wadt, and W. A. Goddard III (Spring 1973) (a more detailed description is given in Ref. 4). An earlier version, referred to as GVBONE, is outlined in Ref. 2.

preserved during the iterations. Each occupied orbital is identified in terms of the shell or correlated pair to which it belongs.

c. *Initial J and K Matrices.* For each open shell, the intial Coulomb and exchange matrices are constructed over the AO basis*

$$J^i_{\mu\nu} = \langle \chi_\mu | J_i | \chi_\nu \rangle$$

$$K^i_{\mu\nu} = \langle \chi_\mu | K_i | \chi_\nu \rangle$$

Likewise an initial h^c matrix is constructed over AOs

$$h^c_{\mu\nu} = \langle \chi_\mu | h | \chi_\nu \rangle + \langle \chi_\mu | \sum_p^{closed} (2J_p - K_p) | \chi_\nu \rangle$$

In these matrices only those $\mu\nu$ elements are evaluated that are required from the subsequent construction of the symmetry block matrices (step *a* below). That is, only if both μ and ν can appear in the expansion of two orbitals of the same symmetry.

5.4.2. The Iterative Process

a. *J and K Matrices over MOs.* The h^c and $\{J_i, K_i\}$ matrices over AOs are transformed to symmetry blocks of occupied orbitals (i.e., only those elements are calculated in which both components belong to the same symmetry).

b. *New Correlated-Pair Coefficients.* For each pair i, the **Y** matrices defined in (130a) are constructed and the elements required for updating (*vide infra*) these matrices are also stored conveniently. Each matrix is diagonalized as in (130b) or solved as in (135) to obtain new pair coefficients $\{\sigma_{ji}\}$. The **Y** matrices for subsequent pairs are updated to reflect the new $\{\sigma_{ji}\}$ for pair i. This process is continued until it converges. Based on the final new pair coefficients $\{\sigma_{ji}\}$, new energy coefficients $\{f_i, a_{ij}, b_{ij}\}$ are defined for subsequent steps.

c. *Orbital Mixing.* Using the matrices over MOs from step a, the sparse **B** matrix of (128) is constructed. Elements connecting orbitals of the different symmetry or connecting orbitals of the same shell are not included. The matrix **B** is diagonalized selecting either the lowest energy root or else the root representing the least change. From the result, new occupied orbitals are defined; however, the h^c and $\{J_i, K_i\}$ over AOs are *not* modified.

---

*Rather than the Coulomb and exchange matrices discussed in the text, the program actually deals with the modified quantities

$$\mathcal{J} \equiv \tfrac{1}{2}(J_i - \tfrac{1}{2}K_i)$$

$$\mathcal{K}_i \equiv \tfrac{1}{4}K_i$$

because these matrices can be constructed much more rapidly using a suitably preprocessed list of two-electron integrals analogous to the approach developed by Raffenetti.[19]

d. OCBSE. For each shell we go through the following sequence. (i) Construct the **F** matrix (over AOs) from the h^c, $\{J_i, K_i\}$ matrices; (ii) transform **F** to MOs including only those symmetry blocks that contain orbitals of this shell; (iii) diagonalize each symmetry block and select roots either on the basis of eigenvalues or else of least change; (iv) define new virtuals for each symmetry affected by (iii); (v) reconstruct new matrices over AOs for this shell (h^c if it is the closed shell or else the appropriate $\{J_i, K_i\}$ if it is an open shell); (vi) proceed to the next shell.

e. If not converged, go back to step *a.*

6. Summary

We have considered herein general procedures of optimizing, self-consistently, the orbitals of Hartree–Fock and generalized valence bond wave functions. In particular, we considered all wave functions with energies that can be expressed as (6),

$$E = = 2 \sum_i f_i h_{ii} + \sum_{i,j} (a_{ij} J_{ij} + b_{ij} K_{ij}) \tag{6}$$

although this technique and formalism are valid for more general cases.

Requiring that the energy be stationary with respect to all orbital variations leads to the general variational condition in (68),

$$\sum_i \langle \delta\phi_i | F_i | \phi_i \rangle = 0 \tag{68}$$

where F_i is the generalized Fock operator,

$$F_i = f_i h + \sum_j (a_{ij} J_j + b_{ij} K_j) \tag{66}$$

Considering variations in which only orbital ϕ_ν changes with all other $n-1$ orbitals kept fixed, (68) leads to

$$\langle \delta\phi_\nu | F_\nu | \phi_\nu \rangle = 0 \tag{76}$$

Ignoring the orthogonality conditions (76), for a complete set of functions, leads to the pseudoeigenvalue equation

$$F_\nu \phi_\nu = \varepsilon_\nu \phi_\nu \tag{136}$$

the more common way of expressing this condition. Although often stated as the variational condition, (76) is a necessary but not sufficient condition for optimum orbitals. Allowing occupied orbitals to mix with each other leads to the condition

$$\langle \phi_\nu | (F_\nu - F_\mu) | \phi_\mu \rangle = 0 \tag{84}$$

for all occupied orbitals μ and ν. Together (76) and (84) form necessary *and* sufficient conditions for a set of orbitals $\{\phi_i\}$ to yield a stationary energy.

Although the variational conditions (76) and (84) tell us whether a given set of orbitals are optimum (i.e., lead to stationary energy), they do not tell us what to do to improve the orbitals when they are not optimum. When using (136), a common approach is to solve

$$F_{\nu_0} \phi_\nu = \varepsilon_\nu \phi_\nu$$

where F is evaluated using the trial functions [this requires the second condition, (84)]. We believe that this is not a reliable approach and instead return to the total energy (the quantity we are trying to minimize), expand it in terms of simultaneous corrections in all orbitals,

$$\phi_i = (\phi_i^0 + \Delta \phi_i)/(1 + \langle \Delta \phi_i | \Delta \phi_i \rangle)^{1/2}$$

keeping terms through second order leading to (94a) with (95) as the condition ensuring orbital orthonormalities. To obtain the optimum orbitals, we require that $(\partial E / \partial \Delta \phi_i) = 0$ for all variations consistent with orbital orthonormalities, leading to an equation analogous to (127),

$$\sum_{(kl)} B_{(ij),(kl)} C_{(kl)} = C_{(ij)} \Delta \varepsilon \tag{137}$$

where B is given by (126) and $C_{(ij)}$ determines the mixing of orbitals ϕ_i and ϕ_j. In addition, to one element for each of the $n(n-1)/2$ rotations, (137) contains an element C_0 corresponding to a renormalization term and the rotations are given by (126b),

$$\lambda_{(ij)} = C_{(ij)}/C_0 \tag{126b}$$

The final orbitals are given as

$$\phi_\nu = (\phi_\nu^0 + \Delta \phi_\nu)/(1 + \langle \Delta \phi_\nu | \Delta \phi_\nu \rangle)^{1/2}$$

where

$$\Delta \phi_\nu = \sum_{i \neq \nu} \zeta_{i\nu} \lambda_{(i\nu)} \phi_i^0 + \sum_{i < \nu} \sum_{j \neq i, \nu} \zeta_{ij} \zeta_{jk} \lambda_{(ij)} \lambda_{(jk)} \phi_i^0 \tag{121}$$

and

$$\zeta_{ij} = -\zeta_{ji} = 1 \qquad \text{for } i > j$$

Since (137) was derived from an energy expression correct through second order in all orbital corrections, then the errors in (137) are of order Δ^2 and hence the orbital corrections $C_{(kl)}$ will converge quadratically.

Unfortunately (137) seems inordinately costly to evaluate and to solve and hence we use a slightly more approximate approach. First we partition the problem into two parts

    1. OCBSE: optimizing each orbital (sequentially) allowing it to mix only with the virtual orbitals.

    2. Orbital mixing: optimizing the mixing of various occupied orbitals with each other.

The OCBSE leads to a simple eigenvalue equation (105b)

$$\bar{\mathbf{F}}_{\nu_0}\mathbf{C}_\nu = \mathbf{C}_\nu\varepsilon_\nu \tag{105b}$$

for the new orbitals (in the space excluding other occupied orbitals).

Including all orbital mixing terms consistently leads to an equation (127), analogous to (137) above. This equation can be simplified to the form (128) at great saving in cost but with only slight penalties in convergence.

For wave functions of the GVB and separated-pair types in which the energy parameters in (6) depend on other expansion coefficients, these coefficients are solved for separately as in (130) or (135).

7. Comparison with Previous Work

As described in Section 4.2, a necessary and sufficient condition for a set of orbitals $\{\phi_i\}$ to be optimal is that the first-order change in the energy be zero for all changes in these orbitals consistent with those constraints (i.e., orbital orthonormality) imposed when deriving the energy expression. For real orbitals this results in the variational expression

$$\sum_i \langle \delta\phi_i | F_i | \phi_i \rangle = 0 \tag{68}$$

Until 1969, the usual approach to this problem was to allow only *one* orbital, say ϕ_ν, to vary at a time keeping the remaining $n - 1$ orbitals fixed. This leads to

$$\langle \delta\phi_\nu | F_\nu | \phi_\nu \rangle = 0 \tag{76}$$

for all variations $\delta\phi_\nu$ consistent with the constraints

$$\langle \delta\phi_\nu | \phi_j \rangle = 0 \qquad \text{for all } \phi_j \tag{78}$$

that is, all $\delta\phi_\nu$ orthogonal to the occupied orbitals. If we substitute $\delta\phi_\nu = \lambda\phi_j$ in (76) [that is, allow $\delta\phi_\nu$ *not* satisfying (78)], the result is

$$\langle \delta\phi_\nu | F_\nu | \phi_\nu \rangle = \lambda \langle \phi_j | F_\nu | \phi_\nu \rangle = \varepsilon_{j\nu} \langle \delta\phi_\nu | \phi_j \rangle$$

where

$$\varepsilon_{j\nu} \equiv \langle \phi_j | F_\nu | \phi_\nu \rangle \tag{138}$$

is referred to as a *Lagrange multiplier*. Thus (76) can be rewritten as

$$\langle \delta\phi_\nu | F_\nu | \phi_\nu \rangle - \sum_j \varepsilon_{j\nu} \langle \delta\phi_\nu | \phi_j \rangle = 0 \tag{139}$$

which is valid for *all* $\delta\phi_\nu$, no constraints. Visualizing $\langle \delta\phi_\nu | F_\nu | \phi_\nu \rangle$ as the derivative of the energy in the direction $\delta\phi_\nu$, we see that the Lagrange multiplier $\varepsilon_{j\nu}$ serves to annihilate that part of the energy derivative arising from varying ϕ_ν into the forbidden direction ϕ_j. In order for (139) to be satisfied for *all* $\delta\phi_\nu$, the $\{\phi_i\}$ must satisfy

$$F_\nu \phi_\nu = \sum_j \varepsilon_{j\nu} \phi_j \tag{140}$$

which can be taken as the fundamental variational equation.[20] Nothing is changed here of course; (140) is equivalent to (76) with (78) and still only ensures that the energy is stationary upon varying only ϕ_ν in the space orthogonal to all occupied orbitals.

Substituting (138) into (140) and rearranging terms leads to the pseudo-eigenvalue equation of Birss and Fraga,[21]

$$R_\nu \phi_\nu = \varepsilon_{\nu\nu} \phi_\nu \tag{141a}$$

where

$$R_\nu \equiv F_\nu - \sum_{j \neq \nu} |\phi_j\rangle\langle\phi_j| F_\nu \tag{141b}$$

"Integrating" (139) leads to a new functional,

$$2I_\nu = E - \varepsilon_{\nu\nu}\langle\phi_\nu|\phi_\nu\rangle - 2 \sum_{j \neq \nu} \varepsilon_{j\nu}\langle\phi_j|\phi_\nu\rangle \tag{142}$$

where E is the energy expression of (6) and the factor $\frac{1}{2}$ with the E arises from the definition of F_ν in (65) and (66). Requiring that the first-order variation $\delta I_\nu^{(1)}$ of (142) be zero for all $\delta\phi_\nu$ leads directly to the correct variational equation (140). Thus, a shortcut to deriving the variational equations is to construct the functional (142) (where the $\varepsilon_{j\nu}$ are yet undetermined), apply the variational principle to obtain (140), and project (140) onto the functions $\{\phi_j\}$ to obtain (138). This sequence of steps is referred to as Lagrange's method of undetermined multipliers. However, it should be kept in mind that the fundamental variational condition is (76) with (78) or equivalently (140) with (138).

Consider now the variational equation for a second orbital ϕ_μ. The equations corresponding to (76)–(78) and (138)–(140) lead to

$$F_\mu \phi_\mu = \sum_j \varepsilon_{j\mu} \phi_j \tag{143}$$

where

$$\varepsilon_{j\mu} = \langle\phi_j|F_\mu|\phi_\mu\rangle \tag{144}$$

which can be integrated to the functional

$$2I_\mu = E - \varepsilon_{\mu\mu}\langle\phi_\mu|\phi_\mu\rangle - 2\sum_{j\neq\mu}\varepsilon_{j\mu}\langle\phi_j|\phi_\mu\rangle \tag{145}$$

We may now consider the question of whether there might be some simple functional, say*

$$2I = E - \sum_{j,k}\varepsilon_{jk}\langle\phi_j|\phi_k\rangle \tag{146}$$

that could be used to derive all n variational equations [such as (140) and (143)]. The answer, in general, is *no*. For example, the variational condition from $\delta\phi_\mu$ requires, letting $j = \mu$ in (138),

$$\varepsilon_{\mu\nu} = \langle\phi_\mu|F_\nu|\phi_\nu\rangle \tag{147}$$

whereas the variational condition from $\delta\phi_\mu$ requires [letting $j = \nu$ in (144)]

$$\varepsilon_{\nu\mu} = \langle\phi_\nu|F_\mu|\phi_\mu\rangle \tag{148}$$

These quantities are not equal in general and hence the functional (146) does not lead to the correct variational equations, e.g., (140) and (143).

Unfortunately many workers have assumed that (146) is the proper starting point for applying the method of Lagrange multipliers to SCF problems. This has lead to some confusion in the literature. Birss and Fraga[21] noticed that the conditions (147) and (148) appear different but believing from (146) that these quantities must be equal, they assumed that

$$\langle\phi_\mu|F_\nu|\phi_\nu\rangle = \langle\phi_\mu|F_\mu|\phi_\mu\rangle$$

or

$$\langle\phi_\mu|(F_\nu - F_\mu)|\phi_\nu\rangle = 0 \tag{149}$$

must automatically be satisfied for the self-consistent solutions of (76) and (78) [all ν]. Das[22] recognized that (149) may not be satisfied and suggested two modifications: (i) symmetrization of the operator R_ν in (141b) and (ii) rotating each orbital pair during the iterative process so as to gaurantee that (149) is satisfied. Levy[25] attempted to incorporate this condition into the psueudoeigen-value problem itself by generalizing the operator R_ν to have the form

$$R_\nu^L \equiv F_\nu - \sum_{j\neq\nu}[|\phi_j\rangle\langle\phi_j|G_{j\nu} + G_{j\nu}|\phi_j\rangle\langle\phi_j|] \tag{150a}$$

---

*Since $\langle\phi_j|\phi_k\rangle = \langle\phi_k|\phi_j\rangle$, we have arbitrarily included the off-diagonal terms twice.

where

$$G_{j\nu} \equiv G_{\nu j} \equiv \lambda F_\nu + (1-\lambda)F_j \qquad \text{for } j > \nu \qquad (150b)$$

and where λ is an arbitrary parameter. Hinze and Roothaan[24] suggested averaging the ε_{ij} from (147) and (148), leading to

$$\varepsilon_{ij} \equiv \langle \phi_i | \tfrac{1}{2}(F_i + F_j) | \phi_j \rangle \qquad (151)$$

[i.e., $\lambda = \tfrac{1}{2}$ in (150b)]. Finally Huzinaga[26] suggested use of an even more general operator

$$R_\nu^H \equiv F_\nu - \sum_{j \neq \nu} [|\phi_j\rangle\langle\phi_j|G_{j\nu} + G_{ij}|\phi_j\rangle\langle\phi_j|] \qquad (152a)$$

where

$$G_{j\nu} \equiv \lambda_{j\nu}F_j + (1-\lambda_{j\nu})F_\nu \qquad (152b)$$

(all j and ν) and where $\lambda_{j\nu}$ is an arbitrary parameter.

It is important to keep in mind that the above formulations are all still based on the variational conditions arising from varying only one orbital at a time. Requiring that (149) also be satisfied was considered as an auxiliary condition which was felt to be necessary since otherwise the variational equations were not consistent with (146).

In 1969 Hunt *et al.*[27]* pointed out that a self-consistent set of orbitals satisfying condition (76) with (78) need not satisfy (149). Thus, they pointed out that the usual functional (146) does not lead to the proper variational equations. However, they soon learned that solving self-consistently for the orbitals satisfying (76) with (78) leads to different solutions depending upon the trial functions.

These difficulties were finally resolved by Goddard *et al.*[30] in 1969 who showed that although a necessary conditon for a set of orbitals to be optimum is that (76) with (78) be satisfied for each orbital; *these conditions do not constitute sufficient conditions.* One must also consider variations in which each orbital is allowed to incorporate character of any other *occupied* orbital as well. As

---

*This is also the paper that proposed the OCBSE method as a simple approach to eliminating the need for Lagrange multipliers or the coupling operators with which they were often replaced. The basic idea in OCBSE of using the current set of occupied and virtual orbitals as the basis in terms of which the SCF equations are solved had been suggested earlier (1961) by Huzinaga[28] (method II) but was apparently never applied. Probably this was because Huzinaga's approach involved carrying out a full transformation over the two-electron integrals to the new basis, a quite impractical procedure to carry out every iteration. Hunt *et al.*[27] showed that one could evaluate the usual Coulomb and exchange operators over atomic orbitals (no integral transformation) just as if there were no constraints and then transform the *final* Fock matrices to the OCBSE basis, a quite practicable procedure. The OCBSE procedure was also rediscovered by Peters[29] in 1972.

discussed in Section 4.2, because of the orthogonality constraint such a variation must allow the other occupied orbital to simultaneously rotate out of the way. Hence one is obligated to allow at least two orbitals to change simultaneously and cannot expect to solve the problem through single-orbital variations alone. Allowing all such variations of ϕ_ν and concurrent mandated variations in all other orbitals., these authors obtained the equation

$$R_\nu \phi_\nu = \varepsilon_{\nu\nu} \phi_\nu \tag{153a}$$

where

$$R_\nu = F_\nu - \sum_{j \neq \nu} |\phi_j\rangle\langle\phi_j| F_j \tag{153b}$$

[note well the presence of F_j in (153b) in place of the F_ν of (141b)]. Projecting (153a) onto ϕ_k we obtain

$$\langle\phi_k|R_\nu|\phi_\nu\rangle = \langle\phi_k|F_\nu|\phi_\nu\rangle - \langle\phi_k|F_k|\phi_\nu\rangle = \langle\phi_k|F_\nu - F_k|\phi_\nu\rangle = 0$$

which is the condition derived in (84.). Comparing with (149) we see that optimizing the mixing of the occupied orbitals with each other does indeed lead to equal off-diagonal Lagrange multipliers, (147) and (148), the condition usually assumed.

Dahl *et al.*[31] showed that starting with (146) *and* (149) leads to the equations derived by Goddard *et al.*[30] and disputed the latter authors' claim of an additional neccessary condition. The error in reasoning by Dahl *et al.* was pointed out in the literature by Albat and Gruen[32] and (153) is recognized[33] as the proper variational equation.

Although (153) are the proper variational conditions, satisfaction of which leads to fully self-consistent orbitals, there are a number of ways of attempting to iteratively solve for these orbitals. One such procedure was proposed by Davidson[34] who based his iterative procedure on a symmetrized form of (153b). In so doing it was noted that some sort of damping was required in order to attain smooth convergence. Wood and Veillard[35] also implemented this equation, using level shifting techniques to overcome convergence difficulties.

In implementing the iterative solutions of pseudoeigenvalue equations such as (141) or (153),

$$R_\nu \phi_\nu = \varepsilon_\nu \phi_\nu \tag{154}$$

the usual approach had been to evaluate the operator R_ν in terms of a set of trial functions $\{\phi_\nu^0\}$ and then to solve equations such as

$$R_{\nu_0} \phi_\nu = \varepsilon_\nu \phi_\nu \tag{155}$$

for the new orbitals; this process being repeated until it converges. In 1970 Hunt et al.[6] pointed out the inconsistency of this approach, showing that in solving for the mixing coefficient for ϕ_ν and ϕ_μ one obtains two different results depending upon whether one starts with the equation for ϕ_ν;

$$R_{\nu_0}\phi_\nu = \varepsilon\phi_\nu$$

or the equation for ϕ_μ,

$$R_{\mu_0}\phi_\mu = \varepsilon_\mu\phi_\mu$$

Hunt et al.[6] resolved this problem by formulating a set of equations based directly upon variation condition (84) to obtain a set of coupled equations,

$$\sum_{(k,l)} B_{(ij)(kl)}\Delta_{kl} = -X_{(ij)}$$

for the linearly independent orbital corrections $\{\Delta_{kl}, k > l\}$. Since these equations neglected only terms to second order in Δ, they lead to quadratic convergence. Hunt et al.[6] then simplified these equations by partitioning the problem into an OCBSE step and an orbital mixing step. For the orbital mixing step they considered independent pairwise rotations as in Section 5.2.1, using the linear Eq. (115c). This approach was the basis of the GVBONE program of Hunt et al.[3]

Our current approach to deriving the variational equations differs from that of Hunt et al.[6] in that we base all considerations upon the total energy expression itself, a starting point that has largely been ignored previously.* The energy is expanded through second order in the orbital corrections (including orthogonality constraints) and the optimum corrections are obtained from the requirement that $(\partial E/\partial\Delta\phi_i) = 0$ for all i. In addition, we have generalized the orbital mixing steps as indicated in Section 5.2.

ACKNOWLEDGMENT

This work was partially supported by a grant (MPS 74-05132) from the National Science Foundation.

---

*Rossi[36] in 1967 suggested solving for SCF orbitals by applying the $n(n-1)/2$ two-by-two rotations among the various orbitals. This formalism was based upon varying the total energy directly. Mukherjee[37] attempted to develop this approach but unfortunately did not retain sufficient terms to satisfy the variational conditions!

References

1. W. A. Goddard III and R. C. Ladner, A generalized orbital description of the reactions of small molecules, *J. Am. Chem. Soc.* **93**, 6750–6756 (1971).
2. R. C. Ladner and W. A. Goddard III, Improved quantum theory of many-electron systems. V. The spin-coupling and optimized GI method, *J. Chem. Phys.* **51**, 1073–1087 (1969).
3. W. J. Hunt, P. J. Hay, and W. A. Goddard III, Self-consistent procedures for generalized valence bond wave functions. Applications H_3, BH, H_2O, C_2H_6, and O_2, *J. Chem. Phys.* **57**, 738–748 (1972).
4. C. C. J. Roothaan, Self-consistent field theory for open shells of electronic systems, *Rev. Mod. Phys.* **32**, 179–185 (1960).
5. F. W. Bobrowicz, Ph.D. thesis, California Institute of Technology (1974).
6. W. J. Hunt, W. A. Goddard III, and T. H. Dunning, Jr., The incorporation of quadratic convergence into open-shell self-consistent field equations, *Chem. Phys. Lett.* **6**, 147–151 1970).
7. W. A. Goddard III, Improved Quantum Ttheory of many-electron systems. II. The basic method, *Phys. Rev.* **157**, 81–93 (1967).
8. B. J. Moss and W. A. Goddard III, Configuration interaction studies on low-lying states of O_2, *J. Chem. Phys.* **63**, 3523–3531 (1975).
9. T. Arai, Theorem on separability of electron pairs, *J. Chem. Phys.* **33**, 95–98 (1960).
10. P.-O. Löwdin , Note on the separability theorem for electron pairs, *J. Chem. Phys.* **35**, 78–81 (1961).
11. G. Levin and W. A. Goddard III, The generalized valence bond description of allyl radical, *J. Am. Chem. Soc.* **97**, 1649–1656 (1975).
12. G. Levin, Ph.D. thesis, California Institute of Technology, April 1974.
13. B. J. Moss, F. W. Bobrowicz, and W. A. Goddard III, The generalized valence bond description of O_2, *J. Chem. Phys.* **63**, 4632–4639 (1975).
14. P.-O. Löwdin and H. Shull, Natural orbitals in the quantum theory of two-electron systems, *Phys. Rev.* **101**, 1730–1739 (1956).
15. J. M. Parks and R. G. Parr, Theory of separated electron pairs, *J. Chem. Phys.* **28**, 335–345 (1958).
16. D. M. Silver, E. L. Mehler, and K. Ruedenberg, Electron correlation and separated pair approximation in diatomic molecules. I. Theory, *J. Chem Phys.* **52**, 1174–1180 (1970).
17. C. C. J. Roothaan, New developments in molecular orbital theory, *Rev. Mod. Phys.* **23**, 69–89 (1951).
18. I. Shavitt, C. F. Bender, A. Pipano, and R. P. Hosteny, The iterative calculation of several of the lowest or highest eigenvalues and corresponding eigenvalues of very large symmetric matrices, *J. Comput. Phys.* **11**, 90–108 (1973).
19. R. C. Raffenetti, Preprocessing two-electron integrals for efficient utilization in many-electron self-consistent field calculations, *Chem. Phys. Lett.* **20**, 335–338 (1973).
20. R. K. Nesbet, Configuration interaction in orbital systems, *Proc. Roy. Soc. London, Ser. A* **230**, 312–321 (1955).
21. F. W. Birss and S. Fraga, Self-consistent-field theory. I. General Treatment, *J. Chem. Phys.* **38**, 2552–2557 (1963).
22. G. Das, Extended Hartree-Fock ground-state wavefunctions for the lithium molecule, *J. Chem. Phys.* **46**, 1568–1579 (1967).
23. B. Levy, Best choice for the coupling operators in the open-shell and multiconfiguration SCF methods, *J. Chem. Phys.* **48**, 1994–1996 (1968).
24. J. Hinze and C. C. J. Roothaan, Multiconfiguration self-consistent field theory, *Prog. Theor. Phys. (Kyoto) Supp.* **40**, 37–51 (1967).
25. B. Levy, Best choice for the coupling operators in the open-shell and multiconfiguration SCF methods, *J. Chem. Phys.* **48**, 1994–1996 (1968).
26. S. Huzinaga, Coupling operator method in the SCF theory, *J. Chem. Phys.* **51**, 3971–3975 (1969).

27. W. J. Hunt, T. H. Dunning, Jr., and W. A. Goddard III, The orthogonality constrained basis set expansion method for treating off-diagonal Lagrange multipliers in calculations of electronic wave functions, *Chem. Phys. Lett.* **3**, 606–610 (1969).
28. S. Huzinaga, Analytical methods in Hartree-Fock self-consistent field theory, *Phys. Rev.* **122**, 131–138 (1961).
29. D. Peters, Simple open-shell SCF molecular orbital computations, *J. Chem. Phys.* **57**, 4351–4353 (1972).
30. W. A. Goddard III, T. H. Dunning, Jr., and W. J. Hunt, The proper treatment of off-diagonal Lagrange multipliers and coupling operators in self-consistent field equations, *Chem. Phys. Lett.* **4**, 231–234 (1969).
31. J. P. Dahl, H. Johansen, D. R. Truax, and T. Ziegler, On the derivation of necessary conditions on Hartree–Fock orbitals, *Chem. Phys. Lett.* **6**, 64–66 (1970).
32. R. Albat and N. Gruen, Examples of known SCF procedures which do not satisfy all necessary conditions for the energy to be stationary, *Chem. Phys. Lett.* **18**, 572–573 (1973).
33. K. Hirao and H. Nakatsuji, General SCF operator satisfying correct variational condition, *J. Chem. Phys.* **59**, 1457–1462 (1973).
34. E. R. Davidson, Spin-restricted open-shell self-consistent-field theory, *Chem. Phys. Lett.* **21**, 565–567 (1973).
35. M. H. Wood and A. Veillard, On convergence guarantees for the multiconfiguration self-conditions for the energy to be stationary, *Chem. Phys. Lett.* **18**, 572–573 (1973).
36. M. Rossi, Variational procedure for open-shell LCAO multideterminant wavefunctions. An approach to the excited-state problem, *J. Chem. Phys.* **46**, 989–996 (1967).
37. N. G. Mukherjee, A variational procedure for the optimization of multi-configuration self-consistent field (MCSCF) orbitals, *Chem. Phys. Lett.* **24**, 441–446 (1974).

5

Pair Correlation Theories

Werner Kutzelnigg

1. Introduction

1.1. One-Electron and Electron-Pair Theories

In the simplest possible description of an n-electron system, one one-electron function (spin-orbital) is associated with each electron and the n-electron wave function is a Slater determinant built up from these spin-orbitals. The one-to-one correspondence between electrons and spin-orbitals gives an acceptable first-order description only for closed-shell and certain open-shell states. A one-electron theory that is applicable in general to open-shell states as well is characterized by assigning sets of electrons to sets of degenerate spin-orbitals, where the number of electrons within one set can be equal to or smaller than the dimension of the irreducible representation spanned by the degenerate set of spin-orbitals. An example is the well-known characterization of an atomic state by its configuration,[1] e.g., for the carbon ground state $1s^2 2s^2 2p^2$, without specifying the m_s and m_l values. (For a general discussion of closed- and open-shell states in the framework of rigorous quantum mechanics, see Refs. 2 and 3.)

An electron-pair theory for a closed-shell state is characterized by as many two-electron functions as there are electron pairs, i.e., $n(n-1)/2$ for an n-electron system, and possibly in addition by n one-electron functions. Some of the two-electron functions can be built from just two one-electron functions. If this holds for all two-electron functions, we have a one-electron theory.

In open-shell states the situation for pair theories is analogous to that for one-electron theories. We associate sets of electron pairs with sets of degener-

Werner Kutzelnigg • Lehrstuhl für Theoretische Chemie, Ruhr-Universität, D463 Bochum, Germany

ate two-electron functions, where the dimension of the electron-pair set is equal to or less than the dimension of the irreducible representation spanned by the set of pair functions. The m_l and m_s values of the pair functions are not specified. For an atomic p^n configuration the pair functions to be considered are of ^3P, ^1D, and ^1S type, and are 9-fold, 5-fold, and 1-fold degenerate, respectively. If $n = 6$, ^3P is occupied by 9 pairs, ^1D by 5 pairs, and ^1S by one pair, the total number of pairs being $\binom{6}{2} = 15$. If $n < 6$ the number of electron pairs associated with one two-electron function is generally smaller than the degeneracy of the latter. This description has, e.g., been used by Steiner.[4]

The very general definition of electron-pair theories given here includes theories based on the two-particle density matrices, as well as those originating from a cluster expansion of the wave function. Second-order perturbation theory of electron correlation automatically leads to a pair theory. There are many more pair theories compatible with the above definition, but we shall in this review concentrate on pair-cluster expansions of the wave function.

1.2. Why Electron-Pair Theories?

There are essentially two reasons why one can expect that electron-pair theories furnish a good approximation for n-electron systems: (1) The Hamiltonian contains only one- and two-particle operators; (2) the Pauli principle prevents three electrons from occupying the same point of space.

If the Hamiltonian contained just one-electron operators, i.e., had the form

$$\hat{H}(1, 2, \ldots, n) = \sum_{i=1}^{n} \hat{h}(i)$$

a one-electron theory would be exact, namely the n-electron wave function Ψ would be a Slater determinant built up from the eigenfunctions of $\hat{h}$.

Unfortunately it is not true that for the Hamiltonian

$$\hat{H}(1, 2, \ldots, n) = \sum_{i=1}^{n} \hat{h}(i) + \sum_{i<j=1}^{n} \hat{g}(i, j) = \hat{H}^{(1)} + \hat{H}^{(2)}$$

an electron-pair theory (in the way defined in Section 1.1) is exact. The situation is not so bad, however, since the average electron interaction can readily be taken into account by a one-electron theory (Hartree–Fock approximation) so that the pair functions have only to take care of that part of the electron interaction which is beyond the average interaction (sometimes referred to as "instantaneous interaction" or "fluctuation").[5]

Since two of three electrons must have the same spin, and since two electrons with the same spin cannot occupy the same point of space[6] (Fermi hole), only two electrons can come close to each other. Therefore, three-particle contributions to the *correlation energy* (i.e., the difference between Hartree–Fock and exact nonrelativistic energy), which are already small because they occur only indirectly (in higher order of perturbation theory), do not contain short-range contributions and are therefore likely to be negligible in many cases.[5,7]

In nuclear physics, pair theories are also used, but the reasons for their validity are different from those for electronic systems. In nuclei it is not true that the Hamiltonian contains only one- and two-particle terms (genuine three-particle potentials are far from neglible), nor is it true that the Pauli principle prevents three particles from coming close to each other. Due to the presence of ordinary spin and isospin, as many as four particles can occupy the same point of space (at least as far as the Pauli principle is concerned).

The reason pair theories seem to work rather well in nuclear theory is the short-range character of the nuclear interaction potentials, mainly the strong repulsion at very short distances (often described by a hard core) that keeps the particles apart.

This difference must be kept in mind when we compare electronic and nuclear pair theories, which have many formal analogies. There is no need for pushing the analogy so far that one tries to define a short-range type of "effective" interelectronic potential ("fluctuation potential")[5] that would play the same role as the real interaction potential in nuclear theory. In any case, the use of a fluctuation potential does not simplify the formalism in any respect and we therefore exclude it from this review.

2. Cluster Expansion of the Wave Function

2.1. The Idea of Separated Pairs

The earliest version of an electron-pair theory is the "separated-pair" approximation introduced by Hurley *et al.*[8-10] The wave function of an *n*-electron state, with *n* even, is approximated by an antisymmetrized product of strongly orthogonal geminals (APSG)[11]

$$\Psi(1, 2, \ldots, n) = \mathscr{A}[\omega_1(1, 2)\omega_2(3, 4) \cdots \omega_{n/2}(n-1, n)] \qquad (1)$$

$$\int \omega_R(1, 2)\omega_S(1, 3)d\tau_1 = 0 \qquad \text{for } R \neq S \qquad (2)$$

The number of pair functions (geminals) used to construct Ψ is only $n/2$ rather than $\binom{n}{2}$ as it is for a general pair theory. No one-electron functions (orbitals) are introduced explicitly, but one can, for singlet-type geminals, of course write

$$\omega_R(1, 2) = c_R\left\{\phi_R(1)\phi_R(2)\frac{1}{\sqrt{2}}[\alpha(1)\beta(2) - \beta(1)\alpha(2)] + u_R(1, 2)\right\} \qquad (3)$$

i.e., express each pair function ω_R as a sum of a one-electron approximation to ω_R and a pair correlation function u_R. When (3) is inserted into (1), one obtains

$$\Psi(1, 2, \ldots, n) = c_0\Big\{\mathscr{A}[\phi_1\alpha(1)\phi_1\beta(2)\phi_2\alpha(3)\cdots\phi_{n/2}\beta(n)]$$

$$+ \sum_{R=1}^{n/2} \mathscr{A}\left[\frac{\phi_1\alpha(1)\phi_1\beta(2)\cdots\phi_{n/2}\beta(n)}{\phi_R\alpha(2R-1)\phi_R\beta(2R)}u_R(2R-1, 2R)\right]$$

$$+ \sum_{R<S=1}^{n/2} \mathscr{A}\left[\frac{\phi_1\alpha(1)\phi_1\beta(2)\cdots\phi_{n/2}\beta(n)}{\phi_R\alpha(2R-1)\phi_R\beta(2R)\phi_S\alpha(2S-1)\phi_S(2S)}\right.$$

$$\left.\times u_R(2R-1, 2R)u_S(2S-1, 2S)\right] + \cdots\Big\} \qquad (4)$$

with

$$c_0 = \prod_{R=1}^{n/2} c_R \qquad (5)$$

The first term in Ψ is just the Slater determinant Φ built up from the $\phi_R\alpha$ and $\phi_R\beta$, the sum over R consists of wave functions obtained from Φ on replacing $\phi_R\alpha\phi_R\beta$ by u_R, i.e., by double substitution, and the sum over R and S contains quadruple substitutions, etc. In terms of annihilation operators $\hat{a}_i$, $\hat{\bar{a}}_i$ for the spin-orbitals $\psi_i = \phi_i\alpha$ and $\bar{\psi}_i = \phi_i\beta$ and creation operators $\hat{b}_R^+$ for the pair correlation functions u_R, the expansion (4) of the APSG function can be written alternatively as

$$\Psi = c_0\Big(\Phi + \sum_{R=1}^{n/2} \hat{b}_R^+\hat{a}_R\hat{\bar{a}}_R\Phi + \sum_{R<S=1}^{n/2} \hat{b}_R^+\hat{b}_S^+\hat{a}_S\hat{\bar{a}}_S\hat{a}_R\hat{\bar{a}}_R\Phi + \cdots\Big) \qquad (6)$$

Keeping in mind that $\hat{a}_k\hat{a}_k = 0$ one can also write

$$\Psi = c_0\Big(1 + \sum_{R=1}^{n/2} \hat{b}_R^+\hat{a}_R\hat{\bar{a}}_R + \frac{1}{2}\sum_{R=1}^{n/2}\sum_{S=1}^{n/2} \hat{b}_R^+\hat{b}_S^+\hat{a}_S\hat{\bar{a}}_S\hat{a}_R\hat{\bar{a}}_R$$

$$+ \frac{1}{3!}\sum_R\sum_S\sum_T \hat{b}_R^+\hat{b}_S^+\hat{b}_T^+\hat{a}_T\hat{\bar{a}}_T\hat{a}_S\hat{\bar{a}}_S\hat{a}_R\hat{\bar{a}}_R + \cdots\Big)\Phi$$

$$= c_0 \exp\Big(\sum_{R=1}^{n/2} \hat{b}_R^+\hat{a}_R\hat{\bar{a}}_R\Big)\Phi \qquad (7)$$

The separated-pair wave function has a number of very nice properties. We note especially the following.[11-13]

a. By minimizing the energy with respect to the APSG ansatz, one can derive effective two-electron Schrödinger equations for the geminal ω_R that can be interpreted as representing two electrons in the effective field of the other electrons.

b. The overall correlation energy $E_{\text{APSG}}^{\text{corr}}$ is, to a high degree of accuracy,* given as the sum of individual pair correlation energies ε_R obtained from independent solutions of effective two-electron equations:

$$E_{\text{APSG}}^{\text{corr}} = \sum_R \varepsilon_R \tag{8}$$

c. A system of n two-electron systems at mutually infinite separation is rigorously described by an APSG function. In particular the total correlation energy is strictly proportional to n.

In practice the APSG ansatz is not satisfactory[14] for the description of atoms or molecules, mainly because it does not account for the so-called interorbital correlation energy and because the strong-orthogonality constraint is too severe. Nevertheless, we can learn one important detail from this wave function, namely the role of the so-called "unlinked clusters."

Let us tentatively break off the expansion (6) of the APSG function after the first sum, i.e., after the pair substitutions with respect to the leading determinant Φ,

$$\Psi_{(2)} = N' \left\{ 1 + \sum_{R=1}^{n/2} \hat{b}_R^+ \hat{a}_R \hat{\bar{a}}_R \right\} \Phi \tag{9}$$

One easily sees[12,15] that for this type of wave function the overall correlation energy is not equal to the sum of the individual pair correlation energies, but rather

$$E_{(2)}^{\text{corr}} = \sum_{R=1}^{n/2} N_R^2 \varepsilon_R \tag{10}$$

$$N_R = (1 + \|u_R\|^2)^{1/2} (1 + \sum_S \|u_S\|^2)^{-1/2}$$

$$\approx \left(1 - \sum_{S(\neq R)} \|u_S\|^2 \right)^{1/2} \approx 1 - \tfrac{1}{2} \sum_{S(\neq R)} \|u_s\|^2 \tag{11}$$

i.e., each pair correlation contribution is multiplied by a factor <1 that depends on the norms of the other pair correlation functions. One can even show that for large n the overall correlation energy is proportional to $\sqrt{n}$, rather than to n (as it should be).† (The proof for a very analogous theorem is given in Section 3.1.) It is very important therefore, to include in the wave function all products of pair substitutions, rather than limit oneself to simple pair substitutions.

---

*In Ref. 13 it is stated explicitly what is meant by "high degree of accuracy." We do not want to elaborate on this point here.
†The author is indebted to Dr. R. Ahlrichs for pointing this out to him.

2.2. The Concept of Cluster Expansions: Analogies from Statistical Mechanics

Consider some quantity A associated with an n-particle system. If it can be expanded in the form

$$A = A_0 + \sum_{i=1}^{n} A_i + \sum_{i<j=1}^{n} A_{ij} + \sum_{i<j<k} A_{ijk} + \cdots \tag{12}$$

such that A_0 is the value of the quantity for statistically independent particles, A_{ij} takes care of two-particle correlation effects, etc., we call (12) a cluster expansion of A. Most of the physically interesting quantities are either additively or multiplicatively separable.[16] Additive separability means that for a supersystem consisting of two *noninteracting* subsystems the quantity A of the supersystem is simply the sum of the corresponding quantities of the subsystems; in the case of multiplicative separability it is the product.

In many cases taking a four-particle cluster A_{ijkl} one cannot tell whether it results from a genuine four-particle correlation effect or whether it contains contributions from two independent pair correlations ("occurring at the same time"), the so-called "unlinked" contributions. However, one can be sure that an additively separable quantity contains only linked contributions. The existence of effects due to simultaneous independent pair correlation in the two subsystems would be in contradiction to additive separability.[16]

Now if A is additively separable one easily sees that $B = \exp(A)$ is multiplicatively separable. The cluster expansion of B is (the A_i, A_{ij}, etc., are just numbers and hence commute)

$$B = B_0 \left(1 + \sum_{i=1}^{n} b_i + \sum_{i<j} b_{ij} + \sum_{i<j<l} b_{ijk} + \cdots \right) \tag{13}$$

If we define

$$a_i = \exp(A_i) - 1 = A_i + \frac{1}{2!} A_i^2 + \frac{1}{3!} A_i^3 + \cdots \tag{14}$$

and the a_{ij}, a_{ijk}, etc., in an analogous way, the b_i, b_{ij}, etc., are given as

$$b_i = a_i$$

$$b_{ij} = a_{ij} + a_i a_j$$

$$b_{ijk} = a_{ijk} + a_i a_{jk} + a_j a_{ik} + a_k a_{ij} + a_i a_j a_k$$

$$b_{ijkl} = a_{ijkl} + a_{ij} a_{kl} + a_{ik} a_{jl} + a_{il} a_{jk}$$

$$+ a_{ijk} a_l + a_{ijl} a_k + a_{ikl} a_j + a_{jkl} a_i$$

$$+ a_{ij} a_k a_l + a_{ik} a_j a_l + a_{il} a_j a_k + a_{jk} a_i a_l$$

$$+ a_{jl} a_i a_k + a_{kl} a_i a_j + a_i a_j a_k a_l \tag{15}$$

Obviously the cluster expansion of B does contain unlinked clusters which are essentially products of the linked clusters of lower orders.

A typical multiplicatively separable quantity of a real gas is its partition function Z (*Zustandssumme*),[17-20] whereas the free energy

$$F = -kT \ln Z \tag{16}$$

is additively separable. The partition function Z is that of an ideal gas Z_0 (ideal means the particles are statistically independent) times a correlation factor (we consider only an atomic gas, Ω is the volume),

$$Z_0 = \frac{(2\pi mkT)^{3n/2}\Omega^n}{n! h^{3n}} \tag{17}$$

$$Z = Z_0 \Omega^{-n} \int \exp[-(1/kT)V(1, 2, \ldots, n)] \, d\tau_1 \cdots d\tau_n \tag{18}$$

If only pair interactions are present, i.e., if

$$V(1, 2, \ldots, n) = \sum_{i<j=1}^{n} v_{ij} \tag{19}$$

one has

$$Z = Z_0 \Omega^{-n} \int \exp\left[-\sum_{i<j} \frac{v_{ij}}{kT}\right] d\tau_1 \cdots d\tau_n \tag{20}$$

The expansion of Z in terms of

$$f_{ij} = \exp\left(-\frac{v_{ij}}{kT}\right) - 1 = -\frac{v_{ij}}{kT} + \frac{1}{2}\frac{v_{ij}^2}{(kT)^2} + \cdots \tag{21}$$

contains both linked clusters such as

$$\Omega^{-3} \int f_{12} f_{23} \, d\tau_1 \, d\tau_2 \, d\tau_3 \tag{22}$$

and unlinked clusters such as

$$\Omega^{-4} \int f_{12} \, d\tau_1 \, d\tau_2 \int f_{34} \, d\tau_3 \, d\tau_4 \tag{23}$$

On taking the logarithm to obtain F, one gets rid of all the unlinked clusters so that all thermodynamic functions have contributions only from linked clusters.[17-20]

The clusters are in a way the response to the interaction. Without interaction there are no correlation effects and in the limit of very small interaction only the direct response matters, i.e., the terms linear in f_{ij} (or v_{ij}) which lead to two-particle clusters in F and to two-particle clusters plus all unlinked products of two-particle clusters in Z. Higher-order unlinked clusters enter only indirectly, but they occur even if no three- and more-particle interactions are present.

In quantum mechanics the energy is additively separable; hence the energy contains only linked clusters.* The wave function is what one may call multiplicatively quasiseparable, i.e., a physically acceptable wave function for a system of two noninteracting subsystems must be the antisymmetrized product of the (strongly orthogonal) wave functions of the subsystems. Like a multiplicative quantity in the proper sense it contains unlinked clusters. One can write Ψ such that the unlinked clusters are formally not present (though in reality they are) by using the exponential ansatz[21]

$$\Psi = e^{\hat{S}} \Phi \tag{24}$$

a special case of which we have seen in the preceding section, and which we shall discuss in detail in the next section. The statement concerning linked clusters in the energy applies strictly to the exact energy, and to approximate energy expressions only if certain conditions are fulfilled. Energy expectation values formed with trial functions may contain unlinked cluster contributions, in particular if the trial wave function is not multiplicatively quasiseparable.

2.3. The Exponential Ansatz of Coester and Kümmel

Any n-electron wave function Ψ can be expanded in terms of the Slater determinants built up from a complete orthonormal set of spin orbitals ψ_i. This configuration interaction (CI) expansion can always be written[22] such that one particular determinant Φ is regarded as unsubstituted and the other determinants are labeled as, e.g., Φ_i^a if the spin-orbital ψ_i contained in Φ is replaced by ψ_a. In this nomenclature Φ_i^a is a singly substituted determinant, Φ_{ij}^{ab} a doubly substituted one, etc. The CI expansion[22]

$$\Psi = c_0 \Phi + \sum_{i,a} c_i^a \Phi_i^a + \sum_{i<j} \sum_{a<b} c_{ij}^{ab} \Phi_{ij}^{ab} + \cdots \tag{25}$$

*We are not giving here the explicit relation between clusters in the wave function and those in the energy. The relation is not as simple as that between partition function and free energy in statistical mechanics and it is different for an expectation-value-type energy expression (Section 5.2) and an expression of the type $\langle \Phi | \hat{H} | \Psi \rangle$ (Section 2.4). Expectation-value-type expressions contain n-particle linked (connected) clusters with $n > 2$ even if Ψ is of the form $e^{\hat{S}_2}\Phi$. An unlinked (disconnected) 4-particle contribution to the energy is, e.g., one which cannot be written as the product of two 2-particle contributions (or a 1- and a 3-particle contribution etc.), particle indices being indices of the ψ_i occupied in Φ. One must bear in mind that the terms "cluster" and "linked" are used by different authors with different meanings and that these terms should only be used when they are precisely defined. In the present literature the word "cluster" is mostly used in the limited sense that it has in diagrammatic perturbation theory.[73] The clusters of the wave functions that play a central part in this review have sometimes been called S-type clusters.[102] The linked-cluster theorem for the energy and other additively separable quantities is more general and fundamental than one might suspect from the very lengthy and complicated proofs of the particular form of this theorem in perturbation theory.

can also be written in second quantization language

$$\Psi = (c_0 + \sum_{i,a} c_i^a \hat{a}_a^+ \hat{a}_i + \sum_{i<j} \sum_{a<b} c_{ij}^{ab} \hat{a}_a^+ \hat{a}_b^+ \hat{a}_j \hat{a}_i + \cdots)\Phi \tag{26}$$

or, if one defines the orbital and pair correlation functions[23]

$$u_i = \frac{1}{c_0} \sum_a c_i^a \psi_a$$

$$u_{ij} = \frac{1}{c_0} \sum_{a<b} c_{ij}^{ab} \frac{1}{\sqrt{2}} [\psi_a(1)\psi_b(2) - \psi_b(1)\psi_a(2)] \tag{27}$$

and the orbital and pair creation operators

$$\hat{b}_i^+ = \frac{1}{c_0} \sum_a c_i^a \hat{a}_a^+$$

$$\hat{b}_{ij}^+ = \frac{1}{c_0} \sum_{a<b} c_{ij}^{ab} \hat{a}_a^+ \hat{a}_b^+ \tag{28}$$

as

$$\Psi = c_0 \left[\Phi + \sum_i \mathscr{A}\left(\frac{\Phi}{\psi_i} u_i\right) + \sum_{i<j} \mathscr{A}\left(\frac{\Phi}{\psi_i \psi_j} u_{ij}\right) + \cdots \right]$$

$$= c_0 \left(1 + \sum_i \hat{b}_i^+ \hat{a}_i + \sum_{i<j} \hat{b}_{ij}^+ \hat{a}_j \hat{a}_i + \cdots \right)\Phi \tag{29}$$

This expansion is in principle exact, if all substitutions up to n-fold are included. One can base a pair theory on this expansion if one truncates it after the pair substitutions, i.e., if one omits $+ \cdots$ in (25), (26), or (29). We see that the truncated expansion (29) is a generalization of $\Psi_{(2)}$ of Eq. (9) and that the truncated expansion (29) will have the same incorrect dependence on the number of electrons as (9), so that it cannot be the appropriate starting point for an acceptable pair theory, at least not for large n. The discussions of Sections 2.1 and 2.2 give us a direct hint for reestablishing the correct n dependence, namely by including the unlinked clusters of pair substitutions, i.e., by using the ansatz[21,23-26]

$$\Psi = e^{\hat{S}}\Phi = e^{\hat{S}_1 + \hat{S}_2}\Phi = \exp\left(\sum_i \hat{b}_i^+ \hat{a}_i + \sum_{i<j} \hat{b}_{ij}^+ \hat{a}_j \hat{a}_i \right)\Phi \tag{30}$$

rather than (29), truncated after the pair terms.

Even if one wants to go beyond a pair theory it is better to use a generalization of (30) rather than (29). This is necessary if one wants to truncate the expansion with n-fold substitutions and if one wants to satisfy the important separability condition that for noninteracting subsystems the total energy is the sum of the energies of the isolated subsystems.

The generalization of (30), namely

$$\Psi = e^{\hat{S}}\Phi = \exp(\hat{S}_1 + \hat{S}_2 + \hat{S}_3 + \cdots)\Phi \tag{31}$$

$$\hat{S}_1 = \sum_i \sum_a d_i^a \hat{a}_a^+ \hat{a}_i$$

$$\hat{S}_2 = \sum_{i<j} \sum_{a<b} d_{ij}^{ab} \hat{a}_a^+ \hat{a}_b^+ \hat{a}_j \hat{a}_i = \tfrac{1}{4} \sum_{i,j} \sum_{a,b} d_{ij}^{ab} \hat{a}_a^+ \hat{a}_b^+ \hat{a}_j \hat{a}_i$$

$$\hat{S}_3 = \sum_{i<j<k} \sum_{a<b<d} d_{ijk}^{abc} \hat{a}_a^+ \hat{a}_b^+ \hat{a}_c^+ \hat{a}_k \hat{a}_j \hat{a}_i \tag{32}$$

has been first proposed by Coester and Kümmel.[21,24] The specialization (30) with $\hat{S}_1 = 0$ was introduced into quantum chemistry by Sinanoğlu. If we compare this linked cluster expansion with the standard cluster expansion (26) we find the following relations between the coefficients c_i^a, etc., of (26) and d_i^a of (32):

$$c_0^{-1} c_i^a = d_i^a$$

$$c_0^{-1} c_{ij}^{ab} = d_{ij}^{ab} + d_i^a d_j^b - d_i^b d_j^a$$

$$c_0^{-1} c_{ijk}^{abc} = d_{ijk}^{abc} + d_i^a d_{jk}^{bc} - d_j^a d_{ik}^{bc} + d_k^a d_{ij}^{bc} - d_i^b d_{jk}^{ac}$$

$$+ d_j^b d_{ik}^{ac} - d_k^b d_{ij}^{ac} + d_i^c d_{jk}^{ab} - d_j^c d_{ik}^{ab} + d_k^c d_{ij}^{ab}$$

$$+ d_i^a d_j^b d_k^c - d_i^a d_j^c d_k^b - d_i^b d_j^a d_k^c + d_i^b d_j^c d_k^a$$

$$+ d_i^c d_j^a d_k^b - d_i^c d_j^b d_k^a \tag{33}$$

The expression for $c_0^{-1} c_{ijkl}^{abcd}$ is already very lengthy.

In working with the $e^{\hat{S}}$ ansatz one can take advantage of the so-called Hausdorff formula

$$e^{-\hat{S}} \hat{X} e^{\hat{S}} = \hat{X} + [\hat{X}, \hat{S}] + \frac{1}{2!}[[\hat{X}, \hat{S}], \hat{S}] + \frac{1}{3!}[[[\hat{X}, \hat{S}], \hat{S}], \hat{S}] + \cdots \tag{34}$$

where $\hat{X}$ is any operator and $[\hat{A}, \hat{B}]$ means the commutator

$$[\hat{A}, \hat{B}] = \hat{A}\hat{B} - \hat{B}\hat{A} \tag{35}$$

2.4. The Hierarchy of the Cluster Equations

We suppose that the chosen Ψ satisfies the Schrödinger equation $\hat{H}\Psi = E\Psi$. If we expand Ψ as

$$\Psi = \sum_\nu c_\nu \Phi_\nu \tag{36}$$

and insert this into the Schrödinger equation we get, after scalar multiplication from the left by Φ_μ (the Φ_ν are supposed to be single Slater determinants* and orthonormal), the system of the so-called CI (*configuration interaction*) equations:

$$\sum_\mu H_{\nu\mu} c_\mu = E c_\nu; \qquad H_{\nu\mu} = \langle \Phi_\nu | \hat{H} | \Phi_\mu \rangle \tag{37}$$

If we write (36) in the form of Eq. (25) we get, since the Hamiltonian contains only one- and two-particle terms, for the line $\nu = 0$, i.e., for $\Phi_\nu = \Phi_0 = \Phi$

$$\langle \Phi | \hat{H} | \Phi \rangle c_0 + \sum_{i,a} \langle \Phi | \hat{H} | \Phi_i^a \rangle c_i^a + \sum_{i<j} \sum_{a<b} \langle \Phi | \hat{H} | \Phi_{ij}^{ab} \rangle c_{ij}^{ab} = E c_0 \tag{38}$$

which can also be written as

$$E = E_0 + \sum_i \varepsilon_i + \sum_{i<j} \varepsilon_{ij} \tag{39}$$

with

$$E_0 = \langle \Phi | \hat{H} | \Phi \rangle$$

$$\varepsilon_i = \frac{1}{c_0} \sum_a \langle \Phi | \hat{H} | \Phi_i^a \rangle c_i^a \tag{40}$$

$$\varepsilon_{ij} = \frac{1}{c_0} \sum_{a<b} \langle \Phi | \hat{H} | \Phi_{ij}^{ab} \rangle c_{ij}^{ab}$$

Equations (39) and (40) contain both the important and trivial statement[22,23] that the knowledge of the $c_0^{-1} c_{ij}^{ab}$ is sufficient to compute the energy.

The lines of (37) with on the r.h.s. $c_\nu = c_i^a$ and $c_\nu = c_{ij}^{ab}$ become

$$\langle \Phi_i^a | \hat{H} | \Phi \rangle c_0 + \sum_j \sum_b \langle \Phi_i^a | \hat{H} | \Phi_j^b \rangle c_j^b + \sum_{j<k} \sum_{b<c} \langle \Phi_i^a | \hat{H} | \Phi_{jk}^{bc} \rangle c_{jk}^{bc}$$

$$+ \sum_{j<k<l} \sum_{b<c<d} \langle \Phi_i^a | \hat{H} | \Phi_{jkl}^{bcd} \rangle c_{jkl}^{bcd} = E c_i^a \tag{41}$$

$$\langle \Phi_{ij}^{ab} | \hat{H} | \Phi \rangle c_0 + \sum_k \sum_c \langle \Phi_{ij}^{ab} | \hat{H} | \Phi_k^c \rangle c_k^c + \sum_{k<l} \sum_{c<d} \langle \Phi_{ij}^{ab} | \hat{H} | \Phi_{kl}^{cd} \rangle c_{kl}^{cd}$$

$$+ \sum_{k<l<m} \sum_{c<d<e} \langle \Phi_{ij}^{ab} | \hat{H} | \Phi_{klm}^{cde} \rangle c_{klm}^{cde}$$

$$+ \sum_{k<l<m<n} \sum_{c<d<e<f} \langle \Phi_{ij}^{ab} | \hat{H} | \Phi_{klmn}^{cdef} \rangle c_{klmn}^{cdef} = E c_{ij}^{ab} \tag{42}$$

---

*They could also be spin- or symmetry-adapted linear combinations of determinants, but we shall not consider this possibility at the moment.

In writing these equations we have not explicitly taken care of the fact that only those matrix elements are different from zero for which the Slater determinants on the left and on the right of $\hat{H}$ differ in no more than two spin-orbitals, which means, e.g., that in the last term of (42) only matrix elements like $\langle \Phi_{ij}^{ab} | \hat{H} | \Phi_{ijmn}^{abef} \rangle$ contribute. One can continue up to the line of (37) where Φ_ν refers to an n-fold substitution.

One sees that in order to calculate the c_i^a from Eq. (41) knowledge of the c_{ij}^{ab} and the c_{ijk}^{abc} is required, while the calculation of the c_{ij}^{ab} from (42) requires knowledge of the c_{ijk}^{abc} and the c_{ijkl}^{abcd}, etc.

One gets a rigorous solution of this hierarchy of the cluster equations (40)–(42) only if one diagonalizes the full CI matrix (37). In doing so one is also sure that one gets an upper bound for the energy. Equation (40) is correct (i.e., furnishes an upper bound to the true E) not only if Ψ is the exact wave function, but also if Ψ represents an eigenvector of the CI matrix.

However, we have not derived the hierarchy of the cluster equations in order to solve them rigorously, but rather as the starting point for approximate treatments. One is tempted to truncate the hierarchy somewhere, e.g., by neglecting higher than double substitutions, but then we would violate the separability condition. In order to satisfy this condition we have to include in the wave function at least those 4-fold substitutions that are unlinked clusters of double substitutions. We must, therefore, transform the hierarchy of the cluster equations to one for the linked-pair cluster expansion (30). Before we do so we consider one simplification that is always possible, rigorously and without loss of generality.

2.5. The Brillouin–Brueckner Condition

As Brenig[23] has pointed out (see also Refs. 22, 27, 28) one can always choose the occupied ψ_i such that all c_i^a $(=d_i^a)$ vanish identically. From (41), imposing the condition $c_i^a = 0$, one gets

$$\langle \Phi_i^a | \hat{H} | \Phi \rangle c_0 + \sum_{j<k} \sum_b c_{jk}^{ab} \langle \Phi_i^a | \hat{H} | \Phi_{jk}^{ab} \rangle + \sum_k \sum_{b<c} c_{ik}^{bc} \langle \Phi_i^a | \hat{H} | \Phi_{ik}^{bc} \rangle$$

$$+ \sum_{\substack{j<k \ b<c \\ (\neq i)}} \sum c_{ijk}^{abc} \langle \Phi_i^a | \hat{H} | \Phi_{ijk}^{abc} \rangle = 0 \tag{43}$$

This condition, sometimes called Brillouin–Brueckner[29] or best-overlap condition,[30] is actually a condition for Φ to be the Slater determinant that has maximum overlap with the exact Ψ.

The Brillouin–Brueckner condition (43) can (ignoring the last term) be regarded as the condition that the expectation value

$$\langle \Phi | \hat{H} | \Phi \rangle$$

with

$$\hat{\bar{H}} = \hat{H}(1 + c_0^{-1} \sum_{i<k} \sum_{a<b} c_{ik}^{ab} \hat{a}_a^+ \hat{a}_b^+ \hat{a}_i \hat{a}_k) \tag{44}$$

is stationary with respect to variation of the ψ_i (occupied in Φ) subject to the normalization of Φ. The corresponding variation principle with $\hat{\bar{H}}$ replaced by $\hat{H}$ leads to the Brillouin condition

$$\langle \Phi_i^a | \hat{H} | \Phi \rangle = 0 \tag{45}$$

for Φ the Hartree–Fock determinant.

At first glance the Brillouin–Brueckner condition and the corresponding variation principle seem perfectly straightforward. On more careful inspection one realizes not only that $\hat{\bar{H}}$ is non-Hermitian but also that $\langle \Phi | \hat{\bar{H}} | \Phi \rangle$ (which is the exact energy) is invariant with respect to simultaneous variation of Φ and the c_{ik}^{ab}, such that a variational principle for Φ can hardly be based on stationarity of $\langle \Phi | \hat{\bar{H}} | \Phi \rangle$. What seems odd with the Brillouin–Brueckner condition is due to the fact that it results from a nonvariational procedure, namely the truncated hierarchy of the cluster equations. (The original derivation was based on a partial summation of perturbation contributions, which does not make a basic difference.) It is more consistent to derive an equivalent of the Brillouin–Brueckner condition in a purely variational framework. We shall see in Section 5.3 that this is in fact possible and that a Hermite form of (43) is an acceptable first approximation to the rigorous condition for the vanishing of single substitutions with respect to Φ.

Since the expectation value $\langle \Phi | \hat{\bar{H}} | \Phi \rangle$ is equal to the exact energy, variational (or rather quasivariational) calculations, based on the minimization of (44), have been referred to as "exact energy Hartree–Fock" methods.[29,31]

Although the expectation value (44) cannot be made stationary with respect to simultaneous variations of Φ and the c_{ik}^{ab}, a stepwise minimization of (44) is possible by first guessing some c_{ik}^{ab}, then obtaining new ψ_i and a new Φ, which can then be the starting point for the calculation of new c_{ik}^{ab}, etc. This iterative procedure, which is now usually called the Brueckner–Hartree–Fock method,[32] is very useful in nuclear theory where the expectation value $\langle \Phi | \hat{H} | \Phi \rangle$ diverges because of the hard-core repulsion, but where $\langle \Phi | \hat{\bar{H}} | \Phi \rangle$ may converge and make a cluster expansion of the wave function possible. In atomic or molecular problems the Brueckner–Hartree–Fock method does not present any noteworthy advantage.

For closed-shell states the "best overlap" or "Brueckner" orbitals do not usually differ much from the Hartree–Fock orbitals, so that one is justified in neglecting single substitutions even if one uses Hartree–Fock orbitals. If one is not allowed to neglect single substitutions it is often easier to include them explicitly rather than obtain the Brueckner orbitals by iteration (see however, Section 5.4).

2.6. The Hierarchy of Equations for Linked Pair Clusters

In the following we assume that single substitutions can be ignored (e.g., because ψ_i are Brueckner spin orbitals). Further, we keep in (41) and (42) only those determinants that are double substitutions or are unlinked clusters of pair substitution with respect to ϕ. In view of (33) we have

$$c_i^a = 0, \qquad c_0^{-1} c_{ij}^{ab} = d_{ij}^{ab}, \qquad c_{ijk}^{abc} = 0 \tag{46}$$

whereas the c_{ijkl}^{abcd} are in analogy with (33), expressed as

$$
\begin{aligned}
c_0^{-1} c_{ijkl}^{abcd} =\ & d_{ij}^{ab} d_{kl}^{cd} - d_{ik}^{ab} d_{jl}^{cd} + d_{il}^{ab} d_{jk}^{cd} - d_{ij}^{ac} d_{kl}^{bd} + d_{ik}^{ac} d_{jl}^{bd} - d_{il}^{ac} d_{jk}^{bd} \\
& + d_{ij}^{ad} d_{kl}^{bc} - d_{ik}^{ad} d_{jl}^{bc} + d_{il}^{ad} d_{jk}^{bc} + d_{ij}^{cd} d_{kl}^{ab} - d_{ik}^{cd} d_{jl}^{ab} + d_{il}^{cd} d_{jk}^{ab} \\
& - d_{ij}^{bd} d_{kl}^{ac} + d_{ik}^{bd} d_{jl}^{ac} - d_{il}^{bd} d_{jk}^{ac} + d_{ij}^{bc} d_{kl}^{ad} - d_{ik}^{bc} d_{jl}^{ad} + d_{il}^{bc} d_{jk}^{ad}
\end{aligned} \tag{47}
$$

Equations (38) and (42) now become

$$\langle \Phi | \hat{H} | \Phi \rangle + \sum_{i<j} \sum_{a<b} \langle \Phi | \hat{H} | \Phi_{ij}^{ab} \rangle d_{ij}^{ab} = E \tag{48}$$

$$\langle \Phi_{ij}^{ab} | \hat{H} | \Phi \rangle + \sum_{k<l} \sum_{c<d} \langle \Phi_{ij}^{ab} | \hat{H} | \Phi_{kl}^{cd} \rangle d_{kl}^{cd}$$

$$+ \sum_{k<l} \sum_{c<d} \langle \Phi_{ij}^{ab} | \hat{H} | \Phi_{ijkl}^{abcd} \rangle c_0^{-1} c_{ijkl}^{abcd} = E d_{ij}^{ab} \tag{49}$$

with $c_0^{-1} c_{ijkl}^{abcd}$ given by (47).

The system of equations (48) and (49) is contained implicitly in Brenig's paper[23] (Brenig introduced additional approximations that led him immediately to a separated-pair theory); it has been worked out in a manageable way using diagram techniques by Čížek et al.[33-36] under the name coupled-pair many electron theory (CP–MET). It differs from the system that one would have obtained by minimization of the energy expectation value $\langle e^{\hat{S}} \Phi | \hat{H} | e^{\hat{S}} \Phi \rangle \langle e^{\hat{S}} \Phi | e^{\hat{S}} \Phi \rangle^{-1}$ with respect to the d_{ij}^{ab}. In fact, in deriving (48) and (49) we have first assumed that the c_{ij}^{ab} and c_{ijkl}^{abcd} are independent variational parameters and, after minimizing the energy with respect to these coefficients, we have imposed the condition (47). Imposing (47) before the minimization would have led to much more complicated equations (see Section 5.1 and 5.2).

Equations (48) and (49) are not variational and this implies that the error in the energy is linear in the error in the wave function with respect to a variational reference wave function, and not quadratic as one is accustomed to from variational wave functions.

A very compact form of the system (48) and (49) is obtained by inserting $\Psi = e^{\hat{S}} \Phi$ into the Schrödinger equation $\hat{H} \Psi = E \Psi$ (which is only rigorous if the exact wave function is of the form $e^{\hat{S}}\Phi$) and multiplying it from the left by $\langle\Phi|$, $\langle\Phi_i^a|$, and $\langle\Phi_{ij}^{ab}|$.[37] For the case $\hat{S} = \hat{S}_2$, to which we limit our attention, we have

$$\langle\Phi|\hat{H}|e^{\hat{S}}\Phi\rangle = \langle\Phi|\hat{H}(1+\hat{S})|\Phi\rangle = \langle\Phi|\hat{H}|\Phi\rangle + \langle\Phi|\hat{H}\hat{S}|\Phi\rangle = E\langle\Phi|e^{\hat{S}}|\Phi\rangle = E$$

(50)

$$\langle\Phi_{ij}^{ab}|\hat{H}e^{\hat{S}}|\Phi\rangle = \langle\Phi_{ij}^{ab}|\hat{H}(1+\hat{S}+\tfrac{1}{2}\hat{S}^2)|\Phi\rangle = E\langle\Phi_{ij}^{ab}|e^{\hat{S}}|\Phi\rangle$$

$$= E\langle\Phi_{ij}^{ab}|1+\hat{S}|\Phi\rangle = E d_{ij}^{ab}$$

(51)

The Brillouin–Brueckner condition is

$$\langle\Phi_i^a|\hat{H}(1+\hat{S})|\Phi\rangle = 0$$

(52)

Alternatively,[38] instead of multiplying the Schrödinger equation by $\langle\Phi|$, $\langle\Phi_i^a|$, and $\langle\Phi_{ij}^{ab}|$ one can multiply it by $\langle\Phi e^{-\hat{S}}|$, $\langle\Phi_i^a e^{-\hat{S}}|$, and $\langle\Phi_{ij}^{ab} e^{-\hat{S}}|$ and then use the Hausdorff formula (34) for $e^{-\hat{S}} H e^{\hat{S}}$. The resulting equations look somewhat different but they are identical to (50) and (51), as the reader may easily verify for himself.

It is noteworthy that there is a partial cancellation of terms in Eq. (49) or its equivalent (51); this we see if we replace E by the explicit expression (50) equivalent to (48)

$$\langle\Phi_{ij}^{ab}|H|\Phi\rangle + \langle\Phi_{ij}^{ab}|\hat{H}\hat{S}|\Phi\rangle + \tfrac{1}{2}\langle\Phi_{ij}^{ab}|\hat{H}\hat{S}^2|\Phi\rangle = \langle\Phi|\hat{H}|\Phi\rangle d_{ij}^{ab} + \langle\Phi|\hat{H}\hat{S}|\Phi\rangle d_{ij}^{ab}$$

(53)

The last terms both on the right-hand side and the left-hand side of (53) contain the "disconnected" contributions

$$\langle\Phi_{ij}^{ab}|\hat{H}|\phi_{ijkl}^{abcd}\rangle d_{ij}^{ab} d_{kl}^{cd} = \langle\Phi|\hat{H}|\Phi_{kl}^{cd}\rangle d_{ij}^{ab} d_{kl}^{cd} \qquad k, l \neq i, j; c, d \neq a, b$$

(54)

which cancel exactly and can be ignored from the very beginning. (It is, by the way, this cancellation that guarantees that the energy is additively separable.) The terms bilinear in the d that remain on the right-hand side are closely related to the EPV (exclusion principle violating) diagrams of pertubation theory.

Kelly[39,40] has proposed a simplification of Eq. (49) which consists of replacing c_{ijkl}^{abcd} in the *ijab* line (i.e., in that line of the system of equations (49) in which, on the right-hand side, one has $E d_{ij}^{ab}$) by $d_{ij}^{ab} d_{kl}^{cd} c_0$ instead of the full expression (47). In the formalism of Eq. (53), Kelly's simplification implies that only the contributions to $\tfrac{1}{2}\langle\Phi_{ij}^{ab}|\hat{H}\hat{S}^2|\Phi\rangle$, which are of the form (54), are considered (not, e.g., $\tfrac{1}{2}\langle\Phi_{ij}^{ab}|\hat{H}|\Phi_{ijkl}^{abcd}\rangle d_{ik}^{ac} d_{jl}^{bd}$), but these cancel with corresponding terms in $\langle\Phi|\hat{H}\hat{S}|\Phi\rangle c_{ij}^{ab}$ such that $\tfrac{1}{2}\langle\Phi_{ij}^{ab}|\hat{H}\hat{S}^2|\Phi\rangle$ completely disappears and that $E = \langle\Phi|\hat{H}|\Phi\rangle + \langle\Phi\hat{H}\hat{S}|\Phi\rangle$ is replaced by an effective energy W_{ij}^{ab} in which the "disconnected" contributions to the correlation energy $\langle\Phi|\hat{H}|\Phi_{kl}^{cd}\rangle$ $(k, l \neq i, j; c, d \neq a, b)$ are missing.

We shall discuss (Section 5.3) this approach in a more general context, together with other attempts to linearize the nonlinear CP–MET system (53). Kelly did not, by the way, start from the $e^{\hat{S}}\Phi$ ansatz, but he came to his approximation by an argument from perturbation theory. He has not actually performed numerical calculations using this approximation.

3. Perturbation Theory of Electron Correlation

3.1. Rayleigh–Schrödinger versus Brillouin–Wigner Perturbation Theory

We assume that we know eigenfunctions Φ_i and eigenvalues W_i of an "unperturbed" Hamiltonian $\hat{H}_0$ that is sufficiently "close" to the exact Hamiltonian $\hat{H}$. The difference $\hat{V}$ between $\hat{H}$ and $\hat{H}_0$ is called the perturbation

$$\hat{H}\Psi_i = E_i\Psi_i, \qquad \hat{H} = \hat{H}_0 + \hat{V}$$
$$\hat{H}_0\Phi_i = W_i\Phi_i, \qquad \hat{H}_\lambda = \hat{H}_0 + \lambda\hat{V} \tag{55}$$

If $\hat{V}$ does not depend explicitly on a "natural" physical perturbation parameter, we can formally multiply $\hat{V}$ by λ, get the solution of the family $\hat{H}_\lambda$ of Hamiltonians, and finally limit our consideration to $\lambda = 1$. Energy E and eigenfunction Ψ of $\hat{H}_\lambda$ (we omit the subscript that labels the eigenfunctions and eigenvalues of $\hat{H}$, usually we are interested in the ground state of $\hat{H}$) are then given as power series in λ

$$E = E^{(0)} + \lambda E^{(1)} + \lambda^2 E^{(2)} + \cdots$$
$$\Psi = \Psi^{(0)} + \lambda\Psi^{(1)} + \lambda^2\Psi^{(2)} + \cdots \tag{56}$$
$$E^{(0)} = W_0, \qquad \Psi^{(0)} = \Phi_0 = \Phi$$

The expressions for the $E^{(k)}$ are given in a compact form if one introduces the reduced resolvent $\hat{G}_0(z)$ of $\hat{H}_0$. Let

$$\hat{H}_0 = \sum_{i=0}^{\infty} W_i\hat{P}_i, \qquad \hat{P}_i = |\Phi_i\rangle\langle\Phi_i| \tag{57}$$

be the spectral representation of $\hat{H}_0$, then the reduced resolvent associated with the lowest eigenvalue W_0 of $\hat{H}_0$ is defined as

$$\hat{G}_0(z) = \sum_{i=1}^{\infty} (z - W_i)^{-1}\hat{P}_i \tag{58}$$

where z is any complex number. We have assumed in (57) and (58) that $\hat{H}_0$ has only a discrete spectrum; the generalization that includes a continuous spectrum is straightforward and does not change the following argument. One easily sees that

$$-\hat{G}_0(W_0)(\hat{H}_0 - W_0) = 1 - \hat{P}_0 \tag{59}$$

This is why G_0 is often symbolically written as

$$G_0(W_0) = \frac{1 - \hat{P}_0}{W_0 - \hat{H}_0} \tag{60}$$

In terms of $\hat{G} = \hat{G}_0(W_0)$ the first $E^{(k)}$ of the Rayleigh–Schrödinger (RS) expansion are given as

$$E^{(1)} = \langle \Phi_0| \hat{V}|\Phi_0\rangle = \langle \hat{V}\rangle$$

$$E^{(2)} = \langle \Phi_0| \hat{V}\hat{G}\hat{V}|\Phi_0\rangle$$

$$E^{(3)} = \langle \Phi_0| \hat{V}\hat{G}(\hat{V} - \langle \hat{V}\rangle)\hat{G}\hat{V}|\Phi_0\rangle$$

$$= \langle \Phi_0| \hat{V}\hat{G}\hat{V}\hat{G}\hat{V}|\Phi_0\rangle - \langle \Phi_0| \hat{V}\hat{G}\hat{G}\hat{V}|\Phi_0\rangle\langle \Phi_0| \hat{V}|\Phi_0\rangle \tag{61}$$

$$E^{(4)} = \langle \Phi_0| \hat{V}\hat{G}(\hat{V} - \langle \hat{V}\rangle)\hat{G}(\hat{V} - \langle \hat{V}\rangle)\hat{G}\hat{V}|\Phi_0\rangle - E^{(2)}\langle \Phi_0| \hat{V}\hat{G}\hat{G}\hat{V}|\Phi_0\rangle$$

The corresponding terms of the Brillouin–Wigner (BW) series are in terms of $\hat{\tilde{G}} = \hat{G}_0(E)$

$$\tilde{E}^{(1)} = \langle \Phi_0| \hat{V}|\Phi_0\rangle = \langle \hat{V}\rangle$$

$$\tilde{E}^{(2)} = \langle \Phi_0| \hat{V}\hat{\tilde{G}}\hat{V}|\Phi_0\rangle$$

$$\tilde{E}^{(3)} = \langle \Phi_0| \hat{V}\hat{\tilde{G}}\hat{V}\hat{\tilde{G}}\hat{V}|\Phi_0\rangle \tag{62}$$

$$\tilde{E}^{(4)} = \langle \Phi_0| \hat{V}\hat{\tilde{G}}\hat{V}\hat{\tilde{G}}\hat{V}\hat{\tilde{G}}\hat{V}|\Phi_0\rangle$$

$E^{(0)}$ and $E^{(1)}$ are the same in the RS and BW expansions, $E^{(2)}$ is formally the same, except that $E^{(2)}$ contains $\hat{G} = \hat{G}_0(W_0)$ and $\tilde{E}_2$ contains $\tilde{G} = \hat{G}_0(E)$; $E^{(3)}$ is even formally different in the two series but looks simpler in BW.

Usually, in comparing the BW and RS series one states that the former has the advantages of being formally simpler, but the disadvantage that knowledge of the perturbed energy E is required so that in most cases an iterative procedure is necessary. For the case of a two-by-two matrix eigenvalue problem, the BW series breaks off after $\tilde{E}^{(2)}$ and is exact. This is also the case for an $n \times n$ matrix with $H_{ij} = H_{ii}\,\delta_{ij}(i, j \neq 0)$.

The main difference between the RS and the BW series is that the latter is *not* an expansion as a power series of the perturbation parameter λ. This is so, since the $\tilde{E}^{(k)}$ (except $\tilde{E}^{(0)}$ and $\tilde{E}^{(1)}$) depend explicitly on the perturbed energy E, and this depends on the perturbation. One can, in principle, eliminate this dependence by expanding E in powers of λ and rearranging the summation, but this would just lead to the RS expansion.[41,42] One may think that the fact that the BW series is not a power series in terms of λ is not crucial for perturbation problems without a natural perturbation parameter, where λ is put equal to 1 anyway. This is true in a way, but the BW series has a serious drawback for the calculation of correlation effects due to its not being a power series expansion. We now illustrate this (for a different demonstration see Refs. 43–45).

Take a system of n noninteracting 2-electron systems, e.g., H_2 molecules at mutually infinite distance. The exact energy of this system is proportional to n. The Hamiltonian $\hat{H} = \hat{H}_0 + \lambda \hat{V}$ is physically meaningful only for $\lambda = 1$ but one can formally also solve the Schrödinger equation for $\lambda \neq 1$, and the resulting energy E must be proportional to n for any λ. Consequently each term $E^{(k)}$ in (56) must be proportional to n. Now compare $E^{(2)}$ (61) with $\tilde{E}^{(2)}$ (62). They differ in $\hat{G} = \hat{G}_0(W_0)$ being replaced by $\hat{\tilde{G}} = \hat{G}_0(E) = \hat{G}_0(W_0 + \Delta E)$.

Before we pursue this discussion we note that two variants of BW perturbation theory are possible.

Variant (a). The energy E that occurs in $\hat{G}_0(E)$ and hence in the whole expansion, is the exact energy, i.e., that of the infinite series, irrespective of the order at which one truncates the series.

Variant (b). If one truncates the series at the kth order of the energy one replaces E in $\hat{G}_0(E)$ by $\sum_{j=0}^{k} \tilde{E}^{(j)}$, i.e., the energy of the truncated series. This allows one to solve the perturbation problem iteratively and one does not need the exact E (which is usually unknown).

We use a basis set that consists of mutually orthogonal subsets that are localized on the individual H_2 molecules. (This is surely possible in the limit $R_{\mu\nu} \to \infty$ for all pairs of H_2 molecules.) Then only those terms in $\hat{G}_0(E)$ or $\hat{G}_0(E_0)$ contribute to $\tilde{E}^{(2)}$ or $E^{(2)}$ that correspond to local single or double substitutions, since all other substituted configurations have vanishing matrix elements of $\hat{V}$ with Φ_0. As a consequence, the energy differences $W_0 - W_i$ in the denominator of nonvanishing contributions to $E^{(2)}$ are all independent of the number of particles and so is any contribution to $E^{(2)}$. The number of terms of one kind is proportional to n and hence $E^{(2)}$ is proportional to n. In $\tilde{E}^{(2)}$ the respective denominators are $E - W_i = \Delta E + W_0 - W_i$. Obviously $W_0 - W_i$ is independent of n but ΔE (essentially the correlation energy) is proportional to n so there is, for large n, a term proportional to n in the denominator and as a consequence $\tilde{E}^{(2)}$ is independent of n.

This argument is only valid for variant (a) of the BW expansion. In variant (b) we cannot presume that ΔE is proportional to n, since the energy E (and hence $\Delta E = E - W_0$) of the truncated BW expansion does not have the correct n dependence. If we note (see Section 3.2) that $E^{(1)}$, which is proportional to n, can always be absorbed into W_0 and hence eliminated, we see that the hypothesis that ΔE of variant (b) is proportional to n leads to the result that $E^{(2)}$ (and similarly the higher terms) is proportional to n^0 (i.e., independent of n) and vice versa. Only the hypothesis that ΔE is proportional to $\sqrt{n}$ is self-consistent. We note that this is the same n-dependence as in a CI expansion with two-particle excitations, but without unlinked clusters.[98]

3.2. Different Variants of RS Perturbation Theory

The most natural way to decompose $\hat{H}$ into $\hat{H}_0$ and $\hat{V}$ is to take $\hat{H}_0$ as the sum of the genuine one-electron operators and $\hat{V}$ the genuine two-electron operators in $\hat{H}$,

$$\hat{H}_0 = \sum_{i=1}^{n} \hat{h}(i) = \hat{H}^{(1)}, \qquad \hat{V} = \sum_{i<j=1}^{n} \frac{1}{r_{ij}} = \hat{H}^{(2)} \qquad (63)$$

This decomposition is particularly convenient for atoms where

$$\hat{h}(i) = -\tfrac{1}{2}\Delta_i - \frac{Z}{r_i} \qquad (64)$$

and where the scale transformation

$$\rho_i = Z r_i \qquad (65)$$

leads to the following Hamiltonian in terms of ρ_i

$$\hat{H} = Z^2 \left[\sum_{i=1}^{n} \hat{h}(i) + \frac{1}{Z} \sum_{i<j} \frac{1}{\rho_{ij}} \right] = Z^2 \hat{H} \qquad (66)$$

$$\hat{H} = \hat{H}_0 + \frac{1}{Z} \hat{V} \qquad (67)$$

The inverse of the nuclear charge $1/Z$ appears to be the natural perturbation parameter and perturbation theory furnishes wave functions and energies for a whole isoelectronic series of atoms (ions) with the same n, but different Z.

The problem of the radius of convergence of the $1/Z$ expansion has received some attention. It has only recently been proven[46] definitely that, for ground states of 2-electron atoms, the series converges for $1/Z < 1/1.98$, i.e., for the He atom, for Li$^+$, etc., but probably not for H$^-$. For atoms with $n \geq 3$ the bound for the radius of convergence is much smaller, and it is unlikely that the series converges at all for neutral atoms with $n \geq 3$. That the second-order results[47] are often rather good seems more or less fortuitous. Nevertheless, the $1/Z$ expansion has considerable merit, in particular it provides some insight into the Z-dependence of the correlation energy in isoelectronic series.[48,49]

If one uses the bare nuclear Hamiltonian as $\hat{H}_0$, the difference between Φ_0 and Ψ is usually very large. One can try to take care of part of the electron interaction by choosing a different $\hat{H}_0$, one in which either the nuclear charge is partially shielded or an effective one-electron operator is added. One then has

$$\hat{H} = \hat{H}_0 + \hat{V}$$

$$\hat{H}_0 = \sum_{i=1}^{n} \hat{h}(i) + \sum_{i=1}^{n} \hat{v}(i) = \hat{H}^{(1)} + \sum_{i=1}^{n} \hat{v}(i) \qquad (68)$$

$$\hat{V} = \sum_{i<j=1}^{n} \frac{1}{r_{ij}} - \sum_{i=1}^{n} \hat{v}(i) = \hat{H}^{(2)} - \sum_{i=1}^{n} \hat{v}(i)$$

The most popular effective one-electron operator is that of Hartree–Fock theory

$$\hat{v}(i) = \sum_{k=1}^{n} [\hat{J}^k(i) - \hat{K}^k(i)] \tag{69}$$

where $\hat{J}^k$ and $\hat{K}^k$ are Coulomb and exchange operators in terms of spin-orbitals. One disadvantage of the choice (68) of $\hat{H}_0$ is that the zeroth-order energy $E^{(0)}$ is not equal to the Hartree–Fock expectation value but equal to the sum of the orbital energies, so that one counts the electron interaction twice. The negative of the interaction energy $\langle \Phi | \hat{H}^{(2)} | \Phi \rangle$ appears then as $E^{(1)}$. One can avoid this by subtracting $\langle \Phi | \hat{H}^{(2)} | \Phi \rangle$ from $\hat{H}_0$, which only shifts the eigenvalues of $\hat{H}_0$. We choose, therefore,

$$\hat{H}_0 = \hat{H}^{(1)} + \sum_{i=1}^{n} \hat{v}(i) - \langle \Phi | \hat{H}^{(2)} | \Phi \rangle$$
$$\hat{V} = \hat{H}^{(2)} - \sum_{i=1}^{n} \hat{v}(i) + \langle \Phi | \hat{H}^{(2)} | \Phi \rangle \tag{70}$$

With this choice $E^{(0)} = \langle \Phi | \hat{H} | \Phi \rangle$ and $E^{(1)} = 0$.

Neither for the decomposition (68) nor for (70) is it usually possible to define a natural perturbation parameter with a direct physical meaning. For perturbation problems without a natural perturbation parameter some ambiguity as to the definition of the orders of the perturbation contribution arises. One can add to $\hat{H}_0$ any operator $\hat{A}$ that has the property[50–52]

$$\hat{A} \Phi_0 = 0 \tag{71}$$

Then $\hat{H}'_0 = \hat{H}_0 + \hat{A}$ satisfies the same eigenvalue equation as $\hat{H}_0$:

$$\hat{H}'_0 \Phi_0 = (\hat{H}_0 + \hat{A}) \Phi_0 = W_0 \Phi_0 \tag{72}$$

The other eigenvalues W'_i ($i \neq 0$) of $\hat{H}'_0$ are, however, not the same as those of $\hat{H}_0$, and consequently the resolvent of $\hat{H}'_0$ differs from that of $\hat{H}_0$, and the kth order perturbation contributions differ as well. Summed up to infinite order the results, of course, have to be the same (provided that either series converges). The operator $\hat{A}$ in the sense of (71) can be chosen as

$$\hat{A}(1, 2, \ldots, n) = \sum_{i=1}^{n} \hat{a}(i) + \sum_{i<j=1}^{n} \hat{b}(i, j)$$

$$\hat{a}(i) = \sum_{p=1}^{\infty} a_p \hat{P}_p(i) - \sum_{k=1}^{n} \hat{a}_k, \qquad \hat{P}_p(i) = |\psi_p(i)\rangle\langle\psi_p(i)| \tag{73}$$

$$\hat{b}(i, j) = \sum_{p,q=1}^{\infty} b_{pq} \hat{P}_p(i) \hat{P}_q(j) - \sum_{k,l=1}^{n} b_{kl}$$

where the ψ_p are the spin-orbitals occupied in Φ. The constants a_p, b_{pq} can be chosen arbitrarily. For any choice of $\hat{A}$ the second-order energy $E^{(2)}$ can be evaluated if one introduces a resolution of the identity between any two operators

$$E^{(2)} = \sum_\nu \sum_\mu \langle\Phi|\hat{V}|\Phi_\nu\rangle\langle\Phi_\nu|\hat{G}|\Phi_\mu\rangle\langle\Phi_\mu|\hat{V}|\Phi\rangle \tag{74}$$

where the Φ_ν are a complete set of eigenfunctions of $\hat{H}_0'$. Since the Φ_ν are eigenfunctions of $\hat{H}_0'$, they are also eigenfunctions of $\hat{G} = \hat{G}_0'(E_0)$ and the terms with $\nu \neq \mu$ vanish; those with $\nu = \mu = 0$ vanish because of the definition of $\hat{G}$. Due to the Brillouin theorem those Φ_ν which are singly substituted with respect to Φ (Φ_i^a) have vanishing matrix elements with $\hat{V}$ (provided that Φ is the Slater determinant of unrestricted Hartree–Fock theory), and higher than doubly substituted determinants do not contribute because $\hat{H}$ contains only one and two particle terms. Hence,

$$E^{(2)} = \sum_{i<j} \sum_{a<b} \langle\Phi|\hat{V}|\Phi_{ij}^{ab}\rangle\langle\Phi_{ij}^{ab}|\hat{G}|\Phi_{ij}^{ab}\rangle\langle\Phi_{ij}^{ab}|\hat{V}|\Phi\rangle \tag{75}$$

With $\hat{A} \equiv 0$ this yields

$$E^{(2)} = \sum_{i<j} \sum_{a<b} \frac{|\langle\Phi|\hat{V}|\Phi_{ij}^{ab}\rangle|^2}{e_a + e_b - e_i - e_j} \tag{76}$$

where the spin-orbitals ψ_i and ψ_a are eigenfunctions of $\hat{F}(i) = \hat{h}(i) + \hat{v}(i)$ with eigenvalues e_i, e_a, respectively. For $\hat{A} \neq 0$ given by (73) we get, instead of (76),

$$E^{(2)} = \sum_{i<j} \sum_{a<b}$$
$$\times \frac{|\langle\Phi|V|\Phi_{ij}^{ab}\rangle|^2}{e_a + e_b + a_a + a_b - e_i - e_j - a_i - a_j + b_{ab} - b_{ij} + \sum_{k=1}^n (b_{ka} + b_{kb} - b_{ki} - b_{kj})} \tag{77}$$

By an appropriate choice of the constants a_p, b_{pq} one can change $E^{(2)}$ almost arbitrarily without affecting the result that one would get for the infinite perturbation series (provided it converges). If one is not worried about this arbitrariness one can choose the a_k and b_{kl} such that $E^{(2)}$ is as close as possible to the full correlation energy. Of course, if one has obtained the optimum $E^{(2)}$ this does not necessarily mean the whole series converges better or that the higher terms have become smaller.

One particular choice of $\hat{A}$ is suggested by the following argument. If one would perform a configuration interaction calculation with Φ and the Φ_{ij}^{ab} as a basis one would have to evaluate the matrix elements $\langle\Phi|\hat{H}|\Phi_{ij}^{ab}\rangle = \langle\Phi|\hat{V}|\Phi_{ij}^{ab}\rangle = \langle ij|\hat{g}|ab\rangle - \langle ij|\hat{g}|ba\rangle$ between Φ and Φ_{ij}^{ab}, the nondiagonal elements $\langle\phi_{ij}^{ab}|\hat{H}|\phi_{kl}^{cd}\rangle$ between the different Φ_{ij}^{ab}, and the diagonal elements $\langle\Phi_{ij}^{ab}|H|\Phi_{ij}^{ab}\rangle$. If one now argues that the off-diagonal elements $\langle \phi_{ij}^{ab} | \hat{H} | \phi_{kl}^{cd} \rangle$ enter the results only indirectly (to "higher order") one can neglect them, which leads to the following approximation for the lowest eigenvalue of the CI-matrix:

$$
\begin{aligned}
E &= \langle \Phi | \hat{H} | \Phi \rangle + \sum_{i<j} \sum_{a<b} \frac{|\langle \Phi | \hat{H} | \Phi_{ij}^{ab} \rangle|^2}{\langle \Phi_{ij}^{ab} | \hat{H} | \Phi_{ij}^{ab} \rangle - \langle \Phi | \hat{H} | \Phi \rangle} \\
&= \langle \Phi | \hat{H} | \Phi \rangle + \sum_{i<j} \sum_{a<b} \frac{|\langle \Phi | \hat{V} | \Phi_{ij}^{ab} \rangle|^2}{e_a + e_b - e_i - e_j + \langle \Phi_{ij}^{ab} | \hat{V} | \Phi_{ij}^{ab} \rangle - \langle \Phi | \hat{V} | \Phi \rangle}
\end{aligned} \tag{78}
$$

One gets this same result in formal second-order RS perturbation theory if one chooses

$$
\hat{A} = \sum_{\nu} (\langle \Phi_{\nu} | \hat{V} | \Phi_{\nu} \rangle - \langle \Phi | \hat{V} | \Phi \rangle) | \Phi_{\nu} \rangle \langle \Phi_{\nu} | \tag{79}
$$

where the sum goes over all possible Slater determinants constructable from the occupied and virtual orbitals of $\hat{H}_0$. In order to achieve (78) one can also choose $\hat{A}$ in the form (73), with

$$
a_k = -\langle k | \hat{v} | k \rangle \tag{80}
$$

$$
b_{kl} = \langle kl | \hat{g} | kl \rangle - \langle kl | \hat{g} | lk \rangle
$$

The variant of RS perturbation theory with $\hat{A} = 0$ was first considered by Møller and Plessett,[53] whereas that with $\hat{A}$ given by (79) has been introduced by Epstein[54] and by Nesbet.[55] Claverie et al.[56] have therefore proposed to call it the Epstein–Nesbet (EN) type perturbation theory. We shall refer to any perturbation calculation based on (72) and (73) as generalized Epstein–Nesbet (GEN) perturbation theory. Note that both the Møller–Plessett (MP) and the EN perturbation theory are variants of the RS perturbation expansion.*

3.3. Decoupling of the First-Order Pairs

The second-order energy $E^{(2)}$ of the GEN expansion (77) can be written as the sum of pair contributions $\varepsilon_{ij}^{(2)}$,

$$
E_2^{(2)} = \sum_{i<j} \varepsilon_{ij}^{(2)} \tag{81}
$$

For the special case of MP perturbation theory ones has in terms of the canonical spin-orbitals (i.e., eigenfunctions of the Hartree–Fock equation)

$$
\varepsilon_{ij}^{(2)} = \sum_{a<b} \frac{|\langle \Phi | \hat{V} | \Phi_{ij}^{ab} \rangle|^2}{e_a + e_b - e_i - e_j} \tag{82}
$$

*See also Notes Added in Proof, p. 188.

The first-order correction to the wave function is

$$\Psi^{(1)} = \sum_{i<j} \sum_{a<b} \frac{\langle \Phi_{ij}^{ab} | \hat{V} | \Phi \rangle}{e_a + e_b - e_i - e_j} \Phi_{ij}^{ab} = \sum_{i<j} \sum_{a<b} c_{ij}^{ab(1)} \Phi_{ij}^{ab} \tag{83}$$

$$c_{ij}^{ab(1)} = \frac{\langle \Phi_{ij}^{ab} | \hat{V} | \Phi \rangle}{e_a + e_b - e_i - e_j} \tag{84}$$

One defines the first-order pair correlation function and the corresponding pair creation operators as

$$u_{ij}^{(1)}(1, 2) = \sum_{a<b} c_{ij}^{ab(1)} \frac{1}{\sqrt{2}} [\psi_a(1)\psi_b(2) - \psi_b(1)\psi_a(2)] \tag{85}$$

$$\hat{b}_{ij}^{+(1)} = \sum_{a<b} c_{ij}^{ab(1)} \hat{a}_a^+ \hat{a}_b^+ \hat{a}_j \hat{a}_i \tag{86}$$

and expresses $\Psi^{(1)}$ in terms of the $u_{ij}^{(1)}$ or the $\hat{b}_{ij}^{+(1)}$ as

$$\Psi^{(1)} = \mathcal{A} \left[\sum_{i<j} \prod_{\substack{k \\ (\pm i,j)}} \psi_k(k) u_{ij}^{(1)}(i, j) \right]$$

$$= \sum_{i<j} \hat{b}_{ij}^{+(1)} \Phi \tag{87}$$

The decoupling of the first-order pairs has been pointed out by Sinanoğlu.[57] Instead of calculating the $u_{ij}^{(1)}$ via their expansion coefficients (84), one can also obtain them in a nonexpanded form by inserting (87) into the Hylleraas variation principle for $\Psi^{(1)}$

$$F(\Psi^{(1)}) = \langle \Psi^{(1)} | \hat{H}_0 - E_0 | \Psi^{(1)} \rangle + 2 \, \mathrm{Re} \langle \Phi | \hat{V} - E_1 | \Psi^{(1)} \rangle$$

$$\delta F(\Psi^{(1)}) = 0 \tag{88}$$

One gets

$$F(\Psi^{(1)}) = \sum_{i<j=1}^{n} f(u_{ij}^{(1)}) \tag{89}$$

$$f(u_{ij}^{(1)}) = \langle u_{ij}^{(1)} | \hat{F}(1) + \hat{F}(2) - e_i - e_j | u_{ij}^{(1)} \rangle$$

$$+ 2 \, \mathrm{Re} \left\langle \frac{1}{\sqrt{2}} [\psi_i(1)\psi_j(2) - \psi_j(1)\psi_i(2)] \Big| \frac{1}{r_{12}} \Big| u_{ij}^{(1)} \right\rangle \tag{90}$$

$[\hat{F}(i)$ is the Hartree–Fock operator, see the line after Eq. (76).] The first-order pair correction functions are those that minimize the functionals (90). When one inserts expansion (85) into (90) and minimizes the $f(u_{ij})$ with respect to the coefficients $c_{ij}^{ab(1)}$ one gets, of course (82–84).

Since the use of pair-natural orbital (PNO) expansions is of great practical importance in pair theories and since the basic idea is most easily understood in the context of the first-order pairs, we now illustrate the PNO expansion method, although it has not been used directly in the present context.

Without loss of generality[8,58-64] we can write any pair function $u_{ij}^{(1)}$ in its natural (or canonical) form in which two natural spin-orbitals $\chi_k^{(ij)}$ and $\chi_k^{\prime(ij)}$, that have the same occupation number, are coupled to the kth natural configuration (the PNOs altogether are an orthonormal set):

$$u_{ij}^{(1)} = \frac{1}{\sqrt{2}} \sum_k c_k^{(ij)} [\chi_k^{(ij)}(1)\chi_k^{\prime(ij)}(2) - \chi_k^{\prime(ij)}(1)\chi_k^{(ij)}(2)] \tag{91}$$

We now minimize $f(u_{ij})$ with respect to both the $\chi_k^{\prime(ij)}$, $\chi_k^{(ij)}$ and the coefficients $c_k^{(ij)}$.

The functional $f(u_{ij})$ is simply the sum of contributions f_k that refer to the kth term in the natural expansion (91); hence one obtains the minimum of f if each f_k has its minimum individually, i.e., if

$$c_k^{(ij)}\hat{F}\chi_k^{(ij)} + (\hat{K}^{ij} - \hat{K}^{ji})\chi_k^{\prime(ij)} = \lambda^{(ij)}\chi^{(ij)}$$
$$c_k^{(ij)}\hat{F}\chi_k^{\prime(ij)} + (\hat{K}^{ji} - \hat{K}^{ij})\chi_k^{(ij)} = \mu^{(ij)}\chi^{\prime(ij)} \tag{92}$$

$$c_k^{(ij)} = \frac{\langle\psi_i\psi_j|\hat{g}|\chi_k^{(ij)}\chi_k^{\prime(ij)}\rangle - \langle\psi_i\psi_j|\hat{g}|\chi_k^{\prime(ij)}\chi_k^{(ij)}\rangle}{\langle\chi_k^{(ij)}|\hat{F}|\chi_k^{(ij)}\rangle + \langle\chi_k^{\prime(ij)}|\hat{F}|\chi_k^{\prime(ij)}\rangle - e_i - e_j} \tag{93}$$

with

$$K^{ij}(1)\psi(1) = \int \psi_i(1)\psi_j(2)\frac{1}{r_{12}}\psi(2)\,d\tau_2 \tag{94}$$

The right-hand sides of (92) are not completely correct because in their derivation we have ignored the fact that the $\chi_k^{(ij)}$ for different k must be orthogonal, and that this has to be taken care of as a constraint of the variation. This would lead to sums like

$$\sum_l \lambda_l^{(ij)}\chi_l^{(ij)} + \sum_l \lambda_l^{\prime(ij)}\chi_l^{\prime(ij)} \tag{95}$$

rather than a single term on the right-hand sides of (92) and thus to a coupling of the different PNOs of one pair.

If we bear in mind that the natural expansion usually converges rapidly (at least as far as the first few terms are concerned), we can take care of the orthogonality of the $\chi_k^{(ij)}$ if we calculate $\chi_1^{(ij)}$ free from constraints, $\chi_2^{(ij)}$ in the subspace orthogonal to $\chi_1^{(ij)}$, etc.

Equations (91)–(95) are formulated in terms of natural *spin*-orbitals. For closed-shell pairs, where $\psi_i = \phi_R\alpha$ and $\psi_j = \phi_R\beta$, $\chi_k^{(ij)} = \phi_k^{(R)}\alpha$, $\chi_k^{\prime(ij)} = \phi_k^{(R)}\beta$, Eqs. (92) and (93) can be cast in the simpler form

$$c_k^{(R)}\hat{F}\phi_k^{(R)} + \hat{K}^i\phi_k^{(R)} = \lambda_k^i\phi_k^i \tag{96}$$

$$c_k^{(R)} = \frac{\langle \phi_R \phi_R | \hat{g} | \phi_k^{(R)} \phi_k^{(R)} \rangle}{2(\langle \phi_k^{(R)} | \hat{F} | \phi_k^{(R)} \rangle - e_R)} \tag{97}$$

In Eqs. (96) and (97) [similarly, in the general case (92) and (93)] the only coupling is between $c_k^{(R)}$ and $\phi_k^{(R)}$. One can first solve (96), with some guess of $c_k^{(R)}$ (usually $c_k^{(R)} = 0$ is a good first guess) for $\phi_k^{(R)}$, then calculate a new $c_k^{(R)}$ by (97) and continue until self-consistency, which is usually reached after 3–4 iterations.

This is the basis of the one-PNO-at-a-time method[63,65,66] for the approximate calculation of the PNOs. Methods of their simultaneous[67,97] calculation have turned out to be somewhat more powerful. PNO expansion and the problem of the direct calculation of PNOs are discussed in detail by W. Meyer.[106] An interesting reduction of the equations for the first-order pairs has recently been discussed by Langhoff.[68]

3.4. Invariance of the Second-Order Energy with Respect to Unitary Transformations, Canonical and Localized Pairs, and Reducible and Irreducible Pairs

As is well known a Slater determinant Φ is invariant with respect to unitary transformations among the occupied spin-orbitals ψ_i. All orders of perturbation theory are invariant in the same sense, provided that $\hat{H}_0$ is invariant. This is the case for the Møller–Plesset type perturbation theory, with $\hat{H}_0$ given by (70), but not for the original or the generalized Epstein–Nesbet perturbation theory, where $\hat{H}_0$ contains in addition the operator $\hat{A}$ defined by (73). In MP, (90) holds for any occupied orthonormal ψ_i.

While in the MP series the sum $E^{(2)} = \sum_{i<j=1}^n \varepsilon_{ij}^{(2)}$ is invariant with respect to unitary transformations among the ψ_i, the individual terms are not invariant and they may turn out to be quite different for, e.g., canonical and localized orbitals. Since the sum is invariant, the choice of a particular unitary transformation is only relevant if one attaches an importance to the individual ε_{ij}; one may, e.g., want to maximize the value of the sum of the intraorbital contributions $\varepsilon_i = \varepsilon_{i\alpha,i\beta}$ or one may want transferability between related molecules.

The second-order energy $E^{(2)}$ of the Møller–Plesset perturbation theory is also invariant with respect to unitary transformations of the pair functions occupied in Φ. If we define a zeroth-order pair function,

$$[i, j] = \frac{1}{\sqrt{2}} [\psi_i(i)\psi_j(j) - \psi_j(i)\psi_i(j)] \tag{98}$$

and the corresponding zeroth-order $(n-2)$ particle function,

$$\langle i, j \rangle = \mathscr{A} \left[\prod_{\substack{k=1 \\ (\neq i,j)}}^n \psi_k(k) \right] \tag{99}$$

(the antisymmetrizer $\mathscr{A}$ being chosen such that it takes care of the normalization as well) then the Schmidt theorem[58] (or Carlson–Keller theorem)[60] tells us that

$$\Phi = \binom{n}{2}^{-1/2} \sum_{i<j=1}^{n} \langle i, j \rangle [i, j] \tag{100}$$

and that the $[i, j]$ are eigenfunctions of the 2-particle density matrix Γ_2 corresponding to Φ. Since the eigenvalue 1 of Γ_2 is $\binom{n}{2}$fold degenerate, one can, instead of the $[i, j]$, choose any linear combinations of them as natural spin geminals

$$\omega_\mu^0 = \sum_{i<j=1}^{n} c_{ij}^\mu [i, j], \qquad \mu = 1, 2, \ldots, \binom{n}{2} \tag{101}$$

and combine them with the appropriate eigenfunctions Ω_μ of Γ_{n-2} to yield

$$\Phi = \binom{n}{2}^{-1/2} \sum_{\mu=1}^{\binom{n}{2}} \omega_\mu^0 (1, 2) \, \Omega_\mu (3, 4, \ldots, n) \tag{102}$$

This expansion is antisymmetric as is (100), though the individual terms in the sum are not. One can easily show that $E^{(2)}$ can be expressed in terms of the ω_μ^0 and the appropriate pair correlation functions $u_\mu^{(1)}$

$$E^{(2)} = \sum_\mu \varepsilon_\mu^{(2)} \tag{103}$$

$$\varepsilon_\mu = \min f(u_\mu^{(1)}) \tag{104}$$

$$f(u_\mu^{(1)}) = \langle u_\mu^{(1)} | \hat{F}(1) + \hat{F}(2) - e_\mu | u_\mu^{(1)} \rangle$$
$$+ 2 \, \mathrm{Re} \left\langle \omega_\mu^0 \left| \frac{1}{r_{12}} \right| u_\mu^{(1)} \right\rangle \tag{105}$$

where

$$e_\mu = \langle \omega_\mu^0 | \hat{F}(1) + \hat{F}(2) | \omega_\mu^0 \rangle \tag{106}$$

Usually the simple antisymmetrized products of orbitals $[i, j]$ span only reducible representations of the invariance group of the Hamiltonian. One can construct such linear combinations ω_μ^0 which transform as irreducible representations of the symmetry group and/or are eigenfunction of $\hat{S}^2$. According to Sinanoğlu[5] we call the former reducible and the latter irreducible pairs. In particular, pairs that are eigenfunctions of $\hat{S}^2$ are called spin adapted. The formalism for these symmetry adapted pairs has been discussed in detail by McWeeny and Steiner.[4,69]

The second-order energy $E^{(2)}$ is invariant with respect to a change from reducible to irreducible pairs only for the Møller–Plesset version of RS-perturbation theory, not, e.g., for the Epstein–Nesbet version.

3.5. Higher Orders of Perturbation Theory

Perturbation theory gives the *exact* answer if one is interested in the first (or a higher) derivative of the energy with respect to a natural perturbation parameter. Perturbation theory gives *good* results if the perturbation is "small" and, therefore, the perturbation series converges rapidly, so that only the first or the first- and second-order contributions to the energy need to be considered. Electron interaction is far from being a small perturbation; it is rather uncertain whether the perturbation series converges at all and there is no *a priori* recipe prescribing the order at which one should stop to get a reliable answer. As long as one limits oneself to the first two orders in the energy, perturbation theoretical calculations are relatively inexpensive, but for higher orders the labor increases tremendously. Perturbation theoretical calculations of electron correlation that include the third order in a rigorous way are very rare and one understands why. While for $E^{(2)}$ only the knowledge of the matrix elements

$$\langle \Phi | \hat{V} | \Phi_{ij}^{ab} \rangle$$

of the interaction potential between the unperturbed Φ and the doubly substituted configurations Φ_{ij}^{ab} is necessary, a third-order treatment requires the knowledge of the elements

$$\langle \Phi_{ij}^{ab} | \hat{V} | \Phi_{kl}^{cd} \rangle$$

as well. However, the same matrix elements are required for a CI calculation in the basis of the Φ and Φ_{ij}^{ab}, which means that, since the diagonalization of the matrix is usually much less expensive than the construction of the matrix elements, a CI calculation or a CEPA calculation that furnishes results much closer to the exact ones (see Section 5.3) needs about the same labor as third-order perturbation theory (for the energy). There are cases where third-order perturbation theory may be preferred nevertheless, e.g., for the calculation of transition or ionization energies, where many perturbation contributions to the two states cancel and need not be evaluated.[70]

Perturbation theory also has some advantages in atomic calculations, where one can work with a continuous rather than discrete basis, and where the integrals over the continuum can be evaluated in a closed form so that problems of basis unsaturation, typical for CI calculations, are circumvented.[71,72] Finally one must admit that for extended systems, e.g., for the theory of the electron gas in a metal, perturbation theory is often the only choice.

Almost all attempts to use perturbation theory beyond second order for the calculation of electron correlation effects are based on a selection of energy contributions for which a bookkeeping in terms of diagrams[73] together with some pictorial "physical" arguments and the use of the linked-cluster theorem [41-43,73] is essential.* Some classes of diagrams are even summed to infinite order, mainly if such a summation can be done in closed form. Obviously in selecting certain diagrams and summing others to infinite order, one completely abandons the basic idea of perturbation theory, namely the expansion in terms of powers of the perturbation. For example, in the theory of an electron gas at high density the so-called ring diagrams turned out to be most important [74] and had to be summed up, whereas in the theory of nuclear matter the so-called ladder diagrams had to be summed to infinite order in order to account for the bulk of correlation effects. [75-78] The need for considering these ladder diagrams is closely connected with the short-range strong repulsion between nucleons, which leads to divergence unless one sums up the ladder diagrams. There are also correlation effects that cannot be accounted for in the framework of perturbation theory, like those related to super-conductivity. [79,80]

It can be shown that the summation of the ladder diagrams to infinite order is equivalent to a variational principle for independent pairs in the effective field of the other particles, which in nuclear theory is called Bethe–Goldstone theory. [75-77] The simplest version of the Bethe–Goldstone theory, the independent pair model, [78] has, for electronic systems, its analogy in the independent electron-pair approximation (IEPA). We will not attempt the perturbative derivation of this essentially nonperturbative method but come back to it later.

The most promising applications of perturbation theory to atomic and small molecular systems are those by Kelly. [71,72] Kelly was able to sum certain two-particle diagrams, so called EPV (Exclusion Principle Violating), to infinite order since they formed a geometrical series. The result of this summation turned out to be equivalent to a shift in the denominator, and identical with the second-order result of Epstein–Nesbet perturbation theory.

Many-body perturbation calculations to third order and, with a Kelly-type partial summation of higher orders, in a finite orbital basis were recently performed by Robb [81] and Silver *et al.* [82] for diatomic hydrides.

A classification of the existing pair theories in the language of many-body perturbation theories has been given by Robb. [83] A review of application of many-body perturbation theory to atoms and molecules has been written by Freed. [84]

In this context one has to mention the PCILO method (*p*erturbative *c*onfiguration *i*nteraction *b*ased on *l*ocalized *o*rbitals) by Malrieu *et al.* [85-89] However, we cannot go into details here.

One remark on perturbation theoretical calculations of correlation energy is necessary and concerns the comparison of the results with those of nonper-turbative calculations. When one speaks about two-particle or three-particle

---

*See the footnote on p. 136.

contributions one does not usually mean the same thing in the two kinds of approaches.

In perturbation theory a contribution

$$\frac{\langle\Phi|V|\Phi_{ij}^{ab}\rangle\langle\Phi_{ij}^{ab}|V|\Phi_{ik}^{ac}\rangle\langle\Phi_{ik}^{ac}|V|\Phi\rangle}{(E-E_{ij}^{ab})(E-E_{ik}^{ac})} \tag{107}$$

would be referred to as a three-particle (or rather three-hole-line) contribution since three indices i, j, k of orbitals occupied in Φ are involved. The variational counterpart of (107),

$$c_{ij}^{ab}\langle\Phi_{ij}^{ab}|V|\Phi_{ik}^{ac}\rangle c_{ik}^{ac} \tag{108}$$

is, however, regarded as a two-particle contribution, specifically a coupling between the two pairs ij and ik. A three-particle contribution in a variational calculation has to involve $c_{ijk}^{abc}\,\Phi_{ijk}^{abc}$ directly. When three-particle contributions are included in a perturbation treatment, this usually means that the coupling between the pairs (like in CEPA, see Section 4.5) is taken care of, whereas in a nonperturbational calculation it means that three-particle clusters are considered, which is a refinement of much higher order (see Section 5.5).

One can define a perturbation scheme[127] which to second-order gives results equivalent to third order and to infinite summation of certain higher-order terms Because of its relation to CEPA-type theories we will come back to it in Section 5.2.

4. The Independent Electron-Pair Approximation (IEPA), the Coupled Electron-Pair Approximation (CEPA), and Related Approximations

4.1. Preliminary Remarks

The independent electron-pair approximation (IEPA) consists of approximating the total correlation energy E^{corr} as a sum of pair contributions ε_μ^{IEPA}, which are calculated independently, from effective two-electron equations. This approximation was suggested by *a priori* theoretical arguments long before it was tested in practical applications. We now know that IEPA is not a very accurate approximation. When it is used together with a nearly saturated basis set it yields, for different systems, between 80% and 120% of the exact correlation energy, and more unfortunate cases may still be found.[90-101] Nevertheless, IEPA reduces the energy error of the Hartree–Fock approximation by an order of magnitude and this reduction of the error may often be decisive for the answer to certain questions. If applied with care IEPA can be very useful in spite of its apparent shortcomings.

There are mainly two facts that lead one to expect that IEPA should work.

a. For separated pairs with strong orthogonality (APSG wave function, see Section 2.1) the total correlation energy is in fact, to a high degree of accuracy, the sum of pair contributions calculated independently; even if strong orthogonality is relaxed this holds rather well.

b. The second-order energy in RS perturbation theory (actually in the MP as well as the EN variant) is exactly the sum of pair contributions calculated individually.

One can conclude from these two facts that IEPA should be good for systems where the intraorbital contributions are responsible for the bulk of correlation energy, and for systems where second-order perturbation theory recovers the main part of the correlation energy.

We shall first "derive" IEPA in a pedestrian way based on a truncated hierarchy of cluster equations, before we discuss how one can go beyond it. Our "derivation" is equivalent to that given by Sinanoğlu[5] in terms of variation of part of the energy functional and, e.g., to that by Freed[102] or Robb[83] starting from perturbation theory.

4.2. Pedestrian Derivation of the Independent Electron-Pair Approximation (IEPA)

We assume that single and triple substitutions with respect to the leading determinant can be neglected and that the quadruple substitutions are exactly unlinked clusters of double substitutions. We can then start from the truncated system of cluster equations (47)–(49), that is essentially due to Brenig.[23]

Let us first assume tentatively that Ψ is an APSG function, i.e., that only the intraorbital double substitutions $c_{R\bar{R}}^{ab}\Phi_{R\bar{R}}^{ab}$ contribute and that the ϕ_a, ϕ_b which occur in $\Phi_{R\bar{R}}^{ab}$ are orthogonal to those that contribute to $\Phi_{S\bar{S}}^{ab}$, and further that the coefficients of quadruple substitutions are given by

$$c_{R\bar{R}S\bar{S}}^{abcd}c_0^{-1} = d_{R\bar{R}}^{ab}d_{S\bar{S}}^{cd} \tag{109}$$

Then (49) reduces to

$$\langle\Phi_{R\bar{R}}^{ab}|\hat{H}|\Phi\rangle + \sum_{c<d}\langle\Phi_{R\bar{R}}^{ab}|\hat{H}|\Phi_{R\bar{R}}^{cd}\rangle d_{R\bar{R}}^{cd}$$
$$+ \sum_{\substack{S\\(\neq R)}}\sum_{c<d}\langle\Phi_{R\bar{R}}^{ab}|\hat{H}|\Phi_{R\bar{R}S\bar{S}}^{abcd}\rangle d_{R\bar{R}}^{ab}d_{S\bar{S}}^{cd} = Ed_{R\bar{R}}^{ab} \tag{110}$$

or with the explicit expression for E, as in (50), and taking care of the cancellation between $\langle\Phi_{R\bar{R}}^{ab}|\hat{H}|\Phi_{R\bar{R}S\bar{S}}^{abcd}\rangle c_{R\bar{R}}^{ab}c_{S\bar{S}}^{ab}$ on the l.h.s. and $\langle\Phi|\hat{H}|\Phi_{S\bar{S}}^{cd}\rangle c_{R\bar{R}}^{ab}c_{S\bar{S}}^{cd}$ on the r.h.s. for $R \neq S$,

$$\langle\Phi_{R\bar{R}}^{ab}|\hat{H}|\Phi\rangle + \sum_{c<d}\langle\Phi_{R\bar{R}}^{ab}|\hat{H}|\Phi_{R\bar{R}}^{cd}\rangle d_{R\bar{R}}^{cd} = (\langle\Phi|\hat{H}|\Phi\rangle + \sum_{c<d}\langle\Phi|\hat{H}|\Phi_{R\bar{R}}^{cd}\rangle d_{R\bar{R}}^{cd}) d_{R\bar{R}}^{ab}$$
$$= E_R d_{R\bar{R}}^{ab} \tag{111}$$

where we have defined

$$E_R = E - \sum_{\substack{S \\ (\neq R)}} \varepsilon_S = \langle \Phi | \hat{H} | \Phi \rangle + \varepsilon_R \tag{112}$$

We see that (111) is identical with the CI equation corresponding to the wave function

$$\Psi_R = c_0 \left(\Phi + \sum_{a<b} d_{R\bar{R}}^{ab} \Phi_{R\bar{R}}^{ab} \right) = c_0 \Phi + \sum_{a<b} c_{R\bar{R}}^{ab} \Phi_{R\bar{R}}^{ab} \tag{113}$$

which represents the correlation within the pair $R\bar{R}$. The equations for the different pairs are in fact, fully decoupled; each pair can be treated independently.

Remember that in Section 2.1 we have quoted that from a variational treatment of the APSG function this very result is obtained to a "high degree of accuracy," whereas from the present derivation it seems to be exact. The explanation of this discrepancy is that conclusions based on a truncated hierarchy of cluster equations can never be exact, except if the exact eigenfunction of the Hamiltonian is of the form of the chosen truncated cluster expansion.

Now we let ourselves be guided by the result for the APSG function in order to derive decoupled-pair equations for a general pair cluster wave function. We, therefore, neglect the pair coupling elements $\langle \Phi_{ij}^{ab} | \hat{H} | \Phi_{kl}^{cd} \rangle$ for $(i, j) \neq (k, l)$, $(a, b) \neq (c, d)$ (which vanish exactly for an APSG wave function) and we assume that like in the APSG case, the main effect of the fourfold substitutions is to cancel the correlation contributions of the other pairs in E, and that one can ignore the fourfold substitutions if one replaces E in analogy with (112) by

$$E_{ij} = \langle \Phi | \hat{H} | \Phi \rangle + \varepsilon_{ij} = E - \sum_{\substack{k<l \\ (\neq i,j)}} \varepsilon_{kl} \tag{114}$$

This leads to

$$\langle \Phi_{ij}^{ab} | \hat{H} | \Phi \rangle + \sum_{c<d} \langle \Phi_{ij}^{ab} | \hat{H} | \Phi_{ij}^{cd} \rangle d_{ij}^{cd} = E_{ij} d_{ij}^{ab} \tag{115}$$

(For a more careful discussion of the cancellation see Sections 2.6 and 5.3.)
One would have obtained (115) directly if one had started from the ansatz

$$\Psi_{ij} = c_0 \Phi + \sum_{a<b} c_{ij}^{ab} \Phi_{ij}^{ab} \tag{116}$$

for the pair ij. Eq. (115) can be interpreted as the condition for the energy expectation value $\langle \Psi_{ij} | \hat{H} | \Psi_{ij} \rangle$ to be stationary and to have the value E_{ij}. One can also formulate this stationarity condition in terms of the pair correlation function

$$u_{ij}(1, 2) = \sum_{a<b} d_{ij}^{ab}[\psi_a(1)\psi_b(2) - \psi_b(1)\psi_a(2)] \tag{117}$$

by requiring that the functional

$$\varepsilon_{ij} \leq f_{ij}(u_{ij}) = \left\{ 2 \, \text{Re} \left\langle \frac{1}{\sqrt{2}}[\psi_i(1)\psi_j(2) - \psi_j(1)\psi_i(2)] \left| \frac{1}{r_{12}} \right| u_{ij}(1, 2) \right\rangle \right.$$
$$\left. + \langle u_{ij}(1, 2)|\hat{H}_{ij}(1, 2)|u_{ij}(1, 2)\rangle \right\}[1 + \|u_{ij}\|^2]^{-1} \tag{118}$$

have its minimum, where the effective two-electron Hamiltonian $\hat{H}_{ij}$ is given by

$$\hat{H}_{ij}(1, 2) = \hat{h}_{ij}(1) + \hat{h}_{ij}(2) + \frac{1}{r_{12}} - F_{ii} - F_{jj} + \left\langle ij \left| \frac{1}{r_{12}} \right| ij \right\rangle - \left\langle ij \left| \frac{1}{r_{12}} \right| ji \right\rangle \tag{119}$$

with

$$\hat{h}_{ij}(1) = \hat{h}(1) + \sum_{\substack{k \\ (\neq i,j)}} [\hat{J}^k(1) - \hat{K}^k(1)] \tag{120}$$

$$F_{ii} = \langle \psi_i | h | \psi_i \rangle + \sum_k \langle \psi_i | \hat{J}^k - \hat{K}^k | \psi_i \rangle \tag{121}$$

All quantities in (117)–(121) are expressed in terms of spin-orbitals.

The approximation based on (115) or equivalently (116) or (117)–(118) has been called many-electron theory (MET) or exact-pair theory by Sinanoğlu,[5] and atomic or molecular Bethe–Goldstone theory by Nesbet.[28] We prefer to use the name independent electron-pair approximation (IEPA) which is likely to give rise to less misunderstanding and controversies than the two other names. If one gives credit to Nesbet and Sinanoğlu for their basic work on electron-pair theories one should not forget another pioneer in this field, namely Szasz.[103]

4.3. The Coupled Electron-Pair Approximation (CEPA)

Since the two approximations on which IEPA is based, (a) neglect of the pair coupling elements and (b) approximate treatment of the cancellation of the nonlinear terms with part of the energy, are independent of each other one can decide to make only one of these approximations. Both numerical results and a theoretical argument that we give in Sections 5.2 and 5.3 suggest that neglect of the pair coupling elements is the more severe approximation. If we make only approximation (b) we are led to the coupled electron-pair approximation (CEPA) (version CEPA-2) proposed by W. Meyer.[97]

The equations for the d_{ij}^{ab} in CEPA differ from the corresponding IEPA equations (115) by the presence of the coupling terms. One now obtains

$$\langle\Phi_{ij}^{ab}|\hat{H}|\Phi\rangle + \sum_{k<l}\sum_{c<d}\langle\Phi_{ij}^{ab}|\hat{H}|\Phi_{kl}^{cd}\rangle d_{kl}^{cd} = E_{ij}d_{ij}^{ab} \tag{122}$$

The CEPA scheme is somewhat intermediate between the IEPA scheme and the coupled-pair many-electron theory (CP–MET) of Čížek (which is close to the pair version of the Coester–Kümmel theory). CEPA takes care of the pair interaction terms ignored in IEPA, but it neglects most of the nonlinear terms that are present in CP–MET. We shall discuss different variants of CEPA in Section 5.3.

4.4. The Spin-Adapted Independent Electron-Pair Approximation

So far we have discussed pair substitutions on the spin-orbital level. We shall now limit our attention to closed-shell states and discuss these in terms of the doubly occupied orbitals ϕ_R $(R = 1, 2, \ldots, n/2)$. In the IEPA discussed, one has to deal with three types of pairs that correspond to substitutions from $\phi_R\alpha\phi_R\beta$, $\phi_R\alpha\phi_S\alpha$, $\phi_R\beta\phi_S\beta$. If one calculates the pair correlation functions u_{ij} for $\phi_R\alpha\phi_S\alpha$ and $\phi_R\alpha\phi_S\beta$ independently, and uses these u_{ij} to construct Ψ either in the form (25) or (30), one will generally not get a Ψ that is a pure singlet function. The requirement that Ψ be a singlet function imposes some relations between the d_{ij}^{ab} which are not satisfied automatically if one treats these pairs independently, except in second-order Møller–Plessett-type perturbation theory (see Section 3.4).

As we have seen, the simple (or "naive") IEPA can be regarded as the recipe to perform $n(n-1)/2$ different variational calculations with the ansatz (116) and to sum up the correlation energies so obtained for all pairs. In the spin-adapted IEPA we replace the simple spin-orbital pairs by spin-adapted pairs; this can be achieved by replacing the wave functions (116) by the following:

$$\Psi_{RR} = N_{RR}\left\{\Phi + \sum_A b_{RR}^{AA}\Phi_{RR}^{A\bar{A}} + \frac{1}{\sqrt{2}}\sum_{A<B} b_{RR}^{AB}(\Phi_{RR}^{A\bar{B}} - \Phi_{RR}^{\bar{A}B})\right\} \tag{123}$$

$$^s\Psi_{RS} = {}^sN_{RS}\left\{\Phi - \frac{1}{\sqrt{2}}\sum_A b_{RS}^{AA}(\Phi_{R\bar{S}}^{A\bar{A}} - \Phi_{RS}^{A\bar{A}})\right.$$

$$\left. - \frac{1}{2}\sum_{A<B} b_{RS}^{AB}(\Phi_{R\bar{S}}^{A\bar{B}} + \Phi_{RS}^{\bar{A}B} - \Phi_{RS}^{\bar{A}B} - \Phi_{RS}^{A\bar{B}})\right\} \tag{124}$$

$$^t\Psi_{RS} = {}^tN_{RS}\left\{\Phi - \frac{1}{2\sqrt{3}}\sum_{A<B} d_{RS}^{AB}(\Phi_{R\bar{S}}^{A\bar{B}} + \Phi_{RS}^{\bar{A}B} + \Phi_{RS}^{\bar{A}B} + \Phi_{RS}^{A\bar{B}} + 2\Phi_{RS}^{AB} + 2\Phi_{RS}^{\bar{A}\bar{B}})\right\} \tag{125}$$

(the N being normalization factors). Here $\Phi_{R\bar{S}}^{A\bar{B}}$ means the Slater determinant obtained from Φ on replacing $\psi_R = \phi_R\alpha$ and $\bar{\psi}_S = \phi_S\beta$ by $\bar{\psi}_A = \phi_A\alpha$ and $\bar{\psi}_B = \phi_B\beta$, where the ϕ_A are spatial orbitals that complement the ϕ_R to a (complete) orthonormal set.

Any one of the functions defined through Eqs. (123)–(125) is a pure singlet, the superscripts s and t refer to the fact that in $^s\Psi_{RS}$ the orbitals ϕ_R and ϕ_S are coupled to a singlet, in $^t\Psi_{RS}$ to a triplet.

The spin-adapted pair correlation energies and total energy for closed-shell states are

$$\varepsilon_{RR} = E_{RR} - \langle\Phi|\hat{H}|\Phi\rangle, \qquad {}^s\varepsilon_{RS} = {}^sE_{RS} - \langle\Phi|\hat{H}|\Phi\rangle, \qquad {}^t\varepsilon_{RS} = {}^tE_{RS} - \langle\Phi|\hat{H}|\Phi\rangle \tag{126}$$

$$E = \langle\Phi|H|\Phi\rangle + \sum_R \varepsilon_{RR} + \sum_{R<S} ({}^s\varepsilon_{RS} + {}^t\varepsilon_{RS}) \tag{127}$$

where E_{RR}, $^sE_{RS}$, $^tE_{RS}$ are the energies corresponding to the wave functions (123)–(125).

The condition for the energy to be stationary with respect to variation of the functions (123)–(125) can, in analogy with (118) and (119), also be formulated as a condition for some functionals f_{RR}, $^sf_{RS}$, $^tf_{RS}$ to be stationary with respect to the variation of the spin-free two-electron correlation functions

$$u_{RR} = \sum_{A\le B} b_{RR}^{AB}(AB)_+, \qquad {}^su_{RS} = \sum_{A\le B} b_{RS}^{AB}(AB)_+, \qquad {}^tu_{RS} = \sum_{A<B} d_{RS}^{AB}(AB)_- \tag{128}$$

where the shorthand notation $(PQ)_\pm$ stands for

$$(PP)_+ = \phi_P(1)\phi_P(2)$$
$$(PQ)_\pm = \frac{1}{\sqrt{2}}[\phi_P(1)\phi_Q(2) \pm \phi_Q(1)\phi_P(2)] \tag{129}$$

To formulate the respective functionals (131) it is convenient to introduce the abbreviations

$$\hat{F} = \hat{h} + \sum_S 2\hat{J}^S - \hat{K}^S$$

$$\hat{H}^R(1) = \hat{F}(1) - 2\hat{J}^R(1) + \hat{K}^R(1) \tag{130}$$

$$\hat{H}_\pm^{RS}(1) = \hat{F}(1) - \hat{J}^R - \hat{J}^S \pm \hat{K}^R \pm \hat{K}^S + \tfrac{1}{2}(\hat{K}^R + \hat{K}^S)$$

$$\varepsilon_{RR} \le f_{RR}(u_{RR}) = \{\langle u_{RR}|\hat{H}^{RR}(1,2)|u_{RR}\rangle + 2\,\mathrm{Re}\langle(RR)_+|\hat{g}|u_{RR}\rangle\}N_{RR}^2$$

$${}^s\varepsilon_{RS} \le {}^sf_{RS}({}^su_{RS}) = \{\langle {}^su_{RS}|\hat{H}_-^{RS}(1,2)|{}^su_{RS}\rangle + 2\,\mathrm{Re}\langle(RS)_+|\hat{g}|{}^su_{RS}\rangle\}{}^sN_{RS}^2 \tag{131}$$

$${}^t\varepsilon_{RS} \le 3\,{}^tf_{RS}({}^tu_{RS}) = \{3\langle {}^tu_{RS}|\hat{H}_+^{RS}(1,2)|{}^tu_{RS}\rangle + 6\,\mathrm{Re}\langle(RS)_-|\hat{g}|{}^tu_{RS}\rangle\}{}^tN_{RS}^2$$

where the effective two-electron operators are defined as

$$\hat{H}^{RR}(1,2) = \hat{H}^{R}(1) + \hat{H}^{R}(2) + \hat{g}(1,2) - 2\langle\phi_R|\hat{F}|\phi_R\rangle + (RR|RR)$$

$$\hat{H}^{RS}_{\pm}(1,2) = \hat{H}^{RS}_{\pm}(1) + \hat{H}^{RS}_{\pm}(2) + \hat{g}(1,2) - \langle\phi_R|\hat{F}|\phi_R\rangle - \langle\phi_S|\hat{F}|\phi_S\rangle \quad (132)$$

$$+ (RR|SS) \mp (RS|SR)$$

In the spin-adapted CEPA one works with the same formalism for the pair functions and expresses the pair interaction terms between spin-adapted pairs. We give more details in connection with the CEPA–PNO method.

One can generalize the above formalism to symmetry-adapted pairs. This is probably useful in atomic calculations, where it has indeed occasionally been used,[93] but has been less useful for molecules.

4.5. The IEPA–PNO and the CEPA–PNO Methods

The IEPA and CEPA schemes in their spin-adapted versions become very powerful when they are combined with the method of the direct calculation of pair-natural orbitals (PNOs).[12,67,97-101] If one expands the pair correlation functions in terms of their PNOs one is led to very small dimensions of the secular equations without losing generality. The rather unpleasant problem of how to select the configurations to be included in a conventional CI is completely circumvented. One can work with a rather large orbital basis without increasing the number of important natural configurations in an unmanageable way. Furthermore, another difficulty of conventional CI, namely the full basis transformation of the two-electron integrals (which is at least a n^5-step), is avoided in the PNO method. The PNO expansion technique is described in detail in Ref. 106. We therefore concentrate here on the most relevant points.

Without loss of generality the spinfree pair correlation functions $^s u_{RR}$, $^s u_{RS}$, and $^t u_{RS}$ can be written in their natural forms

$$u_{RR} = \sum_{k>1} c_k^R \chi_k^R(1) \chi_k^R(2) \tag{133a}$$

$$^s u_{RS} = \sum_{k>2} c_k^{RS} \chi_k^{RS}(1) \chi_k^{RS}(2) \tag{133b}$$

$$^s u_{RS} = \frac{1}{\sqrt{2}} \sum_{k>1} b_k^{RS}[u_k^{RS}(1) v_k^{RS}(2) + v_k^{RS}(1) u_k^{RS}(2)] \tag{133c}$$

$$^t u_{RS} = \frac{1}{\sqrt{2}} \sum_{k>1} d_k^{RS}[u_k^{RS}(1) v_k^{RS}(2) - v_k^{RS}(1) u_k^{RS}(2)] \tag{133d}$$

For $^s u_{RS}$ two alternative forms are indicated: one for pair functions that are totally symmetric with respect to the point group symmetry, the other for a $^s u_{RS}$ that is antisymmetric with respect to at least one symmetry element. One has to use (133c) for an antisymmetric singlet pair only if one insists on u_k and v_k being symmetry adapted. Otherwise by the transformation

$$\chi_{2k-1}^{RS} = \frac{1}{\sqrt{2}}(u_k^{RS} + v_k^{RS})$$

$$\chi_{2k}^{RS} = \frac{1}{\sqrt{2}}(u_k^{RS} - v_k^{RS})$$

(134)

(133c) changes to the form (133b). In this transformed expansion the χ_i^{RS} are not symmetry adapted and χ_{2k-1}^{RS} and χ_{2k}^{RS} have, except for the sign, the same expansion coefficient,

$$c_{2k-1}^{RS} = c_{2k}^{RS} = -\frac{1}{\sqrt{2}} b_k^{RS}$$

(135)

It is convenient to make the identification

$$\phi_R \equiv \chi_1^R = \frac{1}{\sqrt{2}}(\chi_0^{RS} + \chi_1^{RS}) \equiv u_1^{RS}$$

$$\phi_S \equiv \frac{1}{\sqrt{2}}(\chi_0^{RS} - \chi_1^{RS}) \equiv v_1^R$$

(136)

i.e., to count the "strongly occupied" ϕ_R also as PNOs of the respective pairs. This allows us to write the correlated pair functions in exactly the same form as the pair correlation functions in (133) but starting the summation with $k = 1$. The orbitals with subscript 1 are those occupied in the leading determinant Φ.

Insertion of the expansions (133) into the functionals (131) leads immediately to variational equations for the PNOs and their expansion coefficients, in a way similar to that for genuine two-electron systems. For the calculation of the PNOs see Ref. 106.

It is convenient to count the pairs by a single subscript μ or ν and to designate the kth natural configuration of the νth pair as Φ_ν^k. Then the IEPA equations are

$$\langle \Phi_\mu^i | \hat{H} | \Phi \rangle + \sum_k \langle \Phi_\mu^i | \hat{H} | \Phi_\mu^k \rangle d_{\mu,k}^{IEPA} = (E_0 + \varepsilon_\mu^{IEPA}) d_{\mu,i}^{IEPA}$$

(137)

and the CEPA equations

$$\langle \Phi_\mu^i | \hat{H} | \Phi \rangle + \sum_\nu \sum_k \langle \Phi_\mu^i | \hat{H} | \Phi_\nu^k \rangle d_{\nu,k}^{CEPA} = (E_0 + \varepsilon_\mu^{CEPA}) d_{\mu,i}^{CEPA}$$

(138)

The explicit expressions for the coupling elements $\langle \Phi_\mu^i | \hat{H} | \Phi_\nu^k \rangle$ in terms of the PNOs can be found in Ref. 98. The system of equations (138) can be solved in an iterative way, such that only matrices of the dimension of the NO expansion of one pair have to be handled.[98] The PNOs of different pairs are nonorthogonal, but this does not lead to a complication of the matrix elements.

For an analysis of the CEPA results one defines the diagonal and off-diagonal contributions ε_μ and $\Delta\varepsilon_{\mu\nu}$ to the energy:

$$(1+\sum_i |d_{\mu,i}^{CEPA}|^2)\varepsilon_\mu^{CEPA} = \sum_i \langle\Phi_\mu^i|\hat{H}|\Phi\rangle d_{\mu,i}^{CEPA} + \sum_{i,k}\langle\Phi_\mu^i|\hat{H}|\Phi_\mu^k\rangle d_{\mu,i}^{CEPA}d_{\mu,k}^{CEPA} \qquad (139)$$

$$\frac{(1+\sum_i|d_{\mu,i}^{CEPA}|^2)(1+\sum_i|d_{\nu,i}^{CEPA}|^2)}{2+\sum_i|d_{\mu,i}^{CEPA}|^2+\sum_i|d_{\nu,i}^{CEPA}|^2}\Delta\varepsilon_{\mu\nu}^{CEPA} = \sum_{i,k}\langle\Phi_\mu^i|\hat{H}|\Phi_\nu^k\rangle d_{\mu,i}^{CEPA}d_{\nu,k}^{CEPA} \qquad (140)$$

In IEPA the $\Delta\varepsilon_{\mu\nu}$ are neglected. Their sum is hence a measure of the IEPA error

$$\Delta E_{IEPA} = E_{CEPA} - E_{IEPA} \qquad (141)$$

One has to note, however, that the ε_μ are not the same in IEPA as in CEPA, since the expansion coefficients $d_{\mu,i}^{CEPA}$ are readjusted to the interaction. A first approximation to CEPA, called CEPA (1) consists of evaluating the $\Delta\varepsilon_{\mu\nu}$ with the IEPA coefficients, i.e., with the $d_{\mu,i}^{CEPA}$ replaced by the $d_{\mu,i}^{CEPA}$. The energy $E_{CEPA(1)}$ defined in terms of the $d_{\mu,i}^{IEPA}$ is then lowered to E_{CEPA} by allowing a readjustment of the $d_{\mu,k}$. The readjustment energy[98]

$$\Delta E_{CEPA} = E_{CEPA} - E_{CEPA(1)} \qquad (142)$$

is also a quantity useful for an analysis of the correlation energy. Finally one has to look at the difference between the CEPA–PNO and the PNO–CI energies (see Section 4.6) since this is a measure of the importance of the unlinked clusters. It is also an indirect measure of how far the CEPA energy is from a variational bound.*

4.6. IEPA and CEPA for Open-Shell States

In open-shell states the situation is more complicated than for closed-shell states, one reason being that for many types of open-shell states a single Slater determinant is not an acceptable zeroth-order approximation and the cluster expansion of the exact wave function about a Slater determinant Φ is not a good description (see Section 5.5). Even for such open-shell states, which to first order are describable by a single determinant, new problems arise.

For a restricted Hartree–Fock function there are singly substituted configurations for which the Brillouin theorem does not hold, i.e., which have nonzero matrix elements of $\hat{H}$ with Φ, and hence contribute in first order to Ψ. One can avoid this by starting from an unrestricted Hartree–Fock function, but this has other shortcomings. Another complication arises because one has to classify the orbitals into three categories (rather than two, as in the closed-shell

*See Notes Added in Proof, p. 188.

case) namely, those (fully) occupied, those partially occupied, and those unoccupied in Φ. In the carbon ground configuration $1s^2 2s^2 2p^2$ the orbitals $1s$ and $2s$ are obviously fully (i.e., doubly) occupied, whereas $2p$ is only partially occupied. Double substitutions can be classified according to Silverstone and Sinanoğlu[104] as: (a) internal: "excitation" from occupied to partially occupied. The configuration $1s^2 2p^4$ corresponds, e.g., to internal substitution with respect to $1s^2 2s^2 2p^2$; (b) external: "excitation" from occupied or partially occupied to unoccupied orbitals, e.g., $1s^2 2s^2 3s^2$, (c) semiinternal: an example would be $1s^2 2p^3 3s$.

If one chooses symmetry-adapted pairs one has to deal with a large number of different types.[4] Again a combination with the PNO expansion technique is possible and has been worked out in detail.[97] Some of the pertinent formulas can be found in Ref. 105.

4.7. The PNO–CI Method

We were led from the CP-MET equations (47)–(49) to the CEPA scheme by using a simplified expression for the unlinked-cluster contributions to the quadruple substitutions. When one neglects the quadruple substitutions completely, one is led to a system of CI equations limited to double substitutions. When one now performs this CI in terms of the PNOs of the spin-adapted IEPA pairs, one has to solve the system of equations

$$\langle \Phi_\mu^i | \hat{H} | \Phi \rangle + \sum_\nu \sum_k \langle \Phi_\mu^i | \hat{H} | \Phi_\nu^k \rangle d_{\nu,k}^{CI} = E^{CI} d_{\mu,i}^{CI} \tag{143}$$

which has first been used in a treatment limited to intraorbital correlations by Ahlrichs and Kutzelnigg,[12] and on which Meyer's PNO–CI method[97] is based. Equation (143) differs from the CEPA–PNO equation (138) in that $E_0 + \varepsilon_\mu^{CEPA}$ is replaced by the total CI energy E^{CI}.

The advantage of PNO–CI with respect to CEPA–PNO is that it furnishes an upper bound to the energy; its drawback is that it does not have the correct dependence on the number of particles. Since both schemes are formally close it is easy to perform either one of them in the same calculation to get a "variational" and a "separable" energy.

We will not go into details here on the PNO–CI method (see Ref. 106). Let us just stress that the PNO–CI method is a straightforward CI method limited to double substitutions with respect to a leading determinant in a particularly compact and efficient form. An alternative to PNO–CI is the "brute-force" CI with double substitutions in terms of a given basis rather than with NOs. B. Roos[107,108] has programmed a rather powerful method of this kind in which the CI matrix is never explicitly constructed, but where the search for the lowest eigenvalue is, so to speak, done in one step with the calculation of the needed matrix elements. Implicitly in this method very large matrices occur, whereas PNO–CI furnishes results of comparable quality with very small secular equations. A procedure that follows similar lines has been proposed by C. Bender.[128] In the context of CI methods limited to double substitutions one can also mention the first-order wave function method used, e.g., by Schaefer *et al.*[109]

4.8. Nesbet's Hierarchy of *n*th-Order Bethe–Goldstone Equations

Nesbet[110] interprets the independent electron-pair approximation as a step towards the exact solution. If one starts from a single Slater determinant Φ, the first step consists of calculating the orbital correlation energies ε_i from the variational wave function

$$\Psi_i = c_0\Phi + \sum_a c_i^a \Phi_i^a \to E_i = \langle\Phi|H|\Phi\rangle + \varepsilon_i \tag{144}$$

Next, one calculates the pair correlation energies ε_{ij} from

$$\Psi_{ij} = c_0\Phi + \sum_a c_i^a \Phi_i^a + \sum_b c_j^b \Phi_j^b + \sum_{i<j} c_{ij}^{ab} \Phi_{ij}^{ab}$$
$$\downarrow \tag{145}$$
$$E_{ij} = \langle\Phi|H|\Phi\rangle + \varepsilon_i + \varepsilon_j + \varepsilon_{ij}$$

Then the three-particle corrections ε_{ijk} are given through

$$\Psi_{ijk} = c_0\Phi + \sum_a c_i^a \Phi_i^a + \sum_b c_i^b \Phi_j^b + \sum_c c_k \Phi_k^c + \sum_{a<b} c_{ij}^{ab} \Phi_{ij}^{ab}$$
$$+ \sum_{a<c} c_{ik}^{ac} \Phi_{ik}^{ac} + \sum_{b>c} c_{jk}^{bc} \Phi_{jk}^{bc} + \sum_{a<b<c} c_{ijk}^{abc} \Phi_{ijk}^{abc} \tag{146}$$
$$\downarrow$$
$$E_{ijk} = \langle\Phi|H|\Phi\rangle + \varepsilon_i + \varepsilon_j + \varepsilon_k + \varepsilon_{ij} + \varepsilon_{ik} + \varepsilon_{jk} + \varepsilon_{ijk}$$

and so on. Going up to the *n*th order is equivalent to carrying out a full CI calculation. Even the third-order equations are sufficiently difficult that Nesbet has solved them only in very few cases. There is, by the way, no *a priori* requirement that the result of the $(k+1)$-order equation should be better than that of the *k*th order. Too few calculations have been performed to allow general conclusions about the ε_{ijk}. In atoms they are usually positive, i.e., inclusion of the ε_{ijk} reduces the overall correlation energy.

If one analyzes ε_{ijk} one sees that the essential contributions to it are the pair interaction terms

$$\Delta\varepsilon_{ij,ik} = \sum_{a,b} \sum_{c,d} c_{ij}^{ab} c_{ik}^{cd} \langle\Phi_{ij}^{ab}|\hat{H}|\Phi_{ik}^{cd}\rangle$$

of the same type that occur in CEPA, but only for semijoint pairs (one orbital in common), which are the more important ones but which are treated with the wrong coefficients since in calculating the c_{ij}^{ab} only the coupling of Φ_{ij}^{ab} to Φ_{ik}^{cd} and Φ_{kj}^{ef} is taken care of. In addition ε_{ijk} contains pair-triple-coupling terms that are ignored in CEPA.

A variant of Nesbet's hierarchy, in which only the pair–pair coupling terms to third order are considered, has been used by van der Velde and Nieuwpoort.[111,112]

4.9. The Independent-Pair Potential Approximation (IPPA)

Mehler[113] has proposed a method for the calculation of correlation effects, called *independent-pair potential approximation* (IPPA) which consists essentially in performing n variational calculations for an n-electron system, with the trial function

$$\Psi_i = \Phi + \sum_{a<b} \sum_k d_{ik}^{ab} \Phi_{ik}^{ab} \tag{147}$$

i.e., by taking all pairs together that have one orbital index in common. (Mehler has also discussed the inclusion of single substitutions.) The truncated hierarchy of cluster equations corresponding to (147) is

$$E_i = \langle \Phi|\hat{H}|\Psi_i\rangle = \langle \Phi|\hat{H}|\Phi\rangle + \sum_{a<b} \sum_j d_{ij}^{ab} \langle \Phi|\hat{H}|\Phi_{ij}^{ab}\rangle$$

$$= \langle \Phi|\hat{H}|\Phi\rangle + \sum_{j(\neq i)} \varepsilon_{ij} \tag{148}$$

$$\langle \Phi_{ij}^{ab}|\hat{H}|\Phi\rangle + \sum_{\substack{l \\ (\neq i)}} \sum_{c<d} \langle \Phi_{ij}^{ab}|\hat{H}|\Phi_{il}^{cd}\rangle d_{il}^{cd} = E_i d_{ij}^{ab}$$

Equation (148) can be obtained from the CEPA system (122) if one (a) ignores all interaction matrix elements between disjoint pairs $\langle \Phi_{ij}^{ab}|\hat{H}|\Phi_{kl}^{cd}\rangle$, $i \neq k, l; j \neq k, l$; (b) considers only half of the semijoint coupling interaction matrix elements, namely $\langle \Phi_{ij}^{ab}|\hat{H}|\Phi_{il}^{cd}\rangle$ and not $\langle \Phi_{ij}^{ab}|\hat{H}|\Phi_{lj}^{cd}\rangle$ and replaces $E_{ij} = \langle \Phi|\hat{H}|\Phi\rangle + \varepsilon_{ij}$ by $E_i = \langle \Phi|\hat{H}|\Phi\rangle + \sum_{j(=i)} \varepsilon_{ij}$. In this system of equations $|d_{ij}^{ab}| \neq |d_{ji}^{ab}|$, but if one tries to symmetrize Eq. (148), one gets a CEPA-type system in which the semijoint coupling elements enter with a factor $\frac{1}{2}$ and where one replaces E_{ij} by $\langle \Phi|\hat{H}|\Phi\rangle + \varepsilon_{ij} + \frac{1}{2}\sum_{k(\neq i)} (\varepsilon_{ik} + \varepsilon_{kj})$.

5. Toward a Rigorous Treatment of Electron Correlation

5.1. Variational Treatment Based on the exp(S) Ansatz

It is astonishing that genuinely variational treatments of electron correlation have never been applied except for CI calculations limited to double substitutions (e.g., the PNO–CI approach). The inclusion of the unlinked clusters into a CI treatment obviously leads to formidable practical difficulties.

The standard approach would be to start from the ansatz $\Psi = e^{\hat{S}}\Phi$ and to minimize the energy expectation value

$$\frac{\langle\Phi|(e^{\hat{S}})^{\dagger}\hat{H}e^{\hat{S}}|\Phi\rangle}{\langle\Phi|(e^{\hat{S}})e^{\hat{S}}|\Phi\rangle} = \frac{\langle\Phi|(1+\hat{S}^{+}+\frac{1}{2}\hat{S}^{+2}+\cdots)H(1+\hat{S}+\frac{1}{2}\hat{S}^{2}+\cdots)|\Phi\rangle}{\langle\Phi|(1+\hat{S}^{+}+\cdots)(1+\hat{S}+\cdots)|\Phi\rangle}$$

$$= [\langle\Phi|\hat{H}|\Phi\rangle + 2\operatorname{Re}\langle\Phi|\hat{H}\hat{S}|\Phi\rangle + \langle\Phi|\hat{S}^{+}\hat{H}\hat{S}|\Phi\rangle$$

$$+ \operatorname{Re}\langle\Phi|\hat{H}\hat{S}^{2}|\Phi\rangle$$

$$+ \operatorname{Re}\langle\Phi|\hat{S}^{+}\hat{H}\hat{S}^{2}|\Phi\rangle + O(S^{4})][1 + \langle\Phi|\hat{S}^{+}\hat{S}|\Phi\rangle + O(S^{4})]^{-1}$$

$$(149)$$

with respect to $\hat{S}$. [Terms like $\langle\Phi|\hat{H}\hat{S}^{3}|\Phi\rangle$ do not contribute if (as usual) the Hamiltonian contains only 1- and 2-particle operators and Φ is a single Slater determinant. $\hat{S}^{3}\Phi$ is then at least triply excited with respect to Φ and $\hat{H}$ can at most deexcite twofold, so that $\hat{H}\hat{S}^{3}\phi$ is at least singly excited with respect to Φ and hence orthogonal to Φ.] The expansions of both the numerator and the denominator break off at the nth power $\hat{S}$ and $\hat{S}^{+}$ for an n-electron system [if S_1 vanishes at the $(n/2)$ power].

Minimization with respect to $\hat{S}$ leads to a complicated system of nonlinear equations for $\hat{S}$. In the truncated hierarchy of cluster equations $\hat{S}$ is only coupled to $\hat{S}^{2}$, but in the variational equation $\hat{S}$ is coupled to $\hat{S}^{2}$, $\hat{S}^{3}$, etc.

Usually one assumes that $\hat{S}$ is small, i.e., that the absolute value of the coefficients d^{a}_{i}, d^{ab}_{ij}, etc., is small compared to unity. Then it seems reasonable to truncate the expansion of $e^{\hat{S}}$ at some power. The treatment is still variational but it has the wrong n-dependence, which is easily seen for the case where one puts $\Psi = (1+\hat{S})\Phi$ and $\hat{S} = \hat{S}_2$, i.e., where one includes double substitutions only.

Instead of truncating the expansion of Ψ at some power of $\hat{S}$ one can first expand the energy as a power series in $\hat{S}$ and truncate then. This will lead to results that are no longer variational but that appear to have the correct n-dependence. One gets to second order in $\hat{S}$

$$E = (1-\|\hat{S}\Phi\|^{2})\langle\Phi|\hat{H}|\Phi\rangle + 2\operatorname{Re}\langle\Phi|\hat{H}\hat{S}|\Phi\rangle$$

$$+ \langle\Phi|\hat{S}^{+}\hat{H}\hat{S}|\Phi\rangle + \operatorname{Re}\langle\Phi|\hat{H}\hat{S}^{2}|\Phi\rangle \qquad (150)$$

(the term $\operatorname{Re}\langle\Phi|\hat{H}\hat{S}^{2}|\Phi\rangle$ vanishes if $\hat{S}$ does not contain $\hat{S}_1$, and $\hat{H}$ consists of one- and two-particle terms) and to third order in addition,

$$-2\|\hat{S}\Phi\|^{2}\operatorname{Re}\langle\Phi|\hat{H}\hat{S}|\Phi\rangle + \operatorname{Re}\langle\Phi|\hat{S}^{+}\hat{H}\hat{S}\hat{S}|\Phi\rangle \qquad (151)$$

and to fourth order

$$-(\|\hat{S}\Phi\|^4 + \tfrac{1}{4}\|\hat{S}\hat{S}\Phi\|^2)\langle\Phi|\hat{H}|\Phi\rangle - \|\hat{S}\Phi\|^2\langle\Phi|\hat{S}^+\hat{H}\hat{S}|\Phi\rangle$$

$$+\tfrac{1}{4}\langle\Phi|\hat{S}^+\hat{S}^+\hat{H}\hat{S}\hat{S}|\Phi\rangle + \tfrac{1}{3}\,\mathrm{Re}\langle\Phi|\hat{S}^+\hat{H}\hat{S}^3|\Phi\rangle$$

$$-\|S\Phi\|^2\,\mathrm{Re}\langle\Phi|\hat{H}\hat{S}^2|\Phi\rangle \qquad (152)$$

(the last two terms vanish if $\hat{S}_1$ is not present).

5.2. Unitary Ansatz for the Correlated Wave Function

We now propose an alternative formulation of the variational treatment of electron correlation which to the author's knowledge has not been discussed previously in the literature, although the idea of using a unitary ansatz for Ψ can be found in the early work of Primas[114] and Yaris.[115] It goes in fact even back to van Vleck.[129]*

Let us choose

$$\Psi = e^{\hat{\sigma}}\Phi \qquad \text{with } \hat{\sigma} = \hat{T} - \hat{T}^+$$

$$\hat{T} = \tfrac{1}{4}\sum_{i<j}\sum_{a<b} f_{ij}^{ab}\hat{a}_a^+\hat{a}_b^+\hat{a}_j\hat{a}_i$$

$$\hat{\sigma} = \tfrac{1}{4}\sum_{i,j}\sum_{a,b} f_{ij}^{ab}(\hat{a}_i^+\hat{a}_j^+\hat{a}_b\hat{a}_a - \hat{a}_a^+\hat{a}_b^+\hat{a}_j\hat{a}_i) \qquad (153)$$

(The limitation to double substitutions is not essential, but it simplifies the discussion somewhat.) Obviously $\hat{\sigma}$ is anti-Hermitian and hence $e^{\hat{\sigma}}$ is unitary, which means that $\langle\Psi|\Psi\rangle = 1$ if $\langle\Phi|\Phi\rangle = 1$. The expectation value of the Hamiltonian or any other operator is easily written down using the Hausdorff formula

$$\langle\Psi|\hat{H}|\Psi\rangle = \langle\Phi|e^{-\hat{\sigma}}\hat{H}e^{\hat{\sigma}}|\Phi\rangle$$

$$= \langle\Phi|\hat{H}|\Phi\rangle + \langle\Phi|[\hat{H},\hat{\sigma}]|\Phi\rangle + \frac{1}{2!}\langle\Phi|[[\hat{H},\hat{\sigma}],\hat{\sigma}]|\Phi\rangle + \cdots \qquad (154)$$

The series (154) is infinite but it converges like an exponential. The energy to second order in $\hat{T}$ is

$$E = (1 - \|\hat{T}\Phi\|^2)\langle\Phi|\hat{H}|\Phi\rangle + 2\,\mathrm{Re}\langle\Phi|\hat{H}\hat{T}|\Phi\rangle + \langle\Phi|\hat{T}^+\hat{H}\hat{T}|\Phi\rangle$$

$$= (1 - \sum_{i<j}\sum_{a<b}|f_{ij}^{ab}|^2)\langle\Phi|\hat{H}|\Phi\rangle + 2\,\mathrm{Re}\sum_{i<j}\sum_{a<b} f_{ij}^{ab}\langle\Phi|\hat{H}|\Phi_{ij}^{ab}\rangle$$

$$+ \sum_{i<j}\sum_{a<b}\sum_{k<l}\sum_{c<d}\langle\Phi_{kl}^{cd}|\hat{H}|\Phi_{ij}^{ab}\rangle f_{kl}^{cd\,*} f_{ij}^{ab} \qquad (155)$$

To third order one has in addition

$$-\tfrac{3}{4}\|\hat{T}\Phi\|^2\,\mathrm{Re}\langle\Phi|\hat{H}\hat{T}|\Phi\rangle + \tfrac{2}{3}\,\mathrm{Re}\langle\Phi|\hat{T}^+\hat{H}\hat{T}\hat{T}|\Phi\rangle \qquad (156)$$

*See Notes Added in Proof, p. 188.

and the fourth-order term is

$$\tfrac{1}{3}(\|\hat{T}\Phi\|^4 + \tfrac{1}{4}\|\hat{T}^2\Phi\|^2)\langle\Phi|\hat{H}|\Phi\rangle - \tfrac{1}{3}\|\hat{T}\Phi\|^2\langle\Phi|\hat{T}^+\hat{H}\hat{T}|\Phi\rangle$$

$$- \tfrac{1}{3}\langle\Phi|\hat{T}^+\hat{H}\hat{T}^+\,\hat{T}\hat{T}|\Phi\rangle + \tfrac{1}{4}\langle\Phi|\hat{T}^+\hat{T}^+\hat{H}\hat{T}\hat{T}|\Phi\rangle \tag{157}$$

To second order the expansion in powers of $\hat{T}$ (155) agrees formally with that in powers of $\hat{S}$ (150), the third-order contribution in $\hat{T}$ is $\tfrac{2}{3}$ times that in $\hat{S}$, whereas the fourth orders differ by more than just a numerical factor. Obviously the expansion in powers of $\hat{T}$ converges more rapidly than that in powers of $\hat{S}$.

Since the unitary expansion (153) and (154) has not yet been applied practically, we will not go into details here. However, we want to point out (a) that (154) is additively separable (i.e., has the correct n-dependence) in any order of $\hat{\sigma}$ (which is a general property of the Lie-algebraic formulation of many-body quantum mechanics)[114]; (b) [a corollary to (a)] that all disconnected contributions, e.g., contributions like $|f_{ij}^{ab}|^2 f_{kl}^{cd}\langle\Phi|\hat{H}|\Phi_{kl}^{cd}\rangle$ with $(i, j) \neq (k, l)$, $(a, b) \neq (c, d)$ in (156) cancel exactly; (c) that for a two-electron system the sum (154) can be expressed in a simple closed form, namely,

$$E = \cos^2(\|\hat{T}\Phi\|)\langle\Phi\hat{H}|\Phi\rangle + \frac{\sin(\|\hat{T}\Phi\|)\cos(\|\hat{T}\Phi\|)}{\|\hat{T}\Phi\|} 2\,\mathrm{Re}\langle\Phi|\hat{H}\hat{T}|\Phi\rangle$$

$$+ \frac{\sin^2(\|\hat{T}\Phi\|)}{\|\hat{T}\Phi\|^2}\langle\Phi|\hat{T}^+\hat{H}\hat{T}|\Phi\rangle \tag{158}$$

(d) that the f_{ij}^{ab} of (153) differ from the d_{ij}^{ab} of (32) and also from the c_{ij}^{ab} of (38). The relation is for a two-particle system

$$d_{ij}^{ab} = \frac{\tan\|\hat{T}\Phi\|}{\|\hat{T}\Phi\|} f_{ij}^{ab} \tag{159}$$

For $n > 2$ the proportionality constant is a more complicated expression. One generally has

$$d_{ij}^{ab} = f_{ij}^{ab}[1 + O(\|\hat{T}\Phi\|^2)] \tag{160}$$

The coefficients f_{ij}^{ab} that make the energy expression (155) stationary to second order can be obtained from the set of linear equations

$$\langle\Phi|\hat{H}|\Phi_{ij}^{ab}\rangle + \sum_{k<l}\sum_{c<d} f_{kl}^{cd}\langle\Phi_{kl}^{cd}|\hat{H}|\Phi_{ij}^{ab}\rangle = f_{ij}^{ab}\langle\Phi|\hat{H}|\Phi\rangle \tag{161}$$

Formally the same system is obtained with the f_{ij}^{ab} replaced by the d_{ij}^{ab} if one makes (150) stationary. Equation (161) is very close to the CEPA equation (122) with the only difference that $\langle\Phi|\hat{H}|\Phi\rangle$ appears on the r.h.s. instead of $E_{ij} = \langle\Phi|\hat{H}|\Phi\rangle + \varepsilon_{ij}$. The formal solution of (161) only requires a matrix inversion and not a matrix diagonalization.

One is led exactly[127] to the system (161) for the first-order wave function $\Psi^{(1)} = \hat{T}\Phi$ if one performs RS perturbation theory with $\hat{H}_0 = \hat{P}\hat{H}\hat{P} + \hat{Q}\hat{H}\hat{Q} + \hat{R}\hat{H}\hat{R}$, $\hat{P} = |\Phi\rangle\langle\Phi|$, $\hat{Q} = \sum_{i<j}\sum_{a<b}|\Phi_{ij}^{ab}\rangle\langle\Phi_{ij}^{ab}|$, $\hat{R} = 1 - \hat{P} - \hat{Q}$. In this formulation (155) is the energy correct to third order.

5.3 Which is the Best CEPA?

There is strong evidence that, at least as far as atomic and molecular calculations are concerned (the situation in nuclear physics may be different), calculations that go beyond a CEPA-type ansatz are too time consuming for having any chance in practical application, except possibly together with minimum basis sets for very small molecules.[36] This is why CEPA-type theories deserve special attention and a critical comparison of different CEPA approaches is in order.

In this review we have encountered several methods that involve the solution of the system of equations

$$\langle\Phi_{ij}^{ab}|\hat{H}|\Phi\rangle + \sum_{k<l}\sum_{c<d} \langle\Phi_{ij}^{ab}|\hat{H}|\Phi_{kl}^{cd}\rangle \, d_{kl}^{cd} = W_{ij}d_{ij}^{ab}$$

and that differ in the meaning of W_{ij}. A (probably incomplete) list follows.

1. $W_{ij} = E$, the total variational energy, in CI methods limited to double substitutions, especially the PNO–CI method (Section 4.7).

2. $W_{ij} = \langle\Phi|\hat{H}|\Phi\rangle$, the zeroth-order energy, in the method that makes $\langle\Psi|\hat{H}|\Psi\rangle$ stationary to second order in $\hat{S}$ (Section 5.2) and that also occurs in the perturbation scheme[127] mentioned at the end of Section 5.2.

3. $$W_{ij} = \langle\Phi|\hat{H}|\Phi\rangle + \varepsilon_{ij}$$

in Meyer's CEPA (version CEPA-2) (Section 4.3).

4. $$W_{ij} = \langle\Phi|\hat{H}|\Phi\rangle + \varepsilon_{ij} + \tfrac{1}{2}\sum_k (\varepsilon_{ik} + \varepsilon_{kj})$$

in Meyer's CEPA (version CEPA-1).[97]

5. $$W_{ij} = \langle\Phi|\hat{H}|\Phi\rangle + \varepsilon_{ij} + \sum_k\varepsilon_{kj} + \sum_l \varepsilon_{il}$$

in Kelly's CEPA (Section 2.6). Actually Kelly's W is somewhat more complicated, as it contains an additional term that is different for different a, b, but which is probably very small. Robb's approach[81] is close to that of Kelly, but includes still more additional terms.

6. $$W_{ij} = \langle\Phi|\hat{H}|\Phi\rangle + \sum_k \varepsilon_{ik}$$

in Mehler's IPPA (Section 4.8). However, IPPA is not strictly a CEPA-type theory since the disjoint interaction elements are ignored and only half of the interaction elements are considered.

If one discusses these different CEPA methods in terms of the expansion of the energy in powers of $\hat{S}$ or $\hat{T}$, all six variants are correct to second order and they differ in accounting for different parts of third-order contribution. One may wonder whether it is recommended at all to include third-order terms in a basically second-order theory. If one cares for consistency in this sense version 2 above seems the best choice. Version 1 has the merit of being strictly variational but it does not, unlike the other five versions, have the correct n-dependence, which we regard as more important. Versions 3–5 become exact for a supersystem consisting of n noninteracting two-electron systems in a localized description, because then they all reduce to the corresponding systems of an APSG wave function (111). This is a good argument in favor of those versions.

For noninteracting pairs the intersubsystem ε_{ij} vanish and methods 3–5 become indistinguishable. If one regards the correct description of a system of noninteracting pairs as important, one can choose between the two extremes of version 3 (Meyer) or version 5 (Kelly). Kelly's approach bears the closest resemblance to the truncated linked-pair cluster expansion (of Brenig, Coester and Kümmel, and Čížek). From the derivation in Section 2.6 one can conclude that Kelly's W_{ij} is not well balanced since he has fully taken care of the terms $-\langle\Phi|\hat{H}\hat{S}|\Phi\rangle\,d_{ij}^{ab}$ but neglected all contributions of $\frac{1}{2}\langle\Phi_{ij}^{ab}|\hat{H}\hat{S}\hat{S}|\Phi\rangle$, except those that cancel with contributions in the former term. Since the two terms have opposite signs it is better to neglect both of them rather than just one. This argument is more in favor of version 3 or even 1.

An argument against 3 is that it is not invariant with respect to unitary transformation of the strongly occupied spin-orbitals ψ_i. In fact Meyer[97] was led to his CEPA–1 by requiring invariance with respect to transformations between canonical and localized descriptions for a supersystem of noninteracting pairs. It is unfortunately not clear from Meyer's first papers which of the two approaches he would later prefer to call CEPA-1 and CEPA-2. In this review we have always used the term CEPA for the original version that later got the name CEPA-2. Numerical calculations[97] have indicated that in the region of distances that CEPA approaches start to fail, CEPA-2 often gives energies which are too low, while Kelly's CEPA yields energies too high, CEPA-1 seems to be a good compromise. It usually leads to better dipole moment functions than does CEPA-2.[97]

The question of how one should go beyond CEPA cannot be answered in a unique way. The third-order expressions in terms of $\hat{S}$ or $\hat{T}$ and the truncated hierarchy do not lead to the same answer. Two extreme answers concerning the third-order energy correction differ by a factor 2, and the third one is just in between. The answers get much closer to each other if one includes those fourth-order contributions that cancel with third-order contributions as a consequence of the stationarity up to second order. We cannot go into details here, but we mention that one can also compare the different methods in terms of perturbation theory. In these terms the different CEPA methods and the truncated pair cluster hierarchy are essentially correct to third order and contain some diagrams to infinite order. Fourth-order diagrams are partially present and some third-order diagrams are missing. Some improvements of CEPA that are necessary in special cases are discussed in the following sections.

5.4. Single Substitution and Brueckner Orbitals

As mentioned in earlier sections single substitutions Φ_i^a with respect to a Hartree–Fock determinant for closed-shell states are usually not very important as far as ground states are concerned.

One has to include single substitutions if one wants to calculate excited states together with the ground state. For some properties other than the energy, especially second-order properties, single substitutions may be important even if their effect on the energy is negligible.[130] The Φ_i^a do contribute significantly even to the ground-state energy if some Φ_{ij}^{ab} have large expansion coefficients. An example would be the H_2 ground state at large distances, where the coefficients c_1 and c_2 in

$$\Psi = c_1 1\sigma_g(1)1\sigma_g(2) + c_2 1\sigma_u(1)1\sigma_u(2)$$

have nearly the same order of magnitude and opposite sign. If we take $1\sigma_g$ as the Hartree–Fock orbital, then single substitutions like $1\sigma_g(1)2\sigma_g(2)$ have nonnegligible contributions to the energy. If, however, one minimizes $1\sigma_g$ and $1\sigma_u$ for the two-configuration wave function, then by virtue of the generalized Brillouin theorem, single substitutions are negligible.

Single substitutions can be dealt with by a unitary ansatz with

$$\hat{\sigma}_1 = \sum_i \sum_a c_i^a (\hat{a}_a^+ \hat{a}_i - \hat{a}_i^+ \hat{a}_a) = \hat{T}_1 - \hat{T}_1^+ \tag{162}$$

or

$$\hat{\sigma}_1 = \sum_p \sum_q c_p^q \hat{a}_q^+ \hat{a}_p, \qquad c_p^q = -c_q^p \tag{163}$$

(Note that we have agreed to use the subscripts $i, j, \dots$ for occupied, $a, b, \dots$ for virtual, and $p, q, \dots$ for arbitrary orbitals.) The two forms of $\hat{\sigma}_1$ are not identical, but they lead essentially to the same result. Let us choose Eq. (163). Obviously if Φ is a normalized Slater determinant, then

$$\tilde{\Phi} = e^{\hat{\sigma}_1} \Phi \tag{164}$$

is also a normalized Slater determinant. For fixed Φ the Hartree–Fock problem of minimizing the energy with respect to Φ is equivalent to minimizing it with respect to $\hat{\sigma}_1$.

If Φ is sufficiently close to the optimum $\tilde{\Phi}$ one can break off the commutator series after terms in $\hat{\sigma}_1^2$,

$$
\begin{aligned}
\delta E &= \delta \langle \tilde{\Phi} | \hat{H} | \tilde{\Phi} \rangle = \delta \langle \Phi | e^{-\hat{\sigma}_1} \hat{H} e^{\hat{\sigma}_1} | \Phi \rangle \\
&= \delta \langle \Phi | [\hat{H}, \hat{\sigma}_1] + \tfrac{1}{2}[[\hat{H}, \hat{\sigma}_1], \hat{\sigma}_1] | \Phi \rangle \\
&= 2 \operatorname{Re} \sum_p \sum_q \delta c_p^q \langle \Phi | \hat{H} \hat{a}_q^+ \hat{a}_p | \Phi \rangle + 2 \operatorname{Re} \sum_{p,q} \sum_{r,s} c_r^s \delta c_p^q \\
&\quad \times \{ \langle \Phi | \hat{H} \hat{a}_q^+ \hat{a}_p \hat{a}_s^+ \hat{a}_r | \Phi \rangle - \langle \Phi | \hat{a}_q^+ \hat{a}_p \hat{H} \hat{a}_s^+ \hat{a}_r | \Phi \rangle \}
\end{aligned}
\tag{165}
$$

E is stationary if the coefficient of $\delta(c_p^q - c_q^p)$ vanishes; this leads to a condition for an iterative improvement of Φ. When self-consistency is achieved, the Brillouin theorem

$$
\langle \Phi | \hat{H} (\hat{a}_q^+ \hat{a}_p - \hat{a}_p^+ \hat{a}_q) | \Phi \rangle = 0
\tag{166}
$$

or equivalently

$$
\langle \Phi | \hat{H} | \hat{a}_a^+ \hat{a}_i \Phi \rangle = \langle \Phi | \hat{H} | \Phi_i^a \rangle = 0
\tag{167}
$$

holds. We point out that in deriving (166) we have not used the fact that Φ was a single-determinant wave function. The whole argument holds even if Φ is a multiconfiguration function; (165) then offers a means to optimize the orbitals in an MC function, and the condition of self-consistency (166) is the generalized Brillouin condition, probably first derived by Levy and Berthier.[116] [The fact that Φ is a single determinant with the spin-orbitals ψ_i occupied is only used in the transition from (166) to (167).]

In an analogous way one can derive the rigorous variational counterpart of the Brillouin–Brueckner condition, i.e., the condition that the energy expectation value $E = \langle \Phi | e^{-\hat{\sigma}_2} \hat{H} e^{\hat{\sigma}_2} | \Phi \rangle$ cannot be improved on replacing $e^{\hat{\sigma}_2}$ by $e^{\hat{\sigma}_1 + \hat{\sigma}_2}$ with $\hat{\sigma}_1$ given by (162).

Neglecting terms of second and higher order in σ_2 one gets the condition

$$
\operatorname{Re}(\langle \Phi_i^a | \hat{H} + \hat{H} \hat{T}_2 | \Phi \rangle - \tfrac{1}{2} \langle \Phi | \hat{H} \hat{a}_i^+ \hat{a}_a \hat{T}_2 | \Phi \rangle) = 0
\tag{168}
$$

which differs from the classical Brillouin–Bruckner condition mainly in the fact that the real part has to be taken, i.e., that the matrix elements of the non-Hermitian operator $\hat{H}\hat{S}$ in the classical Brillouin–Bruckner condition are to be replaced by those of the Hermitian operator $\tfrac{1}{2}(\hat{H}\hat{S} + \hat{S}\hat{H})$.

5.5. Importance of Higher than Two-Particle Clusters

From numerical experience with atoms and small molecules one may conclude that for closed-shell states not only one-particle clusters (see Section 5.4) but also unlinked three-particle clusters in the wave function are usually negligible, as are the linked four-particle clusters. The next important clusters, after the pair clusters, turn out to be the linked three-particle clusters. This has, e.g., been tested for BH_3 in a minimal basis calculation.[36] One effect for which three-particle clusters are quite important, and that is well understood, applies to the van der Waals interaction between two atoms or molecules in closed-shell states. To illustrate this we consider two He atoms.

The Hartree–Fock wave function can be written as

$$\Phi = |1s_a 1\bar{s}_a 1s_b 1\bar{s}_b| \tag{169}$$

There are two types of double substitutions,*

$$\Phi_{a\bar{a}} = |1p_a 1\bar{p}_a 1s_b 1\bar{s}_b|$$
$$\Phi_{b\bar{b}} = |1s_a 1\bar{s}_a 1p_b 1\bar{p}_b| \tag{170}$$

and

$$\Phi_{ab} = |1s_a 2\bar{p}_a 2p_b 1\bar{s}_b| \tag{171}$$

$\Phi_{a\bar{a}}$ and $\Phi_{b\bar{b}}$ describe the intraorbital correlation and Φ_{ab} the interorbital correlation (dispersion) between two He atoms. To simplify the argument we ignore the fact that p-functions have to be coupled to S-states and that there is a singlet and a triplet Φ_{ab}. We also assume that the pair correlation functions consist of one single term each. Then in IEPA, the correlation energy is just

$$E_{IEPA}^{corr} = \varepsilon_{aa} + \varepsilon_{bb} + \varepsilon_{ab} \tag{172}$$

$$\varepsilon_{aa} = \varepsilon_{bb} = \langle \Phi | \hat{H} | \Phi_{aa} \rangle c_{aa} \tag{173}$$

$$\varepsilon_{ab} = \langle \Phi | \hat{H} | \Phi_{ab} \rangle c_{ab} \tag{174}$$

One finds that ε_{aa} and ε_{bb} vary so little with distance (at least for large R) that omission of ε_{aa} and ε_{bb} leads just to a parallel shift of the potential curve. ε_{ab} is in fact responsible for the dependence of E_{IEPA}^{corr} on the distance. In PNO–CI or CEPA one takes the coupling elements

$$\Delta\varepsilon_{aa,ab} = \Delta\varepsilon_{bb,ab} = \langle \Phi_{aa} | \hat{H} | \Phi_{ab} \rangle c_{aa} c_{ab} \tag{175}$$

into account (the $\varepsilon_{aa,bb}$ are negligible), so that the change of the correlation energy with distance is now determined by

$$(\langle \Phi | \hat{H} | \Phi_{ab} \rangle + 2c_{aa} \langle \Phi_{aa} | \hat{H} | \Phi_{ab} \rangle) c_{ab} \tag{176}$$

---

*We designate as $1p$ those p-AOs that account for the intra-$1s^2$-correlation of one He atom and as $2p$ the p-AOs that account for the intersystem correlation (dispersion).

rather than by (174). Since the two matrix elements in (176) are of comparable magnitude and since c_{aa} is usually negative, (176) is smaller in absolute value than ε_{ab}: the intra–inter orbital coupling reduces the dispersion energy. One may say that (174) represents the dispersion interaction between two "Hartree–Fock He atoms," whereas (176) contains a correction due to the intraatomic correlation of one He atom. To represent the dispersion interaction of two He atoms correctly one should consider not just intersubsystem double substitutions from (169) but rather from

$$
\begin{aligned}
\tilde{\Phi} = \mathscr{A}\{ & (1s_a 1\bar{s}_a + c_{aa} 1p_a 1\bar{p}_a)(1s_b 1\bar{s}_b + c_{aa} 1p_b 1\bar{p}_b) \} \\
= & \Phi + c_{aa} |1s_a 1\bar{s}_a 1p_b 1\bar{p}_b| \\
& + c_{aa} |1p_a 1\bar{p}_a 1s_b 1\bar{s}_b| + c_{aa}^2 |1p_a 1\bar{p}_a 1p_b 1\bar{p}_b|
\end{aligned}
\tag{177}
$$

i.e., in addition to (171) one has to take into account the determinants

$$
\begin{aligned}
\Phi_{abb} &= |1s_a 2\bar{p}_a 2p_b 1\bar{p}_b| \\
\Phi_{aab} &= |1p_a 2\bar{p}_a 2p_b 1\bar{s}_b|
\end{aligned}
\tag{178}
$$

and

$$
\Phi_{aabb} = |1p_a 2\bar{p}_a 2\bar{p}_b 1p_b|
\tag{179}
$$

In view of the smallness of the coefficient c_{aa}^2 the fourfold substitution Φ_{aabb} is less important than the triple substitutions Φ_{abb} and Φ_{aab}. Actual calculations show that the van der Waals coefficient for the He–He interaction[117] (i.e., the coefficient of the $1/R^6$ term) is 1.54 from (174), 1.42 from (176), and 1.46 if (178) is taken into account.

The (opposite) effect of the intra–intercoupling and of triple substitutions is noticeable, though not spectacular. For the Be–Be interaction the corresponding figures[117] are 450, 150, 210, which demonstrate a dramatic effect of intraatomic correlation on the dispersion interaction. The basic reason is the $2s/2p$ near degeneracy in Be which leads to a rather large expansion coefficient of the $2p^2$ configuration. (For an early qualitative discussion of this problem see Refs. 7 and 118; for recent calculations on the He_2 potential curve see Refs. 120 and 121.)

Although the triple substitution that is important here is formally "linked," it is better described as a double substitution with respect to a double substitution. If the ground configuration is $|ijkl|$ and a double substitution is $c_{ij}^{ab}|abkl|$, two possible types of double substitutions from this double substitution are $c_{ij}^{ab}c_{kl}^{cd}|abcd|$, which is classified as unlinked, and $c_{ij}^{ab}c_{bk}^{cd}|acdl|$, which might be called semilinked. It is not hard to see that these semilinked substitutions, which are ignored in the usual linked-pair expansion, ought to be included together with the unlinked contributions to the wave function.

They should be included as

$$c_{bk}^{cd}(\hat{a}_c^+ \hat{a}_d^+ \hat{a}_k \hat{a}_b - \hat{a}_b^+ \hat{a}_k^+ \hat{a}_d \hat{a}_c) \tag{180}$$

into $\hat{\sigma}$. These terms contribute only to third order in $\hat{\sigma}$, one can hence derive equations for the c_{bk}^{cd} that make the third-order contribution to the energy stationary.

If one starts from a multiconfigurational zeroth-order wave function Φ and bases a pair cluster expansion on it, one can include the most relevant semilinked substitutions automatically, if one defines

$$\hat{\sigma}_2 = \tfrac{1}{4} \sum_{p,q,r,s} f_{pq}^{rs} \hat{a}_p^+ \hat{a}_q^+ \hat{a}_s \hat{a}_r, \qquad f_{pq}^{rs} = -f_{rs}^{pq} \tag{181}$$

instead of (153).

5.6. Need for Zeroth-Order Multiconfiguration Wave Functions

In this review we have mainly considered such states which, to zeroth order, are described by a single Slater determinant Φ, i.e., for which the overlap integral between Φ and the exact wave function Ψ, both normalized to unity, is close to 1. For states where this is not the case a cluster expansion about a single Slater determinant is not very useful. Take, e.g., F_2 for large interatomic distances where two Slater determinants Φ_1 and Φ_2 corresponding to the configurations

$$1\sigma_g^2 1\sigma_u^2 2\sigma_g^2 2\sigma_u^2 1\pi_u^4 1\pi_g^4 3\sigma_g^2$$
$$1\sigma_g^2 1\sigma_u^2 2\sigma_g^2 2\sigma_u^2 1\pi_u^4 1\pi_g^4 3\sigma_u^2$$

have nearly the same weight. If one takes the cluster expansion about Φ_1, one has to consider Φ_2 as doubly substituted, hence configurations doubly substituted with respect to Φ_2 are triply or quadruply substituted with respect to Φ_1 but have similar weight to those doubly substituted with respect to Φ_1. All the arguments based on the assumptions that the d_{ij}^{ab} are small and that the d_{ijkl}^{abcd} only have unlinked contributions, are no longer valid. It would seem more appropriate to use the multiconfigurational wave function $\Phi = c_1 \Phi_1 + c_2 \Phi_2$ as Φ, and use this as the starting point for a cluster expansion (see, e.g., Ref. 122). As with the open-shell case[104] one has to distinguish between three classes of orbitals: (a) fully occupied, like $1\sigma_g$, $1\pi_u$, etc.: i,j,k,l; (b) partially occupied, like $3\sigma_g, 3\sigma_u: x, y, z, u, v$; (c) unoccupied: a, b, c, d. To describe correlation, several types of substitutions have to be considered: one-particle substitutions $\Phi_i^a = \hat{a}_a^+ \hat{a}_i \Phi$; Φ_i^x; Φ_x^a; Φ_x^y and two-particle substitutions Φ_{ij}^{ab}; Φ_{ix}^{ab}; Φ_{xy}^{ab}; Φ_{ij}^{ax}; Φ_{ij}^{xy}; Φ_{ix}^{xy}; Φ_{ix}^{yz}; Φ_{xy}^{az};

Φ_{xy}^{zu}. The substitutions to a, b are called external, those to a, x semi-internal, and those to x, y internal.

In the unitary formalism we treat the present case such that $\hat{\sigma}_2$ is divided into two parts, one of which ($\tilde{\hat{\sigma}}_2$) is treated to infinite order and the other ($\hat{\sigma}_2'$) to second order as in CEPA (or possibly to third order). We make the ansatz

$$e^{\hat{\sigma}_2}e^{\tilde{\hat{\sigma}}_2+\hat{\sigma}_1}\phi \tag{182}$$

with $\hat{\sigma}_2'$ of the form (181). $\tilde{\hat{\sigma}}_2$ contains only a very limited number of terms such that $e^{\tilde{\hat{\sigma}}_2}\Phi$ is a linear combination of a sufficiently small number of Slater determinants. One can either determine $\tilde{\Psi}=e^{\tilde{\hat{\sigma}}_2+\hat{\sigma}_1}\Phi$ by a CI calculation or by a multiconfigurational procedure, i.e., iteratively. If $\tilde{\Psi}_n$ is the wave function of the nth iteration cycle, then

$$\tilde{\Psi}_{n+1} = e^{\delta\tilde{\hat{\sigma}}_2+\delta\hat{\sigma}_1}\tilde{\Psi}_n \tag{183}$$

is the MC wave function of the next cycle. One has

$$E_{n+1} = E_n + \langle\tilde{\Psi}_n|[\hat{H}, \delta\tilde{\sigma}_2+\delta\hat{\sigma}_1] + \tfrac{1}{2}[[\hat{H}, \delta\hat{\sigma}_1+\delta\hat{\sigma}_2], \delta\hat{\sigma}_1+\delta\hat{\sigma}_2]|\tilde{\Psi}_n\rangle \tag{184}$$

While the MC–SCF problem is now understood at least in principle, almost no experience at all has so far been compiled about cluster expansions (unitary or not) based on an MC–SCF wave function. It is possibly not sufficient to consider only σ_2'; σ_3' seems to be necessary in many cases, but we cannot go into any details here. Not much has been done on these lines so far. (The work reported in Ref. 123 does not contradict this.) The corresponding problem in perturbation theory was recently discussed.[126]

6. Some Numerical Results

To illustrate some of the points made in this review we outline some results obtained with the SCF–IEPA–PNO, CEPA–PNO, and PNO–CI methods.

First we note that there are "good IEPA molecules," which are rather well described by IEPA, i.e., for which the correction due to the pair interactions are very small, both as far as the total energy and physical properties are concerned. As an example we take the hypothetical MgH_2 molecule[100] in the ground-state configuration $K^2L^8\sigma_g^2\sigma_u^2$ or equivalently $K^2L^8b^2b'^2$, where b means an orbital localized in one MgH bond. The diagonal (ε_μ) and off-diagonal contributions ($\Delta\varepsilon_{\mu\nu}$) to the CEPA valence shell correlation energy in the sense of Eqs. (139)–(140) are given in Table 1 both for the canonical and the localized description. One sees that in the localized case the $\Delta\varepsilon_{\mu\nu}$ are in fact very small (and all positive), so that localized IEPA and CEPA (in the same basis) give nearly the same results ($E_{IEPA}^{corr} = -0.0687$ a.u.; $E_{CEPA}^{corr} = -0.0682$ a.u.), whereas in the delocalized description the $\Delta\varepsilon_{\mu\nu}$ are much larger

Table 1. *Diagonal and Nondiagonal Contributions to* $E_{\text{CEPA}}^{\text{corr}}$
According to Eqs. (139) *and* (140) *for* MgH_2[a]

	Localized[b]			
b	−0.03284	—	—	—
b'	0.00000	−0.03284	—	—
$^1bb'$	0.00003	0.00003	−0.00114	—
$^3bb'$	0.00008	0.00008	0.00006	−0.00169
	Canonical[b]			
σ_g	−0.01285	—	—	—
σ_u	0.00232	−0.01597	—	—
$^1\sigma_g\sigma_u$	−0.00510	−0.00630	−0.02022	—
$^3\sigma_g\sigma_u$	0.00012	0.00002	0.00007	−0.00170

[a] Details of the computation are given in Ref. 100.
[b] All quantities given in atomic units.

and have either sign. The canonical IEPA and CEPA energies differ apprecia-bly ($E_{\text{IEPA}}^{\text{corr}} = -0.0550$ a.u.; $E_{\text{CEPA}}^{\text{corr}} = -0.0685$ a.u.). MgH_2 is a "good IEPA molecule" only in the localized description.

On the CEPA level it does not make a substantial difference whether one has started from a localized or a delocalized description. But IEPA is far from being invariant with respect to a unitary transformation of the occupied orbitals (see also Refs. 124 and 125), in contrast to what holds in second-order perturbation theory (see Section 3.4).

As two other examples we take SiH_4 and Ar (localized) (Table 2); SiH_4 is representative of second row hydrides as far as XH bonds are concerned, and Ar for lone pairs.[100] The XH bonds (b) are better localized and better separated from each other than the lone pairs (n); consequently the intrabond diagonal terms ε_b are larger in absolute value than the ε_n, and for the interbond diagonal terms it is the other way around. More spectacular is the smaller size of the off-diagonal contributions in SiH_4 compared to Ar. The IEPA error ΔE_{IEPA} defined by Eq. (141) is 6.5% of $E_{\text{corr}}^{\text{CEPA}}$ for the valence shell of SiH_4 and 12.8% for the M shell of Ar. The molecules PH_3, H_2S, and HCl fit well in between. One can also conclude from Table 2 that only the $\Delta\varepsilon_{\mu\nu}$ between joint and semijoint pairs are important, those between disjoint pairs (which have no orbital in common) are negligible, but their calculation requires almost no time. All these calculations refer to a particular basis and recover about 70–80% of the valence shell (M-shell) correlation energy.

A similar calculation for the L shells of Ar and Ar^{8+} gave CEPA-correlation energies[100] of −0.1917 a.u. and −0.2154 a.u., respectively. The lower absolute value for Ar is due to the nonavailability of the M-shell AOs for a CI, the so-called exclusion effect.[5] The IEPA error ΔE_{IEPA} is only ~5% for the L shells of Ar and Ar^{8+}, whereas the corresponding error for the L-shell of Ne is ~15%. This is due to the fact that the leading term in the IEPA error is

Table 2. Diagonal and Off-Diagonal Contributions to E_{CEPA}^{corr} in the Localized Representation According to Eqs. (139) and (140)[a]

Type	SiH$_4$[b] multiplicity	Contribution	Type	Ar[b] multiplicity	Contribution
		Diagonal			
b	4	−0.03129	n	4	−0.02296
$^1bb'$	6	−0.00296	$^1nn'$	6	−0.01015
$^3bb'$	6	−0.00457	$^3nn'$	6	−0.01383
		Joint			
$^1bb'/^3bb'$	6	0.00013	$^1nn'/^3nn'$	6	0.00047
		Semijoint			
$b/^1bb'$	12	−0.00002	$n/^1nn'$	12	−0.00069
$b/^3bb'$	12	0.00019	$n/^3nn'$	12	0.00083
$^1bb'/^1bb''$	12	0.00000	$^1nn'/^1nn''$	12	−0.00024
$^1bb'/^3bb''$	24	0.00013	$^1nn'/^3nn''$	24	0.00031
$^3bb'/^3bb''$	12	0.00006	$^3nn'/^3nn''$	12	0.00063
		Disjoint			
b/b'	6	0.00001	n/n'	6	0.00036
$b/^1b'b''$	12	0.00000	$n/^1n'n''$	12	0.00000
$b/^3b'b''$	12	0	$n/^3n'n''$	12	0
$^1bb'/^1b''b'''$	3	0.00000	$^1nn'/^1n''n'''$	3	0.00005
$^1bb'/^3b''b'''$	6	0	$^1nn'/^3n''n'''$	6	0
$^3bb'/^3b''b'''$	3	0	$^3nn'/^3n''n'''$	3	0

[a] Details of the computations are given in Ref. 100.
[b] All quantities in atomic units.

proportional to $1/Z$ in the isoelectronic Ne series, whereas that of E_{IEPA}^{corr} is independent of Z. IEPA turns out to be less good for molecules with triple bonds. For N_2 ΔE_{IEPA} is $\sim 25\%$ of E_{IEPA}^{corr}.

In Table 3 one can find equilibrium distances r_e and force constants k_e for some molecules, computed in different degrees of approximation. Generally, SCF calculations give r_e which is too small and k_e which is too large, whereas the IEPA–PNO results overestimate r_e and underestimate k_e; the PNO–CI

Table 3. Equilibrium Distances (r_e) in a_0 and Force Constants (k_e) in mdyn/Å for the Totally Symmetric Stretching Frequencies of Some Molecules

Molecule	Type	SCF	IEPA–PNO	PNO–CI	CEPA–PNO	Exp.	Reference
BH	r_e	2.313	2.337	2.338	2.342	2.33	100
	k_e	3.40	3.16	3.12	3.06	3.03–306	
NH$_3$	r_e	1.892	1.932	1.907	1.912	1.912	100
	k_e	23.9	20.2	22.6	22.0	21.8	
F$_2$	r_e	2.525	2.781	2.606	2.666	2.68	101
	k_e	8.7	3.6	7.4	5.0	4.8	
N$_2$	r_e	2.020	2.167	2.060	2.078	2.074	101
	k_e	31.0	14.1	26.3	24.1	22.94	

values have errors of the same sign but smaller magnitude than the SCF results, whereas CEPA is usually very good.[97,99–101]

Equilibrium distances and force constants (like inversion and rotation barriers) contain partial information about potential hypersurfaces. CEPA reproduces these hypersurfaces in the neighborhood of the equilibrium distance, where a single Slater determinant Φ has by far the largest weight in Ψ, and where $\|\hat{S}\Phi\| \ll 1$ so that it suffices to go to second order in $\|\hat{S}\Phi\|$.

To obtain the correct dissociation of a bond (see Section 5.6) at large distances two (or more) determinants have the same weight, so that the condition $\|\hat{S}\phi\| \ll 1$ cannot be satisfied for large distances unless Φ is a multiconfiguration function.

Numerical experience on F_2 and N_2 has indicated that CEPA based on a single determinant reproduces the exact potential curves in the region where $|(\Phi, \psi)|^2 \geq 0.8$, i.e., where $\|\hat{S}\Phi\| \leq 0.45$, and $\|\hat{S}\Phi\|^2 \leq 0.2$.

Finally we comment on the transferability of the ε_μ for localized bonds. For the intra-CH bond contributions[101] (computed with the same basis), ε_h are -0.0299, -0.0301, -0.0301, -0.0298 a.u. for C_2H_2, C_2H_4, C_2H_6 (staggered), and CH_4 respectively, the intergeminal-CH bond contribution $\varepsilon_{hh'}$ for the same molecules are -0.0144, -0.0154, -0.0155 a.u., those for intervicinal CH bonds are -0.0013, -0.0014, -0.0011 a.u., respectively. Off-diagonal contributions $\Delta\varepsilon_{\mu\nu}$ have not been compiled to allow conclusions about their transferability.

More detailed numerical data can be found in Ref. 106. What is called CEPA or CEPA–PNO here corresponds to Meyer's CEPA-2. (For the difference between CEPA-1 and CEPA-2, see Section 5.3.)*

ACKNOWLEDGMENT

The author thanks Dr. R. Ahlrichs for valuable comments on a preliminary version of the manuscript and Dr. P. C. Hariharan, Dr. F. Driessler, and H. Diehl for critically reading the final draft. Frl. U. Krupinski had a hard job in typing and correcting several versions of the manuscript.

References

1. E. U. Condon and G. H. Shortley, *The Theory of Atomic Spectra*, Cambridge University Press, New York (1951).
2. W. Kutzelnigg and V. H. Smith, Jr., Open- and closed-shell states in few-particle quantum mechanics. I. Definitions, *Int. J. Quantum Chem.* **2**, 531–552 (1968).
3. V. H. Smith, Jr., and W. Kutzelnigg, Open- and closed-shell states in few-particle quantum mechanics. II. Classification of atomic states, *Int. J. Quantum Chem.* **2**, 553–562 (1968).
4. E. Steiner, Theory of correlated wave functions. II. The symmetry properties of atomic correlated wave functions, *J. Chem. Phys.* **45**, 328–337 (1966).

*See Notes Added in Proof, p. 188.

5. O. Sinanoğlu, Many-electron theory of atoms and molecules. I. Shells, electron pairs versus many-electron correlation, *J. Chem. Phys.* **36**, 706–717 (1962); Many-electron theory of atoms and molecules. II, *J. Chem. Phys.* **36**, 3198–3208 (1962); Many-electron theory of atoms, molecules, and their interactions, *Adv. Chem. Phys.* **6**, 315–412 (1964).

6. E. P. Wigner and F. Seitz, Constitution of metallic sodium I, II, *Phys. Rev.* **43**, 804–810 (1933); **46**, 509–524 (1934).

7. N. R. Kestner and O. Sinanoğlu, Intermolecular-potential-energy curves; theory and calculations on the helium–helium potential, *J. Chem. Phys.* **45**, 194–207 (1966).

8. A. C. Hurley, J. E. Lennard-Jones, and J. A. Pople, The molecular orbitals theory of chemical valency. XVI. A theory of paired electrons in polyatomic molecules, *Proc. R. Soc. London, Ser. A* **220**, 446–455 (1953).

9. J. M. Parks and R. G. Parr, Theory of separated electron pairs, *J. Chem. Phys.* **28**, 335–345 (1958).

10. R. McWeeny, The density matrix in many-electron quantum mechanics. I. Generalized product functions. Factorization and physical interpretation of the density matrix, *Proc. R. Soc. London Ser. A* **253**, 242–259 (1959).

11. W. Kutzelnigg, Direct determination of natural orbitals and natural expansion coefficients of many-electron wave functions. I. Natural orbitals in the geminal product approximation, *J. Chem. Phys.* **40**, 3640–3647 (1964).

12. R. Ahlrichs and W. Kutzelnigg, Direct calculation of approximate natural orbitals and natural expansion coefficients of atomic and molecular electronic wave functions. II. Decoupling of the pair equations and calculation of the pair correlation energies for the Be and LiH ground states, *J. Chem. Phys.* **48**, 1819–1832 (1968).

13. W. Kutzelnigg, Electron correlation and electron pair theories, *Fortschr. Chem. Forsch.* **41**, 31–73 (1973).

14. E. L. Mehler, K. Ruedenberg, and D. M. Silver, Electron correlation and the separated pair approximation in diatomic molecules. II. Lithium hydride and boron hydride, *J. Chem. Phys.*, **52**, 1181–1205 (1970).

15. M. Krauss and A. W. Weiss, Pair correlation in closed-shell systems, *J. Chem. Phys.* **40**, 80–85 (1964).

16. H. Primas, *in: Modern Quantum Chemistry* (O. Sinanoğlu, ed.), Vol. 2, pp. 45–74, Academic Press, New York (1965).

17. H. D. Ursell, The evaluation of Gibb's phase integral for imperfect gases, *Proc. Cambridge Philos. Soc.* **23**, 685–697 (1927).

18. J. E. Mayer, The statistical mechanics of condensing systems. I., *J. Chem. Phys.* **5**, 67–73 (1937).

19. B. Kahn and G. E. Uhlenbeck, Theory of condensation, *Physica* **5**, 399–416 (1938).

20. R. Brout and P. Caruthers, *Lectures on the Many-Electron Problem*, Gordon and Breach, New York (1969).

21. F. Coester, Bound states of a many-particle system, *Nucl. Phys.* **7**, 421–424 (1958).

22. R. K. Nesbet, Brueckner's theory and the method of superposition of configurations, *Phys. Rev.* **109**, 1632–1638 (1958).

23. W. Brenig, Zweiteilchennäherungen des Mehrkörperproblems. I. *Nucl. Phys.* **4**, 363–374 (1957).

24. F. Coester and H. Kümmel, Short range correlations in nuclear wave functions, *Nucl. Phys.* **17**, 477–485 (1960).

25. H. Kümmel, Compound pair states in imperfect Fermi gases, *Nucl. Phys.* **22**, 177–183 (1961).

26. J. da Providencia, Linked graph expansion for the logarithm of the norm of many-body wave functions, *Nucl. Phys.* **44**, 572–578 (1963); Cluster expansion of operator averages for systems of many particles, *Nucl. Phys.* **46**, 401–412 (1963).

27. K. Kumar, Validity of the two-particle approximation in the many-body problem, *Nucl. Phys.* **21**, 99–105 (1960).

28. R. K. Nesbet, Electronic correlation in atoms and molecules, *Adv. Chem. Phys.* **9**, 321–363 (1965).

29. P.-O. Löwdin, Studies in perturbation theory. V. Some aspects on the exact self-consistent field theory, *J. Math. Phys.* **3**, 1171–1184 (1962).

30. W. Kutzelnigg and V. H. Smith, Jr., On different criteria for the best independent-particle-model approximation, *J. Chem. Phys.* **41**, 896–897 (1964).

31. R. A. Krumhout, Exact energy self-consistent field, *Phys. Rev.* **107**, 215–219 (1957).

32. H. Kümmel and J. Q. Zabolitzky, Fully self-consistent Brueckner–Hartree–Fock and renormalized Brueckner–Hartree–Fock calculation for ^{4}He and ^{16}O, *Phys. Rev. C* **7**, 547–552 (1973).

33. J. Čížek, On the correlation problem in atomic and molecular systems. Calculation of wavefunction components in Ursell-type expansion using quantum-field theoretical methods, *J. Chem. Phys.* **45**, 4256–4266 (1966); On the use of the cluster expansion and the technique of diagrams in calculations of correlation effects in atoms and molecules, *Adv. Chem. Phys.* **14**, 35–89 (1969).

34. J. Čížek, J. Paldus, and L. Šroubkova, Cluster expansion analysis for delocalized systems, *Int. J. Quantum Chem.* **3**, 149–167 (1969).

35. J. Čížek and J. Paldus, Correlation problems in atomic and molecular systems. III. Rederivation of the coupled-pair many-electron theory using the traditional quantum chemical methods, *Int. J. Quantum Chem.* **5**, 359–379 (1971).

36. J. Paldus, J. Čížek, and I. Shavitt, Correlation problems in atomic and molecular system. IV. Extended coupled-pair many-electron theory and its application to the borane molecule, *Phys. Rev. A* **5**, 50–67 (1972).

37. H. Kümmel, Theory of many-body wave functions with correlations, *Nucl. Phys. A* **176**, 205–218 (1971).

38. H. Kümmel and H. Lührmann, Equations for linked clusters and the energy variational principle, *Nucl. Phys. A* **191**, 525–534 (1972); Equation for linked clusters and Brueckner–Bethe theory, *Nucl. Phys. A* **194**, 225–236 (1972).

39. H. P. Kelly and A. M. Sessler, Correlation effects in many Fermion systems. Multiple-particle excitation expansion, *Phys. Rev.* **132**, 2091–2095 (1963).

40. H. P. Kelly, Correlation effects in many Fermion systems. II. Linked clusters, *Phys. Rev.* **134A**, 1450–1453 (1964).

41. R. Brout, Variational methods and the nuclear many-body problem, *Phys. Rev.* **111**, 1324–1333 (1958).

42. B. H. Brandow, Compact-cluster expansion for the nuclear many-body problem, *Phys. Rev.* **152**, 863–882 (1966); Linked cluster expansion for the nuclear many-body problem, *Rev. Mod. Phys.* **39**, 771–828 (1967).

43. K. A. Brueckner, Two-body forces and nuclear saturation. III. Details of the structure of the nucleus, *Phys. Rev.* **97**, 1353–1366; Many-body problem for strongly interacting particles. II. Linked cluster expansion, *Phys. Rev.* **100**, 36–45 (1955).

44. H. A. Bethe, Nuclear many-body problem, *Phys. Rev.* **103**, 1353–1390 (1956).

45. J. D. Thouless, *The Quantum Mechanics of Many-Body Systems*, Academic Press, New York (1961).

46. R. Ahlrichs, Convergence of the $1/Z$ expansion, *Phys. Rev. A* **5**, 605–614 (1972).

47. D. Layzer, Z. Horak, M. N. Lewis, and D. P. Thompson, Second-order Z-dependent theory of many-electron atoms, *Ann. Phys.* **29**, 101–124 (1964).

48. J. Linderberg and H. Shull, Electronic correlation energy in 3- and 4-electron atoms, *J. Mol. Spectrosc.* **5**, 1–16 (1960).

49. M. Cohen and A. Dalgarno, The Hartree energies of the helium sequence, *Proc. Phys. Soc. London Ser. A* **77**, 165 (1961).

50. S. T. Epstein, Hartree–Fock Hamiltonians and separable nonlocal potentials, *J. Chem. Phys.* **41**, 1045–1046 (1964).

51. S. T. Epstein, in: *Perturbation Theory and Its Applications in Quantum Mechanics* (C. H. Wilcox, ed.), pp. 49–56, Wiley, New York (1966).

52. E. Steiner, Theory of correlated wavefunctions. III. Alternative initial approximations, *J. Chem. Phys.* **46**, 1717–1736 (1967).

53. C. Møller and M. S. Plessett, Note on the approximation treatment for many-electron systems, *Phys. Rev.* **46**, 618–622 (1934).

54. P. S. Epstein, The Stark effect from the point of view of Schrödinger's quantum theory, *Phys. Rev.* **28**, 695–710 (1926).

55. R. K. Nesbet, Configuration interaction in orbital theories, *Proc. R. Soc. London Ser. A* **230**, 312–321 (1955).

56. P. Claverie, S. Diner, and J. P. Malrieu, The use of perturbation methods for the study of the effects of configuration interaction. I. Choice of the zeroth-order Hamiltonian, *Int. J. Quantum Chem.* **1**, 751–767 (1967).

57. O. Sinanoğlu, Theory of electron correlation in atoms and molecules, *Proc. R. Soc. London Ser. A* **260**, 379–392 (1961).

58. E. Schmidt, Zur Theorie der linearen und nichtlinearen Integralgleichungen. I. Teil: Entwicklung willkürlicher Funktionen nach Systemen vorgeschriebener, *Math. Ann.* **63**, 433–476 (1907).

59. M. Golomb, *in*: *On Numerical Approximation* (R. E. Langer, ed.), pp. 275–327, Wisconsin University Press, Madison (1958).

60. D. C. Carlson and J. H. Keller, Eigenvalues of density matrices, *Phys. Rev.* **121**, 659–661 (1961).

61. A. J. Coleman, Structure of Fermion density matrices, *Rev. Mod. Phys.* **35**, 668–689 (1963).

62. P.-O. Löwdin and H. Shull, Natural orbitals in the quantum theory of two-electron systems, *Phys. Rev.* **101**, 1730–1739 (1956).

63. W. Kutzelnigg, Solution of the two-electron problem in quantum mechanics by direct determination of the natural orbitals. I. Theory, *Theor. Chim. Acta* **1**, 327–342 (1963).

64. W. A. Bingel and W. Kutzelnigg, Symmetry properties of reduced density matrices and natural p-states, *Adv. Quantum Chem.* **5**, 201–218 (1970).

65. R. Ahlrichs, W. Kutzelnigg, and W. A. Bingel, Solution of the two-electron problem in quantum mechanics by direct calculation of the natural orbitals. III. Refined treatment of the helium atom and the helium-like ions. IV. Application to the ground state of the hydrogen molecule in a one-center expansion, *Theor. Chim. Acta* **5**, 289–304, 305–311 (1966).

66. W. Kutzelnigg, *in*: *Selected Topics in Molecular Physics* (E. Clementi, ed.), pp. 91–102, Verlag Chemie, Weinheim (1972).

67. R. Ahlrichs and F. Driessler, Direct determination of pair natural orbitals. A new method to solve the multiconfiguration Hartree–Fock problem for two-electron wave functions, *Theor. Chim. Acta* **36**, 275–287 (1975).

68. P. W. Langhoff, Separation theorem for first-order pair-correlation equations, *Int. J. Quantum Chem.* **7S**, 443–448 (1973).

69. R. McWeeny and E. Steiner, The theory of pair-correlated wave functions, *Adv. Quantum Chem.* **2**, 93–117 (1965).

70. J. P. Malrieu, Cancellation occurring in the calculation of transition energies by a perturbation development of configuration interaction matrices, *J. Chem. Phys.* **47**, 4555–4558 (1967).

71. H. P. Kelly, Correlation effects in atoms, *Phys. Rev.* **131**, 684–699 (1963); Many-body perturbation theory applied to open-shell atoms, *Phys. Rev.* **136B**, 896–912 (1964); Many-body perturbation theory applied to atoms, *Phys. Rev.* **144**, 39–55 (1966); Frequency-dependent polarizability of hydrogen calculated by many-body theory, *Phys. Rev. A* **1**, 274–279 (1970); Applications of many-body diagram techniques in atomic physics, *Adv. Chem. Phys.* **14**, 129–190 (1969).

72. H. P. Kelly, *in*: *Perturbation Theory and Its Application in Quantum Mechanics* (C. H. Wilcox, ed.), pp. 215–241, Wiley, New York (1966).

73. J. Goldstone, Derivation of the Brueckner many-body theory, *Proc. R. Soc. London, Ser. A* **239**, 267–279 (1957).

74. S. K. Ma and K. A. Brueckner, Correlation energy of an electron gas with a slowly varying high density, *Phys. Rev.* **165**, 18–31 (1968).

75. K. A. Brueckner and W. Wada, Nuclear saturation and two-body-forces. Self-consistent solution and the effects of the exclusion principle, *Phys. Rev.* **103**, 1008–1016 (1956).

76. K. A. Brueckner, J. L. Gammel, and H. Weitzner, Theory of finite nuclei, *Phys. Rev.* **110**, 431–445 (1958).

77. H. A. Bethe and J. Goldstone, Effect of a repulsive core in the theory of complex nuclei, *Proc. R. Soc. London Ser. A* **238**, 551–567 (1957).

78. L. C. Gomes, J. D. Walecka, and V. F. Weisskopf, Properties of nuclear matter, *Ann. Phys.* **3**, 241–274 (1958).

79. J. Bardeen, L. N. Cooper, and J. R. Schrieffer, Theory of superconductivity, *Phys. Rev.* **108**, 1175–1204 (1957).

80. N. Bogoljubov, A new method in the theory of superconductivity. I, *Soviet Phys. JETP* **7**, 41–46 (1958).

81. M. A. Robb, Application of many-body perturbation methods in a discrete orbital basis, *Chem. Phys. Lett.* **20**, 274–277 (1973).

82. R. J. Bartlett and D. M. Silver, Many-body perturbation theory applied to electron pair correlation energies. I. Closed-shell first-row diatomic hydrides, *J. Chem. Phys.* **62**, 3258–3268 (1975).

83. M. A. Robb, Pair functions and diagrammatic perturbation theory, *in*: *Computational Techniques in Quantum Chemistry and Molecular Physics* (G. H. F. Diercksen *et al.*, eds.), pp. 435–503, D. Reidel, Dordrecht, Holland (1975).

84. K. F. Freed, Many-body theories of the electronic structure of atoms and molecules, *Annu. Rev. Phys. Chem.* **22**, 313–346 (1971).

85. S. Diner, J. P. Malrieu, P. Claverie, and F. Jordan, Fully localized bond orbitals and the correlation problem, *Chem. Phys. Lett.* **2**, 319–323 (1968).

86. S. Diner, J. P. Malrieu, and P. Claverie, Localized bond orbitals and the correlation problem. I. Perturbation calculation of ground-state energy, *Theor. Chim. Acta* **13**, 1–17 (1969).

87. J. P. Malrieu, P. Claverie, and S. Diner, Localized bond orbitals and the correlation problem. II. Application to π-electron systems, *Theor. Chim. Acta* **13**, 18–45 (1969).

88. S. Diner, J. P. Malrieu, F. Jordan, and M. Gilbert, Localized bond orbitals and the correlation problem. III. Energy up to the third-order in the zero-differential overlap approximation. Application to π-electron systems, *Theor. Chim. Acta* **15**, 100–110 (1969).

89. F. Jordan, M. Gilbert, J. P. Malrieu, and V. Pincelli, Localized bond orbitals and the correlation problem. IV. Stability of the perturbation energies with respect to bond hybridization and polarity, *Theor. Chim. Acta* **15**, 211–224 (1969).

90. C. F. Bender and E. R. Davidson, Correlation energy and molecular properties of hydrogen fluoride, *J. Chem. Phys.* **47**, 360–366 (1967).

91. T. L. Barr and E. R. Davidson, Nature of the configuration-interaction method of *ab initio* calculations. I. Neon ground state, *Phys. Rev. A* **1**, 644–658 (1970).

92. R. K. Nesbet, T. L. Barr, and E. R. Davidson, Correlation energy of the neon atom, *Chem. Phys. Lett.* **4**, 203–204 (1969).

93. A. Weiss, Symmetry-adapted pair correlations in Ne, F^-, Ne^+, and F, *Phys. Rev. A* **3**, 126 (1971).

94. C. M. Moser and R. K. Nesbet, Atomic Bethe–Goldstone calculations of term splittings, ionization potentials, and electron affinities for B, C, N, O, F, and Ne. II. Configurational excitations, *Phys. Rev. A* **6**, 1710–1715 (1972).

95. J. W. Viers, F. E. Harris, and H. F. Schaeffer III, Pair correlations and the electronic structure of neon, *Phys. Rev. A* **1**, 24–27 (1970).

96. D. A. Micha, Many-body contributions to atomic correlation energies, *Phys. Rev. A* **1**, 755–764 (1970).

97. W. Meyer, Ionization energies of water from PNO-CI calculations, *Int. J. Quantum Chem.* **5**, 341–348 (1971); PNO-CI Studies of electron correlation effects. I. Configuration expansion by means of nonorthogonal orbitals, and application to the ground state and ionized states of methane, *J. Chem. Phys.* **58**, 1017–1035 (1973); PNO-CI and CEPA studies of electron correlation effects. II. Potential curves and dipole moment functions of the OH radical, *Theor. Chim. Acta* **35**, 277–292 (1974).

98. R. Ahlrichs, H. Lischka, V. Staemmler, and W. Kutzelnigg, PNO-CI (pair natural orbital configuration interaction) and CEPA-PNO (coupled electron pair approximation with pair natural orbitals) calculations of molecular systems. I. Outline of the method for closed-shell states, *J. Chem. Phys.* **62**, 1225–1234 (1975).

99. R. Ahlrichs, F. Driessler, H. Lischka, V. Staemmler, and W. Kutzelnigg, PNO-CI (pair natural orbital configuration interaction) and CEPA-PNO (coupled electron pair approximation with pair natural orbitals) calculations of molecular systems. II. The molecules BeH_2, BH, BH_3, CH_4, CH_3^-, NH_3 (planar and pyramidal), H_2O, OH_3^+, HF, and the Ne atom, *J. Chem. Phys.* **62**, 1235–1247 (1975).

100. R. Ahlrichs, F. Keil, H. Lischka, W. Kutzelnigg, and V. Staemmler, PNO-CI (pair natural orbital configuration interaction) and CEPA-PNO (coupled electron pair approximation with pair natural orbitals) calculations of molecular systems. III. The molecules MgH_2, AlH_3, SiH_4, PH_3 (planar and pyramidal), H_2S, HCl and the Ar atom, *J. Chem. Phys.* **63**, 455–463 (1975).

101. R. Ahlrichs, H. Lischka, B. Zurawski, and W. Kutzelnigg, PNO-CI (pair natural orbital configuration interaction) and CEPA-PNO (coupled electron pair approximation with pair natural orbitals) calculations of molecular systems. IV. The molecules N_2, F_2, C_2H_2, C_2H_4, and C_2H_6 *J. Chem. Phys.* **63**, 4685–4694 (1975).

102. K. F. Freed, Many-body approach to electron correlation in atoms and molecules, *Phys. Rev.* **173**, 1–24 (1968).

103. L. Szasz, Über die Berechnung der Korrelationsenergie der Atomelektronen, *Z. Natur-forsch.* **15a**, 909–926 (1960); Atomic many-body problem. I. General theory of correlated wave functions, *Phys. Rev.* **126**, 169–181 (1962); Formulation of the quantum-mechanical many-body problem in terms of one- and two-particle functions, *Phys. Rev.* **132**, 936–947 (1963); Pseudopotential theory of atoms and molecules. I. A new method for the calculation of correlated pair functions, *J. Chem. Phys.* **49**, 679–691 (1968).

104. H. J. Silverstone and O. Sinanoğlu, Many-electron theory of nonclosed-shell atoms and molecules. I. Orbital wavefunction and perturbation theory. II. Variational theory, *J. Chem. Phys.* **44**, 1899–1907, 3608–3617 (1966).

105. V. Staemmler and M. Jungen, Application of the independent electron pair approach to the calculation of excitation energies, ionization potentials, and electron affinities of first row atoms, *Theor. Chim. Acta* **38**, 303 (1975).

106. W. Meyer, A recent CI method based on pseudonatural orbitals, this volume, Chapter 11.

107. B. Roos, A new method for large-scale CI calculations, *Chem. Phys. Lett.* **15**, 153–159 (1972).

108. P. Siegbahn and B. Roos, this volume, Chapter 7.

109. H. F. Schaefer III, *Ab initio* potential curve for the $X^3\Sigma_g^-$-state of O_2, *J. Chem. Phys.* **54**, 2207–2211 (1971).

110. R. K. Nesbet, Atomic Bethe–Goldstone equations, *Adv. Chem. Phys.* **14**, 1–34 (1969).

111. G. A. van der Velde and W. C. Nieuwpoort, Generalized Bethe–Goldstone calculations on molecules, *Chem. Phys. Lett.* **13**, 409–412 (1972).

112. G. A. van der Velde, Thesis, Groningen (1974).

113. E. L. Mehler, Independent pair–potential correlated wave functions, *Int. J. Quantum Chem.* **S7**, 437–442 (1973); Orbital correlation effects: The independent pair-potential approximation with application to the ground state and first ionized state of boron hydrides, *Theor. Chim. Acta* **35**, 17–32 (1974).

114. H. Primas, Generalized perturbation theory for quantum mechanical many-particle problems, *Helv. Phys. Acta* **34**, 331–351 (1961).

115. R. J. Yaris, Linked cluster theorem and unitarity, *J. Chem. Phys.* **41**, 2419–2421 (1964); Cluster expansion and the unitary group, *J. Chem. Phys.* **42**, 3019–3024 (1965).

116. B. Levy and G. Berthier, Generalized Brillouin theorem for multiconfiguration S.C.F. theories, *Int. J. Quantum Chem.* **2**, 307–314 (1968).

117. F. Maeder and W. Kutzelnigg, *Ab initio* calculation of van der Waals constants (C_6, C_8, C_{10}) for two-valence-electron atoms, including correlation effects, *Chem. Phys. Lett.* **37**, 285 (1976).

118. H. Margenau and N. R. Kestner, *Theory of Intermolecular Forces*, Pergamon, New York (1969/71).

119. N. R. Kestner, *Chem. Phys.* **3**, 193 (1974).

120. W. Meyer, private communication.

121. B. Liu and A. D. McLean, Accurate calculation of the attractive interaction of two ground state helium atoms, *J. Chem. Phys.* **59**, 4557–4558 (1973).

122. E. Steiner, Theory of correlated wavefunctions. IV. A "configuration interaction plus perturbation" approach, *J. Chem. Phys.* **46**, 1727–1735 (1967).

123. K. Roby, On the theory of electron correlation in atoms and molecules. II. General cluster expansion theory and the general correlated wave function method, *Int. J. Quantum Chem.* **6**, 101–123 (1972).

124. E. R. Davidson and C. F. Bender, Correlation energy calculations and unitary transformations for LiH, *J. Chem. Phys.* **49**, 465–466 (1968).
125. W. Kutzelnigg *in*: *Localization and Delocalization in Quantum Chemistry* (O. Chalvet *et al.*, ed.), pp. 143–153, D. Reidel Dordrecht, Holland (1975).
126. I. Lindgren, The Rayleigh–Schrödinger perturbation and the linked-diagram theorem for a multiconfigurational model space, *J. Phys. B* **7**, 2441 (1974).
127. W. Kutzelnigg, Note on perturbation theory of electron correlation, *Chem. Phys. Lett.* **35**, 283–285 (1975).
128. R. F. Hausman, S. D. Bloom and C. F. Bender, A new technique for describing the electronic states of atoms and molecules—The vector method, *Chem. Phys. Lett.* **32**, 483 (1975).
129. J. H. van Vleck, Sigma-type doubling and electron spin, *Phys. Rev.* **33**, 467–490 (1929).
130. H. J. Werner and W. Meyer, Finite perturbation calculations for the static dipole polarizabilities of the first row atoms, *Phys. Rev.* **13A**, 13–16 (1976).
131. J. S. Binkley and J. A. Pople, Møller–Plessett theory for atomic ground state energies, *Int. J. Quantum Chem.* **9**, 229–236 (1975); J. A. Pople, J. S. Binkley, and R. Seeger, Theoretical models incorporating electron correlation, *Int. J. Quantum Chem.*, to be published.
132. W. Meyer, Theory of self-consistent pairs. An iterative method for correlated many-electron wavefunctions, *J. Chem. Phys.* **64**, 2901–2907 (1976).
133. J. da Providencia and C. M. Shakin, Some aspects of short-range correlations in nuclei, *Ann. Phys.* **30**, 95–118 (1964).
134. W. Meyer and P. Rosmus, PNO-CI and CEPA studies of electron correlation. III. Spectroscopic constants and dipole moment functions for the ground states of the first-row and second-row diatomic hydrides, *J. Chem. Phys.* **63**, 2356–2375 (1975).
135. H. J. Werner and W. Meyer, PNO-CI and PNO-CEPA studies of correlation effects. V. Static dipole polarizabilities of small molecules, *Mol. Phys.* **31**, 855–872 (1976).
136. F. Keil and R. Ahlrichs, Theoretical study of SN_2 reactions. *Ab initio* computation on HF and CI level, *J. Am. Chem. Soc.* **98**, 4787–4793 (1976).
137. K. Hoheisel and W. Kutzelnigg, *Ab initio* calculation including electron correlation of the structure and binding energy of BH_5 and $B_2H_7^-$, *J. Am. Chem. Soc.* **97**, 6970–6975 (1975).
138. F. Keil and W. Kutzelnigg, The chemical bond in phosphoranes. Comparative *ab initio* study of PH_3F_2 and the hypothetical molecules NH_3F_2 and PH_5, *J. Am. Chem. Soc.* **97**, 3623–3632 (1975).
139. R. Ahlrichs, Theoretical study of the H_5^+ system, *Theor. Chim. Acta* **39**, 149–160 (1975).

Notes Added in Proof

To Section 3.2. The Møller–Plessett (MP) version of Rayleigh–Schrödinger perturbation theory for the treatment of electron correlation has recently been applied extensively both to second and third order by Pople and co-workers.[131] It should be mentioned that the second-order MP treatment can be regarded as a first approximation to IEPA and third-order MP as a first approximation to CEPA (see Sections 4.2 and 4.3).

To Section 4.5. Meyer[132] has proposed an improved CEPA method, called the theory of self-consistent pairs (SCP), which is closely related to Ahlrichs' and Driessler's method[67] for two-electron systems. These methods appear to be both more accurate and more economical than the PNO expansion approaches used so far. They achieve this by avoiding any integral transformation, even the partial one required in PNO-CI and CEPA-PNO approaches.

To Section 5.2. The unitary ansatz has occasionally been used in nuclear physics in a different context and a different formalism[133] with rather limited success.

To Section 6. CEPA-PNO calculations on the spectroscopic constants and other ground-state properties of diatomic hydrides[134] gave excellent results that were in some cases more reliable than experimental values. CEPA also proved to be a very good method for the calculation of polarizabilities of atoms[130] and molecules.[135] There are further a number of recent applications of CEPA to problems of chemical interest, like transition states of SN_2 reactions[136] and questions concerning the stability and the geometry of unknown or unstable molecules as, e.g., BH_5,[137] PH_5,[138] or H_5^+.[139]

6

The Method of Configuration Interaction

Isaiah Shavitt

1. Introduction

1.1. Correlation Energy

As the scope of quantum chemistry broadened from the consideration of stable molecules near equilibrium to encompass potential curves and surfaces, transition states, radicals, ions, and excited states, the shortcomings of the Hartree–Fock (HF) approximation for the description of the electronic structure of molecular systems became increasingly evident. The energy error of the restricted HF wave function, i.e., the difference between the HF limit energy (which is the limit approached by restricted self-consistent field calculations as the basis set approaches completeness) and the exact solution of the non-relativistic Schrödinger equation, has been termed the *correlation energy*.[1] It reflects the fact that the HF Hamiltonian contains the average, rather than instantaneous, interelectron potential, and thus neglects the correlation between the motions of the electrons.

The correlation energy is usually a very small fraction of the total energy of a molecular system (0.5% in H_2O, for example), but chemistry is primarily concerned with small energy differences such as those between different electronic states or between different geometries (the binding energy of H_2O, for example, is also 0.5% of the total energy), and these differences may be seriously affected by the correlation error.[1,2] Particularly serious is the fact

Isaiah Shavitt • Battelle Memorial Institute, Columbus, Ohio and Department of Chemistry, The Ohio State University, Columbus, Ohio

that for most molecules the HF approximation becomes progressively less satisfactory as the atoms separate,[1,3] resulting in grossly distorted potential curves and surfaces. Furthermore, the correlation error may vary considerably between different electronic states of the same system. In fact, the principal shortcoming of the conventional definition of correlation energy is that in addition to "dynamical" correlation, which is the concept referred to by the intuitive idea of correlated electronic motion,[4] it also includes nondynamical effects[5,6] due to near degeneracies and rearrangement of electrons within partly filled shells. Thus, in the case of potential curves and surfaces, the incorrect dissociation behavior of most molecules in the HF approximation leads to the counterintuitive result that the correlation energy often increases as the electrons move apart with the separating atoms.[1,7] Such anomalies also appear in the study of excited states,[8] and suggest that some other model, such as an appropriate multiconfigurational wave function, could profitably be used instead of the HF model as a reference point for the definition of correlation energy.[8]

Many schemes have been devised and employed in order to overcome the shortcomings of the Hartree–Fock approach, ranging from relatively minor improvements (unrestricted and extended HF) through moderate extensions (multiconfiguration HF) to more extensive and elaborate methods (perturbation theory methods and configuration interaction). Of these, the configuration interaction method appears to be the most generally applicable, and the most straightforward, though it suffers from serious computational difficulties. Unlike perturbation methods, it can be applied routinely in cases in which the HF wave function is not an adequate zero-order approximation, and thus it can deal in a balanced way with the computation of potential curves and surfaces and with excited states.

1.2. Configuration Interaction

The configuration interaction (CI) method is a straightforward application of the Ritz[9] method of linear variations to the calculation of electronic wave functions. It has been used for both atoms[10] and molecules (either in terms of atomic orbitals,[11] the valence bond method, or in terms of molecular orbitals[12]) since the early days of quantum mechanics, but until electronic computers became available its application was mostly limited to very small systems or to semiempirical treatments. Some of the more ambitious precomputer CI calculations were carried out on the π-electron system of benzene (with some serious approximations) by Parr *et al.*,[13] on some atomic systems by Bernal and Boys[14] and by Boys,[15,16] and on the HF molecule by Kastler.[17] The earliest CI calculations which employed an electronic computer were probably those of Meckler[18] on O_2 (using the computer only for the solution of the matrix eigenvalue problem), of Boys and Price[19] on Cl, Cl⁻, S, and S⁻, and of Boys *et al.*[20] on BH, H_2O, and H_3.

Two characteristics of the CI method make it particularly attractive and important. First, unlike methods which rely on more restricted forms of trial wave functions (such as self-consistent field, separated pairs, etc.), it is capable, *in principle*, of providing accurate solutions of the nonrelativistic, clamped-nuclei Schrödinger equation. Second, it is a general method, applicable *in principle* to any stationary state of an atomic or molecular system. While the qualification "in principle" in the above statements is significant, because of the slow convergence of the CI expansion and of difficulties in dealing with highly excited states, it should be seen in the context of the very rapid improvement in our capabilities for such calculations over the last two decades. Even if it is overoptimistic to expect the same rate of improvement in computer design, computational procedures, and theoretical methods in the near future, it is safe to assume that over the long run the configuration interaction method, or one of its variations described in the present volume, will becomes a standard practical tool for obtaining highly accurate answers to questions involving the electronic structure of small- and medium-size systems.

While the computational difficulties of the CI method are formidable, its conceptual simplicity and generality are very appealing. It results from the application of the variation principle to a trial function which is written as a linear combination of many predetermined terms, each of which is expressed in turn in terms of products of one-electron orbitals. These expansion terms are usually made to satisfy some or all of the boundary and symmetry conditions which apply to the desired final wave function. This can be done for open-shell states almost as easily as for closed-shell systems, for excited states almost as easily as for ground states, and far from equilibrium as easily as near the equilibrium geometry of the system. The very generality of the method can, however, be a disadvantage in some cases, particularly those involving highly excited states. Thus, for example, the restricted form of the SCF wave function makes it possible to use that method to obtain an approximate solution for a core-vacancy state, while it would be very difficult (or even impossible) to prevent a CI wave function for a stationary state embedded in a continuum from acquiring the character of lower states.

The self-consistent field (SCF) method has by now become fairly standardized, at least for closed-shell systems, and several "black box" programs are now available even for the casual user who does not want to learn much about the theory or its implementation. This situation is still far from realization for CI calculations. While very much has been learned in recent years about optimum procedures for various stages of the calculation, the size and complexity of the programs are considerable and many choices and decisions have to be made by the user. It is the aim of this chapter to acquaint the reader with the basic theory and with the various stages of CI calculations, and to discuss the various choices (such as basis sets, orbitals, and configuration functions) and alternative procedures available.

Before proceeding to the formal development of the theory, a few words on the name of the method are in order. The term "configuration interaction" dates from the early days of quantum mechanics, and was taken to imply the cooperative effect of several electronic configurations of a system in stabilizing it, relative to the energy of a single configuration.[21] As recently pointed out by Mulliken,[22] it has often been argued that this name is not appropriate for the CI method as commonly practiced, and alternative terms, such as "superposition of configurations" (SOC) and "configuration mixing" (CM), have been proposed and used by various authors. The term configuration interaction, and particularly the abbreviation CI, are well entrenched, however, and will be used in this chapter.

2. Formalism

2.1. The Linear Variation Method

In this chapter we are concerned with the solution of the time-independent, nonrelativistic, clamped-nuclei, electronic Schrödinger equation,

$$\hat{H}\Psi = E\Psi \tag{1}$$

where the Hamiltonian operator $\hat{H}$ is given by

$$\hat{H} = V_{nn} + \sum_{\mu} \hat{h}_{\mu} + \sum_{\mu < \nu} g_{\mu\nu} \tag{2}$$

with (in atomic units)

$$V_{nn} = \sum_{A < B} Z_A Z_B / r_{AB} \tag{3}$$

$$\hat{h}_{\mu} = -\tfrac{1}{2}\nabla_{\mu}^2 - \sum_{A} Z_A / r_{A\mu} \tag{4}$$

$$g_{\mu\nu} = 1/r_{\mu\nu} \tag{5}$$

The uppercase Latin subscripts represent nuclei (with atomic numbers Z_A, etc.), while the lowercase Greek subscripts indicate electrons, and r_{AB}, etc., stand for the distances between the respective particles. The V_{nn} term, which is constant in the clamped-nuclei model, is often ignored throughout the calculation (except when quoting the final total energy E), since it can be accounted for simply by a shift in the origin of the energy scale. When necessary, we shall distinguish between the different solutions of (1) by means of subscripts on the symbols for the eigenvalues (energies) E and corresponding eigenfunctions (wave functions) Ψ. The numbering of the solutions is such that

$$E_1 \leq E_2 \leq \cdots \tag{6}$$

The method of solution we adopt is based on the variation principle,* which states that the energy functional

$$E[\Psi] \equiv \langle \Psi | \hat{H} | \Psi \rangle / \langle \Psi | \Psi \rangle \tag{7}$$

(using the familiar Dirac bracket notation, with $\langle f | \hat{P} | g \rangle \equiv \int f^* \hat{P} g \, d\tau$) is stationary with respect to all variations in the "trial function" Ψ when (and only when) Ψ coincides with an exact eigenfunction of $\hat{H}$. Furthermore, an approximate solution is obtained when $E[\Psi]$ is stationary with respect to some class of variations in an appropriately chosen restricted form (*ansatz*) of the trial function, and in all cases $E[\Psi]$ is an upper bound to the lowest eigenvalue E_1 of $\hat{H}$.[30] An upper bound to a higher eigenvalue E_p is obtained when it is possible to ensure that Ψ be orthogonal to all *exact* eigenfunctions of $\hat{H}$ corresponding to eigenvalues lower than E_p.

The error in the variational approximation to the eigenvalue is of second order with respect to the error in the corresponding approximation to the eigenfunction. Thus, if we put

$$\varepsilon \Delta \equiv \Psi_{approx} - \Psi_{exact} \tag{8}$$

with Ψ_{approx}, Ψ_{exact}, and Δ all normalized functions and ε a real constant, so that ε is the root-mean-square error in the eigenfunction,

$$\varepsilon^2 = \varepsilon^2 \langle \Delta | \Delta \rangle = \langle \Psi_{approx} - \Psi_{exact} | \Psi_{approx} - \Psi_{exact} \rangle \tag{9}$$

then (see, e.g., Powell and Crasemann[31])

$$E[\Psi_{approx}] = E_{exact} + \varepsilon^2 \langle \Delta | \hat{H} - E_{exact} | \Delta \rangle \tag{10}$$

A bound on ε for the lowest eigenvalue has been derived by Eckart,[30]

$$\varepsilon^2 \leq (E[\Psi_{approx}] - E_1) / (E_2 - E_1) \tag{11}$$

where E_1 and E_2 are the exact lowest two eigenvalues.

The linear variation method,[9] which is at the heart of the CI approach, employs a linear expansion ansatz for the trial function,

$$\Psi = \sum_{s=1}^{n} c_s \Phi_s \tag{12}$$

---

*For general references on the variation principle and on the linear variation method see, for example, Courant and Hilbert,[23] Hylleraas,[24] and Epstein,[25] as well as most texts on quantum mechanics, particularly Pauling and Wilson,[26] Kemble,[27] Messiah,[28] and Slater.[29]

where the Φ_s are predetermined expansion functions and the linear coefficients c_s are the parameters which are varied to make $E[\Psi]$ stationary. As is well known, this leads to the generalized matrix eigenvalue equation

$$\mathbf{Hc} = E\mathbf{Sc} \tag{13}$$

where the elements H_{st} and S_{st} of the $n \times n$ Hermitian matrices $\mathbf{H}$ and $\mathbf{S}$ are defined by

$$H_{st} \equiv \langle \Phi_s | \hat{H} | \Phi_t \rangle \tag{14}$$

$$S_{st} \equiv \langle \Phi_s | \Phi_t \rangle \tag{15}$$

and the column vector $\mathbf{c}$ has the desired coefficients c_s [Eq. (12)] as components. The eigenvalue E of (13) is the value of the functional $E[\Psi]$ when the corresponding eigenvector $\mathbf{c}$ is used in (12). It is generally convenient to choose the expansion functions Φ_s so that they are orthonormal, in which case $\mathbf{S}$ (which is the metric of the expansion basis and, provided the expansion functions are linearly independent, is always positive definite) is a unit matrix, and (13) can be rewritten as a simple eigenvalue equation

$$\mathbf{Hc} = E\mathbf{c} \tag{16}$$

The n linearly independent eigenvectors $\mathbf{c}_p$ of (13) (with components c_{sp} and corresponding eigenvalues E_p) can be chosen to be orthonormal (with respect to the metric $\mathbf{S}$ in the general case),

$$\mathbf{c}_p^\dagger \mathbf{Sc}_q \equiv (\mathbf{c}_p, \mathbf{Sc}_q) \equiv \sum_{s,t=1}^{n} c_{sp}^* S_{st} c_{tq} = \delta_{pq} \tag{17}$$

and will be numbered so that

$$E_1 \le E_2 \le \cdots \le E_n \tag{18}$$

Each eigenvalue E_p of (13) is an upper bound[32,33] to the corresponding eigenvalue of $\hat{H}$, without having to satisfy the more stringent requirement that Ψ_p be orthogonal to all the *exact* eigenfunctions of $\hat{H}$ corresponding to lower eigenvalues. As additional terms are added to the expansion each eigenvalue $E_p^{(n+1)}$ of the $(n + 1)$-term expansion satisfies the inequalities

$$E_{p-1}^{(n)} \le E_p^{(n+1)} \le E_p^{(n)} \tag{19}$$

and as the expansion set $\{\Phi_s\}$ approaches a complete set in the appropriate function space, each eigenvalue $E_p^{(n)}$ of (13) approaches the corresponding eigenvalue E_p of $\hat{H}$ from above.[32] (For error bounds on excited state solutions see also Shull and Löwdin[34] and Weinberger.[35])

2.2. Configuration Functions

Next we turn to the form of the expansion functions Φ_s used in the CI calculations. These functions are often loosely referred to as "configurations," but more appropriate terms which have sometimes been used are "configuration state functions" (CSF),[36] or simply "configuration functions" (CF); the latter term, and corresponding abbreviation, will be used here. They are constructed as linear combinations of products of one-electron functions (spin-orbitals), the linear combinations being chosen so that each Φ_s satisfies all, or at least some, of the symmetry conditions which Ψ is required to satisfy. Chief among these conditions is the permutational antisymmetry requirement, which is easily satisfied by collecting the products of spin-orbitals in Φ_s into Slater determinants.[26,29] A linear combination of Slater determinants is still required, in the general case, in order to satisfy spin and space symmetry conditions. A configuration function which satisfies the spin conditions (i.e., which is an eigenfunction of the total spin operators $\hat{S}^2$ and $\hat{S}_z$ with eigenvalues appropriate for the desired state) is called a *spin-adapted* CF, and if it also satisfies space symmetry conditions it is called *symmetry adapted*.

Since the matrix element H_{st} vanishes if Φ_s and Φ_t belong to different symmetry types, the use of symmetry divides the problem of computing the electronic states of a molecule into separate, smaller problems, one for each symmetry type. The computational labor involved in a CI calculation is mostly proportional to the square of the expansion length, and thus the symmetry separation is very effective in reducing the amount of computation required. Furthermore, making each term in the CI expansion conform to spin and space symmetry conditions enforces these conditions on the final wave function and also helps in establishing the identity of the computed state. Because a symmetry-adapted wave function is automatically orthogonal to all wave functions of other symmetry types, it ensures an upper bound to the energy of an excited state which is lowest for its symmetry type, without having to include lower states in the same calculation.

A convenient way of writing a spin-adapted CF is in the form

$$\Phi_{s\lambda}(\mathbf{x}_1, \mathbf{x}_2, \ldots, \mathbf{x}_N) = \mathscr{A}\,\Xi_s(\mathbf{r}_1, \mathbf{r}_2, \ldots, \mathbf{r}_N)\Theta_\lambda(\sigma_1, \sigma_2, \ldots, \sigma_N) \qquad (20)$$

Here $\mathbf{x}_\mu$ refers collectively to the space ($\mathbf{r}_\mu$) and spin (σ_μ) coordinates of the μth electron, $\mathscr{A}$ is the antisymmetrization operator

$$\mathscr{A} = (N!)^{-1/2} \sum_{\mathscr{P}} \varepsilon(\mathscr{P})\mathscr{P} \qquad (21)$$

[$\varepsilon(\mathscr{P})$ is $+1$ or -1 depending on whether the permutation $\mathscr{P}$ is even or odd, respectively] which operates simultaneously on the space and spin coordinates of the N electrons, Ξ_s is a product of N spatial orbitals $\varphi_i(\mathbf{r})$,

$$\Xi_s(\mathbf{r}_1, \mathbf{r}_2, \ldots, \mathbf{r}_N) = \varphi_{s_1}(\mathbf{r}_1)\varphi_{s_2}(\mathbf{r}_2) \cdots \varphi_{s_N}(\mathbf{r}_N) \tag{22}$$

and Θ_λ is an N-electron spin eigenfunction which is generally a linear combination of products of N elementary one-electron spin functions $\alpha(\sigma)$ and $\beta(\sigma)$. The vector subscript $\mathbf{s}$ on $\Phi_{s\lambda}$ specifies its spatial part Ξ_s (sometimes referred to as a "space configuration") by identifying the N orbitals, φ_{s_i}, in terms of a predetermined ordered list of orbitals $\{\varphi_i\}$. The subscript λ specifies the spin part (or "spin coupling") of $\Phi_{s\lambda}$ by identifying one of a number of linearly independent (and usually orthonormal) N-electron spin eigenfunctions Θ_λ which can be constructed for the desired spin state. Using different spin functions Θ_λ with the same space configuration Ξ_s gives rise, in general, to several linearly independent (and orthogonal, if the Θ_λ are orthogonal) "spin configurations" $\Phi_{s\lambda}$. As is well known, there are limitations on the type of products allowed in the space configuration and on the type of spin function allowed with a particular space configuration, or else the operation of the antisymmetrizer $\mathcal{A}$ on the combined space and spin function will produce a vanishing Φ. Thus, any particular φ_i can appear at most twice in the orbital product Ξ_s, and when it appears twice it imposes a restriction on the allowed spin functions Θ_λ (this will be discussed in more detail in Section 4).

2.3. Orbitals and Basis Functions

The spatial orbitals φ_{s_i} used in constructing any of the "space configurations" Ξ_s are generally selected from a common set of orbitals $\{\varphi_i, i = 1, 2, \ldots, m\}$. In almost all applications these orbitals are chosen to be orthonormal,

$$\langle \varphi_i | \varphi_j \rangle = \delta_{ij} \tag{23}$$

since this greatly simplifies the calculation of the matrix elements H_{st} (and S_{st}, if the CFs are not orthonormal), as discussed below and in Section 5. The orbitals are also chosen, in general, to be of pure symmetry types, i.e., so that each belongs to an irreducible representation of the point group corresponding to the spatial symmetry of the molecule. This facilitates the symmetry adaptation of the CFs, as discussed in Section 4.3.

The nature of the orbital set $\{\varphi_i\}$ is of paramount importance in determining the ultimate success of a CI calculation. First, the one-electron function space spanned by the orbital set determines the best result that can be obtained when all possible CFs which can be constructed from the given set are included in the CI expansion and no further approximations are made. This so-called "full CI" result is invariant with respect to the choice of the individual orbitals used to span the given function space, i.e., it is invariant with respect to any unitary transformation of the orbital set. Second, the particular choice of the individual orbitals within the given space is of crucial importance in determining the convergence rate of the CI expansion. The term "convergence" is used here loosely, referring to the number of CFs needed to achieve results of a particular quality within a given orbital set. It is generally quite impractical to carry out full CI calculations for any but very small orbital sets ($m \leqslant 10$, typically), thus it becomes necessary to select an appropriate subset of all possible CFs, and the likelihood of obtaining satisfactory results with such a selection strongly depends on the choice of the orbitals. The choices of orbitals and configuration functions are discussed more fully in Sections 8 and 9, respectively.

The one-electron function space within which the orbital set is constructed is usually defined by a *basis set*. This generally consists of a set of atom-centered basis functions $\{\chi_p, p = 1, 2, \ldots, m'\}$ ($m \leq m'$) of the same type as those used for SCF and similar calculations. The orbitals are then expressed as linear combinations of the basis functions,

$$\varphi_i = \sum_{p=1}^{m'} \chi_p U_{pi} \qquad (i = 1, 2, \ldots, m) \tag{24}$$

In most cases $m = m'$, but sometimes it is advantageous to use fewer orbitals than basis functions. The choice of a basis set is discussed further in Section 8.1.

2.4. Matrix Elements and Integrals

The principal computational step of a CI calculation is the evaluation of the matrix elements H_{st} [Eq. (14)]. This step, which is discussed in detail in Section 5, is greatly simplified by choosing the orbitals $\{\varphi_i\}$ to be orthonormal. In this case the formula for the matrix element can be written in the form

$$\langle \Phi_s | \hat{H} | \Phi_t \rangle = \sum_{i,j} a_{ij}^{st} h_{ij} + \sum_{i,j,k,l} {}^{\bullet} b_{ijkl}^{st} g_{ijkl} \tag{25}$$

where the *orbital integrals* h_{ij} and g_{ijkl} are defined by

$$h_{ij} \equiv \langle \varphi_i(\mathbf{r}_1) | \hat{h}_1 | \varphi_j(\mathbf{r}_1) \rangle \tag{26}$$

$$g_{ijkl} \equiv \langle \varphi_i(\mathbf{r}_1) \varphi_k(\mathbf{r}_2) | g_{12} | \varphi_j(\mathbf{r}_1) \varphi_l(\mathbf{r}_2) \rangle \tag{27}$$

The coefficients a_{ij}^{st} and b_{ijkl}^{st} depend on the matching of orbitals and spin couplings between Φ_s and Φ_t, but are independent of the nature of the individual orbitals.

The orbital integrals can be obtained in turn from analogous *basis-set integrals,*

$$\bar{h}_{pq} = \langle \chi_p(\mathbf{r}_1)|\hat{h}_1|\chi_q(\mathbf{r}_1)\rangle \tag{28}$$

$$\bar{g}_{pqrs} = \langle \chi_p(\mathbf{r}_1)\chi_r(\mathbf{r}_2)|g_{12}|\chi_q(\mathbf{r}_1)\chi_s(\mathbf{r}_2)\rangle \tag{29}$$

by means of the transformation of Eq. (24):

$$h_{ij} = \sum_{p,q} U_{pi}^* U_{qj} \bar{h}_{pq} \tag{30}$$

$$g_{ijkl} = \sum_{p,q,r,s} U_{pi}^* U_{qj} U_{rk}^* U_{sl} \bar{g}_{pqrs} \tag{31}$$

The transformation of the two-electron integrals, Eq. (31), can require extensive computational resources in the case of large basis sets, and efficient implementation of this step, which is discussed further in Section 3, is of considerable importance.

The evaluation of the basis-set integrals, Eqs. (28) and (29), can also be very time consuming, but this step is also required for other methods which use a basis set, including SCF, and will not be discussed in detail here. Spatial symmetry of the molecule, when it exists, can be very helpful in reducing the number of basis-set integrals which need to be computed directly, and can also be used to great advantage in the integral transformations and in other stages. This will be discussed, where appropriate, in the following sections.

2.5. Outline of Computational Procedure

The computational steps required for a CI calculation can now be outlined as follows:

1. Choose a basis set $\{\chi_p\}$ and compute the basis-set integrals.
2. Choose the orbitals $\{\varphi_i\}$, Eq. (24) (e.g., by an SCF or similar calculation), and transform the basis-set integrals to orbital integrals.
3. Choose and construct a set of symmetry-adapted configuration functions appropriate to the state (or states) under consideration, and compute the Hamiltonian matrix $\mathbf{H}$ in terms of these CFs.
4. Compute the lowest (one or several) eigenvalues and corresponding eigenvectors of $\mathbf{H}$.

This can be followed by a further step in which the one-electron (and perhaps two-electron) reduced density matrices (see Section 7) are computed from the eigenvectors. Together with basis-set integrals for appropriate one-electron (and two-electron) property operators, the density matrices can be used to obtain expectation values for the corresponding electronic properties.

Reduced transition density matrices, involving two different eigenvectors, can also be obtained and used for the calculation of transition moments between two electronic states.

Various additional procedures and iterative loops can be incorporated into the above scheme in order to optimize the choice of orbitals and the selection of configuration functions.

The individual computational steps are discussed in detail in Sections 3–7. The various choices and selections which need to be made are discussed in Sections 8 and 9.

3. Integrals and Integral Transformations

3.1. Notes on Matrix Storage

The matrices which occur in CI calculations, as in most other quantum mechanical calculations, have symmetry properties which can be used to advantage in their storage and handling in the computer. Considering first a real symmetric matrix $\mathbf{A}$ $(A_{ij} = A_{ji})$, or even a complex Hermitian matrix $(A_{ij} = A_{ji}^*)$, such as the one-electron integral matrix $\mathbf{h}$ of Eq. (26), and ignoring for the moment redundancies or blocking due to spatial symmetry, it is obviously enough to compute and store just over half the elements, say those for which $i \geq j$. In order to use computer storage efficiently it is convenient to store such a matrix in a linear array, ordered by rows of the lower triangle,

$$A_{11}, A_{21}, A_{22}, A_{31}, A_{32}, A_{33}, A_{41}, \ldots, A_{nn} \tag{32}$$

In this scheme the position of element A_{ij} $(i \geq j)$ in the linear array is given by its *canonical index*,

$$[ij] \equiv \tfrac{1}{2}i(i-1) + j \qquad (i = 1, 2, \ldots, n; j = 1, 2, \ldots, i) \tag{33}$$

It should be noted that the dimension of the matrix does not enter into the formula for $[ij]$, unlike the case of storage by columns of the lower triangle (or, equivalently, rows of the upper triangle). If elements of $\mathbf{A}$ need to be located randomly, their canonical indices can be determined without multiplication from

$$[ij] = v_i + j \tag{34}$$

where the one-dimensional "locator" array $\mathbf{v}$, given by

$$v_i \equiv [i0] = \tfrac{1}{2}i(i-1) \tag{35}$$

is precomputed and stored. The computation of $\mathbf{v}$ is facilitated by the use of the recurrence

$$v_{i+1} = v_i + i \tag{36}$$

(with $v_1 = 0$). The total storage required for an $n \times n$ array **A** stored in this manner is v_{n+1}.

The two-electron integrals g_{ijkl}, Eq. (27), can be stored in a similar manner using $[ij]$ and $[kl]$ as the row and column indices, respectively, so that the location of $g_{ijkl} (i \geq j, k \geq l, [ij] \geq [kl])$ in the linear array is given by its canonical index $[ijkl] \equiv [[ij][kl]]$. The length of this array (for $i = 1, 2, \ldots, n$, etc.) is $v_{v_{n+1}+1}$, and the length of the array **v** required for locating any integral is v_{n+1}.

In many cases it is possible to take advantage of additional economies in the handling and storage of matrices. Consider first a one-electron integral matrix, like **h** [Eq. (26)], in which the orbitals are symmetry adapted. If the one-electron operator is totally symmetric, as is the case for $\hat{h}$ [Eq. (4)], the integral vanishes unless both orbitals involved are of the same symmetry species (i.e., both belong to the same row of the same irreducible representation). Thus, if the orbitals are ordered by symmetry species, the nonzero part of the matrix takes the form of a sequence of triangular blocks, each of the type **A** discussed above. In the case of orbitals belonging to multidimensional irreducible representations some blocks are redundant and can be omitted. It should also be noted that in some cases involving complex symmetry-adapted orbitals it may be more convenient to store the integral $\int \varphi_i \hat{h} \varphi_j \, dt \equiv \langle \varphi_i^* | \hat{h} | \varphi_j \rangle$, rather than $\langle \varphi_i | \hat{h} | \varphi_j \rangle$, as the (i, j) element of the matrix. In this case some of the nonvanishing blocks will be square and off-diagonal. Whatever the block structure, provided there is no more than one set of contiguous nonvanishing elements in each row of the matrix and if these sets are stored sequentially in one linear array, it is still possible to use a locator array like **v** to find any nonzero element (without having to determine which block it is in), but now the elements of **v** have to be specifically determined to take account of the block structure in such a way that element h_{ij} is in location $v_i + j$ of the linear array for all included elements. Care must be exercised, of course, to avoid attempts to access elements in the omitted vanishing or redundant blocks.

The block structure is usually more complicated for matrices of two-electron integrals [Eq. (27)], since (unless all representations are one-dimensional) the product of two symmetry-adapted orbitals may contain components of more than one symmetry type. Symmetric (lower triangle) and off-diagonal (rectangular in general) blocks will usually occur, and the procedure for locating an element may be rather complicated.

We next consider the case of a symmetric (or Hermitian) matrix which is randomly sparse, like the Hamiltonian matrix **H** [Eq. (14)]. If the degree of sparseness is high enough (typically no more than 20% of the elements of **H** are nonzero) it is advantageous to store nonzero elements only, requiring that each element be identified in some manner. Provided random access is not required, and if the nonzero elements are stored by rows of the lower triangle with the diagonal element (which should never be omitted, even if zero) last in each row, it is enough to indicate the column index t of each element H_{st}. This can be done by packing the column index in the least significant part of the computer word containing H_{st}, or by placing it in a separate integer array having the same structure as the **H** array.[37] The row index s can usually be determined from the looping structure of the program, with the diagonal element signaling the end of the row and triggering the incrementation of s.

Another situation, which occurs particularly for the basis set integrals $\bar{h}_{pq}$ and $\bar{g}_{pqrs}$, is when the elements of an array are obtained in an order which is not systematic in terms of the corresponding indices. When the array is particularly large and has to be stored externally (e.g., on tape) its elements would normally be stored in the order in which they are computed. In this case each element has to be identified fully by listing all its indices (p, q or p, q, r, s), usually packed into one computer word (often with additional information, like a code number indicating any equalities between the indices). This is done, for example, in the POLYATOM program.[38]

It is sometimes necessary to sort a randomly arranged long list of elements into some particular order. If the list is much too long to be sorted completely in the high-speed store of the computer, an ingenious two-pass sorting scheme devised by Yoshimine[39] can be used (see also McLean[40] and Bagus *et al.*[41]).

3.2. Basis-Set Integrals

The calculation of one-electron and two-electron integrals over the basis functions is a common requirement of most methods of wave function calculation for molecules, including the SCF method, and will not be covered here except for some specific comments relating to their use in CI calculations.

The most important of these is the question of accuracy, particularly in the case of large basis sets and situations of near linear dependence of basis functions. The propagation of errors in SCF calculations is usually not very serious, with errors in basis-set integrals being reflected by errors of not much greater magnitude in final energies. This is because the occupied orbitals of an SCF wave function are relatively smooth, with the coefficients U_{pi} [Eq. (24)] expressing them in terms of the basis functions generally being of order of magnitude unity, so that there is no amplification of errors in the transformation of basis-set integrals to integrals over *occupied* SCF orbitals (this step is normally implicit in the construction of the Fock matrix). In CI calculations, on the other hand, we also use the remainder of the basis-set space (the so-called "virtual" space) in order to provide correlating orbitals. Many of these orbitals are highly oscillatory, and in the case of large basis sets and near linear dependence some transformation coefficients U_{pi} (which would generally involve alternating signs) may be one or more orders of magnitude greater than unity. This means that errors in basis-set integrals are greatly amplified in the transformations to integrals over the orbitals, Eqs. (30) and (31), particularly in the case of the two-electron integrals where products of four transformation coefficients occur. Obviously, therefore, it is extremely important to obtain high accuracy, and particularly high internal consistency, in integral evaluations for large basis-set CI calculations. Even then it is often necessary to drop one or more of the most highly oscillatory orbitals from the CI expansion in order to get meaningful results[40] (leading to a situation in which $m < m'$, as mentioned above in Section 2.3). A useful guide to the possible extent of error propagation is provided by the magnitude of the biggest transformation coefficients U_{pi} for the retained orbitals, but the small magnitudes of the CI coefficients c_s [Eq. (12)] for configurations involving the corresponding orbitals should also be taken into account. (See also Löwdin.[42])

Another point to be considered is the construction of symmetry-distinct ("unique") integral lists. Such lists are constructed explicitly in some programs (such as POLYATOM[38]) and implicitly in others in order to avoid the redundant calculation of basis-set integrals which are related by symmetry operations of the molecular point group. The least sophisticated way of constructing such lists is to perform the various symmetry operations on the orbitals in each integral in order to determine if the integral is zero by symmetry or is related to an integral which has already been considered (this check can be facilitated, for example, by a comparison of canonical indices). In POLYATOM this is done individually for each integral (and making use only of equalities, at most with a change of sign, ignoring more complicated linear transformations which connect sets of integrals), and this can be very time consuming in large basis-set calculations with high symmetry. A very simple and effective improvement which can be made in this procedure is to carry out the symmetry analysis for some set of prototype integrals and then expand the resulting list at the time the integrals are computed.

More direct and efficient procedures for determining the smallest set of symmetry-distinct integrals can be obtained by group theoretical methods. This has been done by Davidson[43] as part of his algorithm for transforming the basis-set integrals to integrals over symmetry orbitals.

3.3. Orthogonalization

As noted previously (Section 2.4), the evaluation of matrix elements of the Hamiltonian over the configuration functions is greatly facilitated by constructing the CFs with orthonormal orbitals. If the orbitals $\{\varphi_i\}$ [Eq. (24)] are obtained as solutions of SCF equations, or as iterative natural orbitals (Section 8.3), or in a number of other ways, they are automatically orthonormal. In some other cases (GVB orbitals, different sets of IVOs, etc., discussed further in Section 8.4), they require orthogonalization. Even if the orbitals have been obtained by a scheme which ensures orthogonality, caution is necessary: if the procedure for obtaining the orbitals has not been carried through to adequate accuracy, or if the transformation coefficients are not transmitted to the transformation program with full precision (e.g., if they are read from cards in a format which carries an inadequate number of significant figures), then the orbitals actually used (as distinct from those which were intended to be used) would not be orthonormal to sufficient accuracy. Since the calculation of the Hamiltonian matrix elements assumes orthonormality, the matrix elements would be inaccurate in this case. As already noted, when large basis sets are used, and particularly in the case of near-linear dependence, this can lead to serious errors. Thus it is often advisable to subject the transformation coefficients U_{pi} [Eq. (24)] to an orthonormalization procedure, even if they are already assumed to represent orthonormal orbitals.

Another note of caution: Orbitals obtained from SCF calculations or from natural orbital transformations often have arbitrary phases. This is irrelevant for orbitals which belong to one-dimensional irreducible representations, but carefully controlled phase relationships are important for sets of orbitals which belong as partners to multidimensional irreducible representations. The symmetry adaptation of CFs depends on such phase relationships (i.e., on the transformation properties of sets of degenerate orbitals), and if, for example, a pair of orbitals which are supposed to transform under point-group symmetry operations like (x, y) actually transform like $(x, -y)$, the CFs obtained may not be of the right symmetry species (for a similar problem in another context see Winter *et al.*[44]). Phase control can be accomplished by prescribing the required sign of *one* nonzero transformation coefficient U_{pi} for each orbital φ_i, changing all signs in the ith column, if necessary, to satisfy this condition (for example, for a pair of degenerate e_x and e_y orbitals, a positive sign can be required for a pair of origin-centered p_x and p_y atomic orbitals, respectively).

While not essential, phase control may also be advisable for nondegenerate orbitals, so that CI coefficients may be meaningfully compared between a series of calculations describing a potential curve or surface.

The most commonly used orthonormalization method is the well-known Schmidt procedure, in which the orbitals are orthogonalized sequentially. This method has been criticized as being too arbitrary, since the orbitals are not treated symmetrically and the result depends on the sequence in which the orbitals are processed. For most applications to CI calculations (except perhaps when using valence bond or generalized valence bond orbitals) this is not a serious disadvantage, since a natural order often exists for the orbitals, or they are nearly orthogonal to begin with. When it is desirable to obtain orthonormal orbitals which differ as little as possible from some original set of (nonorthogonal) orbitals, Löwdin's[42,45] method of symmetric orthogonalization can be used.

While the principle of the Schmidt procedure is simple and straightforward, its efficient implementation in the present context is not trivial. A brief

For $i = 1, 2, \ldots, m$:

> Skip to (⋆) if $i = 1$
>
> For $q = 1, 2, \ldots, m'$:
>
> > $$a_q = \sum_{r=1}^{q} U_{ri} s_{qr} + \sum_{r=q+1}^{m'} U_{ri} s_{rq}^*$$
>
> For $j = 1, 2, \ldots, i-1$:
>
> > $$b = \sum_{q=1}^{m'} a_q U_{qj}^*$$
> >
> > For $p = 1, 2, \ldots, m'$:
> >
> > > $$\left[U_{pi} = U_{pi} - b U_{pj} \right.$$
>
> (⋆) $\quad t = \left\{ \mathrm{Re} \sum_{q=1}^{m'} U_{qi}^* \left[2 \left(\sum_{r=1}^{q-1} U_{ri} s_{qr} \right) + U_{qi} s_{qq} \right] \right\}^{-1/2}$
>
> For $p = 1, 2, \ldots, m'$:
>
> > $$\left[U_{pi} = t U_{pi} \right.$$

Fig. 1. Outline of a program for Schmidt orthogonalization. The $m' \times m$ matrix **U** contains the preliminary transformation matrix, which is replaced by the final (orthonormal) transformation matrix. The overlap matrix **s** is assumed to be stored in lower triangular (canonical) form.

outline of a suitable computational algorithm will therefore be given (Fig. 1).

It is assumed that we are given an overlap matrix **s** between a set of basis functions $\{\chi_p, p = 1, 2, \ldots, m'\}$,

$$s_{pq} \equiv \langle \chi_p | \chi_q \rangle \tag{37}$$

and a preliminary transformation matrix **U′** which defines the orbitals $\{\varphi_i', i = 1, 2, \ldots, m\}$ $(m \le m')$ to be orthogonalized in the given order. We proceed through the columns of **U′** in order; for the ith column we first compute

$$a_q = \sum_{r=1}^{m'} U_{ri}' s_{qr} \qquad (q = 1, 2, \ldots, m') \tag{38}$$

(taking into account the symmetric or Hermitian nature of **s** and its storage organization, see Fig. 1), then for $j = 1, 2, \ldots, i-1$ we compute

$$b = \sum_{q=1}^{m'} U_{qj}^* a_q \tag{39}$$

and subtract $b U_{pj}$ from U_{pi}' $(p = 1, 2, \ldots, m')$. Thus, when the nested loops over j and p have been completed, the ith column of **U′** has been replaced by

$$U_{pi}'' = U_{pi}' - \sum_{j=1}^{i-1} U_{pj} \left(\sum_{q=1}^{m'} \sum_{r=1}^{m'} U_{qj}^* U_{ri}' s_{qr} \right) \qquad (p = 1, 2, \ldots, m') \tag{40}$$

These steps (which are omitted for $i = 1$) are equivalent to replacing φ_i' by

$$\varphi_i'' = \varphi_i' - \sum_{j=1}^{i-1} \varphi_j \langle \varphi_j | \varphi_i' \rangle \tag{41}$$

We then compute the normalization factor

$$t = \left[\sum_{q=1}^{m'} (U_{qi}'')^* \sum_{r=1}^{m'} U_{ri}'' s_{qr} \right]^{-1/2} \tag{42}$$

and replace the column by its normalized equivalent,

$$U_{pi} = tU''_{pi} \qquad (p = 1, 2, \ldots, m') \tag{43}$$

We then repeat the whole process for the next column, until the complete matrix $\mathbf{U}$, defining the orthonormal orbitals $\{\varphi_i\}$, has replaced $\mathbf{U'}$.

When carried out in this manner, the computational effort for the whole process is proportional to $m(m')^2$ at most, and storage is required only for $\mathbf{s}$, $\mathbf{U}$, and the one-dimensional array $\mathbf{a}$.

The orbitals $\{\varphi_i\}$ obtained in the orthonormalization procedure are symmetry orbitals if the preliminary orbitals $\{\varphi'_i\}$ have been so chosen. Since orbitals belonging to different symmetry species are automatically orthogonal, the Schmidt process can be carried out separately for each species (i.e., for each set of columns of $\mathbf{U'}$ which define orbitals of a given type). Moreover, if some of the basis functions possess any of the symmetry elements of the molecular point group (e.g., σ and π functions in some molecules) it is often possible to separate $\mathbf{U'}$ (and $\mathbf{U}$) into blocks so that the row and column dimensions of each block are smaller than m' and m, respectively (it is not necessary for the blocks not to have any rows in common). This is particularly useful in the integral transformation stage[46] (Section 3.5).

3.4. Integral Transformations

We next consider the transformation of the basis-set integrals, Eqs. (28) and (29), into integrals over the orthonormal orbitals, Eqs. (26) and (27), formally defined by Eqs. (30) and (31). This will first be discussed without consideration of point-group symmetry, which will be taken up in the next subsection.

The one-electron integral transformation is relatively easy and fast, so we shall concentrate on the two-electron case. The same approach, on a simpler level, can be applied to the one-electron integrals.

A straightforward application of Eq. (31) will involve the computation of $m^4(m')^4$ terms (i.e., m^8 when $m = m'$), which is quite prohibitive for large m. The use of symmetry with respect to index permutations will reduce this by about a factor of 8 [since about $\frac{1}{8}m^4$ distinct elements of $\mathbf{g}$ are needed, and each requires the summation of $(m')^4$ terms]. However, it is easy to see that much redundant computation is involved in this scheme, with inner loops containing computations which are independent of some outer loop indices. A much more practical approach is obtained by breaking up the process into a sequence of four partial summations,

$$a_{pqrl} = \sum_{s=1}^{m'} U_{sl}\bar{g}_{pqrs} \tag{44}$$

$$b_{pqkl} = \sum_{r=1}^{m'} U_{rk}^* a_{pqrl} \qquad (45)$$

$$c_{pjkl} = \sum_{q=1}^{m'} U_{qj} b_{pqkl} \qquad (46)$$

$$g_{ijkl} = \sum_{p=1}^{m'} U_{pi}^* c_{pjkl} \qquad (47)$$

The amount of computation in these sums is proportional to $m(m')^4$ [Eq. (44)], $m^2(m')^3$ [Eq. (45)], $m^3(m')^2$ [Eq. (46)], and $m^4 m'$ [Eq. (47)]. Thus the overall transformation becomes an m^5 rather than an m^8 process, a very considerable improvement.

This partial summation method has been known and used for many years, but has only recently been discussed in the literature.[40,41,46–49] (An m^6 process has been described by Nesbet.[50]) Its implementation is complicated by storage problems due to the very large size of the matrices (for example, if $m = 50$ the matrix $\mathbf{g}$ contains 813,450 distinct elements). The organization of the computer program, in terms of the order of operations, storage organization, and input/output requirements, is of major importance for producing a feasible and economically viable process. The optimum procedure would depend on the characteristics of the computer (storage size, types of auxiliary or external storage available, relative cost of arithmetic and input/output operations) and on the size of the transformation required (as discussed in the Section 3.5, this size may be reduced considerably by symmetry blocking).

An outline of a transformation program for the two-electron integrals, along the lines described by Bender,[46] is given in Fig. 2. It is presented primarily as an example, in order to illustrate some of the techniques involved, since it is limited in the size of matrix it can handle (to about $m = 40$ when user-available storage is 40,000 words) and requires m transfers of the result matrix $\mathbf{g}$ into and out of auxiliary storage (which is not too costly if a large enough auxiliary core storage device, like ECS on CDC machines, is available). More efficient schemes for very large matrices, using direct-access disk files, have been described by McLean,[40] Bagus *et al.*,[41] Diercksen[48] and Elbert[51] (see also the discussion by Pendergast and Fink[49]), but in many cases symmetry blocking can reduce the maximum transformation size to within the capabilities of this type of program.[46]

The program described in Fig. 2 assumes that all the elements of $\bar{\mathbf{g}}$ and $\mathbf{U}$ are real, and requires the original matrix $\bar{\mathbf{g}}$ to be available in canonical order on an external sequential file (tape or disk). The only internal storage required for $\bar{\mathbf{g}}$ is for an input buffer. Each element of $\bar{\mathbf{g}}$ is brought in once, but is used in effect in all eight allowed permutations of its indices (note the two terms in the summation for $\mathbf{b}$ and the four terms for $\mathbf{g}$); since some permutations may not lead to distinct terms if some indices are equal, each $\bar{g}_{pqrs}$ is divided by a

$[mm]=\frac{1}{2}m(m+1)$; $[mmm]=[mm]\cdot m$

Initialize array g to zero in external storage

$[pq]=0$

For $p=1,2,...,m'$:

    Initialize array c_{jkl} ($[jkl]=1,2,...,[mmm]$) to zero

    $e'=1$

    For $q=1,2,...,p$:

        $[pq]=[pq]+1$

        Initialize array b_{kl} ($[kl]=1,2,...,[mm]$) to zero

        If $q=p$ set $e'=2$

        $[rs]=0$

        For $r=1,2,...,p$:

            Initialize array a_l ($l=1,2,...,m$) to zero

            $e=e'$

            $t=r$, but if $r=p$ set $t=q$

            For $s=1,2,...,t$:

                $[rs]=[rs]+1$

                If $s=r$ set $e=e+e$

                If $[rs]=[pq]$ set $e=e+e$

                $f=\bar{g}_{pqrs}/e$ (elements required sequentially)

                For $l=1,2,...,m$:

                    $a_l=a_l+U_{sl}f$

            For $k=1,2,...,m$:

                For $l=1,2,...,k$:

                    $b_{kl}=b_{kl}+U_{rk}a_l+U_{rl}a_k$

        For $j=1,2,...,m$:

            For $k=1,2,...,m$:

                For $l=1,2,...,k$:

                  $c_{jkl}=c_{jkl}+U_{qj}b_{kl}$

For $i=1,2,...,m$:

    For $j=1,2,...,i$:

        Bring in block of g_{ijkl} for current (i,j)

        For $k=1,2,...,i$:

            $h=k$, but if $k=i$ set $h=j$

            For $l=1,2,...,h$:

                $g_{ijkl}=g_{ijkl}+U_{pi}c_{jkl}+U_{pj}c_{ikl}+U_{pk}c_{lij}+U_{pl}c_{kij}$

        Write out block of g_{ijkl} for current (i,j)

Fig. 2. Outline of a program for the transformation of the two-electron integrals. The input matrix $\bar{\mathbf{g}}$ is used in canonical order and is brought in from tape through a buffer. All numbers are assumed real.

symmetry number e which is equal to 1, 2, 4, or 8, depending on index equalities.[46]

    The computation of the intermediate matrices **a**, **b**, and **c** is nested into the outer loops in such a way that only parts of them are needed in the computer at any one time; the indices p, q, and r on these matrices [Eqs. (44)–(47)] are suppressed, since they are not used in determining storage locations. The largest storage array required is for **c**, of length $\frac{1}{2}m^2(m+1)$, next in size being the transformation matrix array **U** (mm') and the arrays for **b** and one block of **g** [$\frac{1}{2}m(m+1)$ each]; the latter contains the partly computed g_{ijkl} for fixed i, j and all k, l. (Obviously, if m is small enough, the whole array **g** can be kept in main storage as it is being computed.) The dependence of storage requirements on

m' is very slight (for the storage of **U** only), so that m' can be much greater than m without exacerbating storage difficulties, but the computation time will be proportional to $m(m')^4$ primarily.

Zero tests for elements of $\bar{\mathbf{g}}$, **U**, **a**, **b**, **c** can be incorporated to advantage into the program if the matrices are sparse, but symmetry blocking would generally be more effective in such cases.

All matrices are usually stored in linear (one-dimensional) arrays, since linear indexing is generally more efficient, as well as more flexible in terms of storage utilization, than two-dimensional array indexing. The indices required to locate elements in the various arrays can be determined by suitable incrementation within the looping structure of the program, as illustrated in Fig. 2 for the canonical indices $[pq]$ and $[rs]$.

3.5. Use of Symmetry

Most small and medium size molecules studied by quantum chemists possess at least some elements of symmetry (though symmetry is often very limited or completely lacking in the study of potential surfaces) and when dealing with a computational step which is proportional to m^5 even one element of symmetry may enable a very significant reduction in the time and storage requirements of a calculation.

As discussed above (Section 3.3) and by Bender,[46] in many cases individual basis functions possess some of the symmetry elements of the molecule, leading to a blocked structure of the transformation matrix **U** and a sparse (or blocked) integral matrix $\bar{\mathbf{g}}$. The ranges of the summations in the transformation process can then be limited appropriately, and in many cases it may be possible to transform blocks of integral matrices independently of each other. Such blocks do not all possess the full symmetry relative to index permutations (this happens when different indices in the block refer to basis functions belonging to different symmetry sets), and a number of modified versions of the procedure described in Fig. 2 are required. These modified procedures have been described in detail by Bender.[46]

A more powerful and general approach to the use of symmetry in the integral transformations is based on carrying out the transformations in two stages: first from basis functions to *primitive symmetry orbitals* (PSO), and then from these to the orthonormal symmetry orbitals which are to be used in the construction of the configuration functions. Primitive symmetry orbitals are defined as the simplest possible linear combinations of basis functions required to form symmetry-adapted functions. The basis functions are divided into as many disjoint subsets as possible, each of which contains a set of functions which transform among themselves under the symmetry operations of the molecular point group. Each PSO is a linear combination of basis functions from just one subset, and the linear combination coefficients are chosen to be 0 and ±1 as far as possible. No attempt is made to orthogonalize or even normalize the PSO at this stage. It is obvious that all the PSOs obtained from any one basis function subset form a disjoint PSO subset, and that the transformation matrix **U** is completely blocked. The integral transformations from basis functions to PSOs can then be done in separate, relatively small blocks, using the techniques described previously.[46] Even these block transformations can still be speeded up if the program can be designed to take advantage of the fact that most nonvanishing transformation coefficients are of unit magnitude.

The techniques just described require the basis-set integrals to be arranged in canonical order by blocks, with all integrals within a block appearing explicitly. This means that the symmetry-distinct ("unique") integral lists have to be expanded to generate all sets of symmetry related integrals and sorted into the required order. This approach is used, for example, in the POLYCAL program[52] (see, e.g., R. M. Stevens[53]).

A more sophisticated and efficient approach would allow the generation of PSO integrals directly from the symmetry-distinct basis-set integrals, without even carrying out an explicit transformation by the methods described above. This approach is particularly easy when applied to the PSO integrals that arise in most types of SCF calculations,[54,55] and can contribute to very substantial economies in the SCF iteration time.[44,56] With an increased programming effort it can be applied to all integrals that occur in CI calculations.[43,55] Davidson's method[43] using double coset decompositions is particularly elegant, and can combine the generation of the symmetry-distinct basis-set integrals and the evaluation of PSO integrals from them in one efficient procedure.

The use of these techniques to obtain integrals over PSOs is usually so efficient that this stage of the transformation is no longer a serious limitation to the use of large basis sets in CI calculations. The second transformation stage, from PSO to the final orbitals $\{\varphi_i\}$, still remains, but the fact that symmetry orbital integrals are blocked, and that each final orbital is made up of PSO of one symmetry species only, allows this transformation to be carried out in blocks by the techniques previously discussed. While each block transformation of two-electron integrals is an m^5 process, the effective m is just the maximum number of orbitals of any one symmetry species. However, before the block transformations can be carried out, the PSO integrals have to be rearranged from the previous order (according to subsets depending on basis function parentage) to canonically ordered blocks according to symmetry species. If the integral lists are very long they can be reordered by the efficient sorting method of Yoshimine.[39–41]

The two-stage transformation approach is also very convenient in the determination of orbitals for CF construction. If SCF, MCSCF, or GVB calculations are used to obtain a set of orbitals (see Section 8), these can be carried out more efficiently using PSO integrals rather than basis function integrals. Iterative loops to obtain natural orbitals (Section 8.3) can also be implemented more efficiently in terms of PSO, since only the second stage of the transformation need to be repeated in each iteration.

4. Construction of Configuration Functions

4.1. Exchange Symmetry and General Considerations

Among all the symmetry properties of the wave function, one which occupies a very special role is exchange symmetry, which requires that the wave function be antisymmetric with respect to permutations of the electrons. While other symmetries (related to spin and space) may be ignored in the construction of the wave function expansion at the cost of an increase in computational effort, no such option realistically exists for the exchange symmetry. All physically meaningful states of electrons (or other fermions) correspond to antisymmetric wave functions, i.e., functions which belong to the antisymmetric irreducible representation of the symmetric group (for the simultaneous permutation of the space and spin coordinates of the electrons). Many physically nonrealizable states which correspond to other representations of the symmetric group would generally have much lower energies than those of the antisymmetric states; unless they can be completely excluded from the expansion, the computed energy will not be an upper bound to the energy of any physically realizable state, and the computed wave function will not be a reasonable approximation for any wave function of such a state.

As discussed above (Section 2.2), antisymmetry can easily be imposed by writing the configuration functions as Slater determinants (first introduced in 1929[57]) or as linear combinations of such determinants. The same functions can be expressed [Eq. (20)] in terms of an antisymmetrizer $\mathscr{A}$ [Eq. (21)] operating on a product of a space function and a spin function. This results in a very compact formalism and does not, in itself, lead to any difficulty in the calculation of matrix elements.

Several considerations should play a role in choosing a method for constructing CFs which satisfy spin and/or space symmetry conditions. Not all of these can be solved optimally and simultaneously in any practical implementation.

First and foremost, the CFs should be constructed in a way which facilitates the computation of the matrix elements of the Hamiltonian [Eq. (14)], this being the most time-consuming step in a large-scale CI calculation. Thus, the topic of the present section is closely related to that of the next section, which discusses the computation of these matrix elements.

Second, it should not be too difficult to obtain a complete and linearly independent set of CFs corresponding to any one configuration (this term is used here in its traditional sense, as just specifying the total occupancy of each set of degenerate orbitals). It is also convenient if the resulting set of CFs is orthonormal at the outset, but if this is not the case, at least it should be easy to carry out the orthonormalization at a later stage.

Third, it is useful if the CFs can be constructed so that important contributions to the wave function from any configuration are concentrated in as few CFs as possible (as in Bunge's method of minimal Hartree–Fock interacting subspaces[58,59]). This can be effective in reducing the length of the CI expansion required for a particular level of accuracy.

Functions adapted to any type of symmetry condition (including spin) can generally be obtained by applying appropriate projection operators to *primitive* (nonadapted) functions, but this often results in complicated expressions and is not necessarily the method of choice. This is the approach generally used for the exchange symmetry; while the antisymmetrizer is not idempotent (a condition which any projection operator should satisfy), since

$$\mathscr{A}^2 = (N!)^{1/2} \mathscr{A} \tag{48}$$

it is proportional to a projection operator

$$\mathcal{O}_{\mathscr{A}} = (N!)^{-1/2} \mathscr{A} = (N!)^{-1} \sum_{\mathscr{P}} \varepsilon(\mathscr{P}) \mathscr{P} \tag{49}$$

(The particular form of $\mathscr{A}$ is chosen so that its operation on a product of orthonormal spinorbitals will result in a normalized Slater determinant.)

4.2. Spin-Adapted CFs

A very extensive literature exists on the construction of spin-adapted CFs and on the computation of Hamiltonian matrix elements between them. Some of the useful reviews in this field are those by Löwdin,[60] Kotani and co-workers,[61,62] Matsen,[63] Pauncz,[64] Harris,[65,66] Löwdin and Goscinski,[67] Musher,[68] Ruedenberg and Poshusta,[69] and Salmon.[70] After a brief survey of the various types of treatment used we shall discuss one particular method in greater detail.

A relatively simple and straightforward approach to the problem of spin adaptation is based on the use of Löwdin's[60,71] spin projection operators.[64–66,72] Such an operator, expressed in product form as

$$\mathcal{O}_S = \prod_{K \neq S} \frac{\hat{S}^2 - K(K+1)}{S(S+1) - K(K+1)}$$

annihilates all unwanted spin components in the function on which it operates, leaving an eigenfunction of the total spin operator $\hat{S}^2$ with eigenvalue $S(S+1)$ (or a null result if the desired component was missing from the original function). The spin eigenfunctions Θ_λ [Eq. (20)] are obtained by operating with $\mathcal{O}_S$ on different *primitive* spin functions θ_λ, each of which is a differently ordered product of p α-factors and q β-factors, with $p+q=N$ and $p-q=2M_S$, M_S being the desired eigenvalue of $\hat{S}_z$. Each spin function may be represented by a *path diagram*[60,64–66] in which successive factors in the corresponding primitive function are described by successive upward slanting (α-factors) or downward slanting (β-factors) links. A linearly independent (but not orthonormal) set of spin functions Θ_λ is obtained by allowing only those path diagrams which do not descend below the baseline. Each Θ_λ can be expressed compactly as a linear combination of primitive functions with the aid of the so-called *Sanibel coefficients*.[72,73]

Some of the most elegant and powerful approaches to the construction of spin-adapted CFs are based on the theory of the symmetric group, and on projection operators derived from its representations. In fact, the requirement that the wave function for a system of electrons be an eigenfunction of the spin operators $\hat{S}^2$ and $\hat{S}_z$ can be viewed instead as a permutational symmetry condition on the spatial part of the wave function. This approach, developed in the work of Weyl,[74] Wigner,[75–77] Dirac,[78] and others,[79–82] predates the more prevalent formulation in terms of spin-orbitals and Slater determinants, and has been extensively revived in recent years.[61–64,67–70,83–97] It combines the treatment of spin adaptation and antisymmetry of the total wave function in one compact formalism, leading to separate permutational symmetry conditions on the spatial and spin parts of the wave function. As pointed out primarily by Matsen,[63,95] Musher,[68,91] and Gallup,[92,93] this allows the construction and use of spinless wave functions in many of the calculations of quantum chemistry.

Another and very powerful group theoretical approach, developed to handle the more complicated symmetry conditions of nuclear structure theory (see, e.g., Moshinsky[98]) and based on the representation theory of the unitary group $U(n)$[99,100] [or of the general linear group $GL(n)$], has recently been adapted for electronic wave function calculations,[101–108] including the treatment of the spatial symmetry for atoms.[101–104] It promises to be a very effective tool in atomic and molecular calculations.

The group theoretical methods for treating the spin conditions allow for great variety in the choice of the specific projection operators and spin functions, and depending on this choice can produce spin-adapted configuration functions equivalent to those obtained by a variety of other methods. For example, the Kotani–Yamanouchi operators[61,62] produce the same functions as those obtained by the successive coupling of individual electron spin functions in the "genealogical construction" (see, e.g., Pauncz[64,109]), while both spin-projected and valence-bond type functions can be produced by means of appropriate Young operators.[63,95,110,111] Several authors have reviewed and compared the different approaches.[64,66–70,95,111,112]

The actual spin functions Θ_λ [Eq. (20)] obtained in the various formalisms can be roughly classified into three general types.[66] These are (a) functions obtained by the use of spin projection operators,[60,64–66,71–73] (b) those obtained by the genealogical construction or the Kotani–Yamanouchi operators,[61,62,64,109,112–114] and (c) valence-bond type functions. We shall concentrate here on the last type.

The valence-bond type spin functions grew out of the work of Heitler and Rumer[115] and of Slater,[116] and have been called Slater bond functions[110] or bonded functions.[117,118] As has been shown by Matsen and others,[63,67,95,110–112] they can also be obtained by the group theoretical approach. Here we shall follow the analysis of Boys and Reeves (see Reeves[117,118] and Sutcliffe[119]). A similar analysis has been developed by Cooper and McWeeny.[120] As is commonly done, we confine our attention to the case of $M_S = S$ (the *principal case*), since the spatial part of the wave function as well as expectation values of spin-independent operators (including the nonrelativistic energy) are independent of M_S. (Spin functions for other values of M_S can be obtained from the principal case functions by means of the stepdown operator $\hat{S}_-$.)

The spin coupling in a CF based on a particular orbital product Ξ_s [Eq. (22)] will be described by a system of parentheses coupling pairs of orbitals in the product. For a given number of electrons N and total spin quantum number S, there will be $\frac{1}{2}N - S$ pairs of coupled orbitals, while the $2S$ orbitals remaining unpaired will be associated with dangling left parentheses. Either a left or a right parenthesis is associated with each orbital, and the coupling between them is determined by the usual conventions for associating nested pairs of parentheses in algebraic expressions. A coupled pair of orbitals $(\varphi_s(\ldots)\varphi_t)$ occupying the μth and νth positions ($\mu < \nu$) in the orbital product are associated with the spin factor

$$2^{-1/2}[\alpha(\sigma_\mu)\beta(\sigma_\nu) - \beta(\sigma_\mu)\alpha(\sigma_\nu)] \tag{50}$$

if $s \neq t$, and

$$[\alpha(\sigma_\mu)\beta(\sigma_\nu)] \tag{51}$$

if $s = t$, while an uncoupled orbital is associated with an α spin factor. For example, for $s \neq t \neq u$,

$$(\varphi_s\varphi_t)(\varphi_u = \mathscr{A}\varphi_s(\mathbf{r}_1)\varphi_t(\mathbf{r}_2)\varphi_u(\mathbf{r}_3)2^{-1/2}[\alpha(\sigma_1)\beta(\sigma_2)\alpha(\sigma_3) - \beta(\sigma_1)\alpha(\sigma_2)\alpha(\sigma_3)] \tag{52}$$

In Eq. (52) the CF has been written in the form $\mathscr{A}\Xi\Theta$ [Eq. (20)]. The same CF can also be written as a linear combination of Slater determinants,

$$(\varphi_s\varphi_t)(\varphi_u = 2^{-1/2}[\mathscr{A}\varphi_s\bar{\varphi}_t\varphi_u - \mathscr{A}\bar{\varphi}_s\varphi_t\varphi_u]$$
$$= 2^{-1/2}[\mathscr{A}\varphi_s\bar{\varphi}_t\varphi_u + \mathscr{A}\varphi_t\bar{\varphi}_s\varphi_u] \tag{53}$$

where unbarred and barred orbital symbols are associated with α and β spin factors, respectively, and where the *serial notation*, in which the μth factor is understood to be a function of the μth electron, has been used. To take another example

$$(\varphi_s\varphi_s)(\varphi_t(\varphi_u\varphi_v)\varphi_w) = \tfrac{1}{2}[\mathscr{A}\varphi_s\bar{\varphi}_s\varphi_t\varphi_u\bar{\varphi}_v\bar{\varphi}_w - \mathscr{A}\varphi_s\bar{\varphi}_s\varphi_t\bar{\varphi}_u\varphi_v\bar{\varphi}_w$$
$$- \mathscr{A}\varphi_s\bar{\varphi}_s\bar{\varphi}_t\varphi_u\bar{\varphi}_v\varphi_w + \mathscr{A}\varphi_s\bar{\varphi}_s\varphi_t\bar{\varphi}_u\varphi_v\varphi_w] \tag{54}$$

As noted previously, there are certain conditions that an orbital product Ξ_s [Eq. (22)] must satisfy if any nonvanishing CFs are to be obtained from it. No orbital may appear more than twice in Ξ_s, and the number d of double occupancies may not exceed $\tfrac{1}{2}N - S$. If CFs are constructed from a permuted form $\mathscr{P}\Xi_s$ of the orbital product, they will be linearly dependent on CFs obtainable from Ξ_s by the rules described below, and thus any set of N orbitals (included allowed repetitions) should be used only in one fixed ordering. In particular, it will be assumed that any repeated orbitals appear in pairs to the left of all singly occupied orbitals in Ξ_s.

A complete and linearly independent set of CFs associated with any orbital product which satisfies the above conditions can be generated by associating p left parentheses and q right parentheses, where

$$p = \tfrac{1}{2}N + S, \qquad q = \tfrac{1}{2}N - S \tag{55}$$

with the orbitals of the product (one parenthesis to each orbital) in all possible ways subject to the rules[117,119] that (a) repeated orbitals are coupled to each other, and (b) there are no dangling right parentheses (i.e., there are more left than right parentheses to the left of any right parenthesis). The second rule is analogous to the path diagram rule (see above), used to ensure linear independence among spin functions constructed by the spin projection operator method. Thus, if each left parenthesis is represented by an upward-slanting link and each right parenthesis by a downward-slanting link in the path diagram, this rule ensures that all diagrams which do not descend below the baseline shall be represented. While the interpretation of each path diagram in the present method is different from that of the spin projection method (see also Salmon[70] and Ruttink[112]), it is clear that the same number of spin functions is produced. This number is given by[64,77]

$$f(N', S) = \binom{N'}{q'} - \binom{N'}{q' - 1} \tag{56}$$

where N' is the number of singly occupied orbitals,

$$N' = N - 2d \tag{57}$$

(d being the number of double occupancies), and

$$q' = q - d = \tfrac{1}{2}N' - S \tag{58}$$

With each spin function Θ_λ we shall associate a *signature* $\pi_\lambda = \pi(\Theta_\lambda)$, defined as the number of pairs of coupled parentheses $(\,)(\,) \cdots (\,)$ which occur before any appearance of a left parenthesis not followed immediately by a right parenthesis in the corresponding coupling scheme (*cf.* Ruedenberg and co-workers[121,122]). In general there may be more than one linearly independent spin function with a given signature for any particular case (N', S).

As an example consider an orbital product of the form $aabcdef$ for a doublet state. In this case $N = 7$, $S = \tfrac{1}{2}$, $d = 1$, $N' = 5$, $q' = 2$, and the number of independent spin functions is $f(5, \tfrac{1}{2}) = 5$. The resulting CFs, with the corresponding signatures, are:

$$
\begin{array}{ll}
(aa)(bc)(de)(f & (\pi = 3) \\
(aa)(bc)(d(ef) & (\pi = 2) \\
(aa)(b(cd)e)(f & (\pi = 1) \\
(aa)(b(cd)(ef) & (\pi = 1) \\
(aa)(b(c(de)f) & (\pi = 1)
\end{array}
$$

The CFs obtained in this manner from one orbital product are normalized (assuming orthonormal orbitals), but are not orthogonal. (CFs obtained from *different* orbital products, i.e., from products which are not just a permutation of each other, are orthogonal to each other if the orbitals are orthogonal.) They can, of course, be orthogonalized by a Schmidt transformation. It is convenient, in any particular CI expansion, to use a single *canonical* list of spin functions, constructed for the case of the greatest number N'_{max} of singly occupied orbitals and ordered according to decreasing signature. (The order of spin functions having the same signature is immaterial at this point, but will be used in the next section to facilitate symmetry adaptation.) In this way, for any particular orbital product with $N' \le N'_{max}$, ordered, as previously specified, with all d double occupancies at the beginning, the first $f(N', S)$ spin functions from the canonical list can be used to construct the complete linearly independent set. Furthermore, if the canonical set is Schmidt orthogonalized in the specified sequence, then all subsets used with the different N' values will be orthonormal. Thus only one Schmidt transformation matrix (which is upper triangular in this case) need be constructed, with the appropriate upper-left submatrix of it being used in any particular case.

It can be shown[123] that with a particular ordering of the spin functions which have the same signature in the canonical set, the resulting Schmidt-orthonormalized spin functions will be of the Serber type,[69,70,80] i.e., they will have definite parity relative to each *geminal transposition*[69] (a transposition of an odd-numbered electron with the following even-numbered one). Thus they will also be appropriate for use in the new formalism developed by Ruedenberg and co-workers[69,121,122] for the construction of spin-adapted CFs and for the computation of matrix elements between them. Furthermore, this particular ordering will be useful in the construction of symmetry-adapted CFs for axial point groups,[124] as discussed further in the next section.

The Boys–Reeves formalism described here provides a reasonable compromise between the different *desiderata* for spin-adapted CFs. As discussed in Section 5, it provides a practical and effective procedure for the calculation of Hamiltonian matrix elements between bonded functions. While the individual bonded functions are not orthogonal to each other, employing a canonical set of spin couplings allows the use of a single, relatively small Schmidt transformation matrix for the orthonormalization of all subsets of nonorthogonal CFs in the expansion. Also (as discussed in Section 5), this allows for the easy evaluation of matrix elements between the *orthonormal* CFs by simple transformations of matrix element blocks involving the nonorthogonal bonded functions. Furthermore, the orbitals in each product can often be ordered in such a way that the important contributions to the wave function can be concentrated in a relatively small number of CFs, since each bonded function can be interpreted in a simple, physically meaningful way so that important spin couplings can be identified. Unlike some of the other schemes, the use of bonded functions does not require the generation and storage of large amounts of information, such as representation matrices of different permutations, which expands very rapidly with an increase in the number of electrons. (For additional discussion of the relative merits of different spin functions see Harris.[66])

4.3. Symmetry-Adapted CFs

The construction of symmetry-adapted CFs (SACFs) is facilitated considerably by the use of symmetry-adapted orbitals. Though it is possible to obtain a SACF by the application of an appropriate group theoretical projection operator[77,125–130] to a spin-adapted CF [Eq. (20)] even if the spatial part Ξ is not a product of symmetry-adapted orbitals (at least in the case of finite symmetry groups, and provided a nonvanishing projection is obtained), the form of the projected function is simplified considerably if symmetry-adapted orbitals are used.

The construction of symmetry-adapted orbitals is a relatively easy problem in most molecular calculations (see, e.g., Wigner,[77] Altmann,[125]

McWeeny,[131] and Flurry[132]). They can often be written down by inspection, and since the number of orbitals is usually not very large the use of the computer is not often required for this step. Automated methods and computer programs are available, however, for the cases in which they are needed.[133-139] SCF orbitals or approximate natural orbitals, which are often used in CI calculations, are usually symmetry adapted, but if they are not (as can be the case in some unrestricted or open-shell calculations) it is generally advantageous to use symmetry-adapted linear combinations in the CI treatment.

For the purpose of discussing the construction of SACFs it is convenient to distinguish four different cases, according to the type of point group involved. In order of increasing complexity they are: (a) Abelian groups ($\mathscr{C}_n$, $\mathscr{C}_{nh}$, $\mathscr{S}_n$, as well as $\mathscr{D}_{2h}$ and all its subgroups), (b) non-Abelian axial groups ($\mathscr{C}_{nv}$, $\mathscr{D}_n$, and $\mathscr{D}_{nh}$ for $n > 2$, $\mathscr{D}_{nd}$, $\mathscr{C}_{\infty v}$, and $\mathscr{D}_{\infty h}$), (c) cubic and icosahedral groups ($\mathscr{T}$, $\mathscr{T}_h$, $\mathscr{T}_d$, $\mathscr{O}$, $\mathscr{O}_h$, $\mathscr{I}$, and $\mathscr{I}_h$), and (d) the orthogonal group in three dimensions $O(3)$ (the atomic symmetry group).

The case of Abelian groups is quite simple, since all irreducible representations (IRs) are one-dimensional and a product of symmetry-adapted orbitals is itself symmetry adapted. Each SACF can be in the form of a single bonded function or any other spin-adapted CF of the type of Eq. (20), $\Phi = \mathscr{A}\Xi\Theta$, with the space part Ξ given as a single orbital product [Eq. (22)]. All that is needed in order to be able to construct SACFs belonging to a given symmetry species is a knowledge of the multiplication table of the IRs. A complication arises, however, in the case of groups with degenerate pairs of complex IRs ($\mathscr{C}_n$, $\mathscr{C}_{nh}$, and $\mathscr{S}_n$, with $n > 2$), since symmetry-adapted orbitals belonging to the complex IRs are necessarily complex. As a result, the transformation matrix **U** [Eq. (24)] and some of the symmetry-orbital integrals [Eqs. (26) and (27)] are complex. It is sometimes thought convenient to forego the separation of the degenerate pairs of symmetry species and use real orbitals which form a basis for the reducible real two-dimensional representations. If this is done it should, however, be remembered that matrix elements between functions belonging to different rows of a *reducible* representation do not necessarily vanish. This reduces the degree of symmetry blocking in the integral matrices, and doubles the length of a CI expansion for a state corresponding to such a representation. The construction of SACFs also becomes more complicated, since the product of orbitals belonging to these reducible representations is not necessarily symmetry adapted, even in the loose sense of belonging to one of the reducible two-dimensional representations. More will be said about this below.

Turning now to the non-Abelian axial point groups, it has been shown[123,124,140] that the use of complex orbitals to span the two-dimensional IRs leads to useful simplifications in this case without requiring complex arithmetic. In fact, the orbital phase factors can be chosen in such a way that all the symmetry-orbital integrals [Eqs. (26), (27)] are real (a condition which cannot always be satisfied for the *Abelian* point groups with complex IRs). This can be achieved, for example, by first constructing real symmetry orbitals and then transforming all pairs of orbitals which belong to the same real IR to the corresponding complex form by the same (complex) transformation. The transformation should be chosen such that each pair of complex orbitals which belong as partners to a two-dimensional IR are the complex conjugate of each other.

The most important simplification that results from the use of complex orbitals is that a symmetry-adapted spatial N-electron function can be written in this case as a linear combination of at most two orbital products, and that when two products are required they are the complex conjugates of each other.[124] Thus each SACF can be written either as a single bonded function of the type discussed in the previous subsection, or as the real or imaginary part of such a function.[124] More specifically, each pair of symmetry orbitals belonging as partners to a two-dimensional IR E_m (or E'_m, E''_m, E_{mg}, E_{mu}) are chosen as eigenfunctions of the highest axial symmetry operation $\hat{C}_n$ (or $\hat{S}_n$) with eigenvalues $\exp(\pm 2\pi im/n)$. A product of such symmetry orbitals is also an eigenfunction of $\hat{C}_n$ (or $\hat{S}_n$), the eigenvalue being of the form $\exp(2\pi iM/n)$, where M is the sum of the individual m-values of the orbitals reduced (modulo n) to the range $-n/2 < M \le n/2$. If M is not equal to 0 or $n/2$ the product is symmetry adapted and belongs to one of the IRs E_M, E'_M, E''_M, E_{Mg}, or E_{Mu}, as the case may be (it is easy to determine which). Otherwise the real and imaginary parts of the product are symmetry adapted and belong to two different one-dimensional IRs; these IRs are A_1 and A_2, etc., if $M = 0$, and B_1 and B_2, etc., if $M = n/2$ (note that the real and imaginary parts are symmetric and antisymmetric, respectively, with respect to a $\hat{\sigma}_v$ reflection in the XZ plane). Similar results are obtained if real symmetry orbitals belonging to the one-dimensional IRs are included in the product.[124] There is a complication in those cases in which the complex conjugate of an orbital product is a permuted form of the same product, but this can be handled[124] by using Serber-type spin eigenfunctions and by arranging the orbitals in the product in such a way that all complex-conjugate pairs of singly occupied orbitals appear in pairs immediately following the doubly occupied orbitals and before any of the other singly occupied orbitals. It then turns out that either the real or the imaginary part of such a product results in a vanishing function when coupled with one of the spin functions under the effect of the antisymmetrizer, while the surviving part provides a symmetry-adapted CF.[124]

The linear molecule symmetry groups $\mathscr{C}_{\infty v}$ and $\mathscr{D}_{\infty h}$ can be handled by the same methods, treating them in terms of $\mathscr{C}_{nv}$ or $\mathscr{D}_{nh}$ subgroups, with n chosen larger than twice the maximum sum of l-values of the orbitals that can occur in any one orbital product in the CFs (where $l\hbar$ is the eigenvalue of the angular momentum operator $\hat{l}_z$ corresponding to the orbital).

The use of complex orbitals is thus seen to be very helpful in the construction of symmetry-adapted CFs for most axial point groups. Unlike the case of real symmetry orbitals (in which quite a few bonded functions may be required to form one SACF), the resulting SACFs have a particularly simple form which facilitates the computation of matrix elements of the Hamiltonian (Section 5.3). The use of complex symmetry orbitals also simplifies the structure of the one-electron and two-electron integral matrices[123] and facilitates the rapid determination of whether any particular integral vanishes by symmetry and, if it does not vanish, where it is located in the stored integral arrays.[123]

It is clear from the foregoing that the use of complex orbitals is also to be preferred for the Abelian axial groups with complex IRs. While it may not be possible to avoid complex arithmetic completely in this case, certain simplifications are possible[124] and most of the advantages seen in the non-Abelian case remain.

More laborious methods are required in the case of the cubic and icosahedral point groups. Of the many papers that can be helpful in treating these cases,[127–130,141] we shall mention in particular those of Gabriel,[142] Wybourne,[143] and Sakata.[144] On the other hand, it is sometimes more convenient to treat a high-symmetry molecule in terms of a lower-symmetry axial subgroup. Thus, for example, computations on methane may be carried out using $\mathscr{C}_{3v}$ or $\mathscr{D}_{2d}$ (if both are used independently this may provide a check on the computer programs). Another interesting approach, applicable in principle to any point group, is to diagonalize the matrices of an appropriate group theoretical operator in terms of the primitive functions.[145–147] This can in fact be used both for spin adaptation (by diagonalizing the matrices of $\hat{S}^2$) and for symmetry adaptation (including the atomic case, using $\hat{L}^2$).

Much work has been done on the important case of atomic (spherical) symmetry, represented by the orthogonal group $O(3)$ (see, for example, Wigner,[77] Condon and Shortley,[21] Racah,[148] Biedenharn and Van Dam,[149] Rotenberg,[150,151] Löwdin,[152] Schaefer and Harris,[146] Bunge and Bunge,[58,153,154] and Munch and Davidson[155]). The most successful methods have been based on vector coupling schemes[21,148,149,156] or on projection operators.[150–154] As previously stated, the newer group theoretical methods based on the unitary or general linear groups[99–108] are capable of treating atomic spatial symmetry together with the spin symmetry in one compact and powerful formalism.[101–104]

5. Hamiltonian Matrix Evaluation

5.1. Projection Operators and Slater Determinants

The reduction of Hamiltonian matrix elements H_{st} (Eq. 14) to linear combinations of orbital integrals [Eq. (25)], sometimes called *projective reduction*,[117,157] is the key step, and usually the most time-consuming step, in large-scale CI calculations. Efficient implementation of this step is of major importance for the feasibility of such calculations. While it is possible to carry out this reduction even when nonorthogonal orbitals are used[158] (though in this case the coefficients a_{ij}^{st} and b_{ijkl}^{st} are not independent of the nature of the orbitals), the computational effort involved is so much greater than in the case of orthonormal orbitals that only the latter will be considered here.

Whether or not the configuration functions used have been obtained with the aid of projection operators, they can often be expressed in terms of such operators acting on a simpler (primitive) function. This applies, in particular, to antisymmetric functions, which can be expressed in terms of an antisymmetrizer [Eq. (21)] acting on a product, or a linear combination of several products, of spin-orbitals [Eqs. (20) and (22)], as well as to spin-adapted and symmetry-adapted CFs. In general, if Ω and Ω' are primitive functions and $\hat{P}$ is a projection operator which commutes with $\hat{H}$, then (because a projection operator is Hermitian and idempotent)

$$\langle \hat{P}\Omega | \hat{H} | \hat{P}\Omega' \rangle = \langle \Omega | \hat{H} | \hat{P}\Omega' \rangle$$
$$= \langle \hat{P}\Omega | \hat{H} | \Omega' \rangle \tag{59}$$

In particular, noting Eqs. (21) and (48),

$$\langle \mathscr{A}\Omega | \hat{H} | \mathscr{A}\Omega' \rangle = (N!)^{1/2} \langle \Omega | \hat{H} | \mathscr{A}\Omega' \rangle$$
$$= \sum_{\mathscr{P}} \varepsilon(\mathscr{P}) \langle \Omega | \hat{H} | \mathscr{P}\Omega' \rangle \tag{60}$$

This reduces the number of terms which can contribute to the matrix element from $(N!)^2$ to $N!$ and results, when Ω and Ω' are each a product of spin-orbitals from a common orthonormal set, in the Slater–Condon rules[21,57] for matrix elements of Slater determinants. Thus, if D and D' are two Slater determinants,

$$D = \mathscr{A}d_1 d_2 \cdots d_N$$
$$D' = \mathscr{A}d_1' d_2' \cdots d_N' \tag{61}$$

the spin-orbitals d_i and d_i' being taken from a common set of orthonormal spin-orbitals, and being arranged so that $d_i = d_i'$ for as many values of i as possible (the maximum coincidence ordering), then noting Eq. (2) we obtain

$$\langle D | \hat{H} | D' \rangle = Q V_{nn} + \sum_i Q_i \langle d_i | \hat{h} | d_i' \rangle$$
$$+ \sum_{i<j} Q_{ij} (\langle d_i d_j | g | d_i' d_j' \rangle - \langle d_i d_j | g | d_j' d_i' \rangle) \tag{62}$$

where $g \equiv g_{12}$, $d_i d_j \equiv d_i(\mathbf{x}_1)d_j(\mathbf{x}_2)$, etc., and

$$Q_{ij...} = \begin{cases} 1 & \text{if } d_k = d'_k \text{ for all } k \neq i, j, \ldots \\ 0 & \text{otherwise} \end{cases} \tag{63}$$

(in particular, $Q = 0$ unless $D = D'$). It is seen that the matrix element vanishes if there are more than two noncoincidences in the spin-orbital sets $\mathbf{d}$ and $\mathbf{d}'$, while it is given by at most two two-electron integrals if there are exactly two noncoincidences (note that an integral such as $\langle d_i d_j | g | d'_i d'_j \rangle$ vanishes unless the spin factors are the same in d_i and d'_i as well as in d_j and d'_j).

Equations analogous to (62) can be obtained for other quantum-mechanical operators. Thus, if $\hat{F}$ is a Hermitian symmetric p-electron operator (which may be spin dependent),

$$\hat{F} = \sum_{\mu_1 < \mu_2 < \cdots < \mu_p} \hat{f}_{\mu_1 \mu_2 \ldots \mu_p} \tag{64}$$

then

$$\langle D | \hat{F} | D' \rangle = \sum_{i_1 < i_2 < \cdots < i_p} Q_{i_1 i_2 \cdots i_p} \sum_{\mathscr{P}} \varepsilon(\mathscr{P}) \langle d_{i_1} d_{i_2} \cdots d_{i_p} | \hat{f} | \mathscr{P} d'_{i_1} d'_{i_2} \cdots d'_{i_p} \rangle \tag{65}$$

where $\hat{f} \equiv \hat{f}_{12\ldots p}$, the serial notation has been used on each side of the Dirac bracket for the p-electron integral, and the sum over $\mathscr{P}$ includes all $p!$ permutations of the p electrons on the right-hand side of the bracket.

Since more complex CFs, such as spin-adapted or symmetry-adapted CFs, can be expressed as linear combinations of Slater determinants, it is possible to compute a matrix element involving such functions as a linear combination of integrals over Slater determinants. Such a procedure would be rather inefficient if each matrix element were computed independently, since the same Slater determinants would usually appear in several CFs. One alternative approach[52,53,145] is to compute the whole Hamiltonian matrix initially in terms of Slater determinants, and then to transform it to a basis of CFs. Another, and more attractive, alternative is to perform this calculation and transformation block by block. A block, in this context, refers to the set of matrix elements which arise from one pair of configurations (where by configuration we mean the set of all CFs corresponding to the same occupancies of sets of degenerate orbitals). When carried out in this way it is possible to take advantage of the properties of spin and symmetry projection operators (even if the CFs have not been derived in terms of projection operators initially), as has been demonstrated by Nesbet[159,160] and by Davidson.[155,161] While even this approach is inherently less efficient than direct methods for the reduction of CF matrix elements to linear combinations of orbital integrals, the relative effectiveness of different procedures ultimately depends very strongly on details of computer program organization and coding. For additional discussions of the computer implementation of Hamiltonian matrix calculations see, e.g., Yoshimine, McLean, and co-workers.[39–41]

5.2. Spin-Adapted CFs

Each method used for the construction of spin-adapted CFs entails a concomitant treatment for the reduction of the corresponding matrix elements. We have already seen how the Slater–Condon rules can be used if the CFs are expressed as linear combinations of Slater determinants. We shall briefly review several other approaches before describing the method which has been used with the bonded functions discussed in Section 4.2.

Considering first the spin projection operator method,[60,64–66,71–73] each CF can be expressed in the form of Eq. (20), with

$$\Theta_\lambda = \mathcal{O}_S \theta_\lambda \tag{66}$$

θ_λ being a primitive spin function. Using the properties of projection operators, it then follows that[66,72]

$$\langle \Phi_{s\lambda} | \hat{H} | \Phi_{t\mu} \rangle = \sum_{\mathcal{P}} \varepsilon(\mathcal{P}) \langle \Xi_s | \hat{H} | \mathcal{P} \Xi_t \rangle \langle \theta_\lambda | \mathcal{O}_S | \mathcal{P} \theta_\mu \rangle \tag{67}$$

Furthermore, it can be shown[66,72] that the spin integral on the right-hand side of (67) is equal to a Sanibel coefficient,

$$\langle \theta_\lambda | \mathcal{O}_S | \mathcal{P} \theta_\mu \rangle = C_j(S, M_S, \tfrac{1}{2}N) \tag{68}$$

where j is the number of α-factors in θ_λ for which there are β factors in corresponding positions of $\mathcal{P}\theta_\mu$. Also, since the spatial products Ξ_s and Ξ_t are constructed from a common set of orthonormal orbitals, only a limited number of permutations $\mathcal{P}$ will give rise to nonzero contributions in (67). The various cases which can arise, depending on double occupancies and noncoincidences in Ξ_s and Ξ_t, have been studied in detail by Harris,[65,66,162] who has also extended the analysis to spin-dependent operators.[163]*

The Sanibel coefficients can be obtained in several ways,[65,72,73,164,165] and some tabulated results are available.[66,73] A very useful feature of this approach is that only a small number of different $C_j(S, M_S, \tfrac{1}{2}N)$ are needed in any one calculation, since S, M_S, and N are fixed by the nature of the problem, and the only allowed values of j are $0, 1, \ldots, \tfrac{1}{2}N - S$ (actually, coefficients for which the last argument is $\tfrac{1}{2}N - 1$, $\tfrac{1}{2}N - 2, \ldots$ may also be required due to

---

*The definition used for the Sanibel coefficients in Refs. 162 and 163 differs from that used elsewhere,[72,164] the third argument being N rather the $\tfrac{1}{2}N$.

double occupancies[162]). The principal case coefficients are particularly simple,[72,164]

$$C_j(S, S, \tfrac{1}{2}N) = (-1)^j \frac{2S+1}{\tfrac{1}{2}N+S+1} \binom{\tfrac{1}{2}N+S}{j}^{-1} \tag{69}$$

Considerable work has also been done on the calculation of matrix elements of spin-adapted CF's based on the symmetric group S_N,[62-67,69,84-97,110,111,121,122,166-171] but much of it is limited to diagonal elements or diagonal blocks (i.e., the set of elements involving the same orbital product on both sides of the operator, but with all possible spin couplings on each side). The choices of group theoretical operators which are equivalent to Löwdin's spin projection operator[89,90,111] or which produce valence-bond type functions[63,67,90,95,110,111] lead to reasonably effective procedures which do not require the storage or generation of large amounts of group theoretical information (see, in particular, Harris,[66] Klein and Junker,[111] and Gallup[168]). On the other hand, the other choices, particularly those which generate the genealogical spin functions, require representation matrices of the symmetric group. Whether these are stored in the computer or computed as needed, their number increases factorially with the number of electrons, and this makes it impractical to use such approaches, in their full generality, for problems involving more than a very few electrons.[66] An alternative method for the reduction of matrix elements involving genealogical spin functions has been developed by Gouyet,[113,114,172,173] using what he calls the occupation-branching number representation, based on Kramers' theory of binary spinor invariants.[174,175]

An interesting new group theoretical approach to the reduction of spin-adapted matrix elements has been developed by Ruedenberg and co-workers.[69,121,122,176] It is based on the use of Serber-type spin functions[80] (which have definite parity with respect to geminal transpositions) and though it requires matrix elements of permutation operators between the spin functions, the formulas appear to be more compact in the case of the Serber functions than in the general case.

Another promising approach, based on expressing the reduction coefficients of two-electron integrals in terms of those of one-electron integrals, has recently been proposed by Wetmore and Segal.[177] It appears to be capable of extension to the direct treatment of symmetry-adapted CFs.

Effective procedures for the reduction of matrix elements have also been developed for the unitary group approach,[101-108] though applications to large-scale CI calculations, particularly with *selected* sets of CFs, remain to be demonstrated.

We turn now to valence-bond type functions. Formulas for diagonal blocks and singlet spin states were first developed by Pauling,[178] using a graphical technique based on Rumer's bond diagrams. An extension to triplet states was given by McLachlan.[179] An improved formulation was given by Shull,[180] and a generalization to any matrix element between bonded functions (Slater bond functions), including those of spin-dependent operators, was made by Cooper and McWeeny.[120] Here we shall describe the alternative and equivalent scheme of Boys and Reeves,[117-119] which has been extended to some spin-dependent operators by Sutcliffe.[119] The Boys–Reeves formalism is particularly amenable to computer implementation, and has been applied in numerous molecular CI calculations.[7,8,20,181] (An effective group theoretical formalism for the same matrix elements has been developed by Matsen, Klein, and co-workers[63,95,110,111]; see also Löwdin and Goscinski,[67] Harris,[66] Heldmann and Schnupp,[90] and Gallup.[168])

Given two bonded functions B and B' of the type described in Section 4.2, the first step in the reduction of the matrix element $\langle B|\hat{H}|B'\rangle$ involves bringing the orbitals in B and B' to a maximum coincidence arrangement. In this reordering, each orbital carries its right or left parenthesis with it, and the new order in both B and B' is such that the spin couplings are preserved, and that coupled orbitals are adjacent as far as possible. The newly ordered sequences of orbitals in B and B' can now be designated $b_1 b_2 \cdots b_N$ and $b'_1 b'_2 \cdots b'_N$, with $b_i = b'_i$ as far as possible. Each orbital b_i or b'_i is now associated with at most two other orbitals (not necessarily different from it), one its paired orbital, b'_i or b_i, respectively, in the other bonded function, and the other, say b_j or b'_k, respectively, the orbital to which it is spin coupled, if any, in the same bonded function. These associations are then chained to form patterns, called *cycles* and *chains*. Each pattern is obtained by following the two types of associations (pairings and spin couplings) alternately until it terminates at both ends by uncoupled orbitals or else it closes upon itself. Closed patterns are called cycles (and this is the only type that occurs for singlet states), while open patterns are called chains, and are further classified as *even* or *odd*, depending on the number of orbital pairs (consisting of one orbital from each bonded function) in them. An even chain has both free ends in the same bonded function, while an odd chain has one end in each function. The total number of chains is equal to $2S$, and the number of even chains is always even (thus there can be no even chains for $S < 1$). Examples of cycle and chain patterns are given in Fig. 3.

The matrix element $\langle B|\hat{H}|B'\rangle$ vanishes if there are more than two noncoincidences, or if there are more than two even chains. Otherwise it is given by the formula (for proof see Reeves[117] or Sutcliffe[119]):

$$\langle B|\hat{H}|B'\rangle = \Gamma\{RQV_{nn} + R\sum_i Q_i\langle b_i|\hat{h}|b'_i\rangle$$

$$+ \sum_{i<j} Q_{ij}[R\langle b_ib_j|g|b'_ib'_j\rangle + q_{ij}\langle b_ib_j|g|b'_jb'_i\rangle]\} \qquad (70)$$

where the notation is similar to that in the Slater–Condon rules formula, Eq. (62), except that the b_i, b'_i refer to orbitals and not to spin-orbitals, the

Bonded Functions	Max. Coincidence Rearrangement	Patterns	Parameters
	i = 1 2 3 4 5 6 7	1 2 3 4 5 6 7	
(ab)(cd) (ad)(bc)	(a b)(c d) (a (b c) d)		$\sigma = \sigma' = 0,\ c = 1,\ J = 0,\ \Gamma = -\frac{1}{2}$ $Q_i = Q_{ij} = 1$ for all i, j $q_{12} = q_{14} = q_{23} = q_{34} = 1,\ q_{13} = q_{24} = -2$
(ab)(c(d (ad)(e(f	(d (a b)(c (d a)(e (f		$\sigma = 1,\ \sigma' = 0,\ c = 0,\ J = 0,\ \Gamma = \frac{1}{2}$ $Q_{34} = 1,\ q_{34} = -1$
(aa)(bc)(d (ab)(cd)(e	(a a)(b c)(d (e (a b)(c d)		$\sigma = \sigma' = 0,\ c = 0,\ J = 1,\ \Gamma = \frac{1}{4}\sqrt{2}$ $Q_1 = Q_{1j} = 1$ for all j $q_{12} = 0,\ q_{14} = 1,\ q_{13} = q_{15} = -2$
(ab)(cd)(e(f (af)(gh)(c(d	(f (a b)(e (c d) (f a)(g h)(c (d		$\sigma = 1,\ \sigma' = 0,\ c = 0,\ J = 0,\ \Gamma = -\frac{1}{4}$ $R = 0,\ Q_{34} = 1,\ q_{34} = 0$
(aa)(bc)(de)(f (ab)(cc)(df)(g	(a a)(c b)(e d)(f (a (c c) b)(g (d f)		$\sigma = \sigma' = 0,\ c = 1,\ J = 2,\ \Gamma = \frac{1}{2}$ $Q_{25} = 1,\ q_{25} = -\frac{1}{2}$

Fig. 3. Examples of cycle and chain patterns for the reduction of matrix elements between bonded functions. The symbols for the parameters are those used in Eqs. (70)–(72). All the Q not specifically listed are zero, and $R = 1$ *unless stated otherwise*. The ticks on some vertical lines in the patterns mark noncoincidences. Note that in the third example $q_{12} = 0$ because $b_1 = b_2$, and in the fourth example the matrix element vanishes.

corresponding integrations being over space coordinates only. The factors $Q_{ij...}$ are given by Eq. (63) (with the d_i replaced by b_i), while

$$R = \begin{cases} 1 & \text{if there are no even chains} \\ 0 & \text{otherwise} \end{cases} \tag{71}$$

The "exchange" integral coefficients q_{ij} are zero if $b_i = b_j$ and/or $b'_i = b'_j$; otherwise they depend on the type of patterns in which i and j occur, as well as on the parity of the positions of i and j within their respective patterns. Writing $p_i = 1$ if i is in an even-numbered position within its pattern and $p_i = -1$ otherwise, the values of q_{ij} for all possible cases are given in Table 1. The overall coefficient Γ is given by[118]

$$\Gamma = (-1)^{\sigma+\sigma'}(-\tfrac{1}{2})^{(N/2)-S-c}2^{J/2} \tag{72}$$

where σ and σ' are the parities of the permutations of the *uncoupled* orbitals in B and B', respectively, required to bring them to their ordering in the cycles and chains patterns ($b_1, b_2, \ldots$ and $b'_1, b'_2, \ldots$), c is the number of cycles, and J is the number of pairs (i, j) in which either $b_i = b_j$ and $b'_i \neq b'_j$, or $b'_i = b'_j$ and $b_i \neq b_j$.

Table 1. Exchange Integral Coefficient q_{ij} in the Boys–Reeves Formalism [Eq. (70)]^a

Patterns containing i and j	$p_i p_j$	q_{ij}
Both in the same cycle or same odd chain	$+1$	-2
	-1	$+1$
In different cycles, or one in a cycle and the other in an odd chain	± 1	$-\frac{1}{2}$
In different odd chains	$+1$	-1
	-1	0
In different even chains	$+1$	$+1$
	-1	-1

^{a} See Refs. 117 and 119. These formulas are for $b_i \neq b_j$, $b_i' \neq b_j'$; otherwise, and in all cases not explicitly listed, $q_{ij} = 0$. The product $p_i p_j$ depends on the relative positions of i and j in their respective patterns (the first and all odd positions in each pattern have $p_i = -1$, while even positions have $p_i = 1$.

Similar formulas for one-electron spin-dependent operators and for two-electron spin–spin and spin–orbit coupling operators (for principal-case functions only) have been given by Sutcliffe.[119]

The formula of Eq. (70) is seen to provide a method for determining the reduction coefficients a_{ij}^{st} and b_{ijkl}^{st} [Eq. (25)] for bonded functions. It is usually advantageous to carry out the calculation of the Hamiltonian matrix **H** in two stages, since the reduction coefficients do not depend on the nature of the orbitals or on numerical parameters and need not be recomputed when orbital exponents or molecular geometry are changed. Also, the computer storage requirements are very different in the determination of the reduction coefficients (when the list of bonded functions is needed) and in the numerical evaluation of the matrix elements (when access to the orbital integrals must be provided). However, rather than terminating the first stage with the production of a list of all nonzero reduction coefficients, a much more compact list is produced if, for the cases of zero and one orbital noncoincidences, the first stage just produces the cycle and chain patterns and a symbol identifying the overall coefficient Γ.

In a typical large-scale CI calculation the majority of Hamiltonian matrix elements (often more than 80%) involve more than two orbital noncoincidences, and thus vanish. (As a rough guide, in an N-electron, large basis-set CI expansion involving all double excitations from a closed-shell configuration, the proportion of nonzero matrix elements is about $8/[N(N-2)]$, this being the probability that two double excitations taken at random are from the same two occupied orbitals.) It is thus important, in the interests of efficiency, to identify zero elements easily and quickly, before any attempt is made to generate cycle and chain patterns for them. The first step in the reduction of a block of matrix elements (all involving the same pair of orbital products in B and B') should thus be the determination of the number of noncoincidences. If this number is greater than two then the whole block can be skipped. Furthermore, the majority of the nonzero elements involve two noncoincidences, and it is worthwhile to make a special effort to provide efficient processing for this case in particular.

Both the reduction stage and the numerical evaluation stage are best carried out block by block, with some of the processing common to all the elements in a block (this includes the determination of the noncoincidences and the identification of all cycles involving the same doubly occupied orbitals in both configurations). Recalling that bonded functions resulting from the same orbital product are not orthogonal in general (Section 4.2), it is convenient to transform each block, after it has been numerically evaluated, to an orthonormal spin-function basis. This can be done by applying the appropriate submatrices of the common Schmidt orthogonalization matrix to the rows and to the columns of the block. Any spin couplings which are to be discarded (because of configuration selection criteria or for reasons of spatial symmetry) can be left out in the transformation, but may not be omitted earlier unless they are not followed by any retained spin couplings. Depending on the method used for the solution of the matrix eigenvalue problem (Section 6), it may also be necessary to rearrange the nonvanishing matrix elements by rows rather than by blocks; this can be done as soon as a complete "slice" (a row of nonzero blocks of the lower triangle of **H**) has been evaluated and transformed.

For another discussion on the implementation of the bonded function formalism the reader is referred to Diercksen and Sutcliffe.[181] Computer programs for the reduction process have been published by Reeves[118] and by Gilson.[182]

5.3. Symmetry-Adapted CFs

There seems to be no effective, generally applicable technique for the *direct* reduction of matrix elements between symmetry-adapted configuration functions (SACFs) for general point groups, though the new method of Wetmore and Segal[177] appears to be capable of easy extension to this case. Most approaches to this problem perform the reduction by first expanding the SACFs in terms of spin-adapted CFs or even in terms of Slater determinants, and are often limited in practical effectiveness to a small number of electrons. The one case for which considerable work has been done is that of atoms (see, e.g., Wigner,[77] Condon and Shortley,[21] Racah,[148] Boys,[183–186] Bunge and Bunge,[154] Sasaki,[156] Harter,[102–104] and Hunter[187]). Boys' vector-coupling scheme has been applied by him[14–16,19] and by Donath.[188] The related approach of Sasaki,[156] based on recoupling transformations, has been applied in large-scale atomic CI calculations by Sasaki and Yoshimine.[189,190] Other large-scale calculations for atoms have been performed by Bunge and Bunge,[58,191,192] using a projection operator formalism,[154] and by Munch and Davidson,[155] using an explicit expansion in terms of Slater determinants, but taking advantage of economies of the same type as those obtained with the use of projection operators, as discussed above.[161]

Turning now to molecules, some work on diagonal elements, based on projection operators, has been done by Simons and Harriman,[193] while approaches similar to some of those for atoms, based on reduction to Slater determinants and the use of coupling coefficients, have been discussed by McWeeny,[131] Griffith,[194] and Wybourne.[143] A specific procedure for diatomic molecules (treating spin-dependent operators) has been developed by Bottcher and Browne,[195] while a method based on the symmetric group and applicable to molecules belonging to axial point groups has been proposed by Walker and Musher.[171] This last method, while performing a direct reduction to orbital integrals, appears to be limited in practical applicability to problems involving a very small number of electrons.

The case of axial point groups has also been treated in detail by Gershgorn and Shavitt,[124] using the spin-adapted CFs based on complex symmetry orbitals discussed in Section 4.3. Since in this formalism each SACF consists of at most two bonded functions, a matrix element between SACFs can involve four bonded-function matrix elements at most. However, since each such SACF can be expressed as the result of a projection operator operating on one bonded function, the use of the projection operator property, Eq. (59), means that no more than two bonded-function matrix elements need be computed to obtain any one SACF matrix element. In fact, since either one of the two right-hand-side expressions in (59) may be used, the only case in which two terms are required is when *both* SACFs of the matrix element consist of two bonded function terms. Furthermore, most of the calculation can be carried out without the use of complex arithmetic,[124] and many economies are possible relative to calculations based on real orbitals.[123]

6. The Matrix Eigenvalue Problem

6.1. General Considerations

The CI energy E and expansion coefficients vector $\mathbf{c}$ are obtained as the appropriate eigenvalue and corresponding eigenvector of the Hamiltonian matrix $\mathbf{H}$ by solving the eigenvalue equation (16). More generally, if the CFs used are not all orthogonal, the *generalized* eigenvalue equation (13) has to be solved. While some of the methods to be discussed can treat the generalized problem directly, it is usually preferable to convert that problem to the simpler version first, by transforming $\mathbf{H}$ to a basis of orthonormal CFs, rather than carry the extra complexity (and data) of the general problem throughout the solution process. The transformation to an orthonormal basis can of course be determined by the Schmidt process (see Section 4.2). The generalized eigenvalue problem has also been discussed by Wilkinson and co-workers[196,197] and by Ford and Hall.[198]

While in general $\mathbf{H}$ can be complex Hermitian, it will be assumed in most of this section that it is real symmetric. The generalization of most solution methods to the complex case is usually straightforward.

The CI eigenvalue problem has some special characteristics which should be taken into account in choosing a method for its solution. One of these, already discussed in Section 5.2, is the sparseness of $\mathbf{H}$. Another is the sheer size of the matrix; the order of the eigenvalue problem in large-scale CI calculations is often in the range 10^3–10^4, or even higher, which precludes the simultaneous storage of the entire matrix in the central memory of the computer. A third characteristic is the dominance of the main diagonal in $\mathbf{H}$ (due to the fact that many electronic states are described reasonably well by a single CF), which typically results in the eigenvector being dominated by one large component. And last, but by no means least, only one or a very few of the lowest eigenvalues, and corresponding eigenvectors, are usually required, so that a complete diagonalization of $\mathbf{H}$ is unnecessary. (While additional higher roots could, in principle, serve as approximations to higher excited states of the system, their validity would be very questionable in most cases, since the basis set and the CFs would have been chosen on the basis of the requirements of the lower states only.)

An appropriate format for the storage and handling of $\mathbf{H}$ has been discussed in Section 3.1. Such a format, in which the matrix is arranged by rows of the lower triangle, with all zero elements omitted and the remaining elements identified by specifying their column index, allows for very efficient processing if the elements are required in the same sequence but is quite impractical if random access to all elements is required. The most useful solution methods would allow the matrix to be kept in an external file (tape or disk), and would require one record at a time to be read into central memory for processing, with only a very limited number of passes through the entire matrix being necessary.

The sparseness of $\mathbf{H}$ is of course very useful in reducing the length of the matrix file and in speeding up the processing, but this can only be used to full advantage if random access is not required and if zero elements are not modified in intermediate stages of the solution. This requirement is not met by most standard methods for the solution of the matrix eigenvalue problem, and such methods are normally practical only for relatively small problems (matrices of order up to about 200–300, depending on central memory size).

The type of method which is most suitable for the CI eigenvalue problem is an iterative method in which the original matrix is not modified in the solution process. An iterative approach which starts with some initial guess for the eigenvector and improves it successively is favored by the fact that reasonably good initial guesses are usually easily available. In particular, the diagonal dominance of **H** implies that relaxation methods in which one component of the trial vector is modified at a time will have adequate convergence properties. Methods of this type, unlike those which subject the original matrix to many successive transformations, have the additional advantage that they do not propagate and accumulate round-off error from one stage to the next. Such iterative methods are discussed in Section 6.3 below. Their principal limitation is that they obtain one (or very few) eigenvectors at a time, and they become progressively less attractive, compared to more standard methods, when more roots are required. In fact, no general method appears to be known at present which is applicable to randomly sparse (or nonsparse) matrices of order $\geq 10^3$, and which makes possible the fast routine calculation of *many* eigenvalues and eigenvectors of such matrices, particularly when the matrices are not diagonally dominant.

6.2. Standard Methods

As is clear from the above discussion, standard diagonalization techniques are only practical, in general, for relatively small matrices, but such matrices do occur in many CI calculations. Even large-scale treatments often involve preliminary smaller calculations for exploratory purposes, for the determination of suitable orbitals, or for CF selection, and some of the standard methods are highly effective in dealing with matrices of order up to about 200, particularly when many roots are required. We shall not attempt to give a complete treatment of such methods, which are covered in detail in several excellent books,[199-201] but shall limit ourselves to some observations on the characteristics of the principal methods and to the listing of some available computer programs for their implementation.

One of the best known matrix diagonalization techniques is the Jacobi method,[199-204] which consists of a sequences of plane (2×2) rotations, each chosen so as to cause one off-diagonal element to vanish. Each rotation changes all the elements in two rows and columns, including any elements which may have been set to zero by previous rotations, thus requiring repeated sweeps through the matrix until all off-diagonal elements are smaller in magnitude than a prescribed tolerance. The Jacobi method has the advantage of simplicity and generality, but is rather inefficient, compared to more modern methods, except for very small matrices. It can be subject to serious accumulation of round-off errors when applied to large matrices. A detailed error analysis has been given by Goldstine, Murray, and von Neumann.[202] Error accumulation has also been discussed by Wilkinson,[200] and methods for the reduction of errors have been described by Corbató[203] and by Rutishauser.[204] Algol and Fortran subroutines have been given by several authors,[201,204,205] including one subroutine for complex Hermitian matrices.[205]

The most widely used diagonalization methods at this time are based on a two-stage process, in which the first stage consists of the transformation of the matrix to *tridiagonal* form.[199–201] In this form all elements except those on the main diagonal and on the two immediately adjacent diagonals (one on each side) are zero. The different methods in this category vary in the way in which they achieve the tridiagonal form and/or the way in which the eigenvalues and eigenvectors of the tridiagonal matrix are obtained. The earliest tri-diagonalization scheme appears to be due to Lanczos,[206] and is particularly suitable for application to large matrices. It does not require repeated modification of the orginal matrix, and can be used for partial tridiagonalization in such a way that the eigenvalues of the tridiagonal submatrix converge to the eigenvalues of the full matrix as the tridiagonalization proceeds (with the eigenvalues of largest magnitude converging first). It can thus be used effectively when only some of the (extreme) roots of a very large matrix are required, but is less efficient than the newer methods of Givens and of Householder when complete tridiagonalization is needed. It is discussed further in the next subsection.

Givens' method of tridiagonalization[200,201,207,208] is based on Jacobi-type plane rotations, but by proceeding down the columns of the lower triangle in order, and using as rotation pivots the elements immediately to the right of those being eliminated, it is possible to bring a matrix of order n to tridiagonal form in exactly $\frac{1}{2}(n-1)(n-2)$ rotations without any eliminated element being changed by subsequent rotations. A detailed error analysis has been given by Givens.[208] Algol and Fortran subroutines for this process have been given by Schwarz *et al.*[201]

A tridiagonalization algorithm which is about twice as fast as Givens' method has been devised by Householder (see Wilkinson[200,209] and others[199,201,210]). It is based on a sequence of $n-2$ orthogonal transformations with matrices of the form $1 - \mathbf{w}_i \mathbf{w}_i^T$ $(i = 1, 2, \ldots, n-2)$, where $\mathbf{w}_i$ is a unit column vector in which the first i components are zero and the remaining components are chosen so as to eliminate the last $n-i-1$ elements in the ith row and column of the matrix of interest. This is the principal method in use today in programs for the efficient computation of eigenvalues and eigenvectors.[201,210–212]

Two basic methods are available[199–201] for finding the eigenvalues of the tridiagonal matrix (which are the same as those of the original matrix). The first of these is the bisection method based on the Sturm sequence of polynomials derived from the tridiagonal matrix[213,214] and the second is the QR procedure[215-218] or one of its modifications.[219-222] The bisection method is particularly convenient and efficient if only some roots, specifically those within a given range of values, are required. The "rational" QR algorithm[221] is most effective if the required roots are near either end of the eigenvalue spectrum of the matrix, while if all eigenvalues are required, the ordinary QR (or QL[220]) algorithm is best, particularly if the eigenvectors are also needed.

Eigenvectors of the tridiagonal matrices can be obtained together with the eigenvalues when the QR method is used.[220,222] Specific eigenvectors can be found efficiently, once the corresponding eigenvalues are known, by an inverse iteration procedure.[200,223,224] The eigenvectors of the original matrix are of course related to those of the tridiagonal form by means of the transformations used to generate that form.

While the entire process for the solution of the matrix eigenvalue problem by means of transformation to tridiagonal form appears complicated and requires an informed choice of the appropriate procedures to be used for the different steps in the process, it has been well researched and several tested and effective programs are available.[199,201,211,212,219,222] We shall mention in particular the excellent set of Algol procedures collected in the book by Wilkinson and Reinsch[199] and the thoroughly tested set of Fortran subroutines based on these procedures in the EISPACK package.[212] A detailed discussion of the optimum path to be followed through these procedures in various cases is included in these publications. We shall also note that the QCPE (Quantum Chemistry Program Exchange) program GIVENS,[211] which is very popular among quantum chemists, actually uses the Householder tridiagonalization method[209] followed by the bisection procedure[213] and inverse iteration[223] to find some or all of the eigenvalues and, if desired, the corresponding eigenvectors.

The computational effort involved in most diagonalization methods is proportional to n^3 (where n is the order of the matrix). In the two-stage processes, the tridiagonalization step is proportional to n^3, while finding a subset of the eigenvalues and the corresponding eigenvectors in the second stage is proportional to n^2 per root. Iterative methods for the direct determination of selected roots without intermediate tridiagonalization, as described in the next section, are proportional to n^2 per root, and are thus preferable in large-scale CI calculations, in which n is very large and the number of roots required is very small.

6.3. Iterative Methods for Large Matrices

A useful review of eigenvalue methods suitable for large and very large matrices has recently been given by Stewart.[225] Here we shall concentrate primarily on two methods which appear to be effective in large-scale CI calculations; these are the method of optimal relaxations (MOR)[226,227] and Davidson's modification[228] of the Lanczos method.[206]

The method of optimal relaxations, as well as several other techniques, such as the gradient method,[229,230] are based on the minimization of the *Rayleigh quotient*. Using the notation of Section 2.1, Eqs. (13)–(18), the Rayleigh quotient $\rho(\mathbf{v})$ of an arbitrary vector $\mathbf{v}$ with respect to the Hermitian matrix $\mathbf{H}$ and the metric $\mathbf{S}$ is defined as

$$\rho(\mathbf{v}) = \frac{\mathbf{v}^\dagger \mathbf{H} \mathbf{v}}{\mathbf{v}^\dagger \mathbf{S} \mathbf{v}} \tag{73}$$

It is stationary (with respect to any variation in $\mathbf{v}$) if, and only if, $\mathbf{v}$ is a solution of the generalized eigenvalue equation (13), and at these stationary points it is equal to the corresponding eigenvalue. The problem of solving the eigenvalue equation is thus equivalent to the problem of finding the stationary points of the Rayleigh quotient. In particular, the lowest (or highest) eigenvalue and corresponding eigenvector can be found by minimizing (or maximizing) $\rho(\mathbf{v})$ with respect to $\mathbf{v}$.

Given some trial vector $\mathbf{v}$ and a "correction vector" $\mathbf{w}$, an improved trial vector

$$\mathbf{v}' = \mathbf{v} + \alpha \mathbf{w} \tag{74}$$

can be found in several equivalent ways. One approach[227] is to solve the eigenvalue problem of order two involving the *interaction matrix*[231]

$$\tilde{\mathbf{H}} = \begin{pmatrix} \mathbf{v}^\dagger \mathbf{H} \mathbf{v} & \mathbf{v}^\dagger \mathbf{H} \mathbf{w} \\ \mathbf{w}^\dagger \mathbf{H} \mathbf{v} & \mathbf{w}^\dagger \mathbf{H} \mathbf{w} \end{pmatrix} \tag{75}$$

and the corresponding metric $\tilde{\mathbf{S}}$ [obtained by replacing $\mathbf{H}$ by $\mathbf{S}$ in (75)]. The two eigenvalues obtained are the values of $\rho(\mathbf{v}')$ corresponding to minimization and maximization, respectively, of the Rayleigh quotient with respect to α, while α itself can be obtained from the corresponding eigenvectors. This technique can easily be generalized to the simultaneous use of several correction vectors.[227,232] Alternatively, α can first be obtained by the direct minimization of $\rho(\mathbf{v}')$ with respect to α, leading to a quadratic equation in α containing the elements of $\tilde{\mathbf{H}}$ and $\tilde{\mathbf{S}}$ in the coefficients.[226,233]

If a general vector $\mathbf{w}$ is used in (74), the calculation of the elements of $\tilde{\mathbf{H}}$ and $\tilde{\mathbf{S}}$ requires the evaluation of scalar products like $\mathbf{w}^\dagger \mathbf{H} \mathbf{v}$, each of which involves at least as many multiplications as the number of nonzero elements in the corresponding matrix. Unless the correction vectors are extremely well chosen, this requires a considerable amount of computation, since very many iterations will normally be needed. One possible choice of $\mathbf{w}$ is along the direction of steepest descent of the Rayleigh quotient at the point $\mathbf{v}$, leading to the *gradient method* of Hestenes and Karush.[229,230] The gradient is easily obtained as

$$\nabla\rho(\mathbf{v}) = 2\mathbf{q}(\mathbf{v})/(\mathbf{v}^\dagger\mathbf{v}) \tag{76}$$

where the *residual vector* $\mathbf{q}(\mathbf{v})$ is given by

$$\mathbf{q}(\mathbf{v}) = (\mathbf{H} - \rho(\mathbf{v})\mathbf{S})\mathbf{v} \tag{77}$$

The calculation of $\mathbf{w}$ (which can be taken equal to $\mathbf{q}$) thus requires the same products $\mathbf{Hv}$ and $\mathbf{Sv}$ that are needed for $\tilde{\mathbf{H}}$ and $\tilde{\mathbf{S}}$. However, the number of iterations normally required in this method is still too large to compete effectively with the relaxation methods discussed below. Alternatively, $\mathbf{w}$ can be chosen along the *conjugate gradient* direction,[234,235] leading to faster convergence than in the gradient method,[236] but the total computational effort still appears to be greater than for relaxation methods in applications to the large, diagonally dominant matrices of CI calculations.

Relaxation methods are based on the successive improvement of the trial vector, one element at a time. Thus, they involve the use of correction vectors $\mathbf{w}$ which are unit basis vectors $\mathbf{e}_p$ [having the elements $(\mathbf{e}_p)_q = \delta_{pq}$]. The minimization of the Rayleigh quotient using the sequence of correction vectors $\mathbf{e}_1, \mathbf{e}_2, \ldots, \mathbf{e}_n$ constitutes one iteration cycle of MOR.[226,227] Each individual step in the minimization requires the multiplication of just one row of $\mathbf{H}$ and of $\mathbf{S}$ by the current trial vector (the diagonal elements $\mathbf{v}^\dagger\mathbf{Hv}$ and $\mathbf{v}^\dagger\mathbf{Sv}$ can be updated from one iteration to the next without the recomputation of $\mathbf{Hv}$ and $\mathbf{Sv}$, as shown by Nesbet[237]). Thus a complete iteration cycle, involving n individual steps, requires about as much computation as one iteration of the gradient method, but experience indicates that the number of such iterations required for convergence is considerably smaller in MOR, at least for the type of matrices involved in large-scale CI calculations.

It can also be shown[228] that the minimization of $\rho(\mathbf{v}')$ relative to α in Eq. (74), with $\mathbf{w} = \mathbf{e}_p$, is equivalent to the solution of

$$q_p(\mathbf{v}') = 0 \tag{78}$$

This is an implicit equation for α, since the corrected vector $\mathbf{v}'$ appears in $\rho(\mathbf{v}')$ in Eq. (77). If, on the other hand, $\rho(\mathbf{v})$ is used instead of $\rho(\mathbf{v}')$ in the calculation of $\mathbf{q}(\mathbf{v}')$, a linear equation for α results,

$$q_p(\mathbf{v}) + (H_{pp} - \rho(\mathbf{v})S_{pp})\alpha = 0 \tag{79}$$

This is in fact the basis of Nesbet's method,[237] which has been used extensively in CI calculations and for which several QCPE programs are available.[238–240] It can be shown that as $\mathbf{v}$ approaches an eigenvector of Eq. (13) the corrections produced by the exact solution of (78) (i.e., by MOR) and by the linear equation (79) become equivalent.

The application of the relaxation methods requires access to one row at a time of **H** and **S**. Since these matrices are often generated in triangular form only (in order to take advantage of their symmetry or Hermiticity) the full row is not usually stored contiguously. Particularly with large matrices stored in external files with zero elements omitted, it is rather difficult to gain access to a complete row without reading through the entire matrix. It is, however, possible to overcome this problem and just use one row at a time of the lower triangle of each matrix, if each nonzero off-diagonal element brought in is used twice, the first time for the calculation of the current correction α and the second for accumulating column sums of the updated vector in preparation for the next iteration.[37,239]

As described so far, the relaxation methods apply only to the lowest (or highest) root of the eigenvalue equation. (While Nesbet's method should apply, in principle, to any root, it is found in practice that convergence to intermediate roots requires trial vectors so close to the desired exact eigenvectors as to be quite impractical in most cases.) Several methods are available, however, to find a few of the lowest, or highest, roots in order.[226,227] The most successful of these appears to be the "root shifting method," in which, after each root is found in turn, the matrix **H** is modified to make the next desired root the lowest eigenvalue of the new matrix. If the first k eigenvalues E_i ($i = 1, 2, \ldots, k$) and corresponding eigenvectors $\mathbf{c}_i$ are known, then the modified matrix $\mathbf{H}^{(k)}$ is defined as

$$\mathbf{H}^{(k)} = \mathbf{H} + \sum_{i=1}^{k} \beta_i (\mathbf{Sc}_i)(\mathbf{Sc}_i)^\dagger \tag{80}$$

This matrix has the same eigenvectors as **H**, while its eigenvalues are $E_i + \beta_i$ for $i \le k$ and E_i for $i > k$. An appropriate choice[226] of the constants β_i will raise all the first k eigenvalues above E_{k+1} and thus allow the calculation of the next root. The explicit calculation and use of $\mathbf{H}^{(k)}$ is not practical, since the modified matrix is no longer sparse, but is not actually necessary. All the required scalar products involving $\mathbf{H}^{(k)}$ can be obtained from corresponding products of **H** by the addition of some easily available quantities[226,227]:

$$\mathbf{a}^\dagger \mathbf{H}^{(k)} \mathbf{b} = \mathbf{a}^\dagger \mathbf{H} \mathbf{b} + \sum_{i=1}^{k} \beta_i (\mathbf{a}^\dagger \mathbf{Sc}_i)(\mathbf{b}^\dagger \mathbf{Sc}_i)^\dagger \tag{81}$$

The choice of the shift constants β_i requires some care,[226] as too large values reduce the diagonal dominance of **H** while too small values cause difficulties due to near degeneracy of the lowest root of $\mathbf{H}^{(k)}$.

The convergence of MOR is generally good for diagonally dominant matrices and well separated eigenvalues. Typically in such cases fewer than 10 iteration cycles are needed to obtain a normalized eigenvector to an accuracy of 10^{-6} (the eigenvalue converges much more rapidly than the eigenvector). In cases of slow convergence two acceleration methods are available. One is extrapolation of the trial vector by the Aitken δ^2-process after every few normal iterations.[226] The other is overrelaxation,[241,242] in which, instead of using the correction $\alpha\mathbf{w}$ in Eq. (74), a larger correction, $\omega\alpha\mathbf{w}$ $(1<\omega<2)$, is applied, where α is the value needed to minimize $\rho(\mathbf{v}')$. The optimum value of the overrelaxation parameter ω varies from case to case and is difficult to determine, but a value around $\omega \approx 1.5$ is often effective. The simultaneous use of several correction vectors[227,232] can also be effective in some cases, particularly when these can be chosen to span the largest off-diagonal element of $\mathbf{H}$. Another method for accelerating the convergence of MOR has been proposed by Feler.[243]

The most serious convergence difficulties of MOR occur in the case of nearly degenerate eigenvalues. One of the possible remedies is the simultaneous calculation of close roots, by the use of several trial vectors, together with one or more correction vectors, in the interaction matrix $\tilde{\mathbf{H}}$, Eq. (75), which is diagonalized in each step.[227] Several variants of this approach are possible, and the details of the most practical and effective procedure are still to be worked out.

Another, and fairly successful, attempt to overcome the convergence difficulties of MOR for nearly degenerate roots is Davidson's[228] modification of Lanczos' algorithm.[206] The latter has already been mentioned (Section 6.2) as a method for the tridiagonalization of a matrix. Complete tridiagonalization by this algorithm requires n multiplications of $\mathbf{H}$ by a vector, equivalent in computational effort to n iteration cycles of MOR, and is thus prohibitive for very large matrices. However, since each such multiplication produces another row and column of the tridiagonal matrix, and since successively larger principal (upper left-hand corner) minors of the tridiagonal matrix have roots which converge to roots of the complete matrix (those of largest magnitude converging first), partial tridiagonalization is feasible and sufficient for the determination of several of the extreme roots of the original eigenvalue equation.

The partial tridiagonal matrices of the Lanczos method are in fact the interaction matrices $\mathbf{B}_k^\dagger\mathbf{H}\mathbf{B}_k$ $(k = 1, 2, \ldots)$ in which $\mathbf{B}_k$ is an $n \times k$ matrix made up of a sequence of orthonormal vectors $\mathbf{b}_1, \mathbf{b}_2, \ldots, \mathbf{b}_k$ which obey a three-term recurrence relation

$$\beta_i\mathbf{b}_{i+1} = (\mathbf{H} - \alpha_i\mathbf{1})\mathbf{b}_i - \beta_{i-1}\mathbf{b}_{i-1} \tag{82}$$

(we have assumed $\mathbf{S} = \mathbf{1}$ in this treatment). The coefficients α_i and β_i are the diagonal and off-diagonal elements, respectively, of the interaction matrix (β_i is determined by the normalization condition on $\mathbf{b}_{i+1}$). The first term on the right of (82) is just the residue vector $\mathbf{q}(\mathbf{b}_i)$, which is proportional to the gradient $\nabla\rho(\mathbf{b}_i)$, so that the vectors $\mathbf{b}_i$ are those which would be obtained by the Schmidt orthonormalization of the sequence of gradients $\mathbf{g}_i$, where $\mathbf{g}_1 = \mathbf{b}_1$ and $\mathbf{g}_{i+1} = \nabla\rho(\mathbf{b}_i)$, or equivalently by the Schmidt orthonormalization of the *Krylov sequence* of vectors $\mathbf{b}_1, \mathbf{H}\mathbf{b}_1, \mathbf{H}^2\mathbf{b}_1, \ldots$. The kth iteration thus provides the best approximation to the eigenvectors within the subspace spanned by the sequence $\mathbf{b}_1, \mathbf{b}_2, \ldots, \mathbf{b}_k$, which is the same subspace as that spanned by the sequence of gradients $\mathbf{g}_1, \mathbf{g}_2, \ldots, \mathbf{g}_k$ or by the Krylov sequence $\mathbf{b}_1, \mathbf{H}\mathbf{b}_1, \ldots, \mathbf{H}^{k-1}\mathbf{b}_1$. Thus the Lanczos algorithm can be viewed as an extension of the gradient method, as well as an extension of the power method (in which the members of the Krylov sequence are used individually as successively better approximations to the eigenvector corresponding to the eigenvalue of largest magnitude).

Davidson argued[228] that since the relaxation methods usually show faster convergence than the gradient method, it would be advantageous to use, as expansion vectors in a Lanczos-like algorithm, vectors having as components the individual corrections of one of the relaxation schemes. Specifically, if $\mathbf{v}_i$ is the approximation to a particular eigenvector obtained in the ith iteration, the $(i+1)$th expansion vector $\mathbf{a}_{i+1}$ has the components

$$(\mathbf{a}_{i+1})_p = q_p(\mathbf{v}_i)/(\rho(\mathbf{v}_i) - H_{pp}) \tag{83}$$

[compare to the Nesbet algorithm, Eq. (79)]. The vector $\mathbf{b}_{i+1}$ is then obtained by orthonormalizing $\mathbf{a}_{i+1}$ to the previous orthonormal expansion vectors $\mathbf{b}_1, \mathbf{b}_2, \ldots, \mathbf{b}_i$. The interaction matrix of order $i+1$ is then diagonalized, and its appropriate eigenvector $\boldsymbol{\alpha}_k$ (the one corresponding in the eigenvalue sequence to the root being sought) is used to determine the next approximation $\mathbf{v}_{i+1}$ to the desired eigenvector of $\mathbf{H}$:

$$\mathbf{v}_{i+1} = \mathbf{B}_{i+1}\boldsymbol{\alpha}_k = \sum_{j=1}^{i+1} \mathbf{b}_j(\alpha_k)_j \tag{84}$$

This approach differs from the original Lanczos algorithm essentially by the inclusion of the denominator in (83). Since the vector sequence $\mathbf{b}_i$ no longer obeys a three-term recurrence relation, and since the interaction matrix is not tridiagonal, the computation is somewhat more cumbersome, but the convergence is much faster than in the Lanczos scheme, at least for diagonally dominant matrices. As in the Lanczos method, nearly degenerate roots do not cause convergence difficulties, but unlike that method, the process has to be repeated for each root sought, since the choice of each correction vector, Eq. (83), depends upon which eigenvector $\boldsymbol{\alpha}_k$ of the interaction matrix is used in Eq. (84). It shares with relaxation methods the dependence of the convergence rate on the diagonal dominance of $\mathbf{H}$, but it shares with the Lanczos method the advantage that the elements of $\mathbf{H}$ need not be used in any specific sequence. In fact, $\mathbf{H}$ may be defined as an operator without giving its explicit matrix representation, which makes this method very useful for those CI techniques which proceed directly from orbital integrals to eigenvectors without forming an explicit $\mathbf{H}$ matrix (see Chapters 7 and 8).

The number of multiplications required for each iteration in the methods described in this section is approximately equal to twice the number of nonzero elements of the lower triangle of the matrix. As the number of iterations required per root is not strongly dependent on matrix size, the computational effort is approximately proportional to n^2 per root sought. In tridiagonalization methods, the main step is proportional to n^3, so that the advantages of the iterative methods increase with n. An exact comparison would depend on the number of roots sought and on the sparseness and diagonal dominance of the matrix, but as a rough guide, whenever the matrix is sufficiently small for the tridiagonalization to be carried out entirely within the high-speed memory of the computer, this approach should be considered as possibly competitive with the large-matrix iterative methods.

7. Density Matrices and Electronic Properties

7.1. Introduction and Definitions

Reduced density matrices of electronic wave functions can be very useful entities in quantum mechanical calculations. They provide a compact representation of essential information present in the wave function; in fact, for Hamiltonians with no higher than two-body forces, the second-order density matrix contains all the physically significant information about the stationary state, and hopefully, if a practical means can be found to obtain this matrix directly, it may obviate the need for the calculation of the many-electron wave function in the study of stationary-state properties. Because of their compactness, reduced density matrices provide powerful means for wave function analysis, including the analysis of the bonding structure in a molecule (for example, population analysis[244]). They make possible detailed comparisons of different calculated wave functions and allow in-depth study of the convergence of a calculation as the basis set and type of wave function are improved. Reduced density matrices facilitate the calculation of molecular electronic properties, particularly those which are represented by expectation values of electronic operators. And finally, they provide for the definition and calculation of natural orbitals (and natural geminals), which are very useful in accelerating the convergence of CI expansions.

In the present context we are interested in density matrices primarily for the calculation of expectation values and for obtaining natural orbitals. This interest focuses primarily on the first-order reduced density matrix and, to a lesser extent, the second-order matrix. Furthermore, for most purposes it will be convenient to deal with spinless (i.e., spin averaged) density matrices. A limited discussion of transition density matrices will be included. More comprehensive treatments can be found in other sources.[71,245-253] A convenient summary of the properties of quantum mechanical density matrices has been given by Smith.[254]

The spin-dependent pth order reduced density matrix $\Gamma_\Psi^{(p)}$ (the p-matrix) corresponding to an N-electron normalized wave function Ψ ($p \leq N$) is defined as

$$\Gamma_\Psi^{(p)}(\mathbf{x}|\mathbf{x}') \equiv \binom{N}{p} \int \Psi(\mathbf{x}, \mathbf{y})\Psi^*(\mathbf{x}', \mathbf{y}) \, d\mathbf{y}, \tag{85}$$

where $\mathbf{x} \equiv (\mathbf{r}, \boldsymbol{\sigma})$ represents the collection of the space $(\mathbf{r})$ and spin $(\boldsymbol{\sigma})$ coordinates of the first p electrons,

$$\mathbf{x} \equiv (\mathbf{x}_1, \mathbf{x}_2, \ldots, \mathbf{x}_p) \tag{86}$$

with $\mathbf{x}_\mu \equiv (\mathbf{r}_\mu, \sigma_\mu)$, while $\mathbf{y} \equiv (\mathbf{s}, \boldsymbol{\tau})$ similarly represents the coordinates of the remaining $q = N - p$ electrons,

$$\mathbf{y} \equiv (\mathbf{x}_{p+1}, \mathbf{x}_{p+2}, \ldots, \mathbf{x}_N) \tag{87}$$

The corresponding spinless p-matrix* $P_\Psi^{(p)}$ is obtained by summing $\Gamma_\Psi^{(p)}$ over the spin coordinates $\sigma_1, \sigma_2, \ldots, \sigma_p$:

$$P_\Psi^{(p)}(\mathbf{r}|\mathbf{r}') \equiv \sum_\sigma \Gamma_\Psi^{(p)}(\mathbf{r}, \boldsymbol{\sigma}|\mathbf{r}', \boldsymbol{\sigma}) \tag{88}$$

The subscript Ψ will be omitted whenever there is no likelihood of confusion. The superscript (p) will be omitted for the 2-matrices, while the 1-matrices will be denoted by the corresponding lowercase Greek letters, $\gamma(\mathbf{x}_1|\mathbf{x}_1')$ and $\rho(\mathbf{r}_1|\mathbf{r}_1')$ for the spin-dependent and spinless matrix, respectively.

Transition density matrices $\Gamma_{\Psi\Psi'}^{(p)}$, $P_{\Psi\Psi'}^{(p)}$, etc., are similarly defined in terms of two normalized wave functions Ψ and Ψ':

$$\Gamma_{\Psi\Psi'}^{(p)}(\mathbf{x}|\mathbf{x}') \equiv \binom{N}{p} \int \Psi(\mathbf{x}, \mathbf{y})\Psi'^*(\mathbf{x}', \mathbf{y}) \, d\mathbf{y} \tag{89}$$

Note that none of the above definitions require Ψ or Ψ' to be an exact or even approximate eigenfunction of $\hat{H}$; they hold for any normalized antisymmetric N-electron functions, such as, for example, the CI expansion functions Φ_s, Eq. (12).

All of the matrices defined so far are continuous matrices (they are in fact the kernels of integral operators), with the trace (Tr) operation defined by means of integration. Specifically,

$$\text{Tr } \Gamma_\Psi^{(p)}(\mathbf{x}|\mathbf{x}') \equiv \int \Gamma_\Psi^{(p)}(\mathbf{x}|\mathbf{x}) \, d\mathbf{x} = \binom{N}{p} \tag{90}$$

with a similar result for Tr $P_\Psi^{(p)}$. If $\{\psi_i, i = 1, 2, \ldots\}$ are a complete, linearly independent (but not necessarily orthonormal) set of spin-orbitals, $\Gamma^{(p)}$ can be expanded uniquely in terms of products of these to give

$$\Gamma^{(p)}(\mathbf{x}|\mathbf{x}') = \sum_{\mathbf{i}, \mathbf{i}'} \Gamma^{(p)}(\mathbf{i}|\mathbf{i}')\psi_{i_1}(\mathbf{x}_1) \cdots \psi_{i_p}(\mathbf{x}_p) \psi_{i_1'}^*(\mathbf{x}_1') \cdots \psi_{i_p'}^*(\mathbf{x}_p') \tag{91}$$

*The symbol P in this context stands for capital ρ rather than capital p.

with $\mathbf{i} \equiv (i_1, i_2, \ldots)$. The discrete matrix $\boldsymbol{\Gamma}^{(p)}(\mathbf{i}|\mathbf{i}')$ is a representation of $\Gamma^{(p)}(\mathbf{x}|\mathbf{x}')$ in the basis $\{\psi_i\}$, and is also referred to as a density matrix. The ordinary matrix trace is not preserved in this representation unless the basis is orthonormal. More generally,

$$\operatorname{Tr} \Gamma^{(p)}(\mathbf{x}|\mathbf{x}') = \sum_{\mathbf{i},\mathbf{i}'} \boldsymbol{\Gamma}^{(p)}(\mathbf{i}|\mathbf{i}') s_{i_1' i_1} s_{i_2' i_2} \cdots s_{i_p' i_p} \tag{92}$$

where the $s_{ij} = \langle \psi_i | \psi_j \rangle$ are overlap integrals. A similar expansion can be made for the spinless matrix $P^{(p)}(\mathbf{r}|\mathbf{r}')$ in terms of a basis of spatial orbitals $\{\varphi_i, i = 1, 2, \ldots\}$, resulting in a representation $\mathbf{P}^{(p)}(\mathbf{i}|\mathbf{i}')$.

7.2. The Calculation of Density Matrices and Electronic Properties

From the definition of $\Gamma^{(p)}$ and the antisymmetry of Ψ it follows that the expectation value of a symmetric p-electron operator $\hat{F}$, Eq. (64), is given by

$$\langle \Psi | \hat{F} | \Psi \rangle = \operatorname{Tr} [\hat{f}_{12\ldots p} \Gamma^{(p)}(\mathbf{x}|\mathbf{x}')]$$

$$= \sum_{\mathbf{i},\mathbf{i}'} \boldsymbol{\Gamma}^{(p)}(\mathbf{i}|\mathbf{i}') \langle \psi_{i_1'} \cdots \psi_{i_p'} | \hat{f} | \psi_{i_1} \cdots \psi_{i_p} \rangle \tag{93}$$

If $\hat{F}$ is spin independent we get

$$\langle \Psi | \hat{F} | \Psi \rangle = \operatorname{Tr} [(\hat{f}_{12\ldots p} P^{(p)}(\mathbf{r}|\mathbf{r}')]$$

$$= \sum_{\mathbf{i},\mathbf{i}'} \mathbf{P}^{(p)}(\mathbf{i}|\mathbf{i}') \langle \varphi_{i_1'} \cdots \varphi_{i_p'} | \hat{f} | \varphi_{i_1} \cdots \varphi_{i_p} \rangle \tag{94}$$

where $\{\varphi_i\}$ is a spatial orbital basis. Similar formulas apply for integrals of the type $\langle \Psi | \hat{F} | \Psi' \rangle$ in terms of transition density matrices. In particular, if Φ_s and Φ_t are two CFs constructed from a basis $\{\varphi_i\}$, and $\hat{F}$ is spin independent, their matrix element F_{st} can be expressed in the form

$$F_{st} \equiv \langle \Phi_s | \hat{F} | \Phi_t \rangle = \sum_{\mathbf{i},\mathbf{i}'} \mathbf{P}_{ts}^{(p)}(\mathbf{i}|\mathbf{i}') \langle \varphi_{i_1'} \cdots \varphi_{i_p'} | \hat{f} | \varphi_{i_1} \cdots \varphi_{i_p} \rangle \tag{95}$$

where $\mathbf{P}_{ts}^{(p)}$ has been written for $\mathbf{P}_{\Phi_t \Phi_s}^{(p)}$. Taking the special case of the Hamiltonian operator, and comparing this with Eq. (25), we find that

$$a_{ij}^{st} = \boldsymbol{\rho}_{ts}(j|i) \tag{96}$$

$$b_{ijkl}^{st} = \mathbf{P}_{ts}(jl|ik) \tag{97}$$

so that the reduction coefficients are nothing but elements of the transition 1-matrix and 2-matrix for the pair of CFs in terms of the given orbital basis. (As pointed out earlier, these elements are independent of the exact nature of the individual orbitals only if the basis is orthonormal.) Thus (see, e.g.,

McWeeny[120,248], any of the methods available for the construction of the Hamiltonian matrix (Section 5) can also be used to determine the matrices $\boldsymbol{\rho}_{st}$ and $\mathbf{P}_{st}$, except that the reduction coefficients in Eqs. (96) and (97) should be derived without the use of Eq. (59) for spatial symmetry projection operators (Section 5.3), since (95) must hold for operators that do not commute with point-group symmetry operations.

Turning our attention now to a CI wave function $\Psi = \sum_s c_s \Phi_s$, its density matrices are easily expressed in terms of the transition density matrices of the CFs,

$$\Gamma_\Psi^{(p)}(\mathbf{x}|\mathbf{x}') = \sum_{s,t} c_t^* c_s \Gamma_{st}^{(p)}(\mathbf{x}|\mathbf{x}') \tag{98}$$

In particular, for the spinless 1-matrix and 2-matrix we have

$$\boldsymbol{\rho}_\Psi(i_1|i_1') = \sum_{s,t} c_t^* c_s a_{i_1'i_1}^{ts} \tag{99}$$

$$\mathbf{P}_\Psi(i_1 i_2|i_1' i_2') = \sum_{s,t} c_t^* c_s b_{i_1'i_1i_2'i_2}^{ts} \tag{100}$$

These two equations provide the means for computing $\boldsymbol{\rho}$ and $\mathbf{P}$ for a CI wave function, once the eigenvector $\mathbf{c}$ has been determined, using the same procedures, and to a large extent the same programs, that had been used in the reduction stage of the construction of $\mathbf{H}$. Similarly, the spin-dependent matrices $\boldsymbol{\gamma}$ and $\boldsymbol{\Gamma}$ can be determined if a reduction procedure for spin-dependent operators[119,120] is available.

The same approach can be used in calculating transition density matrices between two different CI wave functions. If $\Psi = \sum_s c_s \Phi_s$ and $\Psi' = \sum_s c_s' \Phi_s$ are two such functions (the sets of CFs appearing in the two expansions can be the same, as in the case of states of the same symmetry species, or disjoint, as for states of different symmetry, in which case we shall assume a combined numbering scheme for both sets), the transition 1-matrix is obtained from

$$\boldsymbol{\rho}_{\Psi\Psi'}(i_1|i_1') = \sum_{s,t} c_t'^* c_s a_{i_1'i_1}^{ts} \tag{101}$$

with a similar expression for the 2-matrix. Note that it has been assumed that both Ψ and Ψ' have been expressed in terms of the same orbital set $\{\varphi_i\}$; otherwise the calculation of the transition matrices is so difficult as to be generally impractical.

The calculation of stationary-state properties corresponding to expectation values of one- and two-electron operators can proceed as in Eqs. (93) and (94), once the corresponding density matrices have been determined from Eqs. (99) and (100), provided the appropriate orbital integrals $\langle \varphi_{i_1}|\hat{f}|\varphi_{i_1}\rangle$ and $\langle \varphi_{i_1}\varphi_{i_2}|\hat{f}|\varphi_{i_1}\varphi_{i_2}\rangle$ have been evaluated. Normally such integrals would first be evaluated in a nonorthogonal basis $\{\chi_p\}$, the orbitals $\{\varphi_i\}$ being defined in terms of the basis functions by means of a transformation matrix $\mathbf{U}$, Eq. (24). Rather than transform each set of property integrals separately from a basis-set representation to an orbital representation, as in Eqs. (30) and (31), it is more convenient to transform the density matrices once to the basis-set representation so that the basis-set property integrals can be used directly. Taking a one-electron spin-independent property as an example, and writing $\bar{f}_{pq}$ and f_{ij} for the corresponding basis-set and orbital integrals, respectively [compare Eqs. (26)–(31)], substitution of the transformation analogous to Eq. (30) in the 1-matrix analog of Eq. (94) leads to

$$
\langle \Psi | \hat{F} | \Psi \rangle = \sum_{i,i'} \boldsymbol{\rho}(i|i') f_{i'i}
$$

$$
= \sum_{i,i'} \boldsymbol{\rho}(i|i') \sum_{p,p'} U^*_{p'i'} U_{pi} \bar{f}_{p'p}
$$

$$
= \sum_{p,p'} \left[\sum_{i,i'} \boldsymbol{\rho}(i|i') U^*_{p'i'} U_{pi} \right] \bar{f}_{p'p}
$$

$$
= \sum_{p,p'} \bar{\boldsymbol{\rho}}(p|p') \bar{f}_{p'p} \tag{102}
$$

where

$$
\bar{\boldsymbol{\rho}}(p|p') = \sum_{i,i'} \boldsymbol{\rho}(i|i') U^*_{p'i'} U_{pi} \tag{103}
$$

is the 1-matrix in the basis-set representation. Similarly for the 2-matrix,

$$
\bar{\mathbf{P}}(pq|p'q') = \sum_{i,i',j,j'} \mathbf{P}(ij|i'j') U^*_{p'i'} U^*_{q'j'} U_{pi} U_{qj} \tag{104}
$$

Transition properties (such as transition moments) can be computed in an analogous manner, but it should be remembered that the property integral matrices involved, as well as the transition density matrices, do not always possess the same symmetry properties (and hence the same structure) as integral matrices of totally symmetric operators and as single-state density matrices, respectively.

It is thus seen that once the appropriate density matrices have been evaluated, the calculation of first-order electronic properties (i.e., those represented by expectation values) is no more difficult for CI wave functions than for SCF. Second-order properties (those which require a perturbation expansion, such as the polarizability) are much more difficult, and the alternative approach of using a finite perturbation[255,256] (i.e., adding an external field to $\hat{h}$) appears to be more convenient, at least in CI calculations.

In addition to their utility in the computation of electronic properties, the basis-set density matrices are particularly convenient for the analysis of the bonding structure in a molecule. The 1-matrix $\bar{\rho}$ is in fact the basis for the population analysis often used by quantum chemists.

8. Selection of Basis Sets and Orbitals

8.1. Basis Sets

It is a truism that no calculated wave function can be better than the basis set from which it is constructed. CI calculations, in particular, require even greater care in basis-set selection than SCF calculations, since, in addition to the occupied orbitals of the SCF function, the basis also has to provide appropriate correlation orbitals.

The basis-set requirements of a CI calculation begin (but do not end) with those of the SCF treatment. No amount of configuration interaction can remedy defects due to basis-set inadequacies in the underlying SCF calculation. Thus all that has been learned during the last decade about basis sets for molecular SCF calculations (this subject has been discussed in Chapter 1) also applies to CI treatments, as far as the highly occupied part of the orbital space is concerned. But even for this part of the basis some additional considerations apply. In a typical molecular SCF study most of the computation time is used for the evaluation of the basis-set integrals, but this is no longer the case in large-scale CI treatments. Thus, while in SCF calculations the choice of basis set would be dominated by considerations of integral computation time, we can often afford to give less weight to this factor in a CI study, and choose a basis primarily in terms of its expected effectiveness in the more laborious CI stages.

For example, in a recent comparative study of several large basis sets in both SCF and CI calculations on the water molecule,[257] it was found that large Slater-type (STO) basis sets were more effective in lowering the energy of both the SCF and CI wave functions than contracted Gaussian (CGTO) sets of comparable size. However, the basis-set integral evaluation time is an order of magnitude longer for STO than for CGTO sets, making the use of STO bases much less attractive for SCF calculations. In the CI treatments, on the other hand, the STO integral evaluation time was comparable to that of the rest of the calculation, making the use of an STO basis a reasonable choice.

Turning now to the additional basis-set requirements for CI expansions in order to provide an adequate treatment of electron correlation, the principal need here is for basis functions capable of providing good correlation orbitals. These are orbitals which have their greatest amplitude in regions of space in which the occupied SCF orbitals (and thus the electron density) are primarily concentrated, but which have additional nodal surfaces which divide these regions in many and diverse ways. For example, additional nodal surfaces surrounding an atom or group of atoms provide radial (in–out) correlation, while a nodal surface between atoms provides lateral (left–right) correlation. Additional nodal surfaces through atoms or containing the axis of a linear molecule or the plane of a planar molecule provide angular, azimuthal, or up–down correlation, respectively.

An SCF basis for a molecular calculation usually provides some correlation orbitals, specifically the antibonding orbitals, with nodes between the atoms. A high-quality SCF basis which includes polarization functions can often provide an adequate set of correlation orbitals.[258] On the other hand, taking as an extreme example the beryllium atom, even the best restricted Hartree–Fock wave function for the ground state involves s-type orbitals only, so that no basis functions of other types are needed for the SCF stage, while the principal contribution to correlation is angular and is provided primarily by p-type orbitals. Similar considerations apply to π orbitals for H_2 and LiH, and δ orbitals for C_2, N_2, O_2, etc.

It is thus seen that the principal requirement for additional basis functions for the treatment of electron correlation involves functions having similar spatial extent (and thus orbital exponents) as the functions used in the SCF wave function, but having additional nodal surfaces, particularly through the use of higher angular quantum numbers (l). If the aims of a calculation include the adequate treatment of inner-shell correlation (which is usually fairly constant in chemical processes[7] and is normally of little chemical interest), this implies that K-shell p functions must be included, having a radial dependence similar to that of the $1s$ core orbitals. The same considerations apply, of course, to L-shell d functions, etc., but these can often be satisfied by the polarization functions which are normally included in good SCF treatments.

It is also clear that orbital exponents optimized for SCF basis sets are not necessarily optimum for CI treatments. Detailed exponent optimization in CI calculations is so costly, however, that it is rarely done, and no tabulations of optimized CI basis sets, like the many useful tabulations available for SCF, have appeared. Limited optimization, such as for the parameters in an "even tempered" set,[155,259] is sometimes carried out.[189] In most cases, however, we have to be satisfied with an SCF optimized basis, augmented, at best, with several functions designed to provide additional correlation orbitals. Some modification of the contraction scheme of a CGTO basis can sometimes be used to advantage, though, in application to CI.[257]

Finally, special care is also needed in treatments of excited states. These are very frequently included in CI studies, and in order to yield meaningful results they have to use basis sets which take into account the particular requirements of these states, such as functions of higher angular quantum numbers or diffuse functions capable of representing Rydberg or semi-Rydberg (V-state[22]) orbitals.[8]

Further discussion of basis sets for CI calculations can be found in recent reviews by Bagus *et al.*[260] and by Davidson.[258]

8.2. Remarks on the Choice of Orbitals

In a full CI calculation (i.e., one which includes all CFs of the desired symmetry which can be constructed from the orbital set) the computed wave function is invariant to any nonsingular linear transformation of the orbitals. Thus the results of such a calculation depend only on the choice of basis set and not on the particular transformation $\mathbf{U}$, Eq. (24), from basis functions $\{\chi_i\}$ to orbitals $\{\varphi_i\}$ (assuming $m = m'$, i.e., no truncation of the function space in the transformation). Furthermore, in a CI calculation in which, for each level of excitation (promotion) relative to a closed-shell reference configuration, either all or none of the CFs that can be constructed for that level are included in the expansion, the results are invariant to separate nonsingular linear transformations in the occupied space (spanned by the orbitals in the reference configuration) and in the virtual space (spanned by the remaining orbitals), provided the two spaces are orthogonal.

This last statement can be generalized as follows: Let the orbital set $\{\varphi_i\}$ be partitioned into several subsets, with the orbitals in each subset being orthogonal to those in the other subsets. The CFs can then be classified according to the number of electrons they place in each subset (the subset occupancy). Then, if for each class (i.e., each assignment of subset occupancies), either all CFs which can be constructed for that class, or none, are included in the CI expansion, then the CI wave function is invariant to any separate, nonsingular, linear transformations within each of the subsets. (Note that, if pure symmetry states are to be obtained, orbitals which belong as partners to a multidimensional irreducible representation of the molecular point group should be placed in the same subset.) The proof of this statement is obvious, since any transformation of the type specified converts each CF to a linear combination of CFs of the same class.

We shall refer to a wave function of the type discussed in the preceding paragraph as a *full-class CI* wave function. This includes such common cases as full double-excitation CI, as well as the "first-order wave functions" introduced by Schaefer *et al.*[261] (see Section 9.2), and many others, including similar calculations carried out with a truncated orbital set ($m < m'$).[41,53,260]

It is clear that the results of a full-class CI calculation depend only on the subdivision of the basis-set space into subspaces defined by the orbital subsets, and not on the individual orbitals used to span these subspaces. A particular choice of orbitals within each subspace can still have an effect on the detailed form taken by the CI expansion and on the number of CFs making significant contributions to the expansion, but not on the wave function itself or on any of the computed properties. Thus the only orbital optimization that is important in a full-class CI calculation is that of the subdivision of the basis-set space into subspaces. In particular, if a full double-excitation CI calculation is to be undertaken, the use of canonical SCF occupied and virtual orbitals is as effective as any other choice which maintains the SCF boundary between the occupied and virtual subspaces, and about as effective, in general, as any other reasonable choice of orbitals (since the SCF function is usually very similar to any other preferred single-configuration representation, such as the first natural-orbital configuration). It may still be desirable, however, after the wave function has been computed, to attempt to put it in a more compact form, and this may be possible, for example, by a natural-orbital transformation.

A common and useful partitioning of the orbital space is into a core space (which is often left fully occupied in CI calculations), an occupied valence space, a vacant valence space, and the remaining virtual space (which may sometimes be truncated). Some of these spaces may be further divided, in appropriate cases, into σ and π parts, or other, physically motivated divisions. (The valence space is essentially that which would be spanned by the valence-shell basis functions in a minimal basis calculation.)

The principal deficiency of canonical SCF virtual orbitals* for use in CI calculations is that the valence part of the virtual space is often ill-defined and far from optimum, particularly in large basis-set calculations, and most particularly when diffuse functions are included in the basis. As previously stated, optimum correlation orbitals should have most of their amplitude in the same region of space as the electrons which they are meant to correlate. The vacant valence orbitals should thus have their maxima in the same region as the occupied valence orbitals. While this requirement is necessarily satisfied in minimal basis calculations, it is not well satisfied in treatments using large basis sets. The usual SCF effective Hamiltonian represents an $(N-1)$-electron potential when acting on an occupied orbital (because of the cancellation of the Coulomb self-energy term by an equal exchange term in this case), but represents an N-electron potential when acting on a vacant orbital. Thus the low-energy virtual SCF orbitals tend to be more diffuse than the occupied orbitals.[262] In fact, the more diffuse an orbital is, the more likely is its orbital energy to be close to zero. Furthermore, any diffuse part of the basis-set space tends to be distributed to some extent among all the virtual orbitals, blurring any distinction between valence and nonvalence orbitals. This is demonstrated in a particularly striking manner in a recent CI calculation on *trans*-butadiene[8] which included diffuse (Rydberg-type) basis functions and in which expectation values of z^2 were compared for canonical SCF and natural orbitals.

---

*The canonical SCF virtual orbitals referred to here and in subsequent discussions are those obtained in a *finite-basis* matrix-SCF calculation. The *exact* positive-energy eigenfunctions of the Fock operator generally are continuum functions, and are of little use as correlation orbitals.[253]

Various approaches have been used in attempts to overcome these shortcomings. Some of these used the fact that the SCF effective Hamiltonian can be modified in ways which change the individual orbitals without affecting the SCF wave function.[263,264] Alternatively, the virtual orbitals can be determined from a different effective Hamiltonian than the one used for the SCF wave function and explicitly orthogonalized to the occupied orbitals. One objective of the various methods is to obtain orbitals which are appropriate for an $(N-1)$-electron potential (the V^{N-1} potential), as used by Kelly[265] in his many-body perturbation theory treatment of atoms and by Hunt and Goddard[266] for their "improved virtual orbitals" (IVO). Such orbitals correspond more closely to excited-state molecular orbitals[266–268] and do indeed lead to better convergence of CI expansions, particularly when excited states are treated simultaneously with the ground state.[268] Among other methods for choosing virtual orbitals for CI calculations are those which depend on natural orbital (or pseudonatural orbital) transformations or on MCSCF calculations, discussed in the following two subsections, on intermediate small CI calculations,[262] or on the maximization of certain electron repulsion integrals.[269,270]

Another approach is based on the expectation that the use of localized orbitals for both the occupied and virtual subspaces can lead to improved CI convergence, since correlation is largely a local effect and there would be little correlation between electrons in well-separated orbitals.[271–273] The principal disadvantage of this approach in some cases is the loss of orbital symmetry resulting from the localization, but, more importantly, adequate localization of correlation effects is often not possible when several electron pairs are crowded into one atomic valence shell.[274–277]

Most of the above discussion related primarily to single-state calculations at near-equilibrium molecular geometries, and particularly to ground states. Additional considerations are necessary in excited-state and multistate treatments and in the study of potential curves and surfaces.

Considering excited states, a common and convenient practice is to use orbitals based on ground-state SCF calculations. While this simplifies many aspects of the calculation, including the determination of transition moments, it may often be a far from optimum choice. Buenker and Peyerimhoff[278] have found that better excitation energies can be obtained when the CI expansion for each state uses orbitals from an SCF calculation on that state, or at least from a closely related state, rather than using ground-state orbitals for all states. One problem with this approach is that the calculation of transition moments may become prohibitively difficult. More convenient, particularly when computing several states of the same symmetry species, is to use some compromise approach based on orbitals chosen to describe the different states in an equally satisfactory manner. The lack of individual orbital optimization in each state can be compensated for by appropriate choices of the CFs to be included in the CI expansions, as described in Section 9. Such compromise choices are discussed further in the next two subsections.

The calculation of potential curves and surfaces, particularly when including regions far from equilibrium, poses additional problems. The restricted SCF description of most molecules is known to break down as the atoms are pulled apart and bonds are broken.[1,3] This generally applies to any closed-shell (spin-restricted) single-configuration description, whatever the orbitals. Orbitals derived from such a single-configuration wave function would not be expected to provide a balanceed CI treatment for all regions of the surface unless compensated for by including additional CFs. While the use of unrestricted SCF orbitals (based on an open-shell configuration) can overcome the dissociation breakdown problem, as can also be done by some localized orbital descriptions, other problems arise, including loss of orbital symmetry in many cases. This approach, or a similar one based on MCSCF (or GVB) calculations, can however be made to work effectively, and is discussed further in Section 8.4.

8.3. Natural Orbitals

Natural spin-orbitals (NSO) are defined[245] as the eigenfunctions of the spin-dependent first-order density matrix $\gamma(\mathbf{x}|\mathbf{x}')$. For use in CI calculations we are interested in their spin-free counterpart, the *natural orbitals* (NO), defined as the eigenfunctions of the spinless 1-matrix $\rho(\mathbf{r}|\mathbf{r}')$. The expansion of the NOs in terms of a given set of orbitals is obtained by solving the matrix eigenvalue problem (or generalized problem, Eq. (13), if the representation is nonorthogonal) for the corresponding matrix representation $\boldsymbol{\rho}(i|i')$. The eigenvalues obtained are the *occupation numbers* of the corresponding NOs. (For more on natural orbitals and their properties see, for example, the reviews by Davidson[253,279] and Bingel and Kutzelnigg.[252])

The interest in natural orbitals for CI calculations is due to the expectation that CI expansions based on these orbitals should have the fastest convergence, i.e., should require the fewest CFs for a given accuracy. The plausibility of this argument has been demonstrated by Löwdin,[245] but only for the two-electron case is there definite proof[280] that NOs produce the most compact CI expansion possible. While the dramatic improvement in the compactness of the two-electron wave function in terms of NOs does not carry over to the many-electron case,[258,281] there has been accumulating evidence that significantly faster convergence can generally be achieved by the use of natural orbitals or of reasonable approximations thereof.[53,269,282–284] It appears that these orbitals possess the desirable characteristics described in the previous section, particularly in that they provide an effective partitioning of the virtual orbital space into subspaces suitable for correlating the different occupied shells,[258] with the less useful part of the space left over as the lowest-occupancy NOs. Thus NOs allow truncation of the set of orbitals used in the CI expansion with a considerably smaller effect on the result than with a similar truncation in terms of canonical SCF orbitals.[53,283,284]

Another effective way in which NOs can be used to generate a compact expansion is by individual selection of CFs (see Section 9), e.g., on the basis of their estimated energy contributions.[285] This is demonstrated in a recent analysis[286] of one of the CI wave functions from the previously mentioned study of the water molecule.[257] The original wave function was constructed from canonical SCF orbitals computed in a basis set of 39 Slater-type functions, and included all singly and doubly excited CFs relative to the SCF configuration (4120 CFs in all). The NOs of this wave function were obtained, and a series of truncated CI calculations were carried out for both the canonical SCF and natural-orbitals cases, with energy-selected sets of CFs of various length (the energy selection methods are described in Section 9.3). The correlation energy recovered with each type of orbital as a function of the length of the CI expansion is shown in Fig. 4. The faster convergence of the NO-based

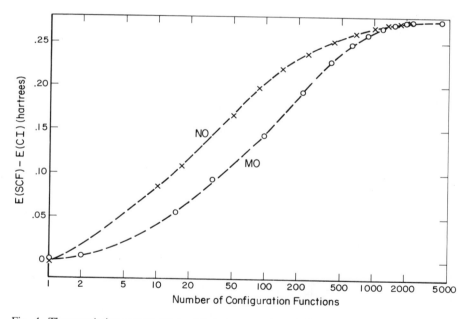

Fig. 4. The correlation energy obtained from energy-selected CI expansions based on canonical SCF orbitals (MO) and on natural orbitals (NO), as a function of the number of CFs included (from Ref. 286, based on the 39-STO CI calculation on H_2O of Ref. 257). The NOs were obtained from the unselected 4120-term expansion. The CFs were included in order of decreasing energy contributions. The points represent actual CI calculations, while the lines represent an approximate interpolation. Note that the horizontal scale is logarithmic. (The energy of the first natural configuration is 0.0014 hartree above the SCF energy, while the 4120-term all-singles-and-doubles CI energy is 0.0005 hartree higher for NOs than for MOs.)

expansion is evident, and is shown more strikingly in Table 2, which gives the number of CFs required in each case in order to recover various percentages of the correlation energy of the original wave function. It is seen that the advantage of NOs over SCF MOs is primarily in the low end of the range of expansion lengths and begins to disappear in this example as the number of CFs exceeds about one-quarter of the number in the unselected wave function.

The principal stumbling block to the use of NOs for CI calculations is that these orbitals are defined in terms of the final wave function, and thus are not known until the calculation is essentially complete. Fortunately, the evidence indicates that any reasonable approximation to the NOs produces almost all of the advantages of the exact NOs. This has led to several practical schemes, one of which is the iterative natural orbital (INO) method.[269,282,285,287] In this procedure a sequence of CI calculations are carried out with a limited set of CFs, each calculation using the NOs of the previous iteration. The success of this scheme requires that an adequate set of CFs be used in the iterations,[258] which can involve a considerable amount of computation. Furthermore, a new integral transformation is required for each iteration, though the use of symmetry (see Section 3.5) can often reduce the effort involved in this step. It appears, however, that the process need not be carried out, in general, beyond two iterations, since one or two iterations produce practically all of the improvement that can be obtained in this method,[282,288,289] provided an adequate set of CFs us used[258] (in fact, the process often begins to diverge after several iterations.[258,289]

Table 2. *Number of Energy-Selected Configuration Functions Required to Recover Various Percentages of the Correlation Energy of the All-Singles-and-Doubles CI Wave Function, When Canonical SCF Orbitals (MO) and Natural Orbitals (NO) Are Used*[a]

| | Number of CFs required[c] | |
Percentage of correlation energy[b]	MO	NO[d]
50	87	32
60	140	50
70	220	79
80	351	147
90	617	362
95	944	697
98	1410	1213
99	1760	1652
100	4120	4120

[a] From Ref. 286, based on the 39-STO CI calculation on H_2O of Ref. 257.
[b] Relative to the correlation energy obtained in the unselected 4120-term expansion.
[c] Obtained by interpolation from the data in Fig. 4.
[d] The NOs were obtained from the unselected CI wave function.

Several other methods for obtaining natural orbitals, or adequate approximations to them, are based on perturbation theory.[283,284,290–293] Particularly simple and easy to carry out is the approach used by Hay,[283] based on the B_k approximation[294] which is discussed in Section 9. *Pseudonatural orbitals* (PNO),[295,296] also called *pair natural orbitals*,[297] which are discussed in Chapter 5, can also be useful in CI calculations.

A problem with the use of NOs for simultaneous calculations on several electronic states is that each state has different NOs, and that these NOs can even differ very considerably from state to state.[8] As in other cases, a compromise approach is called for, and there has been some success in using *average natural orbitals*,[298,299] obtained by diagonalizing the average of the 1-matrices of the states of interest.

An interesting characteristic of natural orbitals is that the *first natural configuration* (FNC), made up of the most strongly occupied NOs, while very similar to the SCF function is found to produce considerably improved values of molecular properties,[300,301] in some cases these values being rather close to extended CI results.[257,286] Another interesting, but unfavorable, property of NO–CI expansions is that they tend to produce Hamiltonian matrices **H** which require considerably more iterations in the solution of the matrix eigenvalue problem (Section 6.3) than for comparable expansions in terms of SCF MOs. This has been traced[286] to the presence of some very large off-diagonal elements in **H**. These elements connect CFs which correlate the inner shells and have extremely large kinetic energy contributions (due to the presence of nodes in the inner shell region). Such large off-diagonal elements would also be bound to have a detrimental effect on the convergence of perturbation expansions based on natural orbitals.[302]

8.4. MCSCF and GVB Orbitals

A multiconfiguration self-consistent field (MCSCF) wave function (discussed in Chapter 3) is a short CI expansion in which both orbitals and expansion coefficients have been optimized. Only part of the basis-set space is normally used in the MCSCF function, this part consisting of a set of orbitals which are usually very similar to the occupied SCF orbitals, augmented by several correlation orbitals. Since these orbitals are optimized in a CI expansion, and thus should possess the desirable characteristics discussed in Section 8.2, it is reasonable to expect that they would also be effective in large-scale CI expansions. The rest of the orbital space which has been left out of the MCSCF wave function has of course not been optimized, and is thus no better than SCF virtual orbitals for the purpose of CI calculations, but if the set of optimized orbitals is large enough to account for the major correlation effects, then a suitable partitioning of the orbital space has been achieved and the remaining orbitals can be limited to a very low occupancy, or truncated altogether, in the CI expansion.

The fact that the correlation orbitals of the MCSCF wave function, unlike the SCF virtual orbitals, are concentrated in the same regions of space as the occupied orbitals has been stressed by Wahl and co-workers[303,304] and by Sabelli and Hinze.[305] Hinze[306] has recently compared the characteristics of MCSCF and large-scale CI techniques, and has concluded that while MCSCF is most effective in providing a compact description of the major correlation effects (and in correctly describing the dissociation process in a molecule[303]), it becomes excessively laborious for obtaining large fractions of the total correlation energy. This is at least partly due to increasing convergence difficulties as the number of CFs in the MCSCF function is increased, related to the fact that with longer expansions the result becomes less sensitive to the form of each orbital. There also are additional difficulties in dealing with excited states. On the other hand, the combination of MCSCF and CI, using a short MCSCF expansion to determine a set of orbitals and then using these orbitals in a large-scale CI expansion, is both feasible and effective for obtaining a quantitative description of the electronic structure of molecules, including the calculation of potential curves and surfaces of ground and excited states. The success of this approach (sometimes coupled with NO techniques[41,260,307]) has been demonstrated by several recent calculations.[307–310]

Unless restricted by symmetry conditions, MCSCF orbitals tend to be localized. This localization is particularly effective for describing molecular dissociation and bond-breaking processes in general, but the removal of symmetry restrictions has also been found to produce improved MCSCF results for normal bound systems.[308,311] For use in CI calculations, however, symmetry restricted orbitals are generally preferable, as they can lead to significant economies and thus can make possible better results for a given level of computational effort.

The generalized valence bond (GVB) wave function, discussed in Chapter 4, is a particular form of an MCSCF wave function in which each chemical bond is described in terms of a pair of localized, nonorthogonal, optimized orbitals. The spatial part of the wave function for each bond is in the "(u, v) form,"[312]

$$\Phi_{uv} = u(\mathbf{r}_1)v(\mathbf{r}_2) + v(\mathbf{r}_1)u(\mathbf{r}_2) \tag{105}$$

rather than the closed-shell "natural-orbital form" of a simple MCSCF function of the optimized double configuration (ODC) type,[313]

$$\Phi_{NO} = c_i\varphi_i(\mathbf{r}_1)\varphi_i(\mathbf{r}_2) + c_j\varphi_j(\mathbf{r}_1)\varphi_j(\mathbf{r}_2) \tag{106}$$

in which φ_i and φ_j are orthonormal. If no symmetry or orthogonality restrictions are placed on the orbitals u and v in (105) then the two optimized functions Φ_{uv} and Φ_{NO} are equivalent, and either gives a correct qualitative description of bond dissociation. However, the (u, v) form has an important advantage, particularly at large internuclear distances, since each of the orbitals u and v would tend to localize on one atom (or molecular fragment, in general), providing a simple description of the dissociation process in terms of a *single*-configuration wave function. Such orbitals would be expected to have beneficial effects on the convergence of a CI expansion, but unfortunately they cannot be used conveniently without orthogonalization. For large internuclear distances u and v are almost orthogonal (since they are localized in different regions of space), and constraining them to be exactly orthogonal (by Schmidt or Löwdin orthogonalization of the optimum GVB orbitals, see Section 3.3) would have little effect on the quality of the (u, v)-form wave function or on the convergence of the CI expansion. Near equilibrium bond length, on the other hand, an orthogonalized (u, v)-form function would be definitely inferior to the two-configuration NO form and may be less effective in providing orbitals for CI. Raffenetti and Kahn[314] have analyzed this problem in detail and have concluded that the orthogonalized (u, v) form would be preferable as a basis for a CI expansion whenever the original GVB orbitals have an overlap integral $\langle u|v \rangle$ smaller in magnitude than $\sqrt{2}-1$, the NO (or ODC) form being preferable otherwise.

Coupled with the previously mentioned improved virtual orbitals[266] (IVO) computed from a V^{N-1} potential obtained by omitting one of the localized GVB orbitals, an orthogonalized GVB and IVO set of orbitals provides a very effective basis for CI expansions, particularly when several electronic states are computed simultaneously and bond dissociation is examined.[268,315]

9. Selection of Configuration Functions

9.1. Excitation Levels, Perturbation Theory, and Cluster Expansions

The number of CFs which can be constructed for a state with N electrons and total spin S out of a set of m orbitals, ignoring spatial symmetry restrictions, is given by Weyl's formula,[74,106,316]

$$n(N, S, m) = \frac{2S+1}{m+1} \binom{m+1}{\frac{1}{2}N-S} \binom{m+1}{\frac{1}{2}N+S+1} \tag{107}$$

For $m \gg N \gg S$ this number is very roughly approximated by

$$n(N, S, m) \approx N^{-2}(2me/N)^N \tag{108}$$

Obviously, full CI calculations (i.e., using all possible configurations) are impractical except for very small systems or when using rather poor basis sets. On the other hand, with well-chosen orbitals, we would hope that only a very small fraction of the CFs would make nonnegligible contributions to the CI wave function. The problem of how to select these CFs prior to a full-fledged calculation is the subject of this section.

The best guidance to the selection process is provided by perturbation theory.[317-320] Using a suitable zero-order wave function, perturbation theory enables us to classify CFs in terms of the order of perturbation in which their contributions first enter the perturbation series, and allows us to estimate their coefficients in the CI wave function and their contributions to the resulting energy.

The simplest and most common analysis is based on the restricted SCF function as the zero-order wave function. All other CFs can then be classified in terms of their excitation (or promotion) level relative to the SCF function; this is defined as the number of electrons promoted from SCF occupied orbitals to virtual orbitals in the given CF. (Some ambiguity may exist in this definition in the case of open-shell states.) Since the Hamiltonian matrix element H_{st} between two CFs Φ_s and Φ_t vanishes if the CFs differ by more than two orbitals, it is clear that the SCF function has vanishing matrix elements with all CFs of excitation levels higher than two. Furthermore, by Brillouin's theorem,[318,319] a closed-shell restricted SCF function has vanishing matrix elements with all singly excited CFs (provided the CFs have been generated from the same basis set used to solve the SCF problem). Thus the only CFs entering the first-order correction to the wave function in the closed-shell case are double excitations, and these serve to determine the energy through the third order. The second-order wave function will include single, triple, and quadruple excitations, which are the only ones having nonvanishing matrix elements with the first-order function. The various blocks of the Hamiltonian matrix **H** can also be classified in this manner in terms of the order in which they first contribute to the perturbation series,[320] and approximate CI methods can be devised which neglect blocks of **H** beyond a certain order.[260,294,320]

The foregoing analysis explains the predominant role that doubly excited CFs generally play in CI expansions. Single excitations are particularly important in open-shell cases,[319] but are also commonly included in closed-shell CI calculations. This is partly due to the very small number of singly excited CFs, but more significantly reflects their important role in the determination of one-electron properties.[300,321] This role can be explained in the following manner.[321] Let the (unnormalized) full CI expansion be written in the form

$$\Psi = \Phi_0 + c_S \Phi_S + c_D \Phi_D + c_T \Phi_T + c_Q \Phi_Q + \cdots \tag{109}$$

where Φ_0 is the SCF function, $c_S \Phi_S$ represents the aggregate of all single-excitation contributions, $c_D \Phi_D$ represents the aggregate of all double-excitation contributions, etc., with $\Phi_0, \Phi_S, \Phi_D, \ldots$ normalized. If we denote the perturbation theory ordering parameter by λ, then for cases in which Brillouin's theorem holds,

$$c_S = \mathcal{O}(\lambda^2) \tag{110}$$

[otherwise $c_S = \mathcal{O}(\lambda)$], and in all cases

$$c_D = \mathcal{O}(\lambda), \ c_T = \mathcal{O}(\lambda^2), \ c_Q = \mathcal{O}(\lambda^2), \ldots \tag{111}$$

The expectation value of an operator $\hat{F}$ can be expanded in the form

$$\langle \Psi | \hat{F} | \Psi \rangle = (1 + |c_S|^2 + |c_D|^2 + \cdots)^{-1} [F_{00} + 2 \operatorname{Re} (c_S F_{0S}) + 2 \operatorname{Re} (c_D F_{0D})$$
$$+ |c_S|^2 F_{SS} + |c_D|^2 F_{DD} + \cdots] \tag{112}$$

where the notation for the matrix elements of $\hat{F}$ is obvious. The only term with a coefficient of order λ (for a closed-shell case) is F_{0D}, but for a one-electron operator this term vanishes since it connects CFs which differ by two orbitals (see Section 5). Thus we are immediately led to the well-known result, first pointed out for the electron density by Møller and Plesset,[317] that while the SCF error in the wave function is of first order in λ, the SCF errors in the one-electron properties, like the SCF error in the energy, are of second order. Furthermore, we find that both Φ_S and Φ_D contribute to the one-electron properties in the same order, λ^2, accounting for the significant effect of singly excited CFs on computed CI one-electron properties.

In actual fact, while formally correct, this analysis somewhat exaggerates the effect of single excitations in closed-shell cases. Not only is the number of singly excited CFs much smaller than the number of doubly excited functions, but it is also generally found that their coefficients are typically much smaller than the squares of the major double-excitation coefficients. Nevertheless, singly excited CFs cannot be left out if good results for dipole moments and other one-electron properties are to be obtained.[300,321]

All the properties discussed above for closed-shell restricted SCF functions, including Brillouin's theorem, also hold for the unrestricted SCF function of any state, but unrestricted SCF orbitals are not generally practical as a basis for CI calculations since they do not satisfy symmetry and equivalence restrictions.[319]

Another source of guidance for the relative importance of different CFs is provided by the various formulations of the cluster expansion approach[4,272,322–329] and the closely related diagrammatic many-body perturbation theory.[330] A detailed account is beyond the scope of this chapter, and only a brief summary of the pertinent results will be given.

The correlation correction to the SCF wave function can be analyzed in terms of cluster contributions, each of which accounts for correlation effects which simultaneously involve all the electrons in a particular subset of the spin-orbitals of the SCF function. These cluster terms, each of which can be described by a certain type of diagram,[324,325] can be classified as either linked (connected, irreducible) or unlinked (disconnected, reducible). An unlinked cluster describes the simultaneous correlation in two or more disjoint subsets of spin-orbitals, and its contribution to the wave function can be entirely determined from a knowledge of the contributions of the smaller, linked clusters contained in it. In fact, the problem can be formulated in such a way that the unlinked terms do not appear explicitly, for example by writing the exact wave function in the form[322-326]

$$\Psi = e^{\hat{S}} \Phi_0 \tag{113}$$

where $\hat{S}$ is a linear combination of linked-cluster excitation operators (the unlinked terms appear in this formulation when the exponential $e^{\hat{S}}$ is expanded as a power series). With the neglect of certain coupling terms (matrix elements of $\hat{H}$ between different linked-cluster terms), each linked-cluster contribution can be calculated separately,[4,272,327-329] and the total correlation energy (as well as other properties) can be obtained simply by adding the separately computed contributions of all the different *linked* clusters. The neglect of the coupling terms, however, is not well justified,[277,326] and can lead to appreciable overestimation of the total correlation energy[277] and to lack of invariance of the result with respect to unitary transformations of the occupied orbitals.[277]

The central result of the cluster analysis of correlation effects is that by far the largest linked cluster contribution is from two-body (electron-pair) terms.[4,272,324-332] This is related to the observation[331] that the close approach of more than two electrons is prevented by the exclusion effect (Fermi hole), since at least two (or more) of them will necessarily have the same spin. A far second among the linked-cluster contributions is provided by the three-body terms,[332] and while one-body terms are important for electronic property calculations[300] (and in open-shell expansions based on a restricted SCF zero-order function[5]), all higher-order linked clusters are negligible for all practical purposes. If however, the analysis is carried out in such a way that the unlinked cluster contributions are displayed separately, rather than being folded into the linked cluster terms,[4,326,331,332] it is found that unlinked clusters made up of linked pairs are of considerable importance, with the four-body unlinked terms (made up of two linked pairs) playing a particularly prominent role. In fact, as the number of electrons increases, the contribution of the unlinked higher-order clusters is expected to increase, until it eventually accounts for most of the correlation energy.[258] (Note that in open-shell states using restricted SCF orbitals, unlinked clusters made up of linked pairs and one-body terms would also be important.)

The various pair-based approaches to the correlation problem, including Sinanoğlu's many-electron theory (MET),[4-6,327,328,331] Nesbet's Bethe–

Goldstone method,[272,275,329] and the independent electron-pair approximation (IEPA) used by Ahlrichs and Kutzelnigg[333] and by Meyer,[296,334] attempt to account for the unlinked cluster contributions to electron correlation by the separate calculation of pair correlation energies, adding these up to provide the total correlation energy. This addition of separately calculated pair contributions automatically incorporates the higher-order unlinked cluster effects, as indicated above, but introduces a serious error due to the neglect of coupling elements between the pairs (pair–pair interactions).[227] With some additional effort, the coupling terms can be included in the calculation, as is done in the coupled-pair many-electron theory (CPMET) of Čížek and Paldus[324–326] and in the coupled electron-pair approximation (CEPA) of Meyer[296,334] (see also Ahlrichs *et al.*[335,336] and Chapter 5).

Another interesting point[332] is that linked three-body clusters enter into the perturbation series for the wave function in the same order (λ^2) as the unlinked four-body terms, while unlinked three-body contributions, like the linked four-body terms, enter in third order. All larger clusters enter in fourth or higher orders. Still, in practice, the unlinked four-body terms are generally more numerous and make a considerably greater contribution than the linked three-body terms, and this contribution increases much more rapidly as the number of electrons is increased.

In a CI expansion of the exact wave function the different clusters are represented by linear combinations of CFs with excitation levels equal to the cluster size.[4,326,331] It is impossible to separate the linked and unlinked parts in the construction of the CFs, each CF containing unlinked effects of all lower-level excitations into which it can be resolved. However, the unlinked cluster contribution to the coefficient of a particular CF can be determined if the coefficients of the appropriate lower-level CFs are known.[4,332,337] It is simply given by the sum of all products of coefficients of lower-level excitations into which it can be resolved [compare Eq. (113)]. This fact, coupled with the knowledge that the only significant contribution of quadruple and higher excitations is by virtue of unlinked clusters made up of linked pairs (at least for closed shells), can be used very effectively in a selection scheme to determine which higher-excitation CFs should be included, after the coefficients of the important double-excitation CFs have been determined.[189,287,337]

The fact that in a CI expansion unlinked cluster contributions can only be accounted for by including quadruple (and higher-order) excitations is one of the principal drawbacks of the CI method. In contrast, such contributions are automatically accounted for, without explicitly computing high-order terms, in some of the cluster-based methods and in many-body perturbation theory. In this sense the CI expansion is much less compact and less efficient than these approaches, and becomes progressively less efficient as the number of electrons increases.[258] A combination of both types of approach, using CI to determine double-excitation coefficients and the cluster expansion ideas to determine the effect of higher-order unlinked terms, can be a practical compromise in many cases. Davidson[258,338] has in fact derived a very simple formula, based on these ideas, for estimating the unlinked quadruple-excitation contribution to the correlation energy ΔE_Q just from the knowledge of the double-excitation correlation energy ΔE_D and the coefficient c_0 of the SCF function in the normalized all-doubles CI expansion:

$$\Delta E_Q = (1 - c_0^2) \Delta E_d \tag{114}$$

The discussion in this section so far has been based on the assumption that the SCF function can serve as a reasonable zero-order approximation on which to base a perturbation series or a cluster expansion. In such cases we see that the most important CFs to include in a CI calculation are double (and perhaps single) excitations, with quadruple and higher even-level excitations becoming more important as the number of electrons increases. Since the number of p-fold excited CFs is roughly proportional[294] to m^p, including any sizable fraction of the higher excitations is usually impractical. The only practical ways to account for these terms appear to be by the preselection of important terms based on double-excitation coefficients,[189,287,337] by the use of extrapolation formulas such as Eq. (114), by cluster-based methods, or by many-body perturbation series.

In cases in which the SCF function is not an adequate zero-order approximation (such as for degenerate or nearly degenerate configurations, for some open-shell cases, and for the calculation of potential curves and surfaces) the situation is more complicated. Both ordinary and many-body perturbation theory, as well as the cluster expansion, can be reformulated to use a more complicated function as the zero-order approximation,[319,339–342] and correspondingly a CI wave function can be constructed using double excitations relative to more than one "root" configuration. This topic is discussed further in the next section.

9.2. Preselection Schemes and Root Configurations

The vast number of CFs which can be generated from even a modest-sized set of orbitals for a system of more than a very few electrons requires careful attention to the construction and pruning of the list of CFs to be included in a CI expansion. The choice of these CFs is often carried out in two stages. The first stage, which we shall call *preselection*, requires the identification of *classes* of CFs which may be important in order for the calculation to provide adequate answers for the problem of interest. The simplest form of preselection is by excitation level relative to the SCF configuration, while more elaborate schemes, for example, can involve preselection by occupancy levels of appropriate orbital subsets, as discussed in Section 8. Since the number of CFs generated by such preselection schemes can still be much too large for practical use in the final CI calculation, and since the majority of CFs so generated may be rather unimportant (see, for example, Fig. 4 and Table 2 in Section 8), a second selection stage often follows, in which individual CFs are examined and screened on the basis of criteria which do not require elaborate calculations. Effective preselection schemes must produce CF lists which are short enough for the screening of individual CFs to be practical and which are not overburdened with too large a proportion of unimportant CFs. At the same time, such schemes should not overlook other CFs which may play a significant role in the problem being studied.

Preselection by excitation level is an obvious choice, but suffers from several disadvantages. Noting the m^p dependence of the number of p-level excited CFs on the number of orbitals m (for $m \gg N$),[294] we are often limited to $p = 2$, or sometimes $p = 3$, for the highest level that can even be conveniently examined. Recalling the importance of unlinked clusters, this often means neglecting many terms which can make significant contributions to the correlation energy and to other properties of interest. (Note however that selected terms of this type may be added after the important double excitations have been identified,[189,337] as discussed in the previous section.) Also, particularly when using large basis sets, preselection by excitation level tends to produce long CF lists in which only a small proportion of the terms survive a screening process. Furthermore, in many cases the SCF function is not a very good zero-order approximation, or is not equally good for the different regions of potential curves or surfaces being investigated, so that preselection by excitation level relative to that function results in bias in computed relative energies.

One effective improvement to the selection by excitation level takes advantage of the subdivision of the orbital space into subspaces of varying characteristics, such as core, valence, etc.,[41,260–262,309] as discussed in connection with the choice of orbitals (Section 8). The selection, or preselection, then proceeds by assigning occupancy levels or limits to the various subspaces, and thus allows us to focus on those correlation effects which are most relevant to the problem being investigated (e.g., dissociation or excitation energies). This method depends for its success on the appropriateness of the orbital space subdivision and on reasonable choices for the occupancy assignments, but in practice these are often done in a somewhat arbitrary manner. If all CFs resulting from the subspace occupancy assignments are used in the CI expansion without further screening, the calculation qualifies as a full-class CI calculation, as defined in Section 8.2, and satisfies the appropriate invariance conditions.

The simplest and most common forms of occupancy-limit preselection involve maintaining full occupancy of inner-shell orbitals and truncating the virtual orbital space. Keeping inner shells fully occupied means complete neglect of inner-shell correlation effects (including intershell contributions), but since these effects are largely insensitive to the chemical environment[7] such neglect is justified in most calculations of energy differences which do not involve core vacancies. Truncation of the virtual space is more arbitrary, however, and may be quite sensitive to the choice of orbitals.[53,262,283,284] One specific possibility, when inner-shell correlation is neglected and the basis set contains more than a minimal set of core functions, is to leave out any core-correlating virtual orbitals, but for this to be possible it may first be necessary to subject the virtual space to a suitable transformation (e.g., to natural orbitals).[260] Another common form of orbital space subdivision for the purpose of CF selection is by orbital energy of canonical SCF orbitals. This often takes the form of allowing just a few of the highest occupied and lowest virtual orbitals to have variable occupancies.[278,343] Alternatively, in studies of π-electron systems the σ orbitals are often left out entirely from the CI part of the calculation,[8,344,345] or just a very small number of them are included.[278,345] The complete neglect of σ orbital participation (including correlation of the electrons in occupied σ orbitals, use of virtual σ orbitals in correlating the electrons in occupied π orbitals, and correlation between σ and π electrons) may, however, seriously affect computed excitation energies and other properties.[8,270,346]

A well-known form of selection by subset occupancy is the *first-order wave function* introduced by Schaefer, Klemm, and Harris,[261] most frequently used in conjunction with the iterative natural orbital (INO) procedure.[282,347] In this approach the virtual orbitals are divided into a valence set and a remainder, and singly and doubly excited CFs which have no more than one electron in the remainder set are included in the calculation (in some cases triple excitations having two electrons in excited valence orbitals and one in the remainder are also included). The importance of the INO procedure in this context is that it is relied upon to produce the optimum partitioning of the virtual space. It is to be emphasized that the first-order wave function is not intended to produce an optimum choice of CFs in terms of the total correlation energy.[289] Instead this approach is aimed[282] at the prediction of reliable unbiased potential energy surfaces. In at least one case,[315] however, the more exhaustive CI techniques described below have proved to be decidedly superior in this regard.

An interesting generalization of preselection by orbital set occupancy assignments is the point system,[348] in which a point value p_i is assigned to each orbital φ_i, and a total point value $P_s = \sum_i n_i^s p_i$ is then derived for each CF, n_i^s being the occupation number of φ_i in the CF Φ_s. The CFs to be included are determined by limits set upon the accepted P_s values. The principal weakness of this scheme is the arbitrariness of the assignment of point values p_i to the orbitals and of limits to the total point values P_s of the CFs, but with suitable choices this can be a very powerful method. A similar idea, but with a more specific choice of point values, using orbital energies for this purpose (so that the total point value of a CF is a rough estimate of its energy expectation value), was used for preselection by Bender and Davidson.[349]

A different approach to the preselection problem is based on the idea of *root* (or *reference*) *configurations*.[59,260,262,268,315,350] This is a generalization of the excitation level criterion, most suitable when several CFs make important contributions to the wave function, and particularly useful for the study of potential curves and surfaces and for multistate calculations. Instead of considering excitation levels defined relative to a single configuration (usually the SCF function), we include all CFs which are obtained by specific excitation levels relative to any one of a small number (typically 2–20) of the more important CFs. Such root CFs (RCFs) have been suggested as a basis for a perturbation calculation by Nesbet,[319] who has also proposed an iterative scheme for their selection. The RCFs would normally include, in addition to the SCF function, any CFs expected to make major correlation contributions, such as the $1s^2 2p^2$ configuration for the ground state of the beryllium atom. In multistate calculations (expanding the wave functions of several states of the same symmetry type in one common set of CFs) the RCFs would include all CFs which make important contributions to any one of the states.[262,268,315] Also, and particularly important for the calculation of potential curves and surfaces, any CF required for the proper description of dissociation, such as the $(1\sigma_u)^2$ configuration for H_2, would be included.[268,315]

The iterative scheme[260,262,268,315,319] for the selection of the RCFs would begin with a physically motivated guess of the important terms, along the lines described above, and these terms would then be used as a basis for the examination of other CFs (usually all single and double excitations relative to all the RCFs) by perturbation theory methods or by any of the other methods for the selection of individual CFs described in the next section. Any CF found to make an important contribution to the wave function, either in terms of energy or, preferably in terms of its estimated coefficient in the CI expansion, would be added to the RCFs, and the process would be iterated until no new important CFs are discovered. In a multistate or potential surface calculation[262,268,307,315] all CFs found to make important contributions to any one of the states at any one of a number of representative points on the surface would be added to the RCFs. A reasonable criterion for the identification of an important CF would normally be[262,315] an estimated CI coefficient of magnitude ≥ 0.1, but better prevention of bias in multistate and potential surface calculations can be obtained by a cumulative contribution criterion,[268,315] as described in the next section.

CI calculations based on root CFs normally include all single- and double-excitation CFs relative to all the RCFs, or use these as a preselected set to be screened on the basis of individual contributions. Potential curves obtained by this selection approach for several states of the BH molecule[315] were found to be very close to corresponding full-CI curves and much better than curves calculated with the same basis set by the first-order wave function method.[351]

9.3. Energy Contributions and Individual CF Selection

The screening and selection of individual CFs is normally carried out on the basis of their estimated contributions to the wave function (as measured by the square of the CI coefficient) or to the energy. Energy contributions are either estimated directly from the second-order energy expression of perturbation theory or derived from estimates of the corresponding CI coefficients. These latter estimates are obtained from preliminary pilot calculations of various types, frequently based on perturbation theory.

The simplest application of perturbation theory (PT) to the estimation of the CI coefficient c_s of a CF Φ_s is through the first-order Rayleigh–Schrödinger PT formula,

$$c_s = \frac{H_{s0}}{H_{00} - H_{ss}} \tag{115}$$

where Φ_0 is the zero-order wave function (with coefficient $c_0 = 1$), normally the SCF function. (We assume in this subsection that the CFs are orthonormal.) This formula, or the corresponding formula for the second-order contribution to the energy,

$$\Delta E_s = \frac{|H_{s0}|^2}{H_{00} - H_{ss}} \tag{116}$$

have been widely used,[14–16,262,270,285,319] and are quite adequate for CF selection, provided Φ_0 is a good zero-order approximation and the principal contribution of Φ_s is through its direct (first-order) interaction with Φ_0.

When a single CF is not a satisfactory zero-order approximation, and particularly when treating potential surfaces, a linear combination of several "primary" CFs, such as the root CFs (RCFs) described in the previous section, can be used as the zero-order function Φ_0.[269,277,294,309,319] The coefficients of this linear combination can be determined from a small CI calculation involving just the primary CFs. The only additional matrix elements required for the testing of each "secondary" CF Φ_s are the elements $H_{st}(t = 1, 2, \ldots, k)$ linking Φ_s to the k primary CFs (which are assumed for convenience to be numbered $1, 2, \ldots, k$) and the diagonal element H_{ss}. This, essentially, is the procedure designated by the symbol* A_k by Gershgorn and

---

*The subscript k was defined in Ref. 294 to be one less than the number of CFs in the primary set, but it is now considered more convenient to define k to equal the number of primary CFs.

Shavitt,[294] to distinguish it from the somewhat more elaborate B_k procedure. The B_k method[294] involves the approximate solution of the CI eigenvalue problem for the primary CFs together with all the secondary CFs (those to be screened) simultaneously. The approximation consists of the neglect of all off-diagonal matrix elements connecting different secondary CFs, so that exactly the same matrix elements are used as in the A_k method. The additional work is primarily the solution of the eigenvalue problem for the approximated Hamiltonian matrix, but since this step is essentially proportional to the number of nonzero elements in the matrix (Section 6.3), and since most of the elements have been neglected, the extra computational effort is not considerable.

The B_k method can also be justified on the basis of perturbation theory.[294] Its principal advantage over the A_k procedure is that it allows the zero-order function (the primary part) to readjust under the influence of the perturbation, and this results in a considerably better approximation to the wave function.[294] It has been used to derive energy contributions for CF selection in several investigations,[294,315] and has also been used for the calculation of approximate natural orbitals.[283,307] However, the A_k method appears to be entirely satisfactory both for CF selection[268] and for obtaining approximate NOs,[269,309] and since it handles the secondary CFs one at a time rather than all together, it avoids serious computer storage problems when the number of secondary CFs is extremely large.

Another procedure, which is intermediate between A_k and B_k, has been used by Buenker and Peyerimhoff.[350] This involves solving a succession of small eigenvalue problems, each containing the primary CFs and just one of the secondary CFs being screened. This too uses the same set of matrix elements as A_k and B_k, and shares with A_k the advantage of handling the secondary CFs one at a time. While it allows readjustment of the zero-order function under the effect of the interaction with just one secondary CF in each case, this cannot be as effective as the readjustment due to the simultaneous interaction with all the secondary CFs. On the other hand it provides a very direct and reliable estimate of each secondary CF's energy contribution, as the difference between the eigenvalue of the corresponding small-CI problem and the energy of the primary set alone. The principal disadvantage of this approach is the need to solve a very large number of small eigenvalue problems. While these many eigenvalue problems can be solved very efficiently, particularly if the primary part of the matrix is diagonalized once at the outset, there seems to be little justification for the extra labor of either this or the B_k method, compared to A_k, just for obtaining energy contribution estimates for CF selection.

The B_k method has also been proposed as an economical substitute for the normal CI calculations,[260,294] saving the labor of computing most of the otherwise required Hamiltonian matrix elements. A considerably larger value of k (the number of primary CFs) is required, however, for this application than for CF selection.[294] Several generalizations of this method have been proposed,[352-354] but these require the use of the full Hamiltonian matrix, and thus do not provide as substantial a saving in the total effort required for the CI calculation. They are obviously too laborious to use for CF selection.

Turning now to the determination of energy contributions, it has already been pointed out how these can be obtained directly from the second-order energy formula, Eq. (116). This formula can also be used with the A_k procedure, taking Φ_0 to be the appropriate linear combination of primary CFs.[268] Two similar formulas can be obtained from (116) by substituting for H_{s0} once or twice, respectively, from Eq. (115):

$$\Delta E_s = H_{0s}c_s \tag{117}$$

$$\Delta E_s = |c_s|^2 (H_{00} - H_{ss}) \tag{118}$$

Obviously, if c_s is taken from Eq. (115), these formulas are entirely equivalent to (116), but they can also be used with coefficients c_s obtained from other sources, such as B_k calculations[294,315] or partial CI calculations,[337,349] in which case their results can differ substantially from those of (116). Equation (117) is the *partial energy* formula used by Bunge.[355] It has the interesting property that the sum of the energy contributions derived from it when the coefficients c_s have been obtained by solving the corresponding eigenvalue equation is equal to the energy eigenvalue; this is obvious when we note that the sum of (117) over s (including $s = 0$) is just the left-hand side of the first equation in the eigenvalue problem, and is thus equal to E [using the normalization $c_0 = 1$, as in Eq. (115)]. A disadvantage of (117) is that in some cases it predicts positive ΔE_s values,[355] even when $H_{00} < H_{ss}$ for $s \neq 0$, giving the false implication that adding the corresponding CF to the CI expansion raises the energy.

Equation (118) has been used for energy contribution assessment by Boys, Shavitt, and co-workers[337,356] and by Bender and Davidson.[349] Its advantage is that it does not contain H_{s0} and can predict a nonzero ΔE_s even for CFs for which the first-order contribution to the wave function vanishes (i.e., $H_{s0} = 0$), provided c_s is obtained from a calculation which takes second-order contributions of Φ_s into account.

A disadvantage of all the above formulas for energy contributions is that they rely on some zero-order approximation and do not treat all CFs equally. This can be particularly important in dealing with potential surfaces, for which the zero-order function may be unequally good at different points on the surface. The exact energy contribution of a CF can of course be obtained by carrying out two separate CI calculations, one containing the CF in question and one excluding it, and taking the difference between the resulting energies. This, however, is impractical in general, but a reasonable alternative can be obtained by using a limited number of the more important CFs when testing each function.[287,350] Another and more practical estimate can be obtained by determining the effect of deleting the CF in question from the CI expansion while leaving the coefficients of all other CFs unchanged. Using the fact that **c** is an eigenvector of **H** results in a very simple energy contribution formula, first derived by Brown,[357]

$$\Delta E_s = (E - H_{ss})|c_s|^2/(\langle\Psi|\Psi\rangle - |c_s|^2) \tag{119}$$

where Ψ is the CI expansion (with arbitrary normalization) containing Φ_s and E is the corresponding variational energy. Both E and the coefficients c_s can be obtained from B_k or other approximate or limited-size calculations. The resemblance of (119) to (118) is obvious, particularly when $|c_s|^2 \ll 1$, but no zero-order function appears in (119), the total energy E taking the place of H_{00}. Thus (119) can be applied equally to all CFs, even including the SCF function or other principal components of the wave function, and can be used very effectively for CF selection in multistate and potential surface calculations.[315]

CF selection on the basis of energy contributions appears to be more appropriate in a variational calculation than selection based on the magnitude of the CI coefficients. As can be seen from Eq. (118) and by examination of published results,[287] these two criteria can lead to significantly different selected lists. In particular, any CF which has a nonnegligible coefficient c_s in spite of having a very high diagonal element H_{ss} must interact quite strongly with Φ_0 [see Eq. (115)] and can have a sizable energy contribution. On the other hand, it appears that selection of terms to be included among the root or primary CFs is best done on the basis of coefficients, since these coefficients would be instrumental in determining the effect of secondary CFs interacting with the zero-order function.

The selection in terms of energy contributions can be carried out in terms of a *threshold* or by a *cumulative selection* criterion. In the threshold method all CFs contributing (in absolute value) less than the specified threshold to all the states in question are dropped.[262,268,350] Typical values of the selection threshold are between 10^{-4} and 10^{-5} hartree, though actual values would depend on the required accuracy and on practicability. Buenker and Peyerimhoff[350,358] have shown that by repeating the last (eigenvalue) stage of the CI calculation with several selection thresholds it is possible to obtain a reliable extrapolation to zero threshold and a considerable improvement in computed potential curves. The extrapolation correction appears to be a fairly constant fraction (usually between 0.6 and 0.9) of the sum of the energy contributions of the CFs which have been left out, the fraction depending on the size of the RCF set.[358]

In a potential surface calculation it is found that the number of selected CFs can vary considerably over the surface,[268,315] and the selection error (the energy loss due to the selection) would then vary as well. Cumulative selection[315] has been evolved to deal with this situation as well as for multistate calculations. In this method CFs are dropped in order of increasing energy contributions until the sum of the contributions of all the dropped CFs reaches a specified limit. In multistate cases the selection can be carried out separately for each state and the selected lists are then merged. The cumulative selection limit would normally be appreciably greater than the value of a threshold criterion (a typical value[315] may be around 10^{-3} hartree) for the same total accuracy. Potential curves obtained by this method for several states of BH were found to be essentially parallel to, and about equally displaced from, corresponding full CI curves.[315]

If properly applied with reasonably tight selection criteria and with an adequate root CF set, the results of such selected CI calculations should be very close to the results of unselected calculations, and should not be overly dependent on the choice of orbitals from which the CFs have been constructed. However, the number of selected CFs, and thus the computational effort, may depend strongly on the appropriateness of the orbital choice.[307]

In summary, it appears that practical and satisfactory CI calculations can be performed in many cases (for small enough molecules) by the appropriate choice of a set of root CFs (which choice can be refined iteratively), and by selecting CFs from all single and double excitations relative to all of the RCFs on the basis of energy contributions obtained from perturbation theory (by an A_k or similar process), using either threshold selection with possible extrapolation or a cumulative selection scheme.

10. Concluding Remarks

As should be obvious from the foregoing, the undertaking of a large-scale CI calculation requires both careful planning and considerable attention to technical details. The planning involves, in particular, the choice of basis set and the determination of the methods to be used for the choice of orbitals and the selection of CFs. These choices have to be made in the light of the type of information that is to be derived from the calculation and the type of scientific questions that are to be answered, but must also be based on computational feasibility. This usually requires some compromises on the size of the basis set and the completeness of the CI treatment, since present facilities and techniques are not usually adequate for a really satisfactory treatment of most problems of interest in chemistry today.

Looking back at the tentative small-scale explorations of two decades ago, impressive strides have been made in the application of the CI method. At the same time it has become evident that the convergence of this method is considerably slower than was tacitly assumed in the earlier days, and that the magnitude of a satisfactory CI treatment increases extremely rapidly with the size of the system being investigated. Thus, even though the limits on the size of practical calculations are continuously being extended by improvements in techniques and in computer capabilities, adequate CI calculations on large molecules are not in sight. At present the CI method can be seen primarily as a tool for the rigorous investigation of small systems (typically containing several small atoms) and for some exploration of medium-size systems. In these areas it can be used to obtain useful and increasingly reliable chemical information, as well as to examine the accuracy of other approaches and to analyze the assumptions and approximations involved in them. The standard CI method described here, as well as some of its modifications described elsewhere in this volume, thus takes its place among other approaches as just one of the tools of quantum chemical investigation. While it cannot offer solutions to all problems, it is a very important, even indispensable, member of the set of methods at our disposal.

Acknowledgments

The author is indebted to the researchers whose work is cited extensively in this chapter, and whose articles, lectures, and discussions contributed greatly to the clarification of the issues discussed here. He owes a particular debt of gratitude to his late teacher, colleague, and friend, Dr. Samuel Francis Boys, a great pioneer and proponent of the configuration interaction method, who has inspired and guided much of the author's work in this field. He is also greatly indebted to Dr. Ernest R. Davidson, with whom many aspects of the CI method were discussed extensively during the writing of this chapter. This work was sponsored by the Battelle Institute Program, Grant No. B-1335-0101.

References

1. P.-O. Löwdin, *Adv. Chem. Phys.* **2**, 207 (1959).
2. H. F. Schaefer III, *The Electronic Structure of Atoms and Molecules. A Survey of Rigorous Quantum Mechanical Results*, Addison-Wesley, Reading, Massachusetts (1972).
3. J. C. Slater, *Phys. Rev.* **35**, 509 (1930).
4. O. Sinanoğlu, *Proc. Nat. Acad. Sci. USA* **47**, 1217 (1961).
5. H. J. Silverstone and O. Sinanoğlu, *J. Chem. Phys.* **44**, 1899 (1966).
6. C. Hollister and O. Sinanoğlu, *J. Am. Chem. Soc.* **88**, 13 (1966).
7. A. Pipano, R. R. Gilman, and I. Shavitt, *Chem. Phys. Lett.* **5**, 285 (1970).
8. R. P. Hosteny, T. H. Dunning, Jr., R. R. Gilman, A. Pipano, and I. Shavitt, *J. Chem. Phys.* **62**, 4764 (1975).
9. W. Ritz, *J. Reine Angew. Math.* **135**, 1 (1909).
10. E. A. Hylleraas, *Z. Phys.* **48**, 469 (1928).

11. S. Weinbaum, *J. Chem. Phys.* **1**, 593 (1933).
12. H. M. James and A. S. Coolidge, *J. Chem. Phys.* **1**, 825 (1933).
13. R. G. Parr, D. P. Craig, and I. G. Ross, *J. Chem. Phys.* **18**, 1561 (1950).
14. M. J. M. Bernal and S. F. Boys, *Philos. Trans. R. Soc. London, Ser. A* **245**, 139 (1952).
15. S. F. Boys, *Proc. R. Soc. London, Ser. A* **217**, 136 (1953).
16. S. F. Boys, *Proc. R. Soc. London, Ser. A* **217**, 235 (1953).
17. D. Kastler, *J. Chim. Phys.* **50**, 556 (1953).
18. A. Meckler, *J. Chem. Phys.* **21**, 1750 (1953).
19. S. F. Boys and V. E. Price, *Philos. Trans. R. Soc. London, Ser. A* **246**, 451 (1954).
20. S. F. Boys, G. B. Cook, C. M. Reeves, and I. Shavitt, *Nature* **178**, 1207 (1956).
21. E. U. Condon and G. H. Shortley, *The Theory of Atomic Spectra*, Cambridge U.P., London (1935).
22. R. S. Mulliken, *Chem. Phys. Lett.* **25**, 305 (1974).
23. R. Courant and D. Hilbert, *Methods of Mathematical Physics*, Vol. 1, Interscience, New York (1953).
24. E. A. Hylleraas, *Mathematical and Theoretical Physics*, Vol. 1, Wiley–Interscience, New York (1970).
25. S. T. Epstein, *The Variation Method in Quantum Chemistry*, Academic Press, New York (1974).
26. L. Pauling and E. B. Wilson, *Introduction to Quantum Mechanics*, McGraw-Hill, New York (1935).
27. E. C. Kemble, *The Fundamental Principles of Quantum Mechanics*, McGraw-Hill, New York (1937).
28. A. Messiah, *Quantum Mechanics*, Vol. II, North-Holland, Amsterdam (1964).
29. J. C. Slater, *Quantum Theory of Matter*, 2nd ed., McGraw-Hill, New York (1968).
30. C. Eckart, *Phys. Rev.* **36**, 878 (1930).
31. J. L. Powell and B. Crasemann, *Quantum Mechanics*, Addison-Wesley, Reading, Massachusetts (1961).
32. E. A. Hylleraas and B. Undheim, *Z. Phys.* **65**, 759 (1930).
33. J. K. L. MacDonald, *Phys. Rev.* **43**, 830 (1933).
34. H. Shull and P.-O. Löwdin, *Phys. Rev.* **110**, 1466 (1958).
35. H. F. Weinberger, *J. Res. Nat. Bur. Stand., Sect. B* **64**, 217 (1960).
36. J. Hinze and C. C. J. Roothaan, *Prog. Theor. Phys. (Kyoto), Suppl. No.* **40**, 37 (1967).
37. I. Shavitt, *J. Comput. Phys.* **6**, 124 (1970).
38. D. B. Neumann, H. Basch, R. L. Kornegay, L. C. Snyder, J. W. Moskowitz, C. Hornback, and S. P. Liebmann, POLYATOM (Version 2), Program 199, Quantum Chemistry Program Exchange, Indiana University, Bloomington, Indiana (1971).
39. M. Yoshimine, *J. Comput. Phys.* **11**, 449 (1973).
40. A. D. McLean, in: *Proceedings of the Conference on Potential Energy Surfaces in Chemistry* (W. A. Lester, ed.), p. 87, Report RA 18, IBM Research Laboratory, San Jose, California (January, 1971).
41. P. S. Bagus, B. Liu, A. D. McLean, and M. Yoshimine, in: *Energy, Structure, and Reactivity* (D. W. Smith and W. B. McRae, eds.), p. 130, Wiley, New York (1973).
42. P.-O. Löwdin, *Adv. Quantum Chem.* **5**, 185 (1970).
43. E. R. Davidson, *J. Chem. Phys.* **62**, 400 (1975).
44. N. W. Winter, W. C. Ermler, and R. M. Pitzer, *Chem. Phys. Lett.* **19**, 179 (1973).
45. P.-O. Löwdin, *J. Chem. Phys.* **18**, 365 (1950).
46. C. F. Bender, *J. Comput. Phys.* **9**, 547 (1972).
47. K. C. Tang and C. Edmiston, *J. Chem. Phys.* **52**, 997 (1970).
48. G. H. F. Diercksen, *Theor. Chim. Acta* **33**, 1 (1974).
49. P. Pendergast and W. H. Fink, *J. Comput. Phys.* **14**, 286 (1974).
50. R. K. Nesbet, *Rev. Mod. Phys.* **35**, 552 (1963).
51. S. T. Elbert, Ph.D. thesis, University of Washington, Seattle (1973).
52. R. M. Stevens, private communication.
53. R. M. Stevens, *J. Chem. Phys.* **61**, 2086 (1974).
54. R. M. Pitzer, *J. Chem. Phys.* **58**, 3111 (1973).

55. R. M. Pitzer, *J. Chem. Phys.* **59**, 3308 (1973).
56. W. C. Ermler and C. W. Kern, *J. Chem. Phys.* **58**, 3458 (1973).
57. J. C. Slater, *Phys. Rev.* **34**, 1293 (1929).
58. A. Bunge, *J. Chem. Phys.* **53**, 20 (1970).
59. A. D. McLean and B. Liu, *J. Chem. Phys.* **58**, 1066 (1973).
60. P.-O. Löwdin, *in*: *Calcul des Fonctions d'Onde Moléculaire* (Colloques Internationaux du Centre National de la Recherche Scientifique, No. 82), p. 23, CNRS, Paris (1958).
61. M. Kotani, K. Ohno, and K. Kayama, *in*: *Encylopedia of Physics* (S. Flügge, ed.), Vol. 37/2, p. 1, Springer, Berlin (1961).
62. M. Kotani, A. Amemiya, E. Ishiguro, and T. Kimura, *Tables of Molecular Integrals*, 2nd ed., Maruzen, Tokyo (1963).
63. F. A. Matsen, *Adv. Quantum Chem.* **1**, 59 (1964).
64. R. Pauncz, *Alternant Molecular Orbital Method*, Saunders, Philadelphia (1967).
65. F. E. Harris, *Adv. Quantum Chem.* **3**, 61 (1967).
66. F. E. Harris, *in*: *Energy, Structure, and Reactivity* (D. W. Smith and W. B. McRae, eds.), p. 112, Wiley, New York (1973).
67. P.-O. Löwdin and O. Goscinski, *Int. J. Quantum Chem., Symp. No.* **3**, 533 (1970).
68. J. I. Musher, *J. Phys. (Paris)* **31**, *Suppl. C4*, 51 (1970).
69. K. Ruedenberg and R. D. Poshusta, *Adv. Quantum Chem.* **6**, 267 (1972).
70. W. I. Salmon, *Adv. Quantum Chem.* **8**, 37 (1974).
71. P.-O. Löwdin, *Phys. Rev.* **97**, 1509 (1955).
72. V. H. Smith, Jr. and F. E. Harris, *J. Math. Phys.* **10**, 771 (1969).
73. R. Manne, *Theor. Chim. Acta* **6**, 116 (1966).
74. H. Weyl, *Gruppentheorie und Quantenmechanik*, Hirzel, Leipzig, Germany (1928) [English translation, *The Theory of Groups and Quantum Mechanics*, Methuen, London (1931), reprinted by Dover, New York].
75. E. Wigner, *Z. Phys.* **40**, 492 (1927).
76. E. Wigner, *Z. Phys.* **40**, 883 (1927).
77. E. Wigner, *Gruppentheorie und ihre Anwendung auf die Quantenmechanik der Atomspektren*, Vieweg, Braunschweig, Germany (1931) [English translation, *Group Theory and Its Application to the Quantum Mechanics of Atomic Spectra*, Academic Press, New York (1959)].
78. P. A. M. Dirac, *Proc. R. Soc. London, Ser. A* **123**, 714 (1929).
79. I. Waller and D. R. Hartree, *Proc. R. Soc. London, Ser. A* **124**, 119 (1929).
80. R. Serber, *J. Chem. Phys.* **2**, 697 (1934).
81. T. Yamanouchi, *Proc. Phys.-Math. Soc. Jpn.* **17**, 274 (1935).
82. T. Yamanouchi, *Proc. Phys.-Math. Soc. Jpn.* **19**, 436 (1937).
83. H. V. McIntosh, *J. Math. Phys.* **1**, 453 (1960).
84. W. A. Goddard III, *Phys. Rev.* **157**, 73 (1967).
85. W. A. Goddard III, *Int. J. Quantum Chem., Symp. No.* **3**, 593 (1970).
86. R. D. Poshusta and R. W. Kramling, *Phys. Rev.* **167**, 139 (1968).
87. J. Gerratt and W. N. Lipscomb, *Proc. Nat. Acad. Sci. USA* **59**, 332 (1968).
88. J. J. Sullivan, *J. Math. Phys.* **9**, 1369 (1968).
89. G. Heldmann, *Int. J. Quantum Chem.* **2**, 785 (1968).
90. G. Heldmann and P. Schnupp, *J. Comput. Phys.* **3**, 208 (1968).
91. J. I. Musher and R. Silbey, *Phys. Rev.* **174**, 94 (1968).
92. G. A. Gallup, *J. Chem. Phys.* **48**, 1752 (1968).
93. G. A. Gallup, *J. Chem. Phys.* **50**, 1206 (1969).
94. D. J. Klein, *J. Chem. Phys.* **50**, 5140 (1969).
95. F. A. Matsen, *J. Am. Chem. Soc.* **92**, 3525 (1970).
96. I. G. Kaplan and O. B. Rodimova, *Int. J. Quantum Chem.* **7**, 1203 (1973).
97. T. K. Lim, *Int. J. Quantum Chem.* **8**, 523 (1974).
98. M. Moshinsky, *Group Theory and the Many-Body Problem*, Gordon and Breach, New York (1968).
99. I. L. Gel'fand and M. L. Tsetlin, *Dokl. Akad. Nauk SSSR* **71**, 825 (1950).
100. I. L. Gel'fand and M. L. Tsetlin, *Dokl. Akad. Nauk SSSR* **71**, 1017 (1950).

101. J. Patera, *J. Chem. Phys.* **56**, 1400 (1972).
102. W. G. Harter, *Phys. Rev. A* **8**, 2819 (1973).
103. W. G. Harter and C. W. Patterson, *Phys. Rev. A* **13**, 1067 (1976).
104. W. G. Harter and C. W. Patterson, *Phys. Rev. A*, in press.
105. F. A. Matsen, *Int. J. Quantum Chem., Symp. No.* **8**, 379 (1974).
106. J. Paldus, *J. Chem. Phys.* **61**, 5321 (1974).
107. J. Paldus, *Int J. Quantum Chem., Symp. No.* **9**, 165 (1975).
108. J. Paldus, *Theor. Chem. Adv. Perspec.* **2**, 131 (1976).
109. R. Pauncz, *Chem. Phys. Lett.* **31**, 443 (1975).
110. F. A. Matsen, A. A. Cantu, and R. D. Poshusta, *J. Phys. Chem.* **70**, 1558 (1966).
111. D. J. Klein and B. R. Junker, *J. Chem. Phys.* **54**, 4290 (1971).
112. P. J. A. Ruttink, *Theor. Chim. Acta* **36**, 289 (1975).
113. J. F. Gouyet, *Phys. Rev. A* **2**, 139 (1970).
114. J. F. Gouyet, *Phys. Rev. A* **2**,, 1286 (1970).
115. W. Heitler and G. Rumer, *Z. Phys.* **68**, 12 (1931).
116. J. C. Slater, *Phys. Rev.* **38**, 1109 (1931).
117. C. M. Reeves, Ph.D. thesis, University of Cambridge (1957).
118. C. M. Reeves, *Commun. ACM* **9**, 276 (1966).
119. B. Sutcliffe, *J. Chem. Phys.* **45**, 235 (1966).
120. I. L. Cooper and R. McWeeny, *J. Chem. Phys.* **45**, 226, 3484 (1966).
121. K. Ruedenberg, *Phys. Rev. Lett.* **27**, 1105 (1971).
122. W. I. Salmon and K. Ruedenberg, *J. Chem. Phys.* **57**, 2776 (1972).
123. A. Pipano and I. Shavitt, unpublished work.
124. Z. Gershgorn and I. Shavitt, *Int. J. Quantum Chem., Symp. No.* **1**, 403 (1967).
125. S. L. Altmann, *in*: *Quantum Theory* (D. R. Bates, ed.), Vol. II, p. 87, Academic Press, New York (1972).
126. M. A. Melvin, *Rev. Mod. Phys.* **28**, 18 (1956).
127. P.-O. Löwdin, *Rev. Mod. Phys.* **39**, 259 (1967).
128. J. Killingbeck, *J. Math. Phys.* **11**, 2268 (1970).
129. A. Gołębiewski, *Mol. Phys.* **20**, 481 (1971).
130. G. A. Gallup, *Int. J. Quantum Chem.* **8**, 267 (1974).
131. R. McWeeny, *Symmetry—An Introduction to Group Theory and Its Applications*, Pergamon, London (1963).
132. R. L. Flurry, Jr., *Theor. Chim. Acta* **31**, 221 (1973).
133. S. Flodmark, SYMPRO, Program 46, Quantum Chemistry Program Exchange, Indiana University, Bloomington, Indiana (1964).
134. S. Flodmark and E. Blokker, *in*: *Group Theory and Its Applications* (E. M. Loebl, ed.), Vol. II, Academic Press, New York (1971).
135. J. R. Gabriel, *J. Comput. Phys.* **2**, 336 (1968).
136. R. Moccia, *Theor. Chim. Acta* **7**, 85 (1967).
137. T. D. Bouman, A. L. H. Chung, and G. L. Goodman, *in*: *Sigma Molecular Orbital Theory* (O. Sinanoğlu and K. B. Wiberg, eds.), p. 333, Yale University Press, New Haven, Connecticut (1970).
138. A. L. H. Chung and G. L. Goodman, *J. Chem. Phys.* **56**, 4125 (1972).
139. T. D. Bouman and G. L. Goodman, GPTHEORY, Program 214, Quantum Chemistry Program Exchange, Bloomington, Indiana (1972).
140. D. Létoquart, *Int. J. Quantum Chem.* **8**, 627 (1974).
141. J. C. Hempel, J. C. Donini, B. R. Hellebone, and A. B. P. Lever, *J. Am. Chem. Soc.* **96**, 1693 (1974).
142. J. R. Gabriel, *J. Chem. Phys.* **51**, 3713 (1969).
143. B. G. Wybourne, *Int. J. Quantum Chem.* **7**, 1117 (1973).
144. I. Sakata, *J. Math. Phys.* **15**, 1702 (1974).
145. R. J. Buenker and S. D. Peyerimhoff, *Theor. Chim. Acta* **12**, 183 (1968).
146. H. F. Schaefer III and F. E. Harris, *J. Comput. Phys.* **3**, 217 (1968).
147. H. F. Schaefer III, *J. Comput. Phys.* **6**, 142 (1970).

148. G. Racah, in: *Ergebnisse der exakten Naturwissenschaften* (G. Höhler, ed.), Vol. 37, p. 28, Springer, Berlin (1965) [in English].
149. L. C. Biedenharn and H. Van Dam (eds.), *Quantum Theory of Angular Momentum*, Academic Press, New York (1965).
150. A. Rotenberg, *J. Chem. Phys.* **39**, 512 (1963).
151. A. Rotenberg, PROP, Program 37, Quantum Chemistry Program Exchange, Indiana University, Bloomington, Indiana (1964).
152. P.-O. Löwdin, *Rev. Mod. Phys.* **36**, 966 (1964).
153. C. F. Bunge and A. Bunge, *J. Comput. Phys.* **8**, 409 (1971).
154. C. F. Bunge and A. Bunge, *Int. J. Quantum Chem.* **7**, 927 (1973).
155. D. Munch and E. R. Davidson, *J. Chem. Phys.* **63**, 980 (1975).
156. F. Sasaki, *Int. J. Quantum Chem.* **8**, 605 (1974).
157. S. F. Boys and G. B. Cook, *Rev. Mod. Phys.* **32**, 285 (1960).
158. R. P. Hosteny and S. A. Hagstrom, *J. Chem. Phys.* **58**, 4396 (1973).
159. R. K. Nesbet, *Ann. Phys. (N.Y.)* **3**, 397 (1958).
160. R. K. Nesbet, *J. Math. Phys.* **2**, 701 (1961).
161. E. R. Davidson, *Int. J. Quantum Chem.* **8**, 83 (1974).
162. F. E. Harris, *J. Chem. Phys.* **46**, 2769 (1967).
163. F. E. Harris, *J. Chem. Phys.* **47**, 1047 (1967).
164. J. E. Harriman, *J. Chem. Phys.* **40**, 2827 (1964).
165. K. Mano, *J. Math. Phys.* **12**, 2361 (1971).
166. R. C. Ladner and W. A. Goddard III, *J. Chem. Phys.* **51**, 1073 (1969).
167. G. A. Gallup, *J. Chem. Phys.* **52**, 893 (1970).
168. G. A. Gallup, *Adv. Quantum Chem.* **7**, 113 (1973).
169. I. L. Cooper and J. I. Musher, *J. Chem. Phys.* **57**, 1333 (1972).
170. I. L. Cooper and J. I. Musher, *J. Chem. Phys.* **59**, 929 (1973).
171. T. E. H. Walker and J. I. Musher, *Mol. Phys.* **27**, 1651 (1974).
172. J. F. Gouyet, *J. Math. Phys.* **13**, 745 (1972).
173. J. F. Gouyet, R. Schranner, and T. H. Seligman, *J. Phys. A* **8**, 285 (1975).
174. H. A. Kramers, *Quantum Mechanics*, North-Holland, Amsterdam (1958) [reprinted by Dover, New York (1964)].
175. H. C. Brinkman, *Applications of Spinor Invariants in Atomic Physics*, North-Holland, Amsterdam (1956).
176. W. I. Salmon, K. Ruedenberg, and L. M. Cheung, *J. Chem. Phys.* **57**, 2787 (1972).
177. R. W. Wetmore and G. A. Segal, *Chem. Phys. Lett.* **36**, 478 (1975).
178. L. Pauling, *J. Chem. Phys.* **1**, 280 (1933).
179. A. D. McLachlan, *J. Chem. Phys.* **33**, 663 (1960).
180. H. Shull, Int. J. Quantum Chem. **3**, 523 (1969).
181. G. H. F. Diercksen and B. T. Sutcliffe, *Theor. Chim. Acta* **34**, 105 (1974).
182. B. R. Gilson, PROJR, Program 218, Quantum Chemistry Program Exchange, Indiana University, Bloomington, Indiana (1972).
183. S. F. Boys, *Proc. R. Soc. London, Ser. A* **207**, 181, 197 (1951).
184. S. F. Boys, *Philos. Trans. R. Soc., Ser. A* **245**, 95 (1952).
185. M. J. M. Bernal and S. F. Boys, *Philos. Trans R. Soc., Ser A* **245**, 116 (1952).
186. S. F. Boys and R. C. Sahni, *Philos. Trans. R. Soc., Ser. A* **246**, 463 (1954).
187. L. W. Hunter, *J. Chem. Phys.* **60**, 2670 (1974).
188. W. E. Donath, *J. Chem. Phys.* **35**, 817 (1961).
189. F. Sasaki and M. Yoshimine, *Phys. Rev. A* **9**, 17 (1974).
190. F. Sasaki and M. Yoshimine, *Phys. Rev. A* **9**, 26 (1974).
191. A. Bunge and C. F. Bunge, *Phys. Rev. A* **1**, 1599 (1970).
192. C. F. Bunge and E. M. A. Peixoto, *Phys. Rev. A* **1**, 1277 (1970).
193. J. Simons and J. E. Harriman, *J. Chem. Phys.* **51**, 296 (1969).
194. J. S. Griffith, *The Irreducible Tensor Method for Molecular Symmetry Groups*, Prentice-Hall, Englewood Cliffs, New Jersey (1962).
195. C. Bottcher and J. C. Browne, *J. Chem. Phys.* **52**, 3197 (1970).

196. R. S. Martin and J. H. Wilkinson, *Numer. Math.* **11**, 99 (1968) [reprinted in Ref. 199, p. 303].
197. G. Peters and J. H. Wilkinson, *SIAM J. Numer. Anal.* **7**, 479 (1970).
198. B. Ford and G. Hall, *Comput. Phys. Commun.* **8**, 337 (1974).
199. J. H. Wilkinson and C. Reinsch, *Linear Algebra* (Vol. II of *Handbook for Automatic Computation*), Springer, New York (1971).
200. J. H. Wilkinson, *The Algebraic Eigenvalue Problem*, Oxford U.P., London (1965).
201. H. R. Schwarz, H. Rutishauser, and E. Stiefel, *Numerical Analysis of Symmetric Matrices*, Prentice-Hall, Englewood Cliffs, New Jersey (1973).
202. H. H. Goldstine, F. J. Murray, and J. von Neumann, *J. Assoc. Comput. Mach.* **6**, 57 (1959).
203. F. J. Corbató, *J. Assoc. Comput. Mach.* **10**, 123 (1963).
204. H. Rutishauser, *Numer. Math.* **9**, 1 (1966) [reprinted in Ref. 199, p. 202].
205. W. E. Baylis, CEIGEN, Program 253, Quantum Chemistry Program Exchange, Indiana University, Bloomington, Indiana (1974).
206. C. Lanczos, *J. Res. Nat. Bur. Stand.* **45**, 255 (1950).
207. W. Givens, *in*: *Simultaneous Linear Equations and the Determination of Eigenvalues* (L. J. Paige and O. Taussky, eds.), p. 117, Nat. Bur. Stand. Appl. Math. Series, No. 29, U.S. Govt. Printing Office, Washington, D.C. (1953).
208. W. Givens, Numerical Computation of the Characteristic Values of a Real Symmetric Matrix, Oak Ridge National Laboratory Report ORNL 1574 (March, 1954).
209. J. H. Wilkinson, *Comput. J.* **3**, 23 (1960).
210. R. S. Martin, C. Reinsch, and J. H. Wilkinson, *Numer. Math.* **11**, 181 (1968) [reprinted in Ref. 199, p. 212].
211. F. Prosser, GIVENS, Program 62.3, Quantum Chemistry Program Exchange, Indiana University, Bloomington, Indiana (1965).
212. B. S. Garbow, *Comput. Phys. Commun.* **7**, 179 (1974).
213. J. M. Ortega, *J. Assoc. Comput. Mach.* **7**, 260 (1960).
214. W. Barth, R. S. Martin, and J. H. Wilkinson, *Numer. Math.* **9**, 386 (1967) [reprinted in Ref. 199, p. 249].
215. J. G. F. Francis, *Comput. J.* **4**, 265 (1961).
216. J. G. F. Francis, *Comput. J.* **4**, 332 (1962).
217. J. M. Ortega and H. F. Kaiser, *Comput. J.* **6**, 99 (1963).
218. H. Bowdler, R. S. Martin, C. Reinsch, and J. H. Wilkinson, *Numer. Math.* **11**, 293 (1968) [reprinted in Ref. 199, p. 227].
219. A. J. Fox and F. A. Johnson, *Comput. J.* **9**, 98 (1966).
220. A. Dubrulle, R. S. Martin, and J. H. Wilkinson, *Numer. Math.* **12**, 377 (1968) [reprinted in Ref. 199, p. 241].
221. C. Reinsch and F. L. Bauer, *Numer. Math.* **11**, 264 (1968) [reprinted in Ref. 199, p. 257].
222. G. W. Stewart, *Commun. ACM* **13**, 365, 369, 750 (1970).
223. J. H. Wilkinson, *Comput. J.* **1**, 90 (1958).
224. G. Peters and J. H. Wilkinson, *in* Ref. 199, p. 418.
225. G. W. Stewart, *in*: *Information Processing 74* (Proceedings of IFIP Congress 74, Stockholm), p. 666, North-Holland, Amsterdam (1974).
226. I. Shavitt, C. F. Bender, A. Pipano, and R. P. Hosteny, *J. Comput. Phys.* **11**, 90 (1973).
227. S. Falk, *Z. Angew. Math. Mech.* **53**, 73 (1973).
228. E. R. Davidson, *J. Comput. Phys.* **17**, 87 (1975).
229. M. R. Hestenes and W. Karush, *J. Res. Nat. Bur. Stand.* **47**, 45 (1951).
230. M. R. Hestenes and W. Karush, *J. Res. Nat. Bur. Stand.* **47**, 471 (1951).
231. M. Clint and A. Jennings, *Comput. J.* **13**, 76 (1970).
232. C. F. Bender, Ph.D. thesis, University of Washington, Seattle (1968).
233. D. K. Faddeev and V. N. Faddeeva, *Computational Methods of Linear Algebra* (English translation), Section 61, W. H. Freeman & Co., San Francisco (1963).
234. W. W. Bradbury and R. Fletcher, *Numer. Math.* **9**, 259 (1966).
235. I. Fried, *J. Sound Vibr.* **20**, 333 (1972).
236. A. Ruhe, *in*: *Eigenwert Probleme* (L. Collatz, ed.), p. 97, Birkhäuser, Basel (1974).
237. R. K. Nesbet, *J. Chem. Phys.* **43**, 311 (1965).

238. H. H. Michels, NESBET, Program 93, Quantum Chemistry Program Exchange, Indiana University, Bloomington, Indiana (1966).
239. I. Shavitt, EIGEN, Program 172.1, Quantum Chemistry Program Exchange, Indiana University, Bloomington, Indiana (1970).
240. M. Raimondi and G. F. Tantardini, NESBET, Program 180, Quantum Chemistry Program Exchange, Indiana University, Bloomington, Indiana (1971).
241. H. R. Schwarz, *Comput. Meth. Appl. Mech. Eng.* **3**, 11 (1974).
242. A. Ruhe, *Math. Comput.* **28**, 695 (1974).
243. M. G. Feler, *J. Comput. Phys.* **14**, 341 (1974).
244. R. S. Mulliken, *J. Chem. Phys.* **23**, 1833 (1955); E. R. Davidson, *J. Chem. Phys.* **46**. 3320 (1967).
245. P.-O. Löwdin, *Phys. Rev.* **97**, 1474 (1955).
246. P.-O. Löwdin, *Phys. Rev.* **97**, 1490 (1955).
247. R. McWeeny, *Proc. R. Soc. London, Ser. A* **253**, 242 (1959).
248. R. McWeeny, *Rev. Mod. Phys.* **32**, 335 (1960).
249. R. McWeeny and Y. Mizuno, *Proc. R. Soc. London, Ser. A* **259**, 554 (1961).
250. A. J. Coleman, *Rev. Mod. Phys.* **35**, 668 (1963).
251. A. J. Coleman and R. M. Erdahl (eds.), Reduced Density Matrices with Applications to Physical and Chemical Systems, Queen's Papers on Pure and Applied Mathematics, No. 11, Queen's University, Kingston, Ontario (1968).
252. W. A. Bingel and W. Kutzelnigg, *Adv. Quantum Chem.* **5**, 201 (1970).
253. E. R. Davidson, *Reduced Density Matrices in Quantum Chemistry*, Academic Press, New York (1976).
254. D. W. Smith, *in* Ref. 251, p. 169.
255. H. D. Cohen and C. C. J. Roothaan, *J. Chem. Phys.* **43**, S34 (1965).
256. J. A. Pople, J. W. McIver, Jr., and N. S. Ostlund, *J. Chem. Phys.* **49**, 2960 (1968).
257. B. J. Rosenberg and I. Shavitt, *J. Chem. Phys.* **63**, 2162 (1975).
258. E. R. Davidson, *in: The World of Quantum Chemistry* (R. Daudel and B. Pullman, eds.), p. 17, Reidel, Dodrecht, Holland (1974).
259. K. Ruedenberg, R. C. Raffenetti, and R. D. Bardo, *in: Energy, Structure, and Reactivity* (D. W. Smith and W. B. McRae, eds.), p. 164, Wiley, New York (1973).
260. P. S. Bagus, B. Liu, A. D. McLean, and M. Yoshimine, *in: Computational Methods for Large Molecules and Localized States in Solids* (F. Herman, A. D. McLean, and R. K. Nesbet, eds.), p. 87, Plenum Press, New York (1973).
261. H. F. Schaefer III, R. A. Klemm, and F. E. Harris, *Phys. Rev.* **181**, 137 (1969).
262. J. L. Whitten and M. Hackmeyer, *J. Chem Phys.* **51**, 5584 (1969); M. Hackmeyer and J. L. Whitten, *J. Chem. Phys.* **54**, 3739 (1971).
263. S. Huzinaga and C. Arnau, *Phys. Rev. A* **1**, 1285 (1970); *J. Chem Phys.* **54**, 1948 (1971).
264. E. R. Davidson, *J. Chem. Phys.* **57**, 1999 (1972).
265. H. P. Kelly, *Phys. Rev.* **136**, B896 (1964).
266. W. J. Hunt and W. A. Goddard III, *Chem. Phys. Lett.* **3**, 414 (1969).
267. N. Björnå, *J. Phys. B.* **6**, 1412 (1973).
268. L. R. Kahn, P. J. Hay, and I. Shavitt, *J. Chem. Phys.* **61**, 3530 (1974).
269. C. F. Bender and E. R. Davidson, *J. Chem. Phys.* **47**, 4972 (1967).
270. J. L. Whitten, *J. Chem. Phys.* **56**, 5458 (1972).
271. J. M. Foster and S. F. Boys, *Rev. Mod. Phys.* **32**, 300 (1960).
272. R. K. Nesbet, *Adv. Chem. Phys.* **9**, 321 (1965).
273. E. Steiner, *J. Chem. Phys.* **54**, 1114 (1971).
274. O. Sinanoğlu and B. Skutnik, *Chem. Phys. Lett.* **1**, 699 (1968).
275. R. K. Nesbet, *Phys. Rev.* **175**, 2 (1968).
276. C. F. Bender and E. R. Davidson, *in* Ref. 251, p. 335.
277. T. L. Barr and E. R. Davidson, *Phys. Rev. A* **1**, 644 (1970).
278. R. J. Buenker and S. D. Peyerimhoff, *J. Chem. Phys.* **53**, 1368 (1970); S. Shih, R. J. Buenker, and S. D. Peyerimhoff, *Chem. Phys. Lett.* **16**, 244 (1972).
279. E. R. Davidson, *Rev. Mod. Phys.* **44**, 451 (1972).

280. P.-O. Löwdin and H. Shull, *Phys. Rev.* **101**, 1730 (1956).
281. G. P. Barnett, J. Linderberg, and H. Shull, *J. Chem. Phys.* **43**, S80 (1965).
282. H. F. Schaefer III, *J. Chem. Phys.* **54**, 2207 (1971).
283. P. J. Hay, *J. Chem. Phys.* **59**, 2468 (1973).
284. A. K. Q. Siu and E. F. Hayes, *J. Chem. Phys.* **61**, 37 (1974).
285. C. F. Bender and E. R. Davidson, *Phys. Rev.* **183**, 23 (1969).
286. I. Shavitt, B. J. Rosenberg, and S. Palalikit, *Int. J. Quantum Chem., Symp. No.* **10**, 33 (1976).
287. C. F. Bender and E. R. Davidson, *J. Phys. Chem.* **70**, 2675 (1966).
288. S. R. Langhoff and E. R. Davidson, *Int. J. Quantum Chem.* **7**, 759 (1973).
289. I. Shavitt, *in*: *Energy, Structure, and Reactivity* (D. W. Smith and W. B. McRae, eds.), p. 188, Wiley, New York (1973).
290. W. Kutzelnigg, *J. Chem. Phys.* **40**, 3640 (1964).
291. E. R. Davidson, *J. Chem. Phys.* **48**, 3169 (1968).
292. R. Albat, *Z. Naturforsch.* **27a**, 545 (1972).
293. V. Kvasnička, *Theor. Chim. Acta* **36**, 297 (1975).
294. Z. Gershgorn and I. Shavitt, *Int. J. Quantum Chem.* **2**, 751 (1968).
295. C. Edmiston and M. Krauss, *J. Chem. Phys.* **45**, 1833 (1966).
296. W. Meyer, *J. Chem. Phys.* **58**, 1017 (1973).
297. R. Ahlrichs and F. Driessler, *Theor. Chim. Acta* **36**, 275 (1975).
298. S. A. Houlden and I. G. Csizmadia, *Theor. Chim. Acta* **30**, 209 (1973).
299. C. F. Bender, private communication.
300. C. F. Bender and E. R. Davidson, *J. Chem. Phys.* **49**, 4222 (1968).
301. A. K. Q. Siu and E. R. Davidson, *Int. J. Quantum Chem.* **4**, 223 (1970).
302. E. R. Davidson, *in*: *Energy, Structure, and Reactivity* (D. W. Smith and W. B. McRae, eds.), p. 179, Wiley, New York (1973).
303. A. C. Wahl, P. J. Bertoncini, G. Das, and T. L. Gilbert, *Int. J. Quantum Chem., Symp. No.* **1**, 123 (1967).
304. P. J. Bertoncini, G. Das, and A. C. Wahl, *J. Chem. Phys.* **52**, 5112 (1970).
305. N. Sabelli and J. Hinze, *J. Chem. Phys.* **50**, 684 (1969).
306. J. Hinze, *in*: *Energy, Structure, and Reactivity* (D. W. Smith and W. B. McRae, eds.), p. 170, Wiley, New York (1973).
307. G. C. Lie, J. Hinze, and B. Liu, *J. Chem. Phys.* **59**, 1872 (1973).
308. P. Dejardin, E. Kochanski, A. Veillard, B. Roos, and P. Siegbahn, *J. Chem. Phys.* **59**, 5546 (1973).
309. S.-I. Chu, M. Yoshimine, and B. Liu, *J. Chem. Phys.* **61**, 5389 (1974).
310. G. D. Gillespie, A. U. Khan, A. C. Wahl, R. P. Hosteny, and M. Krauss, *J. Chem. Phys.* **63**, 3425 (1975).
311. B. Levy, *Int. J. Quantum Chem.* **4**, 297 (1970).
312. H. Shull, *J. Chem. Phys.* **30**, 1405 (1959).
313. G. Das and A. C. Wahl, *J. Chem. Phys.* **44**, 87 (1966).
314. R. C. Raffenetti and L. R. Kahn, unpublished work.
315. R. C. Raffenetti, K. Hsu, and I. Shavitt, *Theor. Chim. Acta* (in press), and unpublished work.
316. J. J. C. Mulder, *Mol. Phys.* **10**, 479 (1966).
317. C. Møller and M. S. Plesset, *Phys. Rev.* **46**, 618 (1934).
318. L. Brillouin, *Les Champs "Self-Consistent" de Hartree et de Fock* (Actualités Sci. Ind. No. 159), Hermann & Cie., Paris (1934).
319. R. K. Nesbet, *Proc. R. Soc. London, Ser. A* **230**, 312 (1955).
320. P. Claverie, S. Diner, and J. P. Malrieu, *Int. J. Quantum Chem.* **1**, 751 (1967).
321. F. Grimaldi, A. Lecourt, and C. Moser, *Int. J. Quantum Chem., Symp. No.* **1**, 153 (1967).
322. F. Coester and H. Kümmel, *Nucl. Phys.* **17**, 477 (1960).
323. J. da Providência, *Nucl. Phys.* **61**, 87 (1965).
324. J. Čížek, *J. Chem. Phys.* **45**, 4256 (1966).
325. J. Čížek, *Adv. Chem. Phys.* **14**, 35 (1969).
326. J. Čížek and J. Paldus, *Int. J. Quantum Chem.* **5**, 359 (1971).
327. O. Sinanoğlu, *Adv. Chem. Phys.* **6**, 315 (1964).

328. O. Sinanoğlu, *Adv. Chem. Phys.* **14**, 237 (1969).
329. R. K. Nesbet, *Adv. Chem. Phys.* **14**, 1 (1969).
330. H. P. Kelly, *Adv. Chem. Phys.* **14**, 129 (1969).
331. O. Sinanoğlu, *J. Chem. Phys.* **36**, 706 (1962).
332. J. Paldus, J. Čížek, and I. Shavitt, *Phys. Rev. A* **5**, 50 (1972).
333. R. Ahlrichs and W. Kutzelnigg, *J. Chem. Phys.* **48**, 1819 (1968).
334. W. Meyer, *Int. J. Quantum Chem., Symp. No.* **5**, 59 (1971).
335. R. Ahlrichs, H. Lischka, V. Staemmler, and W. Kutzelnigg, *J. Chem. Phys.* **62**, 1225 (1975).
336. R. Ahlrichs, F. Driessler, H. Lischka, V. Staemmler, and W. Kutzelnigg, *J. Chem. Phys.* **62**, 1235 (1975).
337. A. Pipano and I. Shavitt, *Int. J. Quantum Chem.* **2**, 741 (1968).
338. S. R. Langhoff and E. R. Davidson, *Int. J. Quantum Chem.* **8**, 61 (1974).
339. C. Bloch and J. Horowitz, *Nucl. Phys.* **8**, 91 (1958).
340. B. H. Brandow, *Rev. Mod. Phys.* **39**, 771 (1967).
341. U. Kaldor, *Phys. Rev. Lett.* **31**, 1338 (1973).
342. I. Lindgren, *J. Phys. B.* **7**, 2441 (1974).
343. S. D. Peyerimhoff and R. J. Buenker, J. Chem. Phys. **49**, 2473 (1968).
344. T. H. Dunning, Jr., W. J. Hunt, and W. A. Goddard III, *Chem. Phys. Lett.* **4**, 147 (1969).
345. S. D. Peyerimhoff and R. J. Buenker, *Theor. Chim. Acta* **19**, 1 (1970).
346. R. J. Buenker and S. D. Peyerimhoff, *Chem. Phys.* **9**, 75 (1975).
347. H. F. Schaefer III and C. F. Bender, *J. Chem. Phys.* **55**, 1720 (1971).
348. K. Morokuma and H. Konishi, *J. Chem. Phys.* **55**, 402 (1971).
349. C. F. Bender and E. R. Davidson, *J. Chem. Phys.* **46**, 3313 (1967).
350. R. J. Buenker and S. D. Peyerimhoff, *Thoer. Chim. Acta* **35**, 33 (1974).
351. P. K. Pearson, C. F. Bender, and H. F. Schaefer III, J. Chem. Phys. **55**, 5235 (1971).
352. G. A. Segal and R. W. Wetmore, *Chem. Phys. Lett.* **32**, 556 (1975).
353. S. Iwata and K. F. Freed, *Chem. Phys.* **11**, 433 (1975).
354. P. J. Fortune and B. J. Rosenberg, *Chem. Phys. Lett.* **37**, 110 (1976).
355. C. F. Bunge, *Phys. Rev.* **168**, 92 (1968).
356. S. F. Boys and I. Shavitt, A Fundamental Calculation of the Energy Surface for the System of Three Hydrogen Atoms, University of Winsconsin Naval Research Laboratory Technical Report WIS-AF-13 (March 4, 1959).
357. R. E. Brown, Ph.D. thesis, Indiana University, Bloomington, Indiana (1967).
358. R. J. Buenker and S. D. Peyerimhoff, *Theor. Chim. Acta* **39**, 217 (1975).

7

The Direct Configuration Interaction Method from Molecular Integrals

Björn O. Roos
and
Per E. M. Siegbahn

1. Introduction

In this paper we will address ourselves to some aspects of the problem of finding accurate solutions to the electronic Schrödinger equation by means of the configuration interaction (CI) method. This method is probably one of the most encouraging for general studies of molecular systems in their ground and excited states, and also for studies of energy surfaces for chemical reactions.

The general scheme for configuration interaction calculations was first proposed by Boys in 1950,[1] who also carried through a ten-configuration calculation on the beryllium atom.[2] Methods similar in philosophy to the CI method had earlier been used by Hartree, Hartree, and Swirles in their study of the oxygen atom[3] and by Hylleraas for the helium atom.[4] The numerical difficulties, however, prohibited further developments of the theory until the electronic computer came into general use and the problem of integral evaluation and the handling of large amounts of numerical data could be efficiently dealt with.

A number of computer programs for *ab initio* calculations on molecular systems on the Hartree–Fock level of approximation were constructed in the

Björn O. Roos and Per E. M. Siegbahn • Institute of Theoretical Physics, University of Stockholm, 113 46 Stockholm, Sweden period 1965–1970 and such calculations are routinely performed today in many laboratories. This development has been of great importance for the application of *ab initio* quantum mechanical methods in studies of a variety of current problems in molecular chemistry and physics, since the Hartree–Fock method is known to yield reliable results for many chemical processes. The Hartree–Fock method can, however, only be used for processes in which the correlation error can be assumed not to vary. Many problems do not belong to this category and it is necessary to include correlation effects in these cases in order to be able to make useful predictions. The most commonly used method for treating these effects is the method of configuration interaction. Today's research in computational quantum chemistry is therefore, to a large extent, centered around the problem of finding efficient tools to use this method in practice.

It is customary to differentiate between two types of correlation effects. The Hartree–Fock model will in certain cases break down due to degeneracies or near degeneracies between several configurations. Such effects frequently occur in calculations of energy surfaces for chemical reactions. They can be dealt with by means of a limited CI expansion which includes all near-degenerate configurations. An extension of the Hartree–Fock method to cases with more than one configuration is then obtained, when the CI expansion coefficients and the orbitals are simultaneously optimized. This procedure constitutes the MC–SCF (multiconfigurational self-consistent-field) method.[3,5–7]

The second effect is the dynamic correlation of the electronic motion. For electrons with parallel spins, this correlation is already to a large extent included in the Hartree–Fock model, through the antisymmetry requirement (Fermi correlation). The main part of the remaining error is, therefore, due to correlation of electron pairs having opposite spins. The fact that electron correlation to a good approximation is described by pair interactions, implies that the dominant part of the correlation energy is obtained with a CI wave function which contains replacements in each pair of electrons. For special reasons single replacements are also important and the wave function describing dynamical correlation effects should, in general, contain the near-degenerate configurations and single and double replacements with respect to all of them.

Such an expansion will in general be very long, since the number of possible double-replacement configurations is large for a reasonably sized molecule orbital basis set. One of the obstacles in using the CI method is therefore the problem of calculating and handling a very large number of matrix elements. In special cases it is possible, however, to avoid the construction of a huge Hamiltonian matrix and instead calculate the CI expansion coefficients directly from the given list of molecular one- and two-electron integrals. This is the philosophy behind the CIMI method[8,9] (direct *configura-*

tion *interaction* from *molecular integrals*) which will be discussed in detail in the following sections. Computer programs for CI calculations using different versions of the CIMI method have been included in the general quantum chemistry program system MOLECULE.[10]

2. The CIMI Method

2.1. General Scheme

In this section general concepts of the configuration interaction scheme are presented in their context, and the basic characteristics of the CIMI method as compared to other methods are described.

The wave function is in the CI method expressed as a linear combination of configuration *state functions* (CSFs), i.e.,

$$\Psi = \sum_\mu C_\mu \Phi_\mu \tag{1}$$

where the CSFs are linear combinations of Slater determinants having correct spin and space symmetry.

The expansion coefficients C_μ are obtained by applying the Rayleigh–Ritz variation principle

$$\sum_\nu (H_{\mu\nu} - ES_{\mu\nu})C_\nu = 0 \tag{2}$$

with

$$H_{\mu\nu} = \langle \Phi_\mu | \hat{H} | \Phi_\nu \rangle$$

$$S_{\mu\nu} = \langle \Phi_\mu | \Phi_\nu \rangle$$

The eigenvalues obtained from (2) are upper bounds to the true eigenvalues of the Hamiltonian.[11] The actual solution of the equation system (2) is, in practice, divided into several parts, which will be briefly described here.

The Slater determinants, which are used to construct the CSFs, are built from an ordered set of one-electron functions (orbitals). These orbitals are in turn usually obtained as linear combinations of a given set of atom-centered basis functions (the LCAO expansion). Therefore, the starting point of a calculation is the selection of an atomic basis set. The choice of types of functions ($1s$, $2p$, $3d$, etc.) on the different atomic centers and their exponents will be very important for the accuracy of the calculation. Properties not included in the basis set, for example the diffuse nature of an excited state, will obviously not be recovered even in an extended CI calculation. In some standard cases there is enough experience available to make the choice of basis set almost a technical problem. In general, however, a deep insight into the chemical problem is necessary, because in practice one has to work far from the ideal complete basis set.

Once the basis set is chosen all possible one- and two-electron integrals are calculated. This is the only integral evaluation in the calculation. The rest will be pure matrix operations.

The one-electron space spanned by the basis functions is usually divided into two parts. Using the terminology of Sinanoğlu et al.[12] we shall call these spaces internal and external. Ideally, the internal space should include the occupied HF space and, in addition, all orbitals which are needed to describe degeneracy and near-degeneracy effects. The external space is then used to describe the effects of dynamical correlation. The internal space is usually obtained from a HF calculation, or in the case of near degeneracies, from an MC–SCF calculation. For reasons which will become clear later it is of interest to keep the internal space as small as possible. Often the use of the full external space leads to a CI expansion which is too large to be manageable. Therefore, a reduction of the external space may be necessary. This can, for example, be achieved by a transformation to natural orbitals, which have certain optimum convergence properties.[13,14] A method for obtaining approximate natural orbitals will be demonstrated in a later section.

A configuration basis set is obtained by building Slater determinants with orbitals from the internal and external set. In order to keep down the number of configurations some sort of selection can be made. One way to achieve this is to use the concept of interacting space.[15] Starting from a set of reference-state configurations, in which only internal orbitals are occupied, one includes, in the CI expansion, only configurations which have a nonvanishing Hamiltonian matrix element with at least one of the reference CSFs. This subset is called the first-order interacting space. In the general case this means that only configurations which correspond to single or double replacements in the reference states are taken into account. This gives fewer configurations than the more general concept of one or two electrons outside the internal set. Present experience indicates that the use of the first-order interacting space with an appropriate choice of reference states is an ideal approach for CI calculations. The main problem will then be to design methods that could accomplish this goal in an efficient way. At present this problem is not solved for the general case and is an area of increased research.

Configurations corresponding to three or more replacements appear, to a good approximation, only in the form of unlinked clusters.[16] It is possible to include the effect of these configurations in an approximate way, for example, by means of the coupled electron-pair approximation.[17]

For each configuration we can, in general, form a number of linearly independent Slater determinants by different choices of the spin part of the orbitals. Configuration state functions are formed by taking the linear combinations of these determinants which are eigenfunctions of the total spin angular momentum operators. A number of methods for this spin projection are available and the reader is referred to the literature for more details.[18]

One of the bottlenecks with the CI method seems to be the slow convergence of the CI expansion. Even if the optimum properties of the natural orbitals are used to reduce the length of the expansion, the number of CSFs easily exceeds 10^4. The possibility to obtain accurate wave functions with this method has therefore been limited to systems containing few electrons. The CIMI method has been designed to overcome this problem and make possible CI calculations with high accuracy on larger systems as well. The method has so far been restricted to some special choices of reference states but has the potential to treat more general cases.

In order to explain what the CIMI method is we will now go back to the step in the calculation where the integrals are computed and the internal and external orbitals determined. The next step would be to set up the Hamiltonian matrix for the selected n-electron space. For CSFs constructed from orthogonal orbitals, these matrix elements are simple expressions involving one- and two-electron integrals over these orbitals.[18] The first step is therefore a transformation of all integrals, from the atomic basis to our selected orthogonal orbital basis. Having performed this transformation and having selected the CSFs, we then need the explicit expressions that resolve the Hamiltonian matrix elements in terms of integrals and coupling coefficients. Due to the number of matrix elements this is a formidable computational problem. Many presently used CI techniques have their main limitations in this step. When these formulas, the so-called symbolic matrix elements, have been created, it is a seemingly straightforward problem to construct the Hamiltonian matrix. An example from a calculation on the H_3 system[40] will, however, point out some of the computational difficulties. This calculation involved 20,704 CSFs, which made up a matrix of 28×10^6 nonzero elements, constructed from 3.7×10^5 integrals. Everyone familiar with such computations will realize the difficulty in constructing a matrix of that dimension.

Once the matrix is constructed the root(s) of interest should be extracted from the secular equation. For problems with more than a few hundred configurations this cannot be done with methods involving rotations of the matrix such as the Jacobi or the Givens–Householder methods. Instead, special iteration or perturbation techniques have to be applied. Both of these schemes basically lead to a multiplication of the Hamiltonian matrix with a trial wave function in the form of a vector $\mathbf{C}^{(k-1)}$ generating a vector $\boldsymbol{\sigma}^{(k)}$:

$$\boldsymbol{\sigma}^{(k)} = \mathbf{H}\mathbf{C}^{(k-1)} \qquad (3)$$

A new trial wave function is constructed from $\boldsymbol{\sigma}^{(k)}$ through simple algebra, and the procedure is repeated until convergence. These iteration and perturbation schemes will be discussed in more detail in a later section. The key of the CIMI method lies in the simple expression (3). As the Hamiltonian matrix elements are linear combinations of one- and two-electron integrals, it must obviously be possible, in principle, to formulate expressions for the vector $\boldsymbol{\sigma}^{(k)}$ directly in terms of these integrals, thereby avoiding all reference to the huge Hamiltonian matrix. A formula representing the increment to the element i of the vector $\boldsymbol{\sigma}$ from a particular integral $(ab|cd)$ of the type

$$\Delta\sigma_i^{(k)} = A(ab|cd)C_j^{(k-1)} \qquad (4)$$

where A is a coupling coefficient depending on the spin-symmetry of the involved CSFs, will therefore form the basis for the CIMI method.

In the following sections the crucial coupling coefficients A and the diagonalization schemes will be discussed in somewhat more detail.

2.2. Coupling Coefficients

In a general case Eq. (4) may be very complicated. The problem that has to be solved for a chosen well-defined CI scheme, is to break up (4) into pieces and to find easy to use algorithms for each piece. There may be many ways of attacking this problem,[51] but we will only discuss the methods used in the MOLECULE-CI package. The integrals are divided into types and each type of integral contributes to the interaction between classes of configuration states. The definition of integral type is connected with the division of the orbitals into groups, according to their role in the CI scheme. In a general CI we will divide the orbitals into three groups. The first group contains orbitals that are fully occupied in all reference states. Only excitations *from* these orbitals are possible. The second group contains the rest of the internal orbitals. Excitations both to and from these orbitals are included. The final third group contains all the external orbitals. By definition only excitations *to* these orbitals are possible. The treatment of a fourth group containing the uncorrelated orbitals is trivial and will not be discussed further.

A two-electron integral is defined through four orbitals. One of the two factors that determine the type of integral is the division of these four orbitals into the three different groups. The other factor is the internal relationship between the indices of orbitals belonging to the same group. There are, as shown in Table 1, fourteen different integral types with all four orbitals belonging to the same group. With a closed-shell reference state there are 53 different integral types. If the reference state has one open shell, 46 new integral types appear due to the introduction of the second internal orbital group. If this group is increased by just one other orbital another 59 integral types are added making a total of 148 types. For each integral type, algorithms defining the interacting classes of configuration states have to be found. This is usually no serious problem for each type, but these algorithms may be

Table 1. The 14 Different Types of Integrals (ab/cd) Obtained When All Orbitals Belong to the Same Group

Type number	Index relations	Type number	Index relations
1	$a = b = c = d$	8	$c = d, b < c$
2	$a = b = c$	9	$a = c$
3	$b = c = d$	10	$b = d$
4	$a = b, c = d$	11	$b = c$
5	$a = c, b = d$	12	$b > c$
6	$a = b$	13	$b < c, b > d$
7	$c = d, b > c$	14	$b < d$

complicated to use. Also the large number of different types set up limits to when this procedure is useful in practice. One way to reduce this number is to reduce the number of orbital groups. When all possible configuration states are formed (complete CI) all orbitals are treated equally and therefore form only one group, the second in our definition. In this case there will consequently be only fourteen different integral types (see Table 1). A complete CI in the internal space only, reduces the number of orbital groups from three to two. The first internal group is now included in the second. The thereby introduced extra configurations could be deleted by use of the index list (discussed in a later section). The size of the index list will, however, increase rapidly and form a new limitation to the possible size of the calculation. A different way of simplifying the work required in using the CIMI method is to consider only certain classes of configuration states in the reference space, for example, only closed-shell states. This will, besides reducing the number of integral types, also make the algorithms for each integral simpler than they would be in the general case.

Returning to Eq. (4), a specified interaction between two configuration state functions through an integral type requires a coupling coefficient. This coefficient will depend on the spin coupling of the interacting configuration states and the permutation that lines up equal orbitals in these configuration states. A brute force method for obtaining the coupling coefficients was used for the case of complete CI on three valence electrons. A representative list of configuration states was fed into a computer program, which constructed formal matrix elements of the Hamiltonian operator.[20] A search for a selected integral type was then made through the list of matrix elements. Three indices will, in this case, define a particular configuration and six indices will therefore define the interaction. Four of these indices, two from each configuration, will be the same as the indices of the integral and will here be denoted as the fixed indices. If the integral contains four different indices, the remaining third index for each configuration will be called a running index. The classes of interacting configurations are therefore generated by letting the running index loop over all n orbitals. As there are n^4 integrals, full CI for three electrons will be an n^5

process. Similarly, it is easily seen that for complete CI with m electrons the number of operations will be proportional to n^{m+2}.

For double replacements from a single reference state each replacement is defined by four indices, and the interaction is described by eight indices. Four of these, two from each configuration, have to be fixed indices defining the integral, and the remaining two running indices for the configurations have to match. Simple algorithms for the interacting classes of configurations can therefore be constructed with two fixed indices and two running indices. This will also be demonstrated in a later section.

Once these algorithms have been determined we are able to express the coupling coefficients in (4) in a formal way through general expressions. Two different ways of describing the coupling coefficients have been used. One is the straightforward way, as illustrated in Table 2 for complete CI on three electrons. These coupling tables are directly obtainable through simple computer programs. For a particular type of integral Table 2 shows both the algorithm used to describe the classes of interacting configuration states and the coupling coefficient that should be used when the running index loops over all orbitals. This is precisely what is needed to use Eq. (4) in practice, and is a useful description but gives no hints to any possible generalization of the formulas. Another method has been used for the case of all single and double replacements from a single closed-shell reference state. Through a few coupling tables it is possible to formulate an expression for the increment to σ which is general for all types of integrals. This formula and the coupling tables will be described later in the section on the CIMI method for closed-shell systems. However, in spite of the generality of this formula it is the other procedure that generates the most useful description in a particular case.

Table 2. *Full CI for Three Electrons. Coupling Coefficient Tables for Integral $(ab|cd)$ Where $c > b > d$ (Type 13 of Table 1)[a]*

Spin coupling		d		b		c		a	
			Interaction elements $(kda^{(1,2)}\|kbc^{(1,2)})^b$						
$(1\|1)$	$+\frac{1}{2}$	0	$-\frac{1}{2}$	$-\sqrt{6}/2$	-1	$+\sqrt{6}/2$	$-\frac{1}{2}$	0	$+\frac{1}{2}$
$(1\|2)$	$-\sqrt{3}/2$	$-\sqrt{2}$	$-\sqrt{3}/2$	0	0	0	$+\sqrt{3}/2$	$-\sqrt{2}$	$+\sqrt{3}/2$
$(2\|1)$	$-\sqrt{3}/2$	0	$+\sqrt{3}/2$	$+\sqrt{2}/2$	0	$+\sqrt{2}/2$	$-\sqrt{3}/2$	0	$+\sqrt{3}/2$
$(2\|2)$	$-\frac{1}{2}$	0	$-\frac{1}{2}$	0	$+1$	0	$-\frac{1}{2}$	0	$-\frac{1}{2}$
			Interaction elements $(kca^{(1,2)}\|kdb^{(1,2)})^b$						
$(1\|1)$	$+1$	$+\sqrt{6}/2$	$+\frac{1}{2}$	0	$-\frac{1}{2}$	0	$+\frac{1}{2}$	$-\sqrt{6}/2$	$+1$
$(1\|2)$	0	0	$+\sqrt{3}/2$	0	$-\sqrt{3}/2$	$-\sqrt{2}$	$-\sqrt{3}/2$	$+\sqrt{3}/2$	0
$(2\|1)$	0	$+\sqrt{2}/2$	$+\sqrt{3}/2$	$-\sqrt{2}$	$+\sqrt{3}/2$	0	$-\sqrt{3}/2$	0	0
$(2\|2)$	$+1$	0	$-\frac{1}{2}$	0	$-\frac{1}{2}$	0	$-\frac{1}{2}$	0	$+1$

The top of the value columns is headed $k \to$.

[a] The entries in columns 2–10 correspond to various values of the running index k, e.g., for the entries under d the running index k equals d; for the entries in the next column (between d and b) we have $d < k < b$; etc.

[b] k is the running index and the superindex denotes spin coupling.

2.3. Diagonalization Schemes

As was pointed out earlier, extraction of roots and energies of equation system (2) has to be done either by iterative techniques or by perturbation theory, due to the usually long expansion (1), running into several thousand configuration states. These two approaches are essentially equivalent computationally, but formally differ slightly. Both of them can basically be made to require the construction of a vector $\boldsymbol{\sigma}$ according to (3). We should also mention here that some of the diagonalization techniques most frequently used today[21,22] cannot be used in their original form in connection with the CIMI method as they require a differently constructed $\boldsymbol{\sigma}$. We will start the discussion of diagonalization procedures with the perturbation techniques, and especially treat the particular scheme that has been found most efficient in connection with the CIMI method, even in comparison with general iterative techniques.

The perturbation equations are obtained by making the usual splitting of the Hamiltonian,

$$\hat{H} = \hat{H}_0 + \hat{V} \tag{5}$$

and the expansions of the wave function and energy

$$\Psi = \sum_n \Psi^{(n)} \tag{6}$$

$$E = \sum_n \varepsilon_n \tag{7}$$

and then substituting these expressions into the Schrödinger equation. Collecting terms of order k, leads to

$$(E_0 - \hat{H}_0)\Psi^{(k)} = \hat{V}\Psi^{(k-1)} - \sum_{n=0}^{k-1} \varepsilon_{k-n}\Psi^{(n)} \tag{8}$$

If a simple expression is wanted for the components of $\Psi^{(n)}$ an $\hat{H}_0$ is needed which gives a simple inverse of $(E_0 - \hat{H}_0)$. A diagonal representation of this operator is therefore an advantage. If we make an expression of the perturbation functions into orthogonal configuration state functions $|\Phi_\mu\rangle$, a diagonal $(E_0 - \hat{H}_0)$ is, for example, obtained by

$$\hat{H}_0 = \sum_\mu |\Phi_\mu\rangle\langle\Phi_\mu|\hat{H}|\Phi_\mu\rangle\langle\Phi_\mu| \tag{9}$$

where the sum is over all configuration state functions in the basis set. In component form, (8) is then written as

$$C_\mu^{(k)} = \frac{1}{E_0 - H_{\mu\mu}}\left[\sum_{\nu=1}^N V_{\mu\nu}C_\nu^{(k-1)} - \sum_{n=0}^{k-1} \varepsilon_{k-n}C_\mu^{(n)}\right] \tag{10}$$

The first term inside the brackets is recognized as being basically equal to the vector component σ_μ defined by (3). The expressions for the perturbation energies obtained simultaneously as the perturbation functions $\Psi^{(k)}$ are harder to derive, and the reader is referred to Ref. 23 for details. The final formulas are

$$\varepsilon_{2k-1} = \sum_{\mu=1}^{N} C_\mu^{(k-1)} \sum_{\nu=1}^{N} V_{\mu\nu} C_\nu^{(k-1)} - \sum_{m=1}^{k-1} \sum_{n=1}^{k-1} \varepsilon_{2k-1-m-n} \sum_\mu C_\mu^{(m)} C_\mu^{(n)}$$

$$\varepsilon_{2k} = \sum_{\mu=1}^{N} C_\mu^{(k)} \sum_{\nu=1}^{N} V_{\mu\nu} C_\nu^{(k-1)} - \sum_{m=1}^{k} \sum_{n=1}^{k-1} \varepsilon_{2k-m-n} \sum_\mu C_\mu^{(m)} C_\mu^{(n)}$$

(11)

The components σ_μ are found also in these expressions. Equations (10) and (11) can be used together with the linear expressions (6) and (7) until convergence is reached. A better procedure [24] is to solve the secular equation in terms of the perturbation functions $\Psi^{(k)}$. As the vectors $\sigma^{(k)}$, according to (3), are constructed in each iteration, a single scalar product of the previously obtained perturbation functions will automatically yield the full Hamiltonian matrix in terms of these functions. These matrix elements may also be written in closed form,[23]

$$\langle \Psi^{(p)} | \hat{H} | \Psi^{(q)} \rangle = \langle \Psi^{(p)} | \hat{H}_0 | \Psi^{(q)} \rangle + \varepsilon_{p+q+1} + \sum_{k=1}^{p} \sum_{l=1}^{q} \varepsilon_{p+q+1-l-k} \langle \Psi^{(k)} | \Psi^{(l)} \rangle$$

(12)

After n iterations a secular problem of dimension $(n-1)$ is easily set up and solved by standard techniques. This method, in Table 3, is referred to as nonlinear variation perturbation theory. Other ways of obtaining improved convergence on the energy involve using the $[N, N-1]$ and $[N, N]$ Padé approximants to extrapolate the perturbation energies. For more details on this approach, including closed expressions for the approximants, the reader is referred to Refs. 24 and 25. Finally, the variational energy from a linear summation of the perturbation functions is also easily calculated from simple closed formulas.[23] This method is shown in Table 3, and is referred to as linear

Table 3. *Convergence Properties of Different Perturbative Methods for Solving the CI Equations*[a]

Iteration	Linear perturbation theory (ε_{2n})	Linear variation perturbation theory	Nonlinear variation perturbation theory	$[N, N-1]$ Padé approximation	$[N, N]$ Padé approximation
2	−0.413239	−0.242832	−0.267611	−0.337680	−0.356097
4	−0.239193	−0.279266	−0.313136	−0.342975	−0.386712
6	−0.137370	−0.133265	−0.320340	−0.324810	−0.325122
8	−2.105398	−0.180377	−0.320674	−0.324273	−0.310087
9	−2.758177	−0.056128	−0.320685	−0.319864	−0.322604
10	7.954382	0.047972	−0.320688	−0.321207	−0.320586

[a] The example illustrated refers to a calculation on the HF molecule with $d(\text{HF}) = 4.0$ a.u. and 971 configurations.

variation perturbation theory. All these methods have been compared to each other and results from a representative case [a calculation on HF with d (HF) = 4.0 a.u.] are shown in Table 3. The nonlinear variation perturbation approach has always been found superior, and convergence to six decimal places in the correlation energy for a ground state is usually obtained in 6–7 iterations. Very rarely are up to 10 iterations needed. The iterative techniques described below have shown inferior convergence properties. However, a certain flexibility in both the iterative and the perturbation techniques has not yet been fully explored, so the question of which is the most efficient diagonalization method in connection with the CIMI method is still not completely answered. In perturbation theory this flexibility appears in the choice of $\hat{H}_0$. The matrix elements $\langle \Phi_\mu | \hat{H} | \Phi_\mu \rangle$ appearing in (9) can be replaced by arbitrary constants α_μ, making up a still-diagonal $\hat{H}_0$ of the form

$$\hat{H}_0 = \sum_\mu |\mu\rangle \alpha_\mu \langle \mu| \tag{13}$$

There does not seem to be any theoretically based best choices of $\hat{H}_0$. By diagrammatic arguments the choice (9) has been inferred based on inherent summation to higher order. However, this does not seem to be connected with faster convergence properties. On the contrary, better convergence has so far been obtained choosing α_μ equal to diagonal elements of a sum of Fock operators

$$\hat{H}_0 = \sum_\mu |\mu\rangle \langle \mu | \hat{F} | \mu \rangle \langle \mu| \tag{14}$$

where $\hat{F} = \sum_{i=1}^n \hat{f}(i)$ and $\hat{f}$ is a Hartree–Fock type operator.

Finally, we should mention that the variation perturbation approach is also well suited for convergence on excited states.[26] However, this has still not been used with the CIMI method. A similar approach has recently been suggested by Davidson[27] and seems to compete well even with optimally constructed iterative techniques.

An iterative solution of an equation system of the type

$$\mathbf{A} \cdot \mathbf{C} = 0 \tag{15}$$

starts with the division of the matrix $\mathbf{A}$ into two parts

$$\mathbf{A} = \mathbf{A}' + \mathbf{A}'' \tag{16}$$

Moving $\mathbf{A}' \cdot \mathbf{C}$ to the right-hand side and operating to the left with $\mathbf{A}''^{-1}$ sets up the general iterative scheme

$$\mathbf{C}^{(k)} = -\mathbf{A}''^{-1} \cdot \mathbf{A}' \cdot \mathbf{C}^{(k-1)} \tag{17}$$

which is then carried till convergence. In order for this scheme to be practically useful $\mathbf{A}''$ should be diagonal, so that the inverse is easily constructed. Therefore, $\mathbf{A}''$ is taken as the whole or part of the diagonal elements of $\mathbf{A}$. In the former case, as applied to the equation system (2) with an orthogonal basis set, we get, in component form,

$$A''_{\mu\nu} = (H_{\mu\mu} - E)\,\delta_{\mu\nu}$$

$$A'_{\mu\nu} = H_{\mu\nu}(1 - \delta_{\mu\nu})$$

which leads to

$$C^{(k)}_{\mu} = \frac{1}{E^{(k-1)} - H_{\mu\mu}} \left[\sum_{\nu=1}^{N} H_{\mu\nu} C^{(k-1)}_{\nu} - H_{\mu\mu} C^{(k-1)}_{\mu} \right]$$

or more commonly written as

$$\Delta C^{(k)}_{\mu} = \frac{1}{E^{(k-1)} - H_{\mu\mu}} \left[\sum_{\nu=1}^{N} H_{\mu\nu} C^{(k-1)}_{\nu} - E^{(k-1)} C^{(k-1)}_{\mu} \right] \tag{18}$$

where the energy is simultaneously updated by

$$E^{(k-1)} = \sum_{\mu=1}^{N} C^{(k-1)}_{\mu} \sum_{\mu=1}^{N} H_{\mu\nu} C^{(k-1)}_{\nu} \Big/ \sum_{\mu=1}^{N} [C^{(k-1)}_{\mu}]^2 \tag{19}$$

The basic operation (3) is recognized in both (18) and (19). Various modifications of the updating form (18) can be obtained by choosing the splitting differently in (16). A more general division of A than was done in (18) leads to the denominator $E^{(k-1)} - H_{\mu\mu} - \alpha'_{\mu}$ instead, where α'_{μ} can be chosen to speed up the convergence [no other change in (18) is made by this modification]. This flexibility corresponds to the equivalent freedom of choosing $\hat{H}_0$ in the perturbation technique. Similar techniques of shifting denominators have been used in Hartree–Fock[28] and MC–SCF calculations.[29,30]

2.4. Approximate Natural Orbitals

We mentioned earlier that in many cases a truncation of the external orbital space is desired in order to make the CI expansion shorter. With a given truncation the so-called natural spin orbitals[13] will, in a certain respect, be the optimal choice as an orbital set. The natural spin orbitals are defined as the orbitals that diagonalize the first-order reduced density matrix which, with a CI expansion of type (1), has the form

$$\rho_1(x'_1; x_1) = \sum_{\mu} \sum_{\nu} C^*_{\mu} C_{\nu} \rho_1(\mu\nu | x'_1 x_1) \tag{20}$$

where

$$\rho_1(\mu\nu | x'_1 x_1) = n \int \Phi^*_{\mu}(x'_1 x_2 \cdots x_n) \Phi_{\nu}(x_1 x_2 \cdots x_n)\, d\tau_2 \cdots d\tau_n$$

is called the transition density matrix. The density matrix may be represented in terms of the spin-orbitals $\psi_p(x)$, according to

$$\rho_1(x', x) = \sum_{p,q} D_{pq} \psi_p^*(x') \psi_q(x) \tag{21}$$

where the sum runs over all orbitals in the internal and external set. The useful theorem that can be rigorously proved for the natural spin-orbitals[14] states that the expansion of a trial wave function Φ in a natural spin-orbital basis minimizes the least-square deviation Δ:

$$\Delta = \langle \Psi - \Phi | \Psi - \Phi \rangle$$

where Ψ is the exact wave function.

As the first-order reduced density matrix and consequently the natural orbitals are constructed from the exact wave function, some approximation to this matrix has to be used in practice. Various methods for constructing approximate density matrices have for some time been used by different groups.[31–33] Some of these schemes are essentially based on perturbation theory. The components of the external part of the density matrix can be constructed to second order from a wave function known only to first order. As can be seen from (8) the expressions for the coefficients of the first-order wave function are simple and can therefore be easily constructed. The Hamiltonian matrix elements needed are those that connect the external space with the reference space and, depending on the $\hat{H}_0$ chosen, possibly also the diagonal elements. In this context we should remind the reader that a result obtained from finite-order perturbation theory, as in this case the wave function to first order, is only well defined with respect to a certain splitting of the Hamiltonian into $\hat{H}_0 + \hat{V}$. In fact, by choosing $\hat{H}_0$ according to (13) any result for the wave function and consequently the density matrix can be obtained due to the freedom of choice of α_μ. The usefulness of approximate natural orbitals constructed by perturbation theory therefore has to be empirically judged. Orbitals that seem to give reasonable convergence properties are obtained by choosing $\hat{H}_0$ according to (9) or (14). We have used the latter $\hat{H}_0$. For the case of a single closed-shell reference state, a simple formula for the *external elements* of the first-order density matrix in terms of molecular integrals are obtained to second order:

$$D_{ab}^{(2)} =$$

$$\sum_{i,j}^{occ} \sum_{c}^{ext} \frac{2(ai|cj)(bi|cj) + 2(aj|ci)(bj|ci) - (ai|cj)(bj|ci) - (aj|ci)(bi|cj)}{(\varepsilon_c + \varepsilon_a - \varepsilon_i - \varepsilon_j)(\varepsilon_c + \varepsilon_b - \varepsilon_i - \varepsilon_j)} \tag{22}$$

where a, b, and c denote external orbitals, i, j internal orbitals, and ε are the orbital energies. This formula is essentially equivalent to that obtained by diagrammatic techniques and by Green's function theory.[33] By ordering the integrals in a specific order, the density matrix can be constructed directly from one pass through the integrals. This technique is similar in spirit to the CIMI method itself, where the long list of Hamiltonian matrix elements are avoided in going from $C^{(n-1)}$ to $\sigma^{(k)}$. Here the long list of wave function coefficients is avoided in going from the integrals to the short list of density matrix elements. The method is possible to extend and keep efficient to at least some more general cases with a multiconfigurational reference space. The by far most time-consuming step in the construction of these approximate natural orbitals will be the required transformation of the integrals. By using the fact that only certain integrals are needed in (22) we can reduce the transformation time substantially. Further, the transformation should optimally proceed in the following order:

$$(pi|rs) = \sum_q C_{iq}(pq|rs)$$

$$(pi|rj) = \sum_s C_{js}(pi|rs)$$

(23)

$$(ai|rj) = \sum_p C_{ap}(pi|rj)$$

$$(ai|bj) = \sum_r C_{br}(ai|rj)$$

where the same convention as in (22) has been used for the indices. The rate-determining step will be the first summation in (23). If the number of internal orbitals is N_{int} and the total number of orbitals is N_{tot}, the ratio between the number of operations here and the corresponding full transformation is N_{int}/N_{tot}, which will usually be a small number. The number of operations in the other three summations in (23) will be reduced even further.

Convergence on the correlation energy with different orbital sets is illustrated in Fig 1 from a study on the water molecule. A basis set of essentially

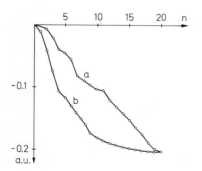

Fig. 1. Correlation energy for H_2O (equlibrium geometry) as a function of the number of external orbitals. Curve a is obtained with virtual canonical HF orbitals ordered by increasing orbital energy; curve b with natural or approximate natural orbitals ordered by decreasing occupation numbers.

double zeta plus polarization quality was used and the $1s$ orbital on the oxygen was left uncorrelated. Three sets of orbitals are compared, the canonical SCF orbitals (COs) ordered with respect to orbital energies, the approximate natural orbitals (ANOs) obtained from (22), and the natural orbitals (NOs) obtained from the converged CI calculation with all canonical orbitals. The latter two sets are ordered with respect to occupation numbers. As could have been expected from the above theorem, we see that the NO expansion converges much quicker on the correlation energy than the CO expansion (the ANO expansion yields results so similar to those obtained with the NO expansion that they cannot be distinguished on the scale used in Fig. 1). This is not obvious, however, as the theorem states something about the overlap and not the energy. If the core has been correlated, for example, selection principles other than high occupation numbers would have to have been considered.[34] What is demonstrated in Fig. 1, is that if a certain number of orbitals are taken out from the orbital set when correlating the valence shells, less correlation energy is lost if the orbitals are ANOs than if they are COs. However, in practice the interest is not so much in the correlation energy itself as in a particular chemical property, for example, an equilibrium bond distance. What is wanted in this case is an *equal* loss of correlation energy for *different* distances. For an ordinary chemical bond a loss of valence shell correlation energy will usually be proportional to the loss of accuracy on the bond distance. For a van der Waals interaction, however, good results can be obtained by neglecting the intraatomic correlation completely, though this is orders of magnitude larger than the interaction energy itself.[35,36] A different truncation for intra- and interatomic correlation is therefore recommended in this case.[56] In general, a division of a problem into different chemical effects may allow a more severe truncation of the orbital set and thereby yield a more efficient calculation.

3. Application to Some Different Types of CI Expansions

3.1. A Closed-Shell Single-Determinant Reference State

The direct CI method from molecular integrals was first applied to systems with one closed-shell reference state,[8] including in the CI expansion all single- and double-replacement states with respect to the reference state:

$$\Psi = \Phi_0 + \sum_i \sum_a C_{i \to a} \Phi_{i \to a} + \sum_{i>j} \sum_{a>b} C_{ij \to ab} \Phi_{ij \to ab} \qquad (23)$$

where Φ_0 is the reference state function, Φ_i^a represents a single-replacement state, and Φ_{ij}^{ab} a double-replacement state. Indices $i, j, \ldots$ are, as before, used to denote orbitals occupied in Φ_0 (internal orbitals), while $a, b, \ldots$ are used for the remaining orbital space (external orbitals). The state functions should be spin eigenfunctions corresponding to a singlet state.

Spin eigenfunctions can be obtained in a number of different ways. The state functions used in the CI expansion (23) have been chosen to have the following form:

$$\Phi_i^a = \sqrt{2}\hat{O}[\ldots \varphi_i \bar{\varphi}_a \ldots] \tag{24a}$$

$$\Phi_{ii}^{aa} = [\ldots \varphi_a \bar{\varphi}_a \ldots] \tag{24b}$$

$$\Phi_{ij}^{aa} = \sqrt{2}\hat{O}[\ldots \varphi_i \bar{\varphi}_a \ldots \varphi_a \bar{\varphi}_j \ldots] \tag{24c}$$

$$\Phi_{ii}^{ab} = \sqrt{2}\hat{O}[\ldots \varphi_a \bar{\varphi}_b \ldots] \tag{24d}$$

$$^1\Phi_{ij}^{ab} = -\sqrt{3}\hat{O}[\ldots \varphi_i \varphi_a \ldots \bar{\varphi}_b \bar{\varphi}_j \ldots] \tag{24e}$$

$$^2\Phi_{ij}^{ab} = \hat{O}\{[\ldots \varphi_i \bar{\varphi}_a \ldots \varphi_j \bar{\varphi}_b \ldots] + [\ldots \varphi_a \bar{\varphi}_i \ldots \varphi_j \bar{\varphi}_b \ldots]\} \tag{24f}$$

where the square brackets are used to denote a normalized determinant. Only the replacements are shown explicitly. $\hat{O}$ is a spin-projection operator for singlet states:

$$\hat{O} = \prod_{S \neq 0} \left[1 - \frac{\hat{S}^2}{S(S+1)} \right] \tag{25}$$

The state functions (24a)–(24f) should belong to the totally symmetric representation of the molecular point group. The treatment of symmetry is considerably simplified by allowing only point groups with one-dimensional irreducible representations. For some molecules this will mean that only a subgroup of the full symmetry group can be used. The state functions are then automatically symmetry adapted, provided that symmetry-adapted one-electron orbitals are used. The condition for a configuration state $\Phi_{ijk\ldots}^{abc\ldots}$ to belong to the same representation as reference state Φ_0 of arbitrary symmetry is that the product of the orbitals in the replacement belongs to the totally symmetric representation Γ_1. For one-dimensional groups this means that

$$\gamma_i \times \gamma_j \times \gamma_k \times \cdots \times \gamma_a \times \gamma_b \times \gamma_c \times \cdots = \Gamma_1 \tag{26}$$

where γ_ν is the irreducible representation for orbital ν.

Using the state functions given above we can construct 21 different types of matrix elements of the Hamiltonian $\hat{H}$. All these elements can be written as linear combinations of one- and two-electron integrals. To simplify these expressions the Hamiltonian is first partitioned into a Hartree–Fock operator $\hat{F}$

and a fluctuation potential $\hat{V}$:

$$\hat{H} = \hat{F} + \hat{V} \tag{27}$$

where

$$\hat{F} = \sum_{i=1}^{2m} \hat{f}(i)$$

with

$$\hat{f}(i) = \left\{ \hat{h}(i) + \sum_{j=1}^{m} [2\hat{J}_j(i) - \hat{K}_j(i)] \right\} \tag{28}$$

and

$$\hat{V} = \sum_{i<j} (1/r_{ij}) - \sum_{i=1}^{2m} \sum_{j=1}^{m} [2\hat{J}_j(i) - \hat{K}_j(i)] \tag{29}$$

where $\hat{h}$ is the one-electron operator and $\hat{J}_j$ and $\hat{K}_j$ are the ordinary Coulomb and exchange operators; m is the number of doubly occupied orbitals in the reference state Φ_0.

As was pointed out in the preceding section, the most crucial point in a CI calculation is the evaluation of the quantity

$$\sigma_\mu = \sum_\nu (H_{\mu\nu} - H_{00}\,\delta_{\mu\nu}) C_\nu \tag{30}$$

By using (27) this quantity can be written as

$$\sigma_\mu = \sum_\nu (F_{\mu\nu} - F_{00}\,\delta_{\mu\nu}) C_\nu + \sum_\nu (V_{\mu\nu} - V_{00}\,\delta_{\mu\nu}) C_\nu \tag{31}$$

where $F_{\mu\nu}$ and $V_{\mu\nu}$ are matrix elements of the operators $\hat{F}$ and $\hat{V}$, respectively. Characteristic for the CIMI method is the resolution of the second sum in (31) in terms of molecular two-electron integrals. Define the integrals

$$f_{pq} = \int \varphi_p^*(1)\hat{f}(1)\varphi_q(1)\,dV_1 \tag{32}$$

$$(pq|rs) = \int \varphi_p^*(1)\varphi_q(1)(1/r_{12})\varphi_r^*(2)\varphi_s(2)\,dV_1\,dV_2 \tag{33}$$

The vector $\boldsymbol{\sigma}$ can be written as a sum over these one- and two-electron integrals multiplied with appropriate coupling coefficients which depend only on the spin couplings for the corresponding configuration state functions. For a double-replacement state ($\mu = ij \to ab$) we obtain the following contribution to

$\boldsymbol{\sigma}_\mu$ from the interaction with all double-replacement states:

$$\Delta\sigma_{ij\to ab} = \sum_c j_2(\mu, \nu)\{f_{bc}C_{ij\to ac} + f_{ac}C_{ij\to cb}\}$$

$$-\sum_k j_1(\mu, \nu)\{f_{kj}C_{ik\to ab} + f_{ik}C_{kj\to ab}\}$$

$$+\sum_{k\geq l} \{j_1(\mu, \nu)(ik|jl) + k_1(\mu, \nu)(il|kj)\}C_{kl\to ab}$$

$$+\sum_{c\geq d} \{j_2(\mu, \nu)(ac|bd) + k_2(\mu, \nu)(ad|bc)\}C_{ij\to cd}$$

$$+\sum_{k,c} \{j_3(\mu, \nu)(ac|ik) + k_3(\mu, \nu)(ai|ck)\}C_{kj\to cb}$$

$$+\sum_{k,c} \{j_4(\mu, \nu)(bc|jk) + k_4(\mu, \nu)(bj|ck)\}C_{ik\to ac}$$

$$+\sum_{k,c} \{j_5(\mu, \nu)(bc|ik) + k_5(\mu, \nu)(bi|ck)\}C_{kj\to ac}$$

$$+\sum_{k,c} \{j_6(\mu, \nu)(ac|jk) + k_6(\mu, \nu)(aj|ck)\}c_{ik\to cb} \tag{34}$$

The coupling constants $j_i(\mu, \nu)$ and $k_i(\mu, \nu)$ depend only on the spin coupling appropriate for the state functions Φ_μ and Φ_ν, and on the signature of permutations which bring equal indices into equal positions. They take simple numerical values (Table 4).

From the interaction with single-replacement states we obtain the following contribution:

$$\Delta\sigma_{ij\to ab} = i_1(\mu)f_{bj}C_{i\to a} + i_2(\mu)f_{aj}C_{i\to b}$$

$$+i_1(\mu)f_{bi}C_{j\to a} + i_2(\mu)f_{ai}C_{j\to b}$$

$$+\sum_c \{i_1(\mu)(ac|bj) + i_2(\mu)(aj|bc)\}C_{i\to c}$$

$$+\sum_c \{i_1(\mu)(ai|bc) + i_2(\mu)(ac|bi)\}C_{j\to c}$$

$$-\sum_k \{i_1(\mu)(ai|jk) + i_2(\mu)(aj|ik)\}C_{k\to b}$$

$$-\sum_k \{i_1(\mu)(bj|ik) + i_2(\mu)(bi|jk)\}C_{k\to a} \tag{35}$$

Table 4. Coupling Constants for Interactions between Double Replacement CSFs in the CICS Case [a]

$J_1(\mu, \nu)$

μ	1	2	3	4	5
1	1	$\frac{1}{2}\sqrt{2}$	—	—	—
2	$\sqrt{2}$	1	—	—	—
3	—	—	1	$\frac{1}{2}\sqrt{\frac{3}{2}}$	$\frac{1}{2}\sqrt{\frac{1}{2}}$
4	—	—	$\sqrt{\frac{3}{2}}$	1	0
5	—	—	$\sqrt{\frac{1}{2}}$	0	1

$K_1(\mu, \nu)$

μ	1	2	3	4	5
1	0	$\frac{1}{2}\sqrt{2}$	—	—	—
2	0	1	—	—	—
3	—	—	0	$\frac{1}{2}\sqrt{\frac{3}{2}}$	$\frac{1}{2}\sqrt{\frac{1}{2}}$
4	—	—	0	$\frac{1}{2}$	$\sqrt{3}/2$
5	—	—	0	$\sqrt{3}/2$	$-\frac{1}{2}$

$J_2(\mu, \nu)$

μ	1	2	3	4	5
1	1	—	$\frac{1}{2}\sqrt{2}$	—	—
2	—	1	—	$\frac{1}{2}\sqrt{\frac{3}{2}}$	$\frac{1}{2}\sqrt{\frac{1}{2}}$
3	$\sqrt{2}$	—	1	—	—
4	—	$\sqrt{\frac{3}{2}}$	—	1	0
5	—	$\sqrt{\frac{1}{2}}$	—	0	1

$K_2(\mu, \nu)$

μ	1	2	3	4	5
1	0	—	$\frac{1}{2}\sqrt{2}$	—	—
2	—	0	—	$\frac{1}{2}\sqrt{\frac{3}{2}}$	$\frac{1}{2}\sqrt{\frac{1}{2}}$
3	0	—	1	—	—
4	—	0	—	$\frac{1}{2}$	$\sqrt{3}/2$
5	—	0	—	$\sqrt{3}/2$	$-\frac{1}{2}$

$J_3(\mu, \nu)$

μ	1	2	3	4	5
1	-4	$-2\sqrt{2}$	$-2\sqrt{2}$	$-\sqrt{3}$	-1
2	$-2\sqrt{2}$	-2	-2	$-\sqrt{\frac{3}{2}}$	$-\sqrt{\frac{1}{2}}$
3	$-2\sqrt{2}$	-2	-2	$-\sqrt{\frac{3}{2}}$	$-\sqrt{\frac{1}{2}}$
4	$-\sqrt{3}$	$-\sqrt{\frac{3}{2}}$	$-\sqrt{\frac{3}{2}}$	$-(3+p)/4$	$\sqrt{3}(p-1)/4$
5	-1	$-\sqrt{\frac{1}{2}}$	$-\sqrt{\frac{1}{2}}$	$\sqrt{3}(p-1)/4$	$-(1+3p)/4$

$K_3(\mu, \nu)$

μ	1	2	3	4	5
1	2	$\sqrt{2}$	$\sqrt{2}$	$\sqrt{3}(1-p)/2$	$(1+3p)/2$
2	$\sqrt{2}$	1	1	$\sqrt{6}(1-p)/4$	$\sqrt{2}(1+3p)/4$
3	$\sqrt{2}$	1	1	$\sqrt{6}(1-p)/4$	$\sqrt{2}(1+3p)/4$
4	0	0	0	0	0
5	2	$\sqrt{2}$	$\sqrt{2}$	$\sqrt{3}(1-p)/2$	$(1+3p)/2$

$J_4(\mu, \nu)$

μ	1	2	3	4	5
1	0	0	0	0	0
2	$-2\sqrt{2}$	-2	-2	$-\sqrt{\frac{3}{2}}$	$-\sqrt{\frac{1}{2}}$
3	$-2\sqrt{2}$	-2	-2	$-\sqrt{\frac{3}{2}}$	$-\sqrt{\frac{1}{2}}$
4	$-\sqrt{3}$	$-\sqrt{\frac{3}{2}}$	$-\sqrt{\frac{3}{2}}$	$-(3+p)/4$	$\sqrt{3}(p-1)/4$
5	-1	$-\sqrt{\frac{1}{2}}$	$-\sqrt{\frac{1}{2}}$	$\sqrt{3}(p-1)/4$	$-(1+3p)/4$

$K_4(\mu, \nu)$

μ	1	2	3	4	5
1	0	0	0	0	0
2	$\sqrt{2}$	1	1	$\sqrt{6}(1-p)/4$	$\sqrt{2}(1+3p)/4$
3	$\sqrt{2}$	1	1	$\sqrt{6}(1-p)/4$	$\sqrt{2}(1+3p)/4$
4	0	0	0	0	0
5	2	$\sqrt{2}$	$\sqrt{2}$	$\sqrt{3}(1-p)/2$	$(1+3p)/2$

$J_5(\mu, \nu) = J_6(\mu, \nu)$

μ	1	2	3	4	5
1	0	0	0	0	0
2	0	0	0	0	0
3	0	0	0	0	0
4	$-\sqrt{3}$	$-\sqrt{\frac{3}{2}}$	$-\sqrt{\frac{3}{2}}$	$-(3+p)/4$	$\sqrt{3}(p-1)/4$
5	-1	$-\sqrt{\frac{1}{2}}$	$-\sqrt{\frac{1}{2}}$	$\sqrt{3}(p-1)/4$	$-(1+3p)/4$

$K_5(\mu, \nu) = K_6(\mu, \nu)$

μ	1	2	3	4	5
1	0	0	0	0	0
2	0	0	0	0	0
3	0	0	0	0	0
4	$\sqrt{3}$	$\sqrt{\frac{3}{2}}$	$\sqrt{\frac{3}{2}}$	$3(1+p)/4$	$\sqrt{3}(1-3p)/4$
5	-1	$-\sqrt{\frac{1}{2}}$	$-\sqrt{\frac{1}{2}}$	$-\sqrt{3}(1+p)/4$	$(3p-1)/4$

[a] The indices (μ, ν) refer to the different types of double-replacement CSFs by Eqs. (24b)–(24f)[52] and p is the signature of the permutation which brings equal indices into equal positions in the two configurations.

Table 5. Coupling Coefficients for the Interactions between Double- and Single-Replacement CSFs in the CICS Case[a]

μ	$i_1(\mu)$	$i_2(\mu)$
1	$\sqrt{2}/4$	$\sqrt{2}/4$
2	$\frac{1}{2}$	$\frac{1}{2}$
3	$\frac{1}{2}$	$\frac{1}{2}$
4	0	$\sqrt{\frac{3}{2}}$
5	$\sqrt{2}$	$-\sqrt{\frac{1}{2}}$

[a] The indices (μ) refer to the five state functions defined in Eqs. (24b)–(24f).

with a corresponding relation for the contributions from double-replacement states to $\sigma_{i \to a}$. The coupling coefficients $i_1(\mu)$ and $i_2(\mu)$ are found in Table 5. Finally, the interaction between single-replacement states gives the contribution

$$\Delta\sigma_{i \to a} = \sum_b f_{ab}C_{i \to b} - \sum_j f_{ij}C_{j \to a} + \sum_{j,b} \{2(ai|bj) - (ab|ij)\}C_{j \to b} \qquad (36)$$

Thus we have replaced the sum over configurations with a much shorter sum over the one- and two-electron integrals. Computationally, the construction of the vector $\boldsymbol{\sigma}$ is accomplished during the sequential reading of the integral list. Each integral is identified with respect to its type as defined in Eqs. (34)–(36). It is then multiplied with a coupling constant and the contribution is added to the appropriate σ_μ terms. The necessary programming effort will strongly depend on the number of different types of two-electron integrals, since each integral type requires its own specific loop structure.

The CIMI method for closed-shell systems has been implemented as the MOLECULE-CICS program. The overall flow scheme for this program is shown in Fig. 2. The actual CI calculation is preceded by an integral evaluation and an SCF-calculation. This calculation provides the orbital set used for the construction of the configuration state functions. Usually either the canonical orbitals or approximate natural orbitals are used as constructed according to the method described in the preceding section. Other possibilities can easily be implemented, e.g., to use localized orbitals.

A transformation of the two-electron integrals has to precede the actual CI calculation. The atomic two-electron integrals are, in MOLECULE, computed directly over a symmetry-adapted orbital basis. The transformation can therefore also be symmetry blocked. This greatly reduces the necessary computational effort, since the time needed is proportional to the fifth power of the maximum number of orbitals in one irreducible representation. The transfor-

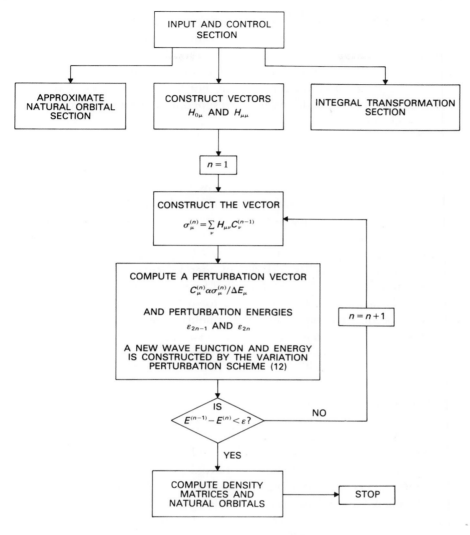

Fig. 2. Flow chart for the MOLECULE-CICS program.

mation is performed using the method given by Yoshimine,[37] which makes heavy use of direct access. An efficient two-electron transformation is an important part of a CI calculation since the limit in the size of the atomic orbital basis is to a large extent set by this step.

The actual CI calculation starts with a computation of the vectors $H_{0\mu}$ and optionally $H_{\mu\mu}$ for later use. This part also gives the second-order perturbation energy. The corresponding first-order CI vector may be used as a starting vector for the iterative search for an eigenvector and eigenvalue of the Hamiltonian.

The possible perturbation (or iteration) schemes have been previously discussed in detail. The by far most time consuming part in these schemes is the construction of the vector $\boldsymbol{\sigma}$. To efficiently work with the vectors $\boldsymbol{\sigma}$ and $\mathbf{C}$ it will be necessary to quickly identify the nonzero components from the four integers, i, j, a, and b, of the replacement. For this reason an index vector is constructed. A position in this vector corresponding to the double replacement $ij \rightarrow ab$ is given by

$$N[i(i-1)/2+j]+a(a-1)/2+b$$

where N is the number of orbital pairs in the external set. This position contains either a reference to the first spin state for this configuration or a zero. If a zero is found, the corresponding state functions have been excluded from the CI expansion. This provides an easy way of excluding all configurations which are not allowed by symmetry. In the construction of the index vector it is also possible to impose other conditions on the CI expansion, e.g., to exclude all intramolecular replacements in studies of intermolecular interactions.

The iteration procedure is controlled by a convergence test which can optionally be made on either the CI vector or the correlation energy. The iteration procedure is ended when the threshold is reached, and the calculation is completed with the construction of the first-order density matrix and the natural orbitals.

3.2. Expansion in Single Determinants

The extension of the CIMI method to the general open-shell case is a difficult problem mainly due to the complicated dependence of the constants A in (4) on the spin coupling. One way to avoid most of these difficulties is to remove the spin coupling completely and instead work with single determinants. The main drawback of such a procedure will be the increased number of independent terms in the CI expansion. Due to spin symmetry, relationships exist between the coefficients for different determinants, and this can be used to make the construction of the σ vector more efficient. If single determinants are used it is, on the other hand, possible to relax the condition of pairing electrons into the same molecular orbitals and instead allow different space orbitals for different spin. This will destroy the relationships between the coefficients but it can be assumed that it will increase the convergence of the CI expansion. A determinant obtained from an *u*nrestricted *H*artree–*F*ock (UHF) calculation would in this case be allowed as the reference state. A CI program constructed from these principles, here denoted CISD, uses, like the CICS program, a basis set of all single and double replacements out of one single determinant.

The CI basis set will then consist of the following types of determinants, where we use the sign convention of ordering α-spins first:

The reference state

$$\Phi_0 = [\ldots i_\alpha j_\alpha \ldots \bar{i}_\beta \bar{j}_\beta \ldots] \tag{37}$$

Single replacements

$$\Phi_{i_\alpha}^{a_\alpha} = [\ldots a_\alpha j_\alpha \ldots \bar{i}_\beta \bar{j}_\beta \ldots] \qquad (\alpha \text{ excitations})$$

$$\Phi_{i_\beta}^{a_\beta} = [\ldots i_\alpha j_\alpha \ldots \bar{a}_\beta \bar{j}_\beta \ldots] \qquad (\beta \text{ excitations}) \tag{38}$$

Double replacements

$$\Phi_{i_\alpha j_\alpha}^{a_\alpha b_\alpha} = [\ldots a_\alpha b_\alpha \ldots \bar{i}_\beta \bar{j}_\beta \ldots] \qquad (\alpha\alpha \text{ excitations})$$

$$\Phi_{o_\beta l_\beta}^{a_\beta b_\beta} = [\ldots i_\alpha j_\alpha \ldots \bar{a}_\beta \bar{b}_\beta \ldots] \qquad (\beta\beta \text{ excitations}) \tag{39}$$

$$\Phi_{i_\alpha j_\beta}^{a_\alpha b_\beta} = [\ldots a_\alpha j_\alpha \ldots \bar{i}_\beta \bar{b}_\beta \ldots] \qquad (\alpha\beta \text{ excitations})$$

In the special case of a closed-shell reference state the determinants (37)–(39) will span the same space as the projected configurations used in the CICS program. The spin operator $\hat{S}^2$, when operating on such determinants, will not generate any function outside the space. The space is "closed" under operations with $\hat{S}^2$, which will guarantee a CI expansion with correct spin symmetry. This property is used in the calculation of nuclear spin–spin coupling constants, as described later, where a correct triplet wave function is needed in the space spanned by (37)–(39) with a closed-shell reference state. For open-shell cases and when different orbitals for different spin are used, a wave function based on this type of expansion will not have the correct spin symmetry. This error can, however, be expected to be small in most cases. A rough idea of the accuracy of the determinantal expansion is obtained by noting that for equal orbitals for different spin the first-order interacting space (configurations interacting directly with the reference state) is included in the space spanned by all single and double replacements.

The length of the expansion is, in open-shell cases, about a factor of three longer than an expansion in projected first-order interacting CSFs, and of approximately the same length as an expansion in all projected configurations. An expansion in determinants with different orbitals for different spin is advantageous especially in those cases where a single RHF determinant is not a good representation of the wave function, as is usually the case when a molecule dissociates. An illustration of this for the HF molecule will be discussed later. The main disadvantage with different space orbitals is the increased number of molecular two-electron integrals. The list of such integrals will be four times longer, which makes the integral transformation *three* times slower, and the CI calculation *four* times slower. An efficient transformation program is consequently even more important for the CISD than for the CICS program. A modification of the CICS transformation to include the possibility of different orbitals for different spin is included in the CISD package. The actual CI program has essentially the same general structure as has been described for the CICS case, including the index list and special routines for each integral type. With the restriction of paired spin-orbitals removed, the number of integral types has increased, and the program is correspondingly larger. The coupling coefficients will, however, simply be plus or minus one. This makes it possible to write the contributions to σ without reference to any coupling tables. We obtain the following contributions to σ from interactions between double-replacement determinants:

$$\Delta\sigma_{i_\mu j_\nu \to a_\mu b_\nu} = \sum_\kappa \sum_{c_\kappa} (-1)^p [f^\nu_{b_\nu c_\kappa} C_{i_\mu j_\nu \to a_\mu c_\kappa} \delta_{\kappa\nu} + f^\mu_{a_\mu c_\kappa} C_{i_\mu j_\nu \to c_\kappa b_\nu} \delta_{\kappa\mu}]$$

$$+ \sum_\kappa \sum_{k_\kappa} (-1)^{p+1} [f^\nu_{k_\kappa j_\nu} C_{i_\mu k_\kappa \to a_\mu b_\nu} \delta_{\kappa\nu} + f^\mu_{i_\mu k_\kappa} C_{k_\kappa j_\nu \to a_\mu b_\nu} \delta_{\kappa\mu}]$$

$$+ \sum_{k_\mu > l_\nu} [(i_\mu k_\mu | j_\nu l_\nu) - (i_\mu l_\nu | k_\mu j_\nu) \delta_{\mu\nu}] C_{k_\mu l_\nu \to a_\mu b_\nu}$$

$$+ \sum_{c_\mu > d_\nu} [(a_\mu c_\mu | b_\nu d_\nu) - (a_\mu d_\nu | c_\mu b_\nu) \delta_{\mu\nu}] C_{i_\mu j_\nu \to c_\mu d_\nu}$$

$$+ \sum_\kappa \sum_{k_\kappa c_\kappa} (-1)^p [(a_\mu i_\mu | c_\kappa k_\kappa) - (a_\mu c_\mu | i_\mu k_\kappa) \delta_{\kappa\mu}] C_{k_\kappa j_\nu \to c_\kappa b_\nu}$$

$$+ \sum_\kappa \sum_{k_\kappa c_\kappa} (-1)^p [(b_\nu j_\nu | c_\kappa k_\kappa) - (b_\nu c_\kappa | j_\nu k_\kappa) \delta_{\kappa\nu}] C_{i_\mu k_\kappa \to a_\mu c_\kappa}$$

$$+ \sum_{k_\mu c_\nu} (-1)^p [(b_\nu i_\mu | c_\nu k_\mu) \delta_{\mu\nu} - (b_\nu c_\nu | i_\mu k_\mu)] C_{k_\mu j_\nu \to a_\mu c_\nu}$$

$$+ \sum_{k_\nu c_\mu} (-1)^p [(a_\mu j_\nu | c_\mu k_\nu) \delta_{\mu\nu} - (a_\mu c_\mu | j_\nu k_\nu)] C_{i_\mu k_\nu \to c_\mu b_\nu} \tag{40}$$

where i, j, and k denote internal spin-orbitals, a, b, and c denote external spin-orbitals, and the indices μ, ν, and κ denote spin α or β, so that f^ν is the α-spin Fock matrix if $\nu = \alpha$, or β-spin if $\nu = \beta$. p is the number of permutations necessary to line up equal spin-orbitals. For the interaction between single and double replacements we obtain

$$\Delta\sigma_{i_\mu j_\nu \to a_\mu b_\nu} = f^\nu_{b_\nu j_\nu} C_{i_\mu \to a_\mu} - f^\nu_{a_\mu j_\nu} C_{i_\mu \to b_\nu} \delta_{\mu\nu}$$

$$- f^\mu_{b_\nu i_\mu} C_{j_\nu \to a_\mu} \delta_{\mu\nu} + f^\mu_{a_\mu i_\mu} C_{j_\nu \to b_\nu}$$

$$+ \sum_{c_\mu} [(a_\mu c_\mu | b_\nu j_\nu) - (a_\mu j_\nu | c_\mu b_\nu) \delta_{\mu\nu}] C_{i_\mu \to c_\mu}$$

$$+ \sum_{c_\nu} [(a_\mu i_\mu | b_\nu c_\nu) - (c_\nu a_\mu | b_\nu i_\mu) \delta_{\mu\nu}] C_{j_\nu \to c_\nu}$$

$$+ \sum_{k_\nu} [(a_\mu j_\nu | i_\mu k_\nu) \delta_{\mu\nu} - (a_\mu i_\mu | j_\nu k_\nu)] C_{k_\nu \to b_\nu}$$

$$+ \sum_{k_\mu} [(b_\nu i_\mu | j_\nu k_\mu) \delta_{\mu\nu} - (b_\nu j_\nu | i_\mu k_\mu)] C_{k_\mu \to a_\mu} \tag{41}$$

Finally, for the interaction between single-replacement determinants the contributions are given by

$$\Delta\sigma_{i_\mu \to a_\mu} = \sum_{b_\mu} f^\mu_{a_\mu b_\mu} C_{i_\mu \to b_\mu} - \sum_{j_\mu} f^\mu_{i_\mu j_\mu} C_{j_\mu \to a_\mu}$$

$$+ \sum_\kappa \sum_{j_\kappa b_\kappa} [(a_\mu i_\mu | b_\kappa j_\kappa) - (a_\mu b_\kappa | i_\mu j_\kappa) \delta_{\kappa\mu}] C_{j_\kappa \to b_\kappa} \tag{42}$$

The possibility to generalize the CISD method to cases with many reference determinants is in principle straightforward. A severe practical problem is the length of the expansion. A possible simplification would be to treat only the correlation between electrons with different spin. In the present scheme this corresponds to including only replacements of the $\alpha\beta$ type. Besides shortening the expansion, such a procedure also leads to a much shorter number of integral types and also to considerable simplifications in the formulas for the contributions to the σ vector. In spite of these simplifications an expansion of this type may be able to yield the major part of the structure-dependent correlation energy.

3.3. Complete CI Expansions

A particularly well-suited class of problems for the CIMI method is configuration interaction in the space of all possible spin- and space-allowed configurations, here denoted as complete CI. This type of wave function is of value in two quite different types of problems. It has been used as an investigating tool for problems where excited surfaces are of interest.[38,39] Usually it is appropriate in these cases to use a small basis set so that the Hamiltonian matrix can be diagonalized with conventional rotation methods like the Givens or Jacobi method. For this type of problem, with a short CI expansion, there is no advantage in using the CIMI method which does not yield a Hamiltonian matrix. The second type of problems where complete CI can be efficiently used is in accurate investigations on systems with few valence electrons. The CIMI methods has in these cases great advantages since it is relatively easy to set up the formulas for the calculation of the σ vector. For two valence electrons the CICS or CISD programs can be directly used. A special program (CI3) for the case of complete CI on three-electron systems has recently been constructed[40] and a similar four-electron program is presently being developed. Programs for complete CI on more than four electrons would probably not be so useful since the number of configurations soon becomes prohibitive.

The main reason for the relative simplicity in applying the CIMI method for complete CI problems is the small number of integral types which occur in this case. Only the fourteen types of two-electron integrals given in Table 1 will appear in the equations determining the σ vector. The precise way in which one of these integral types is handled in the program is illustrated in Table 2. As mentioned in the section about the coupling coefficients, this description of the contributions to the σ vector is preferable from the programming point of view. The more compact description used for the CICS and the CISD programs is however possible to use also for the CI3 program, but the coupling tables will have a more complicated structure in this case.

For three electrons forming a doublet state, it is possible to write down three different types of configuration state functions. If two space orbitals are equal we get

$$\Phi_{aab} = [\varphi_a \bar{\varphi}_a \varphi_b] \tag{43a}$$

If all orbitals are different two linearly independent doublet state functions can be constructed:

$$\Phi_{abc}^{(1)} = \frac{1}{\sqrt{6}}\{2[\varphi_a \bar{\varphi}_b \varphi_c] - [\varphi_a \varphi_b \bar{\varphi}_c] - [\bar{\varphi}_a \varphi_b \varphi_c]\} \tag{43b}$$

$$\Phi_{abc}^{(2)} = \frac{1}{\sqrt{2}}\{[\varphi_a \varphi_b \bar{\varphi}_c] - [\bar{\varphi}_a \varphi_b \varphi_c]\} \tag{43c}$$

With these definitions for the CSFs a part of the general contribution to a σ element takes the form

$$\Delta\sigma_{abc} = \sum_{d \geq e} \{j_1(\mu, \nu)(bd|ce) + k_1(\mu, \nu)(be|cd)\}C_{ade}$$

$$+ \sum_{d \geq e} \{j_2(\mu, \nu)(ad|ce) + k_2(\mu, \nu)(ae|cd)\}C_{dbe}$$

$$+ \sum_{d \geq e} \{j_3(\mu, \nu)(ad|be) + k_3(\mu, \nu)(ae|bd)\}C_{dec} + \cdots \tag{44}$$

where contributions in terms of single sums have been omitted. The coupling coefficients j_i and k_i, which in this case depend on the permutation of indices $a-e$ in a more complicated way, are given in Table 6. To be able to set up the general expressions given in this table four parameters (r, s, t, and u) had to be introduced. This makes the usefulness of the table even more doubtful.

A fairly detailed timing analysis of the CI3 program has been made for a calculation on H_3.[40] It was found that the CPU time was shared roughly equal between index handling and the actual multiplications according to (4). As expected very little time was spent identifying and sending the integrals to the appropriate subroutines, and in the diagonalization routine once the σ vector had been constructed. The most efficient way to speed up a calculation on a particular system is to reduce the number of integrals that significantly contribute to the energy, as the time needed is directly proportional to the number of integrals. One way of doing this is to delete orbitals after a transformation to

Table 6. Coupling Coefficients for the Interaction Elements Defined in Eq. (44)[a]

μ	ν=1	2	3	μ	ν=1	2	3
	$j_1(\mu,\nu)$				$k_1(\mu,\nu)$		
1	-1	—	—	1	0	—	—
2	$(u\sqrt{\tfrac{3}{2}}+2)$	$(1-3r/2+rs)$	—	2	0	$(rs+3s/2-1)$	—
3	$(u\sqrt{\tfrac{1}{2}}+3)$	$r\sqrt{3}/2$	$(1-3r/2)$	3	$3-u\sqrt{2}$	$-s\sqrt{3}/2$	$(1-3s/2)$
	$j_2(\mu,\nu)$				$k_2(\mu,\nu)$		
1	$-u$	—	—	1	0	—	—
2	$\sqrt{\tfrac{3}{2}}$	$(1-r/2)$	—	2	$-\sqrt{\tfrac{3}{2}}$	$(r/2-1)$	—
3	$u\sqrt{\tfrac{1}{2}}$	$\sqrt{3}/2$	$(1-3r/2)$	3	$u\sqrt{\tfrac{1}{2}}$	$\sqrt{3}/2$	$(1-3r/2)$
	$j_3(\mu,\nu)$				$k_3(\mu,\nu)$		
1	0	—	—	1	0	—	—
2	$-u\sqrt{\tfrac{3}{2}}$	$(rs-3r/2+1)$	—	2	0	$(3s/2-1-rs)$	—
3	$u\sqrt{\tfrac{1}{2}}$	$-r\sqrt{3}/2$	$(1-3r/2)$	3	$3-u\sqrt{2}$	$s\sqrt{3}/2$	$(1-3s/2)$

[a] The indices (μ,ν) refer to the CSFs defined in (43a)–(43c). The parameters used are defined according to:

$$r = \begin{cases} 0 & \text{no permutation} \\ 1 & \text{otherwise} \end{cases} \qquad s = \begin{cases} 0 & \text{two permutations} \\ 1 & \text{otherwise} \end{cases}$$

$$t = \begin{cases} -1 & \text{index 2 and 3 permuted} \\ 0 & \text{no permutation} \\ 1 & \text{index 1 and 2 permuted} \end{cases} \qquad u = \begin{cases} -1 & d=e \\ 1 & \text{otherwise} \end{cases}$$

natural orbitals. Methods in which the number of small integrals can be directly maximized are of great interest but have not yet been investigated.

4. Examples of CI Calculations Performed with the CIMI Method

In this section we shall briefly discuss some CI studies in which the CIMI method has been applied. The discussion will be concentrated around the trends of the results and the reader is referred to the separate articles for more details.

The CIMI method has so far been applied to three different types of CI expansions; a closed-shell single-determinant reference state including all single- and double-replacement state functions (a CSSRS-expansion); a single determinant UHF reference state including all single- and double-replacement determinants without spin projection (an OSSRS-expansion), and finally complete CI expansion (including all CSFs obtainable from a given orbital set) for three-electron systems.

4.1. Studies on Closed-Shell Systems

The closed-shell version of the CIMI method has been implemented as the MOLECULE-CICS programs. Its main use is in calculations on molecules in their ground electronic states and on molecular interactions of moderate energy (e.g., hydrogen bond energies, etc.). Very weak van der Waals interactions require, however, a more accurate wave function, also including higher-order replacements, if results with an accuracy of a few degrees Kelvin are desired.

Apart from missing certain types of configurations, the accuracy of the wave function is determined by the length of the expansion, which for a given symmetry of the system under consideration, depends on the size of the orbital set. In applications with the CIMI method the limitations to this length are usually set by the dimension of the available processor storage and not so much by time requirements. It is necessary to have random access to two CI vectors [$C^{(k)}$ and $C^{(k-1)}$ in Eq. (10)] and the index vector simultaneously. Thus two floating point vectors and one integer vector have to be kept in the processor storage. The total storage requirements are determined by the number $2N_{CI} + N_{conf}$, where N_{CI} is the length of the expansion and N_{conf} is the number of *possible* configurations:

$$N_{conf} = \tfrac{1}{4} n_{int}(n_{int}+1)\, n_{ext}(n_{ext}+1) + n_{int} \cdot n_{ext}$$

where n_{int} and n_{ext} are the sizes of the internal and external orbital sets, respectively. A recent calculation on the water dimer had $n_{int} = 8$ and $n_{ext} = 58$. This gave $N_{conf} = 62,060$ and $N_{CI} = 56,268$. Approximately 1200 kbytes of IBM processor storage was used in this case. On the other hand, calculations with N_{CI} around 10,000 can be done with a storage size of less than 500 kbytes (the storage requirements can be reduced further if multiple passes over the integral list are allowed). Timing is another bottleneck in all types of CI calculations. The CIMI method minimizes the number of necessary multiplications needed to build the Hamiltonian matrix, as the coupling coefficients A in (4) can be factorized out of matrix elements of similar type. It is, however, necessary to construct the nonzero elements of the Hamiltonian matrix in every iteration. The total computation time will, therefore, be proportional to the number of iterations and it becomes important to use quickly convergent iteration procedures. Using the variation perturbation technique described earlier typically 6–8 iterations are necessary to converge to around 10^{-6} a.u. in the correltion energy for a closed-shell case with one dominant configuration. Figure 3 shows how the CPU time for one iteration in the MOLECULE-CICS program varies as a function of the number of CSFs. The I/O time is very short (CPU efficiency around 90% in typical cases) and significantly differs on this point from conventional CI methods. We conclude, on the basis of the data for core space and timing, that CIMI calculations with up to 10^4 CSFs require only a moderate amount of computer time and can be done with a limited core

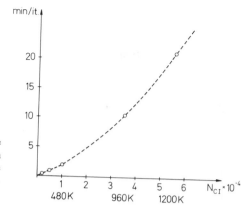

Fig. 3. Timing for the CICS program. The figure shows the CPU time for one iteration (computer: IBM 360/91) as a function of the number of configurations. Core storage requirements (in bytes) are also indicated.

space. On the large computers available today a CI expansion of 10^5 CSFs is possible and would not be prohibitively time consuming.

What accuracy can we expect to obtain with a CI expansion of the closed-shell type? One limitation is related to the incorrect dissociation behavior of the closed-shell determinant. At large internuclear separations several configurations have comparable weights and an ideal CI expansion would include single and double replacements with respect to all these configurations. The missing configurations corresponding to higher-order replacements can thus be expected to be strongly structure dependent. The possibility of using the CSSRS expansion in calculations of energy hypersurfaces is, therefore, restricted to cases which dissociate correctly in the HF approximation. This means essentially that only interactions between closed-shell systems can be studied. Calculations of portions of the energy hypersurfaces close to the equilibrium geometry should, however, be less affected by the limitation to a single reference state, since the additional configurations usually become important only at larger internuclear separations. Such calculations have been performed on a number of different molecules using a variety of Gaussian basis sets. Some of the results are collected in Table 7. The importance of an appropriate basis-set choice is clear from these results. Calculations not including polarization functions will, in general, not yield reliable equilibrium distances, as is illustrated by the calculations on H_2O and HF. These calculations were made with a basis set of essentially "double zeta" quality. Calculated distances improve considerably when polarization functions are added to this basis set and differ in all cases, except acetylene, less than 0.005 Å from the experimental values. This is very gratifying since very large basis sets cannot be used for other than small molecular systems. In general, we expect the single reference state approximation to give a better description of the near-equilibrium energy hypersurface for single bonded systems than for multiple bonds, as illustrated by the results on acetylene. The additional π bonds which are present in the latter case are usually weaker than the σ bond, and

Table 7. Equilibrium Geometries Obtained in CI Calculations Using a CSSRS Expansion

Molecule	Basis set[a]	Coordinate	SCF results[b]	CI results[c]	Observed	Reference
NH_3	⟨N/4, 2, 1/H/3, 1⟩	d(NH)	1.004	1.016	1.012	
		∢HNH	105.5°	102.5°	106.7°	46
H_sO	⟨O/5, 3/H/2⟩	d(OH)	0.954	0.979	0.957	48
		∢(HOH)	112.4°	112.4°	104.5°	
H_2O	⟨O/5, 4, 1/H/3, 1⟩	d(OH)	0.944	0.960	0.957	
		∢HOH	105.3°	103.8°	104.5°	43
H_3O^+	⟨O/5, 4, 1/H/3, 1⟩	d(OH)	0.959	0.972	—	
		∢HOH	113.5°	111.6°	—	43
HCN	⟨N, C/6, 3, 1/H/3, 1⟩	d(HC)	1.063	1.068	1.066	
		d(CN)	1.137	1.153	1.153	49
CH_4	⟨C/4, 2, 1/H/2, 1⟩	d(CH)	1.082	1.092	1.088	47
HF	⟨F/4, 2/H/2, 1⟩	d(FH)	0.916	0.932	0.917	45
C_2H_2	⟨C/4, 2, 1/H/2, 1⟩	d(CC)	—	1.211	1.207	
		d(CH)	—	1.074	1.061	50
C_2H_4	⟨C/4, 2, 1/H/2, 1⟩	d(CC)	—	1.337	1.337	
		d(CH)	—	1.091	1.086	
		∢HCH	—	117.2°	117.3°	50

[a] A basis set of contracted Gaussian functions is used in all cases. Only the contracted set is given.
[b] Distances in angströms.
[c] The expansion includes all single and double replacement CSFs, except from the core orbitals.

configurations containing antibonding molecular orbitals of π type can therefore be expected to have a larger weight than the corresponding σ-antibonding configurations. Higher-order replacement states, therefore, become more important in this case.

Another question is whether the results shown in Table 7 will persist if even larger basis sets are used. Some indication for this being the case is given by the results obtained by Meyer for the first and second row hydrides,[41] using extended basis sets. Errors in calculated equilibrium distances were in all cases found to be less than 0.005 Å. These results were obtained with the PNO–CI method, which is essentially parallel to the CSSRS expansion. Inclusion of higher-order replacements in an approximate way diminished the error to around 0.002Å.

The large basis-set calculations of Meyer also gives an indication of how much correlation energy it is possible to obtain with a CSSRS expansion. Thus for H_2O 90% of the empirically estimated value was obtained,[42] which is probably close to the limiting value. Examples of correlation energies obtained with smaller basis sets on the ammonia molecule are given in Table 8. The importance of polarization functions is again clear. Less than 50% of the correlation energy is obtained, if they are not included. A CI expansion based on a "double zeta plus polarization" basis yields, on the other hand, around 70% of the valence electron correlation energy. Polarization functions describe angular correlation, which is an important and structure-dependent part of the total correlation energy. The inclusion of such functions in the basis set is therefore essential if an accurate description of the chemical bond is desired. Further extensions of the basis set will only slowly improve the result.

Correlation effect on interaction energies between closed-shell systems are usually small in magnitude. Typical effects are induced dipole–dipole interactions (van der Waals interactions) and charge delocalization. The first effect increases the magnitude of the correlation energy and consequently increases the binding energy. The interaction energy is dominated by this term in the case of weakly interacting closed-shell molecules. The second effect usually leads to a decrease in magnitude of the correlation energy caused by the effective charge seen by the electrons. A typical example is H_3O^+. A recent calculation,[43] using the CSSRS expansion, predicted the correlation energy for this ion to be 1.5 kcal/mol smaller in magnitude than for the water molecule.

Changes in the intramolecular correlation energy can also give important contributions to the interaction energy between closed-shell systems. Such changes can be caused by polarization or by structural changes (e.g., changes in bond distances). CSSRS calculations on the water dimer give an illustration of the importance of polarization and dispersion effects on the binding energy. The correlation energy was found to increase the binding energy with 0.9 kcal/mol (from 5.14 to 6.05 kcal/mol), or 18% in this case.[43] A considerable shortening of the oxygen–oxygen bond from 2.99 to 2.92 Å was also observed (a linear hydrogen bond structure was assumed in these calculations). Strongly interacting systems ($\Delta E > 10$ kcal/mol) will in some cases undergo structural changes which in turn influence the correlation energy. The incorrect behavior of a closed-shell Hartree–Fock wave function at dissociation limits will, for example, give rise to appreciable correlation effects when bond distances are changed. The interaction between the ion OH^- and H_2O gives an illustration of this effect. This system has recently been studied both on the SCF and the CI level.[44] The SCF calculation of the equilibrium geometry of the system $[H_a-O_a-H_bO_bH_c]^-$ gave a value of 1.02 Å for the distance R (H_bO_b)

Table 8. Correlation Energy for NH_3 Obtained with a CSSRS Expansion for Different Basis Sets

Basis set[a]	Number of external orbitals	Number of CSFs	Correlation energy,[b] a.u.
$\langle N/4, 2/H/2\rangle$	11	875	-0.1328
$\langle N/5, 3/H/3\rangle$	17	2,271	-0.1424
$\langle N/4, 2, 1/2, 1\rangle$	25	2,690	-0.1909^c
$\langle N/5, 3, 1/3, 1\rangle$	32	4,335	-0.2033^c
$\langle N/7, 6, 1/2, 1\rangle$	40	10,800	-0.2428

[a] Contracted Gaussian functions. The primitive basis set is $(N/10, 6, 1/H/5, 1)$.
[b] Estimated exact value is -0.328 a.u.
[c] $1s$ core electrons frozen.

Table 9. Binding Energies for Some Closed-Shell Interacting Systems[a]

System	SCF	CI	Total[b]	Reference	Experimental
$H^+—OH_2^c$	174.3	172.8	167.5	43	166 ± 2
$Li^+—OH_2^c$	36.1	34.9	32.9	43	34.0
$F^-—H_2O^c$	24.2	26.2	23.0	43	23.3
$H_2O—H_2O^c$	5.1	6.1	5.1	43	~5
$HO^-—H_2O^c$	24.5	28.1	—	44	25
$HO^-—H_2O^d$	34	39	—	45	25
$F^-—HF^d$	53	56	—	45	37

[a] Units are kcal/mol.
[b] Including a correction for the change in zero-point vibrational energy.
[c] Basis set: $\langle O, F/5, 4, 1 \rangle$, $\langle Li/5, 2 \rangle$, $\langle H/3, 1 \rangle$.
[d] Basis set: $\langle O, F/4,2 \rangle$, $\langle H/2, 1 \rangle$.

which is considerably longer than the value 0.94 Å obtained in an equivalent calculation on a free water molecule.[43] This is a considerable weakening of the $O_b H_b$ bond and we can expect that the correlation energy will lead to a further weakening, due to increasing near degeneracy effects leading to a larger hydrogen bond energy, an even longer distance $R(H_b O_b)$ and to a decrease in the proton transfer barrier. Such results were also obtained on the CI level. Thus $R(H_b O_b)$ was computed to be 1.08 Å. The binding energy which on the SCF level of approximation was calculated to be 24.5 kcal/mol, increased to 28.1 kcal/mol when correlation energy was taken into account. The proton transfer barrier simultaneously decreased from 1.4 to 0.2 kcal/mol.

A compilation of some calculated values for interaction energies between closed-shell systems is given in Table 9. The results obtained for $H_3 O_2^-$ and FHF$^-$ with the smaller basis set (essentially "double zeta") illustrate the importance of choosing an appropriate basis set for calculations of this type. Diffuse functions have to be used in order to obtain an adequate description of the negative ions F$^-$ and OH$^-$. If such basis functions are not added an excessively large binding energy will be obtained.

4.2. Calculations Based on Expansion in Single Determinants

An important advantage with the OSSRS expansion as compared to the CSSRS expansion is the higher flexibility of the reference state which is introduced by allowing different space orbitals for different spins. With this type of expansion it becomes possible to use the CIMI method for studies of open-shell systems like radicals, etc. A part of the energy surface for the reaction $H + H_2O$ has recently been studied with the CISD program.[48] It is, however, also possible to use this method for closed-shell systems in cases where the RHF approximation corresponds to a saddle point instead of a true minimum. This will, for example, occur in dissociation processes. To illustrate the usefulness of the OSSRS expansion in this case we have carried out a set of comparative calculations on the HF molecule. The atomic basis set used consisted of 11s, 6p, and 1d Gaussian functions contracted to 5s, 3p, and 1d for fluorine and a 5s, 1p set contracted to 3s 1p for hydrogen.[53] The fluorine d exponent was chosen to be 1.1 and the hydrogen p exponent to be 0.7, which is optimal for H_2. The results of these calculations are illustrated in Fig. 4, which shows the dissociation curve in four different approximations. The two upper curves have been obtained in the *restricted* (RHF) and the *unrestricted* (UHF) Hartree–Fock approximations. The different space orbitals give enough flexibility to allow the UHF curve to dissociate properly into a hydrogen atom and a spin-polarized fluorine atom. The characteristic breakdown of the RHF

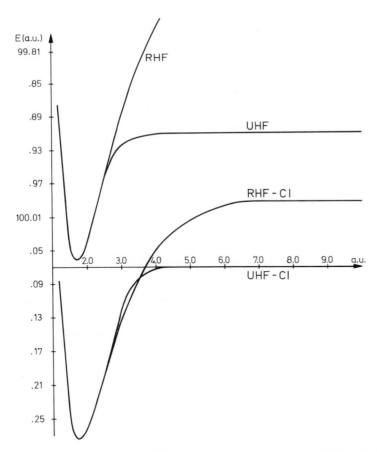

Fig. 4. Potential curves for HF calculated with the restricted (RHF) and unrestricted (UHF) methods and with CI expansions based on either an RHF or an UHF reference state (energies in atomic units).

approximation starts occurring at around 1.35 Å, corresponding to an energy of about 2.7 eV above the energy minimum, which occurs at a bond distance of 0.90 Å. The lowest vibrational frequencies will thus not be affected by the erratic behavior of the RHF curve. Outside 1.35 Å, the UHF solution yields more and more localized orbitals whereas the RHF orbitals remain delocalized. This is the usual behavior of the RHF solution. In fact very few systems dissociate to well-defined species in the RHF approximation. The system $H^+ + F^-$ represents a self-consistent solution to the RHF equations, but there also exists a solution which corresponds to a mixture of different ionic states and which has a lower energy. This solution is closely related to the first of the two closed-shell determinants which are needed for a proper description of the dissociated system.

The two lower curves in Fig. 4 are obtained from calculations with all single and double replacements out of the RHF and UHF determinants, respectively. Although somewhat surprising, the incorrect dissociation of the RHF approximation still persists in the RHF–CI curve, in spite of the fact that the two configurations needed for correct dissociation of a delocalized bond are included in the CSSRS expansion. The hydrogen molecule obviously dissociates correctly in the CSSRS approximation. We can use first-order perturbation theory in order to obtain a qualitative understanding of the RHF–CI curve in Fig. 4. A qualitatively correct potential curve for a single bonded system like HF is obtained with a two-configurational wave function of the type

$$\Phi_0 = C^I[\sigma\sigma] + C^{II}[\sigma^*\sigma^*] = C^I\Phi^I + C^{II}\Phi^{II} \tag{45}$$

where σ is the bonding and σ^* the corresponding antibonding molecular orbital (leaving out all equal orbitals). At infinite distances the coefficients take the values $C^I = -C^{II} = 1/\sqrt{2}$ and the wave function (45) then becomes equivalent to a localized description with one of the electrons on H and the other on F. In order to describe the correlation of the remaining electrons, all single- and double-replacement CSFs with respect to the two reference configurations should be added. Concentrating on the most important double replacements we can write the resulting wave function (at infinite HF distance) as

$$\Psi = \frac{1}{\sqrt{2}}(\Phi^I - \Phi^{II}) + \sum_{ij,ab} C^I_{ij\to ab}\Phi^I_{ij\to ab} + \sum_{ij,ab} C^{II}_{ij\to ab}\Phi^{II}_{ij\to ab}$$

where $\Phi^I_{ij\to ab}$ and $\Phi^{II}_{ij\to ab}$ denotes double-replacement states with respect to Φ^I and Φ^{II}. Since

$$\langle\Phi_0|\hat{H}|\Phi^I_{ij\to ab}\rangle = \langle\Phi_0|\hat{H}|\Phi^{II}_{ij\to ab}\rangle$$

we obtain, to first order in perturbation theory,

$$C^I_{ij\to ab} = C^{II}_{ij\to ab} = C_{ij\to ab}$$

Notice that Φ^I and Φ^{II} as well as the corresponding pairs of replacement states become degenerate at infinity. The total energy, to second order, is given by

$$E = \langle \Phi_0 | H | \Psi \rangle = E_0 + 2 \sum_{ij,ab} C^I_{ij \to ab} \langle \Phi_0 | H | \Phi^I_{ij \to ab} \rangle = E_0 + E_{corr} \qquad (47)$$

where the second term represents the correlation energy. It is clear from this expression that a CI expansion of the CSSRS type with double replacements form Φ^I only, will give an energy of $E_0 + \frac{1}{2} E_{corr}$. This is nicely illustrated in Fig. 4 where the RHF–CI curve approaches the midpoint between the UHF and UHF–CI curve when the distance becomes infinite.

In conclusion, the RHF–CI approach seems, in general, to be valid only in a region less than 0.5 Å away from the equilibrium distance, whereas the UHF–CI method seems to be able to generate quite useful potential curves for chemical reactions. The calculation on HF presented here thus yields an equilibrium r_e value of 0.92 Å and a dissociation energy $D_e = 5.6$ eV. The experimental values are 0.917 Å and 6.1 eV. On the UHF level these quantities were calculated to be 0.90 Å and 4.1 eV. Experimentally, the difference in correlation energy between F and HF can be estimated to 1.6 eV. Most of this structure-dependent part of the correlation energy is thus recovered in the UHF–CI calculation.

4.3. Three-Electron Complete CI Calculations

A recent study of the full potential surface for the H_3 system [19] will serve as an example of a calculation made with the CI3 program. This calculation is also illustrative of the type of reasoning which lies behind the final choice of the one-and many-particle basis sets.

The energy surface for linear H_3 had earlier been computed to high accuracy,[54] The absolute errors in the energies computed in this study were smaller than 1 kcal/mole with a much higher relative accuracy. The purpose of the CI3 calculations was to also predict the nonlinear part of the surface to a similar relative accuracy, with a reasonable effort in terms of computer time. The one-particle basis used for the linear case consisted of four s-, three p-, and two d-type functions of Slater type. For the nonlinear system, contracted Gaussian functions, CGTOs, were chosen since the integral evaluation is much more efficient for such functions with the computer programs presently available.

Comparative calculations on the hydrogen molecule with several different basis sets were made, in order to investigate possible truncations which could be made without any significant loss of accuracy. Deletion of one set of p functions (two instead of three) was found to cause a loss of accuracy in the (numerically computed) vibrational frequencies of the order of 10 cm^{-1}. The corresponding change obtained with one set of d functions deleted was around $1\ cm^{-1}$ This was considered more acceptable and four s-, three p-, and one d-type CGTOs were consequently chosen as the one-particle basis set.

The first step in a determination of an appropriate many-particle basis was to investigate the possibility of deleting molecular orbitals. For this purpose approximate natural orbitals were determined, both by perturbation theory and from a limited CI expansion. The latter orbitals were found to be slightly superior. Truncations in this orbital basis were made by only keeping orbitals with occupation numbers larger than specified thresholds. Test calculations at the transition state of H_3 led to a final choice of $30a'$ and $13a''$ orbitals. Complete CI calculations were then made in this space, using the cɪ3 program. Several points along the minimal energy path (which is linear) were calculated in order to finally check the validity of the truncations against the larger STO calculations. The two sets of calculations give minimal energy paths which are parallel to around 0.01 kcal/mole in a region of at least two atomic units around the barrier.

The timing (CPU time on an IBM 360/195) for each step of the calculation on a representative nonlinear point on the energy surface is as follows:

1.	Integral evaluation	5.72 min
2.	SCF calculation	3.48 min
3.	Full transformation	7.78 min
4.	Limited CI	5.37 min
5.	Truncated transformation	4.52 min
6.	Complete CI	19.45 min

This adds up to a total time of 46.32 minutes for each point on the surface. Around one hundred points had to be calculated in order to uniquely define the most important part of the surface. A detailed discussion of the results obtained in this study of H_3 will be presented elsewhere,[19] The cɪ3 program has also been used for studies of the reaction $H + Li_2$.[55]

4.4. Second-Order Properties

Experimentally, molecules are studied by observing their response to external perturbations, for example, electric and magnetic fields. The properties of a molecule are divided into classes according to which order of perturbation theory the corresponding operator will make a contribution to the energy. First-order properties are consequently calculated simply by taking the expectation value of the operator over the undisturbed wave function. For a one-electron operator h, as the dipole moment, this is achieved by a direct multiplication of the first-order density matrix D_{pq} and the corresponding one-electron integral matrix J_{pq} in the same basis

$$\langle h \rangle = \sum_{p,q} D_{pq} I_{pq} \tag{48}$$

For a first-order N-electron property the corresponding procedure will require the Nth order reduced density matrix, and is consequently much more difficult to set up. The second-order properties are hard to calculate for another reason. They will, by definition, require the calculation of a disturbed wave function. Such a calculation will, in general, be as complicated as the calculation of the correlated wave function itself. The first-order wave function is, according to (8), defined through the differential equation

$$(\hat{H}_0 - E_0)|\psi^{(1)}\rangle = (E_1 - \hat{V})|\psi^{(0)}\rangle \tag{49}$$

By the usual expansion of $|\psi^{(0)}\rangle$ and $|\psi^{(1)}\rangle$ in a configuration basis

$$|\psi^{(0)}\rangle = \sum_k C_k^{(0)}|\Phi_k\rangle$$

$$\tag{50}$$

$$|\psi^{(1)}\rangle = \sum_l C_l^{(1)}|\Phi_l'\rangle$$

where $|\Phi_l'\rangle$ is orthogonalized to $|\psi^{(0)}\rangle$ by

$$|\Phi_l'\rangle = (1 - \hat{P})|\Phi_l \tag{51}$$

with $\hat{P} = |\psi^{(0)}\rangle\langle\psi^{(0)}|$, and multiplying from the left by $\langle\Phi_m'|$ we transform the differential equation to an inhomogeneous equation system

$$\sum_l C_l^{(1)}[\langle\Phi_m'|\hat{H}_o|\Phi_l'\rangle - E_o(\delta_{lm} - C_l^{(0)}C_m^{(0)})] = -\langle\Phi_m'|\hat{V}|\psi^{(0)}\rangle \tag{52}$$

The correlated undisturbed wave function $|\psi^{(0)}\rangle$ will be an exact eigenfunction to $\hat{H}_o$ with the correlated energy as the eigenvalue if we define

$$\hat{H}_o = \hat{P}\hat{H}\hat{P} + (1 - \hat{P})\hat{H}(1 - \hat{P}) \tag{53}$$

with the projection operator $\hat{P}$ defined as above. $\hat{H}$ is the undisturbed total Hamiltonian. The matrix elements of $\hat{H}_o$ in the basis $|\Phi'\rangle$ are related to the matrix elements of the undisturbed Hamiltonian $\hat{H}$ in the basis $|\Phi\rangle$ by

$$\langle\Phi_m'|\hat{H}_o|\Phi_l'\rangle = \langle\Phi_m'|\hat{H}|\Phi_l'\rangle = \langle\Phi_m|\hat{H}|\Phi_l\rangle - E_0 C_l^{(0)} C_m^{(0)} \tag{54}$$

The perturbation operator $\hat{V}$ contains besides the real perturbation $\hat{v}$ the additional terms of $\hat{H}$ not included in $\hat{H}_0$ so that

$$\hat{V} = \hat{P}\hat{H} + \hat{H}\hat{P} - 2\hat{P}\hat{H}\hat{P} + \hat{v} \tag{55}$$

The first three operators on the right-hand side will, however, not have any matrix elements between the basis sets $|\Phi'\rangle$ and $|\Phi\rangle$. The equation system will therefore take the simple form

$$\sum_l C_l^{(1)}[\langle \Phi_m|\hat{H}|\Phi_l\rangle - E_0\,\delta_{lm}] = -\sum_k C_k^{(0)}[\langle \Phi_m|\hat{v}|\Phi_k\rangle - E_1\delta_{km}] \tag{56}$$

The same equation system would have been obtained without the orthogonalization (27), which shows that $\psi^{(1)}$ is not entirely unique. The desired property, however, is defined through the second-order energy contribution, which for both basis sets is given by

$$E_2 = \sum_k \sum_l C_k^{(0)}C_l^{(1)}(\langle \Phi_k|\hat{v}|\Phi_l\rangle - E_1\,\delta_{kl}) \tag{57}$$

where $|\psi^{(0)}\rangle$ is assumed as normalized.

The set of equations (56) can be solved by similar iteration schemes as were described in the section diagonalization procedures. Computationally, solving (56) again basically means construction of a vector $\sigma^{(k)}$ in iteration k, according to (3), and is therefore well suited to be handled by the CIMI method. The vector σ, however, has to be modified by the terms on the right-hand side, which depends on the pertubation operator but not on the undetermined coefficients $C_l^{(1)}$. These terms can be computed once and for all in the beginning of the calculation and stored.

The procedure described above has been used for the calculation of the nuclear spin–spin coupling constants observed in the splitting of NMR lines. Three different physical effects are responsible for this splitting: the electron orbital effect, the dipolar interaction between the electron and nuclear spin, and the Fermi contact interaction. For coupling between protons the latter effect is strongly dominant and is described by the operator

$$\hat{v} = \frac{16\pi\beta\hbar}{3}\sum_N \gamma_N \sum_k \delta(r_{kN})\hat{S}_k \cdot \hat{I}_N \tag{58}$$

where $\delta(r_{kN})$ is the Dirac delta function at nucleus N, $\hat{S}_k$ denotes the spin of electron k, and $\hat{I}_N$ the spin of nucleus N. The nuclear spin–spin coupling appears in the second-order energy as a contribution of the type

$$E'_{NN} = hJ'_{NN}\hat{I}_N \cdot \hat{I}'_N \tag{59}$$

where J'_{NN} is called the coupling constant between nucleus N and N'. The operator $\hat{v}$, according to (58), couples the singlet wave function $|\psi^{(0)}\rangle$ to triplet configurations $|\Phi'_m\rangle$, according to (52), which will require matrix elements of the undisturbed Hamiltonian over triplet configurations. As the MOLECULE-CICS program is restricted to singlet CSFs the CISD program has to be used. This program, however, has no explicit spin projection, but will automatically yield a correct triplet configuration in the space used. This means that for a double replacement from a closed-shell reference state we have to use all *six* possible determinants in order to obtain $|\psi^{(1)}\rangle$ as a true spin eigenfunction.

The CIMI method for nuclear spin–spin coupling has been applied to some smaller molecules. The calculation on H_2[57] with extended basis sets has an accuracy of a few percent for the coupling constant and is illustrative of the accuracy that can be obtained with this approach. For the H_2O molecule fairly nice agreement with experiments was also reached, whereas the results for NH_3, CH_4, C_2H_4, and C_2H_4 were more disappointing.[58] The difficulty in describing the cusp of the wave function at the nuclear center with Gaussian functions seems to be reponsible for this problem. Various ways of obtaining a better stability of the electron density at the nuclei are now being investigated. One way might be the use of Slater functions which have the right cusp behavior.

5. Concluding Remarks

We have, in this contribution, discussed in some detail the characteristics of the direct configuration interaction method. The merits of this method lie primarily in the possibility to use very long CI expansions, and thus obtain very accurate wave functions. The method is most easily adapted in the case of a single-determinant reference state including the full first-order subspace in the CI expansion. Examples have been given with a RHF and UHF reference state, where it has been shown that the direct method leads to very efficient algorithms for the determination of the CI coefficients.

The main drawback of the method is the lack of generality. It is not possible today to see how it could be generalized to arbitrary open-shell cases and to a more general selection of configurations. Special and very useful extensions are, however, possible. One rather straightforward way to generalize the method is to allow the reference state to be a linear combination of closed-shell determinants and include the first-order subspace with respect to all these determinants. Such a wave function gives a correct description of singlet states in which the electrons can be pairwise coupled to singlets. It then becomes possible to treat properly, for example, dissociation of single bonds and excited singlet states. Also the case of one odd electron outside a closed shell can be treated by means of a virtual coupling to a noninteracting electron. A computer program which can use this more general type of CI expansion is presently being developed. Future work on the CIMI method will also be related to the problem of including open-shell configurations in the reference state.

References

1. S. F. Boys, Electronic wave functions, I. A general method of calculation for the stationary states of any molecular system, *Proc. R. Soc. London, Ser. A* **200**, 542–554 (1950).

2. S. F. Boys, Electronic wave functions. II. A calculation for the ground state of the beryllium atom, *Proc. R. Soc. London, Ser. A* **201**, 125–137 (1950).

3. D. R. Hartree, W. Hartree, and B. Swirles, Self-consistent field, including exchange and superposition of configurations with some results for oxygen, *Philos. Trans. R. Soc. London, Ser. A* **238**, 229–247 (1939).

4. E. A. Hylleraas, Neue Berechnung der Energie des Heliums im Grundzustande, sowie des tiefsten Terms von Ortho-Helium, *Z. Phys.* **54**, 347–366 (1929).

5. A. P. Jucys, On the Hartree–Fock method in multi-configuration approximation, *Adv. Chem. Phys.* **14**, 191–206 (1969) (atoms).

6. G. Das and A. C. Wahl, Extended Hartree–Fock wavefunctions: Optimized valence configurations for H_2 and Li_2, optimized double configurations for F_2, *J. Chem. Phys.* **44**, 87–96 (1966) (diatomic molecules).

7. T. L. Gilbert, Multiconfigurational self-consistent-field theory for localized orbitals. I. The orbital equation, *Phys. Rev. A* **6**, 580–600 (1972). A thorough discussion of the subject can also be found in Ref. 30.

8. B. Roos, A new method for large scale CI calculations, *Chem. Phys. Lett.* **15**, 153 (1972).

9. B. Roos and P. Siegbahn, in: *Chemical and Biochemical Reactivity, the Jerusalem Symposia on Quantum Chemistry and Biochemistry, VI*, The Israel Academy of Sciences and Humanities, Jerusalem (1974).

10. J. Almlöf, B. Roos, and P. Siegbahn, the Program System MOLECULE, Reference Manual (to be published).

11. J. K. L. MacDonald, Successive approximations by the Rayleigh–Ritz variation method, *Phys. Rev.* **43**, 830–833 (1933).

12. H. J. Silverstone and O. Sinanoğlu, Many-electron theory of nonclosed-shell atoms and molecules. I. Orbital wave function and perturbation theory, *J. Chem. Phys.* **44**, 1899–1907 (1966); II. Variational theory, *J. Chem. Phys.* **44**, 3608–3617 (1966).

13. P.-O. Löwdin, Quantum theory of many-particle systems. I. Physical interpretations by means of density matrices, natural spin-orbitals, and convergence problems in the method of configurational interaction, *Phys. Rev.* **97**, 1474–1489 (1955).

14. A. J. Coleman, Structure of fermion density matrices, *Rev. Mod. Phys.* **35**, 668–689 (1963).

15. A. D. McLean and B. Liu, Classification of configurations and the determination of interacting and noninteracting spaces in configuration interaction, *J. Chem. Phys.* **58**, 1066–1078 (1973).

16. O. Sinanoğlu, Many-electron theory of atoms and molecules. I. Shells, electron pairs versus many-electron correlations, *J. Chem. Phys.* **36**, 706 (1962).

17. W. Meyer, PNO–CI studies of electron correlation effects. I. Configuration expansion by means of nonorthogonal orbitals, and application to the ground state and ionized states of methane, *J. Chem. Phys.* **58**, 1017–1035 (1973).

18. R. McWeeny and B. T. Sutcliffe, *Methods of Molecular Quantum Mechanics*, Academic Press, London (1969).

19. B. Liu and P. Siegbahn, An accurate three-dimensional potential surface for H_3, to be published.

20. C. Bender, private communication.

21. R. K. Nesbet, Algorithm for diagonalization of large matrices, *J. Chem. Phys.* **43**, 311–312 (1965).

22. I. Shavitt, C. F. Bender, A. Pipano, and R. P. Hosteny, The iterative calculation of several of the lowest or highest eigenvalues and corresponding eigenvectors of very large symmetric matrices, *J. Comput. Phys.* **11**, 90–108 (1973).

23. P.-O. Löwdin, Studies in perturbation theory IX. Upper bounds to energy eigenvalues in Schrödinger's perturbation theory, *J. Math. Phys.* **6**, 1341–1353 (1965).

24. E. Brändas and O. Goscinski, Variation-perturbation expansions and Padé-approximants to the energy, *Phys. Rev. A* **1**, 552–560 (1970).

25. O. Goscinski and E. Brändas, in: *Theory of Electronic Shells of Atoms and Molecules*, Report from the Vilnius International Symposium, 1969, Mintis, Vilnius (1971).

26. R. J. Bartlett and E. J. Brändas, Reduced partitioning procedure in configuration interaction studies. II. Excited states, *J. Chem. Phys.* **59**, 2032–2042 (1973).
27. E. R. Davidson, The iterative calculation of a few of the lowest eigenvalues and corresponding eigenvectors of large real-symmetric matrices, *J. Comput. Phys.* **17**, 87–94 (1975).
28. V. R. Saunders and I. H. Hillier, A "level-shifting" method for converging closed shell Hartree–Fock wave functions, *Int. J. Quantum Chem.* **7**, 699–705 (1973).
29. V. R. Saunders and M. F. Guest, in: *Proceedings of SRC Atlas Symposium No. 4, "Quantum Chemistry—the State of the Art"* (V. R. Saunders and J. Brown, eds.), Atlas Computer Laboratory, Chilton Didcot, Oxfordshire (1975).
30. J. Hinze, MC-SCF. I. The multi-configurational self-consistent-field method, *J. Chem. Phys.* **59**, 6424–6432 (1973).
31. P. S. Bagus, B. Liu, A. D. McLean, and M. Yoshimine, in: *Wave Mechanics the First Fifty Years* (W. C. Price, S. S. Chissik, and T. Ravensdale, eds.), pp. 88–98, Butterworth & Co., London (1973).
32. C. Edminston and M. Krauss, Pseudonatural orbitals as a basis for the superposition of configurations. I. He_2^+, *J. Chem. Phys.* **45**, 1833–1839 (1966).
33. W. P. Reinhardt and J. D. Doll, Direct calculation of natural orbitals by many-body perturbation theory: Application to helium, *J. Chem. Phys.* **50**, 2767–2768 (1969).
34. P. Dejardin, E. Kochanski, A. Veillard, B. Roos, and P. Siegbahn, MC–SCF and CI calculations for the ammonia molecule, *J. Chem. Phys.* **59**, 5546–5553 (1973).
35. P. Bertoncini and A. C. Wahl, Ab initio calculation of the helium–helium $^1\Sigma_g^+$ potential at intermediate and large separations, *Phys. Rev. Lett.* **25**, 991–994 (1970).
36. H. F. Schaefer, D. R. McLaughlin, F. E. Harris, and B. J. Alder, Calculation of the attractive He pair potential, *Phys. Rev. Lett.* **25**, 988–990 (1970).
37. M. Yoshimine, The use of direct access devices in problems requiring the reordering of long data lists, IBM Corp. Tech. Rep. RJ 555 (1973).
38. F. E. Harris and H. H. Michels, Open-shell valence configuration-interaction studies of diatomic and polyatomic molecules, *Int. J. Quantum Chem.* **1S**, 329–338 (1967).
39. H. F. Schaefer and F. E. Harris, Ab initio calculations on 62 low-lying states of the O_2 molecule, *J. Chem. Phys.* **48**, 4946–4955 (1968).
40. P. Siegbahn. in: *Proceedings of SRC Atlas Symposium No. 4, "Quantum Chemistry—the State of the Art"* (V. R. Saunders and J. Brown, eds.), Atlas Computer Laboratory, Chilton Didcot, Oxfordshire (1975).
41. W. Meyer, in: *Proceedings of SRC Atlas Symposium No. 4, "Quantum Chemistry—the State of the Art"* (V. R. Saunders and J. Brown, eds.), Atlas Computer Laboratory, Chilton Didcot, Oxfordshire (1975).
42. W. Meyer, Ionization energies of water from PNO–CI calculations, *Int. J. Quantum Chem.* **5S**, 341–348 (1971).
43. G. H. F. Diercksen, W. P. Kraemer, and B. Roos, SCF-CI Studies of correlation effects on hydrogen bonding and ion hydration, The systems: H_2O, $H^+ \cdot H_2O$, $Li^+ \cdot H_2O$, $F^- \cdot H_2O$ and $H_2O \cdot H_2O$, *Theor. Chim. Acta* **36**, 249–274 (1975).
44. G. H. F. Diercksen, W. P. Kraemer, and B. Roos, SCF-CI studies of the equilibrium structure and proton transfer barrier in $H_3O_2^-$, *Theor. Chim. Acta* **42**, 77–82 (1976).
45. A. Støgård, A. Strich, J. Almlöf, and B. Roos, Correlation effects on hydrogen-bond potentials. SCF-CI calculations for the systems HF_2^- and $H_3O_2^-$, *Chem. Phys.* **8**, 405–411 (1975).
46. J. Kowalewski and B. Roos, Large configuration interaction calculations of nuclear spin–spin coupling constants. III. Vibrational effects in ammonia, *Chem. Phys.* **11**, 123–128 (1975).
47. K. Niblaeus, private communication.
48. K. Niblaeus, B. Roos, and P. Siegbahn, An UHF-CI study of the H_3O radical, to be published.
49. P. K. Pearson, G. L. Blackman, H. F. Schaefer III, B. Roos, and U. Wahlgren, HNC molecule in interstellar space? Some pertinent theoretical calculations, *Astrophys. J.* **184**, L19–L22 (1973).
50. A. D. McLean and P. Siegbahn, to be published.

51. R. F. Hausman, Jr., S. D. Bloom, and C. F. Bender, A new technique for describing the electronic states of atoms and molecules—the vector method, *Chem. Phys. Lett.* **32**, 483–488 (1975).

52. The original tables published in Ref. 8 were unfortunately incorrect. The authors are grateful to Dr. N. C. Handy for pointing out a number of these errors.

53. F. B. van Duijneveldt, Gaussian basis sets for the atoms H–Ne for use in molecular calculations, IBM Research Report RJ 945 (1971).

54. B. Liu, *Ab initio* potential energy surface for linear H_3, *J. Chem. Phys.* **58**, 1925–1937 (1973).

55. P. Siegbahn and H. F. Schaefer III, Potential energy surface for $H + Li_2 \rightarrow LiH + Li$. I. Ground state surface from large scale configuration interaction, *J. Chem. Phys.* **62**, 3488–3495 (1975).

56. B. Liu and A. D. McLean, Accurate calculation of the attractive interaction of two ground state helium atoms, *J. Chem. Phys.* **59**, 4557–4558 (1973).

57. J. Kowalewski, B. Roos, P. Siegbahn, and R. Vestin, Large configuration interaction calculations of nuclear spin–spin coupling constants. I. HD molecule, *Chem. Phys.* **3**, 70–77 (1974).

58. J. Kowalewski, B. Roos, P. Siegbahn, and R. Vestin. Large configuration interaction calculations of nuclear spin–spin coupling constants. II. Some polyatomic molecules, *Chem. Phys.* **9**, 29–39 (1975).

8

A New Method for Determining Configuration Interaction Wave Functions for the Electronic States of Atoms and Molecules: The Vector Method

R. F. Hausman, Jr.
and
C. F. Bender

1. Introduction

At least one procedure has been developed for accurately describing the ground and excited electronic states of atoms and molecules. The method, termed configuration interaction (CI), uses expansion techniques for correcting the SCF wave function ϕ_o,

$$\psi = C_o\phi_o + \sum_{i=1}^{N} C_i\phi_i \equiv \mathbf{C}^\dagger\boldsymbol{\phi} \tag{1}$$

The correction functions, or configurations, ϕ_i are usually linear combinations of Slater determinants. The expansion coefficients are determined by the

R. F. Hausman, Jr. and C. F. Bender • Lawrence Livermore Laboratory, University of California, Livermore, California

variation principle which requires solution of an eigenvalue problem

$$H_{op}\psi = E\psi \tag{2}$$

Standard CI methods determine the coefficients c_i by simply diagonalizing the "interaction" or Hamiltonian matrix **H**,

$$\mathbf{HC} = \mathbf{EC} \tag{3}$$

The matrix elements of H are

$$(\mathbf{H})_{pq} = \langle \phi_p | H_{op} | \phi_q \rangle \tag{4}$$

This method is computationally simple to implement, and there have been many CI calculations on electronic states of atoms and molecules. The status of such calculations has been recently reviewed,[1] and relative accuracies of 0.1–0.3 eV are attainable today.

For polyatomic molecules, most accurate CI studies have been limited to the ground states of small systems ($20e^-$ and $Z < 10$). The main reason for this limitation is that CI expansions are slow to converge, that is, large numbers (1,000–10,000) of terms are required to obtain the above-mentioned accuracy. The problem arises, not in manipulation of the wave function, but rather in the formation and diagonalization of the interaction matrix. Clearly an N-term wave function will lead to an interaction matrix with N^2 elements. Thus, for reasonable N the complete matrix often cannot be stored in fast computer storage and standard diagonalization methods cannot be used. Low-lying excited states present two additional problems: (1) Methods for extracting higher eigenvalues have not been developed to the same degree as those for the lowest eigenvalue and (2) even more configurations are required to describe the excited states. Studies involving determination of polyatomic potential energy surfaces have been greatly hampered by the requirement of large numbers of energy calculations to adequately describe a surface. Therefore, if CI methods are to be successfully used to study potential energy surfaces, very rapid methods are required for calculating the energy and associated wave function.

Over the past two decades, great progress has been made in solving these problems. The number of terms (configurations) has been reduced using a number of methods. Often certain correlation effects are present throughout the process and their effect on relative energies is minimal, e.g., inner-shell ($1s^2$) correlation corrections. Studies have shown that certain configurations which are highly excited relative to the SCF function are unimportant in the wave function. Thus generation by excitation level with elimination of higher-order excitations can reduce the number of configurations. Recently, more sophisticated configuration enumeration techniques have been developed.[2] Perturbative methods have been used to estimate[3-5] the energy contribution of a configuration, and unimportant configurations can be eliminated. A unique application of perturbative selection has been recently presented by Buenker and Peyerimhoff.[6]

The number of configurations can be indirectly reduced by proper choice of the unoccupied molecular orbitals. The lowest virtual SCF orbitals are sometimes localized at great distances from the occupied orbitals. Thus, the virtual orbitals may be essentially zero where the correlation error in the SCF wave function is greatest. A number of methods have been suggested to "contract" the virtual orbitals.[3,7,8] Other methods seek to estimate the natural orbitals of the system.[9–12] Meyer's method looks very encouraging but involves nonorthogonal orbitals which could have dramatic effects on the determination of the expansion coefficients in Eq. (2). Probably the "best" orbitals are determined by the multiconfiguration self-consistent field[13] (MC–SCF) or the iterative natural orbital[14] (INO) methods. Both methods iteratively determine the optimum orbitals for a CI wave function. Unfortunately there are usually additional computational requirements.

Even with reduced numbers of configurations, the fact remains that the expansion coefficients in Eq. (2) must be rapidly determined. The logistic problems associated with construction, manipulation, and diagonalization of the H matrix in Eq. (2) have been solved in part by theoretical analysis and development of innovative computer algorithms. Rapid methods for computing matrix elements have been developed.[15] Recently two promising methods have appeared[16,17] but have not yet been applied computationally. Fortunately the H matrix is sparse; in some calculations only 5%–10% of the matrix elements are nonzero. Computer techniques have been developed to take advantage of this sparseness. Specifically designed diagonalization procedures have been developed to extract the lowest few eigenvalues.[18–20] Further reductions in computing the many points required for a potential energy surface have been realized by using formula tape or point-by-point configuration selection techniques. In the former case, all the formulas for nonzero Hamiltonian matrix elements are derived and stored compactly on the formula tape. This dramatically reduces the computer requirements for construction of the H matrix for each point on the surface. On the other hand, rapid configuration selection methods have been developed which can be applied before construction of the H matrix. This method could lead to irregular surfaces, depending on the selection criteria.

Recently a new method for large-scale CI calculations was introduced by Roos.[21] The expansion coefficients are obtained directly from the two-electron integrals. Thus the formation, manipulation, and diagonalization of the Hamiltonian matrix was eliminated. This method has been successfully applied to a number of molecular systems. The method has been developed to construct singlet wave functions which include single and double excitations relative to a closed-shell single configuration. In addition a code has been developed[22] to generate full CI wave functions for the three-electron case.

The method is indeed powerful, but there are two notable disadvantages. First, the singlet restriction dramatically limits the systems which can be studied *and* precludes excited states. Second, the formalism required to extend the level of excitation beyond single and double replacements is extremely complicated.

The conventional approach to nuclear shell model calculations has been recently reviewed by McGrory.[23] Here, as in most chemical studies, the Hamiltonian matrix is formed and diagonalized. The second quantized form of the Hamiltonian operator was used by Talman[24] to study ^{10}Ne. An important step in attacking large problems was taken by Whitehead,[25] who used the second quantized representation of the Hamiltonian and an iterative diagonalization technique to produce wave functions involving as many as 100,000 Slater determinants.

The Whitehead approach combined with the impressive computer times attained by Roos has suggested to us a general approach to generating large CI wave functions. Our *vector method*[26] (VM) does not require formation, manipulation, and diagonalization of the Hamiltonian matrix. Wave functions and eigenvalues are generated directly and are constructed from configurations, not necessarily simple Slater determinants. The generality of the VM allows study of any spin state, and wave functions can include any level of excitation relative to any number of configurations. As will be seen later, the computing times depend linearly on the number of terms in the wave function and these times are already competitive with, and in some cases faster than, standard CI methods.

2. The Vector Method

In the vector method the operator equation is solved directly for the expansion coefficients $\mathbf{d}_E$,

$$H_{op}\mathbf{d}_E^+\boldsymbol{\phi} = E\mathbf{d}_E^+\boldsymbol{\phi} \tag{5}$$

In all applications to date, $\boldsymbol{\phi}$ has been a set of Slater determinants; later in this paper we will show that linear combinations of Slater determinants can be used instead. The key step in solving Eq. (5) is replacement of the normally used nonrelativistic electronic Hamiltonian operator

$$H_{op} = -\sum_i \left(-\tfrac{1}{2}\nabla_i^2 + \sum_A \frac{Z_A}{r_{iA}}\right) + \sum_{i>j} \frac{1}{r_{ij}} \tag{6}$$

with the second quantized or occupation number representation

$$H_{op} = \sum_{\alpha\beta\gamma\delta} \langle\alpha\beta|H|\gamma\delta\rangle a_\alpha^+ a_\beta^+ a_\delta a_\gamma \tag{7}$$

In Eq. (6) A is an atom with nuclear charge Z_A and i, j represent the electrons.

The sum in Eq. (7) runs over all possible orthonormal spin-orbitals, and a_k and a_k^+ are the normal fermion annihilation and creation operators, respectively. They obey the following relationships:

$$[a_\alpha^+, a_\beta^+]_+ = a_\alpha^+ a_\beta^+ + a_\beta^+ a_\alpha^+ = 0 \tag{8a}$$

$$[a_\alpha, a_\beta]_+ = 0 \tag{8b}$$

$$[a_\alpha, a_\beta^+]_+ = \delta_{\alpha\beta} \tag{8c}$$

Slater determinants are used because they can be written as a product of N single-particle creation operators,

$$D_p\{\alpha(1)\beta(2)\omega(N)\} = Q_p a_\alpha^+ a_\beta^+ \cdots a_\omega^+|0\rangle \tag{9}$$

where Q_p is a normalization constant and $|0\rangle$ is the vacuum state. The matrix elements in Eq. (7) are

$$\langle\alpha\beta|H|\gamma\delta\rangle = \frac{1}{N-1}\langle\alpha|H_1|\gamma\rangle\langle\beta|\delta\rangle + \tfrac{1}{2}\langle\alpha\beta|H_2|\gamma\delta\rangle \tag{10}$$

where

$$\langle\alpha|H_1|\gamma\rangle = \left\langle\alpha\left|\tfrac{1}{2}\nabla_1^2 - \sum_A \frac{Z_A}{r_{1A}}\right|\gamma\right\rangle \tag{11a}$$

and

$$\langle\alpha\beta|H_2|\gamma\delta\rangle = \left\langle\alpha\beta\left|\frac{1}{r_{12}}\right|\gamma\delta\right\rangle \tag{11b}$$

Note that the above integrals are simply combinations of the integrals required to compute Hamiltonian (interaction) matrix elements.

The equivalence of the two representations can be demonstrated by evaluating matrix elements between Slater determinants. The formulas for Slater determinants which differ by 0, 1, 2 spin-orbitals are identical and matrix elements involving greater differences are zero in both representations.

In practice a different form of the operator is used,

$$H_{op} = \sum_{\substack{\alpha>\beta \\ \delta>\gamma \\ (\alpha\beta)\geq(\gamma\delta)}} \langle\alpha\beta|H'|\gamma\delta\rangle\{a_\alpha^+ a_\beta^+ a_\delta a_\gamma + a_\gamma^+ a_\delta^+ a_\beta a_\alpha\} \tag{12}$$

here $(\alpha\beta)$ denotes the pair $\alpha\beta$. The above form reduces the number of two-body matrix elements and the Hermiticity of the operator can be used. The matrix elements become somewhat more complicated:

$$\langle\alpha\beta|H'|\gamma\delta\rangle = \frac{1}{N-1}\{\langle\alpha|H_1|\gamma\rangle\langle\beta|\delta\rangle + \langle\beta|H_1|\delta\rangle\langle\alpha|\gamma\rangle$$

$$- \langle\alpha|H_1|\delta\rangle\langle\beta|\gamma\rangle - \langle\beta|H_1|\gamma\rangle\langle\alpha|\delta\rangle\}$$

$$+ \langle\alpha\beta|H_2|\gamma\delta\rangle - \langle\alpha\beta|H_2|\delta\gamma\rangle \tag{13}$$

The operator Eq. (5) is solved in an iterative fashion. For an approximate wave function ψ with corresponding energy E,

$$H_{op}\psi = E\psi + \phi \tag{14}$$

The wave function is then corrected;

$$\psi' = \psi + \lambda\phi \tag{15}$$

when the correction λ is selected to guarantee convergence to one of the eigenvalues.

Iterative diagonalization methods are ideally suited to sequentially correct an approximate wave function and can be applied to matrices *or* operators. In the operator case, matrix multiplication is replaced by the operator rules. While all eigenvalue algorithms are in some sense iterative, we shall use the term *iterative method* to denote those algorithms which do not entail modification of the matrix elements. The term *direct method* denotes algorithms like those of Householder,[27] which modify the elements of the matrix as they proceed. Two important classes of iterative methods can be derived from the Rayleigh quotient and the Krylov sequence.

The simplest example of the former class is perhaps the gradient method for the lowest (or highest) eigenvalue. Having selected an appropriate correction vector ϕ, the scalar factor λ is determined by requiring the derivative of the Rayleigh quotient to vanish. The Rayleigh quotient is

$$\rho(\lambda) \equiv \langle\psi'|H\psi'\rangle / \langle\psi'|\psi'\rangle \tag{16}$$

and the derivative with respect to λ is

$$\frac{dP(\lambda)}{d\lambda} = \frac{1}{\langle\psi'|\psi'\rangle^2} \{2\langle\psi|\psi\rangle\langle\phi|H\psi\rangle - 2\langle\phi|\psi\rangle\langle\psi|H\psi\rangle$$

$$+ \lambda[2\langle\psi|\psi\rangle\langle\phi|H\phi\rangle - 2\langle\phi|\phi\rangle\langle\psi|H\psi\rangle]$$

$$+ \lambda^2[2\langle\phi|\psi\rangle\langle\phi|H\phi\rangle - 2\langle\phi|\phi\rangle\langle\phi|H\psi\rangle]\} \tag{17}$$

This method, although simple to use, is very slow to converge and cannot easily be applied to eigenvalues other than the lowest (or highest) ones.

The simplest algorithm derivable from the Krylov sequence is the power method. The n-term Krylov sequence in the operator H and wave function ψ is

$$K^n = \{\psi, H\psi, H^2\psi, \ldots, H^{n-1}\psi\}$$

The power method merely computes the limit of this sequence which, when normalized, is the eigenfunction corresponding to the largest eigenvalue of H.

This is easily seen by expanding ψ in the eigenfunctions $\mathscr{E}_i$.

$$\psi = \sum_{i=1}^{N} a_i \mathscr{E}_i, \qquad a_i \equiv \langle \psi | \mathscr{E}_i \rangle \tag{18}$$

and

$$H\mathscr{E}_i = E_i \mathscr{E}_i \tag{19}$$

Then, assuming that $E_1 \geq E_2 \geq \cdots \geq E_n \geq 0$, we have

$$\lim_{n \to \infty} H^n \mathscr{E}_1 = \lim_{n \to \infty} \sum_{i=1}^{N} a_i H^n \mathscr{E}_i$$

$$= \lim_{n \to \infty} E_1^n \left\{ a_1 \mathscr{E}_1 + \sum_{i=2}^{N} a_i \left(\frac{E_i}{E_1} \right)^n \mathscr{E}_i \right\}$$

$$= a_1 E_1^n \mathscr{E}_1 \tag{20}$$

The above method illustrates two difficulties common to this class of algorithms. First, convergence is apt to be quite slow, being governed essentially by E_2/E_1. Second, an inappropriate choice of ψ may further retard convergence (for example, if a_1 vanishes the method, in principle, cannot converge to $\mathscr{E}_1$).

The Lanczos algorithm[28] generates a tridiagonal representation of H by sequential orthogonalization of the Krylov sequence. The same representation may be generated by modified Gram–Schmidt orthogonalization of the Krylov sequence, but the procedure is numerically less stable. In the Lanczos algorithm, at each iteration a new wave function is added to a sequence satisfying

$$H[\psi_1, \psi_2, \ldots, \psi_i] = [\psi_1, \psi_2, \ldots, \psi_i]^T T_i \tag{21}$$

where T_i is tridiagonal. The eigenvalues of T_i converge monotonically to the eigenvalues of H, the largest and smallest converging most rapidly. The algorithm will terminate "prematurely" if H has degenerate eigenvalues or if ψ_1 is inappropriately chosen. In either case, the eigenvalues of T_i will have converged to some subset of the spectrum of H.

The method we prefer was recently developed by Davidson.[20] This method sequentially determines a number of eigenvalues and is related to both the gradient and Lanczos methods. Our calculations indicate that the Davidson algorithm converges more rapidly than other algorithms to the lowest (1–4) eigenvalues. There is some indication, however, that the Lanczos algorithm may be competitive for higher (5 or greater) eigenvalues.

It is important to note that not all possible Slater determinants need be included. If ϕ contains a subset of Slater determinants,

$$H_{op} \mathbf{d}^+ \phi = \mathbf{e}^+ \phi + \mathbf{R} \tag{22}$$

where $\mathbf{R}$ contains all Slater determinants *not* included in $\boldsymbol{\phi}$, then the energy for the subset is given by

$$E_\phi = \langle \mathbf{d}^+\boldsymbol{\phi}|H_{op}\mathbf{d}^+\boldsymbol{\phi}\rangle/\langle \mathbf{d}^+\boldsymbol{\phi}|\mathbf{d}^+\boldsymbol{\phi}\rangle \tag{23a}$$

$$= \langle \mathbf{d}^+\boldsymbol{\phi}|\mathbf{e}^+\boldsymbol{\phi}+\mathbf{R}\rangle/\langle \mathbf{d}^+\boldsymbol{\phi}|\mathbf{d}^+\boldsymbol{\phi}\rangle \tag{23b}$$

$$= \langle \mathbf{d}^+\boldsymbol{\phi}|\mathbf{e}^+\boldsymbol{\phi}\rangle/\langle \mathbf{d}^+\boldsymbol{\phi}|\mathbf{d}^+\boldsymbol{\phi}\rangle \tag{23c}$$

since

$$\langle \mathbf{R}|\boldsymbol{\phi}\rangle = 0 \tag{24}$$

Therefore, only Slater determinants in $\boldsymbol{\phi}$ are required to calculate the corresponding energy, E_ϕ.

Selections or subsets of Slater determinants can be generated in two fashions. First, orbitals can be frozen, e.g., configurations involving excitations of certain orbitals are not included. For example, inner-shell correlation effects can be neglected. This is done by excluding certain types of two-body matrix elements. If f denotes the orbitals to be frozen and o the remaining orbitals, two-body matrix elements of the following form are included:

$$\begin{aligned}
&\langle ff'|H'|f''f'''\rangle \\
&\langle of|H'|o'f'\rangle \\
&\langle of|H'|f'o'\rangle \\
&\langle fo|H'|o'f'\rangle \\
&\langle fo|H'|f'o'\rangle \\
&\langle oo'|H'|o''o'''\rangle
\end{aligned} \tag{25}$$

A second method of selection is by excitation level. Slater determinants are generated with specified levels of excitation relative to a list of Slater determinants.

An attractive feature of the vector method is configuration generation. If ψ_o is a collection of Slater determinants,

$$H_{op}\psi_o$$

generates single and double excitations which have nonzero matrix elements with at least one of the Slater determinants in ψ_o. Thus H_{op}^K will generate up to $2K$-fold excitations. The symmetry of the initial wave function will be preserved provided the two-body matrix elements reflect that symmetry.

3. Computational Details

Two computer codes have been developed to carry out VM calculations. Both codes utilize integrals generated by SCREEPER, which is a general molecular code developed at LLL over the past six years. SCREEPER contains all necessary integral programs for contracted Cartesian Gaussian basis sets. Two SCF programs are included. A large standard CI program has been developed which uses formula tape techniques. All codes required to automatically carry out INO[3] calculations have also been included in SCREEPER.

Both VM codes begin with a guess to the wave function, a set of Slater determinants. Correct symmetry is guaranteed by taking appropriate linear combinations. The Hamiltonian operator is then applied and a (sometimes) larger set is generated subject to excitation level restrictions. The approximate wave function is corrected and the process is repeated until the energy has converged.

The first VM code (NSM) was developed (by R. Hausman) to carry out nuclear shell model calculations. We have used the same code to study molecular systems. The code has some very unique features; up to sixty spin-orbitals and electrons are allowed, and two diagonalization procedures are available, Lanczos and Davidson. Wave functions containing up to 30,000 Slater determinants can be constructed. The time-consuming step of operating with the Hamiltonian on a wave function has been minimized by using sophisticated vector techniques.

The second code (VM) was initially developed to better understand the method. Similar features have been included. Sixty spin-orbitals are allowed, only the Davidson diagonalization procedure is available, and the maximum number of Slater determinants has been reduced to 10,000. Two important features have been included which dramatically affect the computer time. First, the hermiticity of the two-body matrix elements is used. This reduces the number of two-body matrix elements and also the computer execution time by a factor of two. Probably the most significant feature is the inclusion of formula tape techniques. Once the list of Slater determinants has stabilized, i.e., application of H_{op} generates no new acceptable Slater determinants, then the formulas for applying

$$H_{op}\mathbf{d}_i^+\boldsymbol{\phi} = \mathbf{d}_o^+\boldsymbol{\phi} \tag{26}$$

are saved. Subsequent iterations, higher roots and different geometries then require considerably less computer time. The formula tape contains two-body matrix elements $(\alpha\beta|H'|\gamma\delta)$, two lists of indices $\alpha\beta$ and $\delta\gamma$, and a list of coefficients $C(\alpha\beta, \delta\gamma)$. For Slater determinants, the $C(\alpha\beta, \delta\gamma)$ are ± 1. For a given two-body matrix element, its contribution to all elements of $\mathbf{d}_o$ is computed. Symbolically

$$\mathbf{d}_o(\delta\gamma) = \mathbf{d}_o(\delta\gamma) + \mathbf{d}_i(\alpha\beta)\langle\alpha\beta|H'|\gamma\delta\rangle C(\alpha\beta, \delta\gamma) \tag{27a}$$

and using the Hermiticity of the Hamiltonian operator

$$\mathbf{d}_o(\alpha\beta) = \mathbf{d}_o(\alpha\beta) + \mathbf{d}_i(\delta\gamma)\langle\alpha\beta|H'|\gamma\delta\rangle C(\alpha\beta, \delta\gamma) \tag{27b}$$

The above steps are particularly amenable to vector techniques which are available in recently developed state-of-the-art computing machinery (CDC-STAR, TI, CRAY-I).

4. LiH Calculation

The calculation was undertaken for checking purposes and timing comparisons. A double-zeta plus polarization contracted Gaussian basis set, employed in earlier calculations,[29] was used. Only one point on the potential energy curve was studied ($R = 3.08$ bohr). This was the equilibrium internuclear separation for the SCF function. The normal restricted Hartree–Fock occupied (1σ and 2σ) and Davidson's IVO[7] unoccupied orbitals were used in the calculation. Three wave functions were constructed, the SCF function plus (1) all single and double excitations from the 2σ, (2) all single and double excitations from both 1σ and 2σ, and (3) full CI. Table 1 gives the number of electronic configurations (Space), linearly independent spin configurations (Spin), and the number of Slater determinants (SD) used in the three calculations. The number of terms vary by almost two orders of magnitude, allowing realistic timing comparisons.

The energy results for the various wave functions are given in Table 2. The energies agree to six figures beyond the decimal point with standard CI energies. One interesting and possibly misleading result is the energy lowering due to inner-shell effects (0.00882 hartrees). This small amount is probably due to the fact that the basis set is inappropriate for describing inner-shell correlation. Particularly absent is the angular correlation due primarily to π orbitals with large exponents; such orbitals were *not* included in the basis set. The same rationale should be used to explain the small energy decrease due to the inclusion of triple and quadruple excitations (0.00028 hartrees).

Table 1. Numbers of Terms in Wave Functions for LiH Calculation

Wave function	Space	Spin	SD
1. SCF	1	1	1
2. SCF+single and double replacement of 2σ orbital	48	48	82
3. SCF+all single and double replacements of 1σ and 2σ orbitals	136	162	357
4. Full configuration interaction	1002	1353	3033

Table 2. Energy Results for LiH Calculation

Wave function[a]	Total energy, hartrees	Correlation energy, 10^{-3} hartrees
1	-7.98262	0
2	-8.00451	21.89
3	-8.01333	30.71
4	-8.01361	30.99

[a] See Table 1 for definition of the wave functions.

The timings of the two VM codes (NSM and VM) are given in Table 3. Standard configuration interaction timings (from SCREEPER) are also included. Since both SCREEPER and VM use formula tape techniques, the timings for a second point on the potential curve (or higher root) have also been included. Clearly the formula tape method, as already determined for standard CI wave functions, is superior. The reduction is due to the elimination of duplicate logic. Another feature of the timing is that the time to create the VM formula tape is nearly a linear function of the number of Slater determinants, while the creation of the CI formula tape is nearly a quadratic function of the number of spin eigenfunctions. Although our CI formula tape code is known to be inefficient, we do not expect the quadratic dependence to be reduced by more efficient coding. Timing for the nth point on the potential curve is nearly the same for the two formula tape codes, and the slight difference probably reflects different diagonalization convergence criteria and diagonalization methods. SCREEPER uses a modified Nesbet technique[18] and VM uses the Davidson technique.

Table 3. Timing Comparisons for LiH Calculation[a]

Code	Point number	Wave function[b]		
		2	3	4
NSM	1	16	66	236
	N	16	66	236
VM	1	1.5	6.2	78.1
	N	0.4	1.4	13.1
SCREEPER	1	1.3	7.9	278.1
	N	0.5	1.2	19.2

[a] Time in 7600 CPU sec.
[b] See Table 1 for definition of the wave functions.

5. H_2O Double-Zeta Calculation

This calculation was also undertaken for checking purposes and timing comparisons. In this case the results of Hosteny *et al.*[30] could be used to compare our vector method with standard CI codes. The double-zeta Gaussian basis set was the "optimally contracted" $[4s2p/2s]$ basis set of Dunning.[31] Only one geometry was studied ($R_{OH} = 1.8111$ bohr and $\theta_{HOH} = 104.45°$). Three different wave functions were studied, the SCF plus replacements through (1) doubles, (2) triples, and (3) quadruples. In all wave functions, inner-shell correlation effects were neglected (i.e., frozen $1a_1$). The numbers of spin eigenfunctions and Slater determinants are given in Table 4. Again the number of terms varies sufficiently to make timing comparisons.

The total energy and correlation energy for the SCF and three CI wave functions are reported in Table 5. Again the energy results agree with the published results to within 0.000002 hartrees. Table 6 gives timings for the published work and the two VM codes. Unfortunately it is difficult to compare our timings with Hosteny *et al.*, because they used a CDC 6400 and our times are for a CDC 7600. Benchmark computer runs indicate that the CDC 7600 is 10–20 times the speed of the CDC 6400.[32] The NSM code is 1–10 times slower than the standard CI code used by Hosteny *et al.* The advantage is even greater for a second point on a potential energy surface since the calculation for the standard CI is reduced by a factor of two. The timing results for the VM code are very encouraging. Although the first point on the surface will require a little more computer time (1–2 times) the subsequent points will be computed *faster than the standard CI (2–6 times*)! We have analyzed the timing using the function

$$f(t) = AN^x \tag{28}$$

where N is the number of terms in the wave function (spin eigenfunctions for standard CI and Slater determinants for VM). The analysis gave

$$X \simeq 1.6$$

for standard CI and

$$X \simeq 0.9$$

for VM. We therefore expect even greater timing advantages (over standard CI) as the number of terms increases.

This test calculation clearly demonstrated that VM and standard CI produce the same wave functions and that by using formula tape techniques, the VM will require less computer time for generating potential energy surfaces. This is extremely important to us because of our Laboratory's (LLL) need for accurate descriptions of excited states and potential energy surfaces.

Table 4. Numbers of Terms for Wave Function in
Double-Zeta H_2O Calculation

Wave function	Spin	SD
1. SCF + single and double replacements	224	523
2. 1 + triple replacements	1,558	4,559
3. 2 + quadruple replacements	6,779	22,005

Table 5. Energy Results for H_2O Double-Zeta
Calculation

Wave function[a]	Total energy, hartrees	Correlation energy, 10^{-3} hartrees
SCF	−76.009256	0
1	−76.135406	126.150
2	−76.136479	127.223
3	−76.142248	132.992

[a] See Table 4 for definition of the wave functions.

Table 6. Timing Comparison for H_2O Double-Zeta
Calculation

Method	Wave function[a]		
	1	2	3
Hosteny *et al.*[b,c] (total)	45	943	9262
Code	23	487	5177
Form-H matrix	17	307	2870
Lowest eigenvalue	5	149	1215
Less code	22	456	4085
NSM (total)[d]	25.9	176	1039.3
VM (total)[d]	7.17	128.11	—
Less formula tape	0.80	7.12	—

[a] See Table 4 for definition of the wave functions.
[b] See Ref. 30.
[c] CDC 6,400 CPU sec.
[d] CDC 7,600 CPU sec.

6. Vertical Transition Energies in Water

This was our first calculation on a system of chemical interest (an application). Water was selected because of the wealth of experimental data.[32–36] In some cases a puzzling feature has been observed in the 4.5 eV region. In addition, there have been a number of theoretical studies of this system.[37–45] Our main goal in this study was to demonstrate the VM could be used to study excited states. This was, therefore, a test of the method and particularly a test of the diagonalization procedure.

Two basis sets were used in the study. The first was a 4-Gaussian expansion[46] of a Slater's rule[47] single-zeta ($O_{1s} = 7.7$, $O_{2s,2p} = 2.275$, $H_{1s} = 1.2$) basis set augmented with an s and p set ($\alpha = 0.028$) for describing the Rydberg states.[48] The second basis set used the Dunning double-zeta contracted Gaussian[31] set augmented with the same Rydberg s and p set. In both calculations, the ground state SCF orbitals were used. Since only vertical transitions were of interest, one geometry ($\theta_{HOH} = 104.45$, $R_{OH} = 1.8111$ bohr) was studied. In all, nine singlet states of water ($4-^1A_1$, $1-^1A_2$, $3-^1B_1$, $1-^1B_2$) were studied using a number of different wave functions. The states and dominant configurations are given in Table 7. For the single-zeta basis set the $1a_1$ and $2a_1$ orbitals were "frozen," i.e., configurations involving excitations from these orbitals were not allowed. The effect of different levels of excitation was studied. In cases where more than one eigenvalue was possible, excitation from all dominant (Table 7) terms were included. For each symmetry ($^1A_1$, $^1A_2$, $^1B_1$, $^1B_2$) three different wave functions were generated; the dominant configurations and excitations through (1) doubles, (2) quadruples, and (3) sextuples. For the double-zeta calculation only the $1a_1$ was frozen, and two wave functions were determined for the different states; the dominant configurations and excitations through (1) singles and (2) doubles. The numbers of Slater determinants in the various wave functions are given in Table 8.

All calculations were carried out with the NSM code. The timings for the calculations are given in Table 9. Again the time varies nearly linearly with the number of Slater determinants.

Table 7. Singlet Electronic States of H_2O:C_{2v} Symmetry

State	Electronic configuration	States
Ground	$(1a_1)^2(2a_1)^2(3a_1)^2(1b_2)^2(1b_1)^2$	$^1A_1$
Rydberg	$(1a_1)^2(2a_1)^2(3a_1)^2(1b_2)^2 1b_1 3s_R$	$^1B_1$
	$(1a_1)^2(2a_1)^2(3a_1)^2(1b_2)^2 1b_1 3p_R$	$^1A_1, {}^1A_2, {}^1B_1$
	$(1a_1)^2(2a_1)^2(1b_2)^2(1b_1)^2 3a_1 3s_R$	$^1A_1$
	$(1a_1)^2(2a_1)^2(1b_2)^2(1b_1)^2 3a_1 3p_R$	$^1A_1, {}^1B_1, {}^1B_2$

Table 8. Number of Slater Determinants in Water Calculation

Wave function	$^1A_1$	$^1A_2$	$^1B_1$	$^1B_2$
Basis set 1, frozen $1a_1$ and $2a_1$				
1. ϕ_0 (dominant configurations)	4	1	3	1
2. ϕ_0+excitations through doubles	1051	516	1088	468
3. ϕ_0+excitations through quadruples	7920	6218	7996	6250
4. ϕ_0+excitations through sextuples	11149	10952	21001	11101
Basis set 2, frozen $1a_1$				
1. ϕ_0	4	1	3	1
2. ϕ_0+excitations through singles	306	116	304	98
3. ϕ_0+excitations through doubles	6371	2844	6332	2388

The energies and effects of different excitation levels for the single-zeta basis set are shown in Table 10. The absolute accuracy is not impressive, due to the basis set. It appears the basis set described the excited Rydberg states better than the ground state; the differences between Rydberg states is in better agreement with the experiment than the vertical transition energies to the ground state. An important outcome of the calculation is seen in the effects of higher excitations. First, if results to within 0.001 hartree are acceptable, only single and double replacements relative to all dominant configurations need be included. Clearly no higher than quadruple excitations have to be included. This encouraging result will be useful to both standard CI and VM application calculations.

The vertical transition energies for the various low-lying singlet states of water are given in Table 11. The energies reported in the table were computed

Table 9. Timings for H_2O Calculations: NSM Code[a]

Wave function	$^1A_1$	$^1A_2$	$^1B_1$	$^1B_2$
Basis set 1				
Wave function 2	387+	20	81	20
Wave function 3	683	150	430	211
Wave function 4	1145	328	770	328
Basis set 2				
Wave function 2	207	21	144	20
Wave function 3	5967	326	5188	274

[a] Using here the old version of NSM code.

Table 10. *Effect of Different Excitation Levels*

Symmetry	Root	Wave function 4, hartrees	$\Delta(2-4)$,[a] 10^{-3} hartrees	$\Delta(3-4)$,[a] 10^{-3} hartrees
$^1A_1$	1	-75.56989	2.364	0.001
$^1A_1$	2	-75.30550	1.125	0.005
$^1A_1$	3	-75.28712	1.024	0.003
$^1A_1$	4	-75.20569	1.315	0.004
$^1A_2$	1	-75.32366	1.038	0.005
$^1B_1$	1	-75.39132	1.098	0.004
$^1B_1$	2	-75.30774	0.899	0.003
$^1B_1$	3	-75.22222	0.626	0.002
$^1B_2$	1	-75.23186	0.772	0.003

[a] $\Delta(i-4)$ denotes difference between wave function i and 4.

Table 11. *Vertical Transition Energies (eV) for Low-Lying Singlet States of Water*

State	SCF[a]	CI[a]	McKoy[b]	VM Wave function 1	VM Wave function 2	Experimental[a]
$^1A_1$-1	0	0	0	0	0	0
-2	—	9.8	9.5	8.9	9.6	9.7
-3	—	10.3	9.6	9.6	10.2	10.2
-4	—	—	—	11.5	12.1	—
$^1A_2$-1	8.3	9.2	9.0	9.0	9.4	9.1
$^1B_1$-1	6.5	7.3	7.2	6.9	7.4	7.4
-2	—	9.9	9.5	9.2	10.1	10.0
-3	—	11.7	—	11.1	11.9	—
$^1B_2$-1	10.3	11.2	—	11.2	11.7	—

[a] Ref. 44.
[b] Ref. 43.

using the two wave functions for the double-zeta basis set. The results are in good agreement with previous calculations and available experimental results. Our calculations also agree with previous theoretical studies in that we do not find any transitions in the 4.5 eV region.

This calculation showed the unimportance of higher excitation levels in CI wave functions and demonstrated that the vector method could be successfully applied to problems of chemical interest. One problem arose during the course of the calculation: it was difficult to determine singlet states only. Because Slater determinants were used, the iterative diagonalization procedures used often determined the low-lying triplets in addition to the singlets. Had spin eigenfunctions been used this would not have been a problem.

7. Improvement of the Vector Method

Clearly the above calculations demonstrate that the vector method is already competitive with standard CI techniques, both in computer times and the size of problem which can be studied. The question arises then, how can the method be improved? Besides innovative computer code modification, particularly to formula tape sections, there are two improvements which should have dramatic effects on the computer time and the size of problem.

First, the use of configurations which are eigenfunctions of all operators which anticommute and commute with the Hamiltonian operator will be considered. Simply using spin eigenfunctions rather than Slater determinants will reduce the number of terms dramatically. This improvement is possible since the configurations can always be expressed as linear combinations of Slater determinants; hence annihilation–creation operations can be carried out. This would be required during formation of the formula tape. Once constructed all subsequent operations would refer to configurations.

The second improvement is to use "spinless" two-body matrix elements. This is possible by rewriting the Hamiltonian operator as

$$H_{op} = \sum_{\alpha\beta\gamma\delta} \langle \alpha\beta|H|\gamma\delta \rangle a_\alpha^+ a_\beta^+ a_\delta a_\gamma \tag{29}$$

$$= \sum_{\substack{i \geq j \\ l \geq k \\ (ij) \geq (kl)}} \{[F_{ijkl}Q_{ijlk} + J_{ijkl}Q_{ijkl}] + (1 - \delta(ij)(kl))[F_{lkji}Q_{lkij} + J_{lkji}Q_{lkji}]\} \tag{30}$$

where now the indices i, j, k, l refer to spinless orthonormal orbitals. The spin is concealed in Q_{ijkl},

$$Q_{ijlk} = \sum_{p,q} a_{ip}^+ a_{jq}^+ a_{lq} a_{kp} \tag{31}$$

and the two-body matrix elements become somewhat more complicated,

$$F_{ijkl} = \langle ij|H|kl \rangle + (1 - \delta ij)(1 - \delta kl)\langle ji|H|lk \rangle$$

$$J_{ijkl} = (1 - \delta ij)\langle ji|H|kl \rangle + (1 - \delta lk)\langle ij|H|lk \rangle$$

and

$$\langle ij|H|kl \rangle = \frac{1}{N-1}\langle i|H_1|k \rangle \langle j|1 \rangle + \tfrac{1}{2}\langle ij|H_2|kl \rangle \tag{32}$$

This form of the Hamiltonian operator thus preserves the hermiticity feature and reduces the number of two-body matrix elements by approximately 16. Thus the formula tape length will be considerably shorter.

We are currently developing a new vector method code (SVM) which will include the above-mentioned features. Calculations will proceed in two steps: (1) a geometry-independent step and (2) calculation of wave functions and corresponding energies. Potential energy surfaces can then be generated inexpensively since step (1) will only be required once. The first step will involve generating the desired configurations and then symbolically determining the formula tape. Only the group multiplication table for the general geometry (e.g., C_S, C_{2v}, etc.) need be used. The calculation step is very similar to the standard CI step. First the integrals are computed, an SCF calculation is made to define orthonormal molecular orbitals, the one- and two-electron integrals are transformed to the orthogonal space, and the two-body matrix elements are formed. Then the formula tape is used to determine the wave functions and energies.

8. Conclusions and Discussion

A review of the attributes of VM shows some very useful features. The configuration list is very simple to generate. Calculations have demonstrated that computing times are nearly a linear function of the number of terms in the wave function. Formula tape techniques, similar to those used in standard CI, significantly decrease the time for higher roots and/or multiple geometry calculations.

The *most* outstanding feature of VM is that no large matrices are formed, manipulated, or diagonalized. Because of this fact, VM calculations are already competitive with or faster than standard CI. The outlined improvements should yield computer times which are one to two orders of magnitude faster than standard CI.

Recall, CI techniques were developed over the past two decades while we have developed VM during the past year!

ACKNOWLEDGMENT

This work was performed under the auspices of the United States Energy Research & Development Administration.

References

1. H. F. Schaefer III, *The Electronic Structure of Atoms and Molecules: A Survey of Rigorous Quantum Mechanical Results*, Addison Wesley, Reading, Massachusetts (1972).
2. H. F. Schaefer III and C. F. Bender, Multiconfiguration wave functions for the water molecule, *J. Chem. Phys.* **55**, 1720 (1971).
3. C. F. Bender and E. R. Davidson, A natural orbital based energy calculation for He–H and Li–H, *J. Phys. Chem.* **70**, 2675 (1966).
4. Z. Gershgorn and I. Shavitt, An application of perturbation theory ideas in configuration interaction calculations, *Int. J. Quantum Chem.* **2**, 751 (1968).
5. L. R. Kahn, P. J. Hay, and I. Shavitt, Theoretical Study of curve crossing; *ab initio* calculations on the four lowest $^1\Sigma^+$ states of LiF, *J. Chem. Phys.* **61**, 3530 (1974).
6. R. J. Buenker and S. D. Peyerimhoff, Individualized configuration selection in CI calculations with subsequent energy extrapolation, *Theor. Chim. Acta* **35**, 33 (1974).
7. E. R. Davidson, Selection of the proper canonical Roothaan–Hartree–Fock orbitals for particular applications. I. Theory, *J. Chem. Phys.* **57**, 1999 (1972).
8. C. F. Bender and H. F. Schaefer III, New theoretical evidence for the nonlinearity of the triplet ground state of methylene, *J. Chem. Phys.* **55**, 4798 (1971).
9. C. F. Bender and E. R. Davidson, A theoretical calculation of the potential curves of the Be–Be molecule, *J. Chem. Phys.* **47**, 4972 (1967).
10. W. Meyer, PNO-CI studies of the electron correlation effects. I. Configuration expansion by means of nonorthogonal orbitals, and application to the ground state and ionized states of methane, *J. Chem. Phys.* **58**, 1017 (1973).
11. A. K. Q. Siu and E. R. Davidson, A study of the ground state wavefunction of carbon monoxide, *Int. J. Quantum. Chem.* **4**, 223 (1970).
12. A. K. Q. Siu and E. F. Hayes, Configuration interaction procedure based on the calculation of perturbation theory natural orbitals: Application to H_2 and LiH, *J. Chem. Phys.* **61**, 37 (1974).
13. D. R. Hartree, W. Hartree, and B. Swirles, Self-consistent field, including exchange and superposition of configurations, with some results for oxygen, *Philos. Trans. R. Soc. London, Ser. A* **238**, 229 (1939).
14. C. F. Bender and H. F. Schaefer, Electronic splitting between the $^2B_1$ and $^2A_1$ states of the NH_2 radical, *J. Chem. Phys.* **55**, 4798 (1971).
15. A. Pipano and I. Shavitt, The use of complex symmetry orbitals in large scale molecular configuration interaction calculations (preprint).
16. J. Paldus, Group theoretical approach to the configuration interaction and perturbation theory calculations for atomic and molecular systems, *J. Chem. Phys.* **61**, 5231 (1974).
17. W. G. Harper and C. W. Patterson, A unitary calculus for electronic orbitals (preprint submitted to *Phys. Rev.*).
18. R. K. Nesbet, Algorithm for diagonalization of large matrices, *J. Chem. Phys.* **43**, 311 (1965).
19. I. Shavitt, C. F. Bender, A. Pipano, and R. P. Hosteny, The iterative calculation of several of the lowest or highest eigenvalues and corresponding eigenvectors of very large symmetric matrices, *J. Comput. Phys.* **11**, 90 (1973).
20. E. R. Davidson, The iterative calculation of a few of the lowest eigenvalues and corresponding eigenvectors of large real-symmetric matrices, *J. Compnt. Phys.* **17**, 87 (1975).
21. B. Roos, A new method for large-scale CI calculations, *Chem. Phys. Letts.* **15**, 153 (1972).

22. P. Seigbahn and H. F. Schaefer III, Potential energy surfaces for $H + Li_2 \rightarrow LiH + Li$ ground state surface from large scale configuration interaction, *J. Chem. Phys.* **62**, 3488 (1975).

23. J. B. McGrory, The twobody interaction in nuclear shell model calculations, *in: The Two-body Force in Nuclei* (Proceedings of the Gulf Lake Conference, 1971) (S. M. Austin and G. M. Crawley, eds.), Plenum Press, New York (1972).

24. J. D. Talman, Shell model calculation of the even parity states of ^{20}Ne, Crocker Nuclear Laboratory Report CNL-UCD-21, April 1964, Crocker Nuclear Laboratory, University of California, Davis, California.

25. R. R. Whitehead, A numerical approach to nuclear shell model calculations, *Nucl. Phys. A* **182**, 290 (1972).

26. R. F. Hausman, Jr., S. D. Bloom, and C. F. Bender, A new technique for describing the electronic states of atoms and molecules—the vector method, *Chem. Phys. Lett.* **32**, 483 (1975).

27. J. H. Wilkinson, *The Algebraic Eigenvalue problem*, Oxford University Press, London (1965).

28. C. Lanczos, An iteration method for the solution of the eigenvalue problem of linear differential and integral operators, *J. Res. Nat. Bur. Stand.* **45**, 255 (1950).

29. P. A. Kollman, C. F. Bender, and S. Rothenberg, Theoretical prediction of the existence and properties of the lithium hydride dimer, *J. Am. Chem. Soc.* **94**, 8016 (1972).

30. R. P. Hosteny, R. R. Gilman, T. H. Dunning, Jr., A. Pipano, and I. Shavitt, Comparisons of the slater and contracted Gaussian basis sets in SCF and CI calculations on H_2O, *Chem. Phys. Lett.* **7**, 325 (1970).

31. T. H. Dunning, Jr., Gaussian basis functions for use in molecular calculations. I. Contraction of $(9s\,5p)$ atomic basis sets for the first row atoms, *J. Chem. Phys.* **53**, 2823 (1970).

32. P. K. Pearson, private communication.

33. E. N. Lassettre and A. Skerbele, Generalized oscillator strengths for 7.4 eV excitation of H_2O at 300, 400, and 500 eV kinetic energy. Singlet–triplet energy differences, *J. Chem. Phys.* **60**, 2464 (1974).

34. F. W. E. Knoop, H. H. Brongersma, and L. J. Oosterhoff, Triplet excitation of water and methanol by low-energy electron-impact spectroscopy, *Chem. Phys. Lett.* **13**, 20 (1972).

35. S. Trajmar, W. Williams, and A. Kupperman, Detection and identification of triplet states of H_2O by electron impact, *J. Chem. Phys.* **54**, 2274 (1971).

36. K. Watanabe and M. Zelikoff, Absorption coefficients of water vapor in the vacuum untraviolet, *J. Opt. Soc. Am.* **43**, 753 (1953).

37. R. P. Hosteny, A. P. Hinds, A. C. Wahl, and M. Krauss, MC SCF calculations on the lowest triplet state of H_2O, *Chem. Phys. Lett.* **23**, 9 (1973).

38. J. A. Horsley and W. H. Fink, *Ab initio* calculations of some low-lying excited states of H_2O, *J. Chem. Phys.* **50**, 750 (1969).

39. K. J. Miller, S. R. Mielczarek, and M. Krauss, Energy surface and generalized oscillator strength of the ^1A Rydberg state of H_2O, *J. Chem. Phys.* **51**, 26 (1969).

40. W. J. Hunt and W. A. Goddard III, Excited states of H_2O using improved virtual orbitals, *Chem. Phys. Lett.* **3**, 414 (1969).

41. R. A. Gangi and R. F. W. Bader, Study of the potential surfaces of the ground and first excited singlet states of H_2O, *J. Chem. Phys.* **55**, 5369 (1971).

42. R. F. W. Bader and R. A. Gangi, The lowest singlet and triplet potential surfaces of H_2O, *Chem. Phys. Lett.* **6**, 312 (1970).

43. D. M. Bishop and Ay-Ju A. Wu, An investigation of the $^1B_1$ excited states of water, *Theo. Chim. Acta (Berlin)* **21**, 287 (1971).

44. D. Yaeger, V. McKoy, and G. A. Segal, Assignments in the electronic spectrum of water, *J. Chem. Phys.* **61**, 755 (1974).

45. R. J. Buenker and S. D. Peyerimoff, Calculation of the electronic spectrum of water (preprint 1974).

46. W. J. Hehre, R. F. Stewart, and J. A. Pople, Atomic electron populations by molecular orbital theory, *J. Symp. Faraday Soc.* **2**, 15 (1968).

47. H. Eyring, J. Walter, and G. E. Kimball, *Quantum Chemistry*, John Wiley and Sons, Inc., New York (1944).

48. N. W. Winter, private communication (1973).

9

The Equations of Motion Method: An Approach to the Dynamical Properties of Atoms and Molecules

Clyde W. McCurdy, Jr.,
Thomas N. Rescigno,
Danny L. Yeager,
and Vincent McKoy

1. Introduction

This chapter is concerned with the equations of motion method as a many-body approach to the dynamical properties of atoms and molecules. In a wide range of spectroscopic experiments one is primarily concerned with just dynamical properties. These dynamical properties include excitation energies and oscillator strengths in optical spectroscopy, the dynamic or frequency-dependent polarizability in light scattering studies, photoionization cross sections, and elastic and inelastic electron scattering cross sections. These experiments probe the response of an atom or molecule to some external perturbation. If one is concerned with these properties one should develop a formalism which aims

Clyde W. McCurdy, Jr., Thomas N. Rescigno, Danny L. Yeager, and Vincent McKoy • Arthur Amos Noyes Laboratory of Chemical Physics, California Institute of Technology, Pasadena, California 91125

directly at these properties. Of course this idea is not novel. For example, one might try to calculate the appropriate Green's functions whose poles, and residues at these poles, are directly the excitation energies and transitions densities, respectively. One could also attempt to solve the time-dependent Schrödinger equation directly, e.g., in the time-dependent Hartree–Fock approximation. The approach to these dynamical properties of atoms and molecules which we will discuss is based on the equations of motion formalism as suggested by Rowe.[1] This is a very practical formalism based on the equations of motion for excitation operators defined as operators that convert one stationary state of a system into another state.

The basic procedure of this method is to derive the equations of the motion which the excitation operator $O_\lambda^\dagger$ for state $|\lambda\rangle$, defined by

$$O_\lambda^\dagger|0\rangle = |\lambda\rangle$$
$$O_\lambda|0\rangle = 0 \tag{1}$$

satisfies. In Eq. (1), $|0\rangle$ and $|\lambda\rangle$ are the ground and excited states respectively. Clearly the operators $O_\lambda^\dagger$ and O_λ of Eq. (1) satisfy these equations of motion

$$[H, O_\lambda^\dagger]|0\rangle = \omega_\lambda O_\lambda^\dagger|0\rangle$$
$$[H, O_\lambda]|0\rangle = -\omega_\lambda O_\lambda|0\rangle \tag{2}$$

However, starting from Eq. (2) it can also be shown straightforwardly that $O_\lambda^\dagger$ satisfies the equation of motion,[1]

$$\langle 0|[O_\kappa, H, O_\lambda^\dagger]|0\rangle = \omega_\lambda \langle 0|[O_\kappa, O_\lambda^\dagger]|0\rangle \tag{3}$$

where ω_λ is the excitation energy of the state $|\lambda\rangle$ and the double commutator is defined as

$$2[A, B, C] = [A, [B, C]] + [[A, B]C] \tag{4}$$

Equation (3) is the formal equation of motion for the excitation operator $O_\lambda^\dagger$ derived by Rowe.[1] The usefulness of this approach for studying electronic problems of atoms and molecules depends, of course, on whether we can find accurate and practical solutions of this equation for these systems. In recent papers[2,3] we have developed hierarchy of solutions to these equations which we have applied to a range of problems in atomic and molecular physics. These applications include the calculation of excitation energies and transition moments of electronic transitions in diatomic[4] and polyatomic[5-7] molecules, photoionization cross sections by numerical analytic continuation techniques, and cross sections for the inelastic scattering of low- and high-energy electrons by molecules.

For the pedagogical purposes of this article, we now emphasize several formal features of the equations of motion method previously discussed by Rowe.[1] This is necessary not only to identify the possible advantages of this method but also to put our solutions to the equations of motion in the right perspective relative to direct methods for solving the Schrödinger equation. It is important to realize that the equations of motion for $O_\lambda^\dagger$, Eq. (3), are formally exact. However, one obviously cannot obtain exact solutions to these equations, and approximations must be made in two aspects of the solution of these equations. First, the linear combination of simpler operators, e.g., $a_i^\dagger a_j$ and $a_i^\dagger a_j^\dagger a_k a_l$, in terms of which $O_\lambda^\dagger$ is expanded, must be truncated and then some approximations must be made for the ground state wave function $|0\rangle$ in evaluating the expectation value in Eq. (3). The truncation of the simpler operators which define $O_\lambda^\dagger$ is equivalent to selecting those excitations relative to the ground state which are important in generating the excited state. Hence $O_\lambda^\dagger$ can include single excitations (a particle–hole pair) and double excitations (two particle–hole pairs) relative to the ground state. This ground state wave function is, of course, correlated. Similar choices must be made in a configuration interaction study of the stationary states $|0\rangle$ and $|\lambda\rangle$. Next some approximate ground state wave function must be used to evaluate the ground state expectation value of the double commutator $[O_\lambda, H, O_\lambda^\dagger]$. Here the method has a definite advantage over the direct solution of the Schrödinger equation where one must evaluate the expectation value of $\langle 0|O_\lambda H O_\lambda^\dagger|0\rangle$, since for these excitation operators $O_\lambda^\dagger$ the double commutator $[O_\lambda, H, O_\lambda^\dagger]$ is always an operator of particle rank two lower than the product $O_\lambda H O_\lambda^\dagger$. This reduction in particle rank of the operators is a very practical advantage when evaluating their ground state expectation values, since for some approximate ground state wave function we can expect better accuracy in the expectation value of the operator of lower rank. A simple example of the reduction of particle rank of operators through commutators would be

$$[x, [H, x]] = 1 \tag{5}$$

which follows from $[x, p_x] = i$ and $[H, x] = -ip_x$.

Another feature of the equations of motion method is the definition of the ground state as the vacuum of the excitation operators O_λ, i.e., Eq. (1). Formally, one should use, in Eq. (3), the ground state wave function $|0\rangle$ which also satisfies the equation $O_\lambda|0\rangle = 0$, a requirement which leads to an iterative procedure for solving these two equations. In our solutions to the equations of motion[2] we developed a scheme for solving these two equations self-consistently. This procedure converges very rapidly, usually requiring two or three iterations. A simpler procedure would be to determine the ground state wave function through a more direct method, e.g., Rayleigh–Schrödinger perturbation theory, and to use this wave function directly in Eq. (3) ignoring the condition $O_\lambda|0\rangle = 0$. In fact this procedure works very well[4] and does so because of the double commutators in the equation of motion. By lowering the particle rank of the operators through double commutators, we have made the required expectation value insensitive to approximations in $|0\rangle$. This same observation also explains an apparent inconsistency in the derivation of the *random phase approximation* (RPA) from the equations of motion for $O_\lambda^\dagger$. We will see that even though an uncorrelated ground state is used to evaluate the expectation value in Eq. (3), ground state correlations are implicitly assumed in the excitation process.

A low-order solution of the equations of motion for $O_\lambda^\dagger$ leads to the random phase approximation. In this order of approximation the method is hence equivalent to the Green's function method and time-dependent Hartree–Fock theory.[8] However, general relationships between the Green's function method and the equations of motion formalism have not been studied. It does seem clear that it is much simpler to obtain higher-order solutions of the equations of motion, Eq. (3), than to extend the Green's function method systematically beyond the RPA.[2] It is important to realize that the equation of motion in the form of Eq. (3) removes the arbitrariness of the linearization procedure previously used to solve the equations of motion for the excitation operator.[8] In fact the RPA was often criticized on the basis of this linearization of the equations of motion. In this method one solves the basic equation satisfied by an excitation operator, i.e., $[H, O_\lambda^\dagger] = \omega_\lambda O_\lambda^\dagger$ by assuming a form for $O_\lambda^\dagger$ and neglecting all terms resulting from the commutator which cannot be put in this form.

In the next section we outline the derivation of our solution to the equations of motion and discuss the applications of these equations to predicting the excitation energies and oscillator strengths of electronic transitions in molecules. In Section 4 we show how equations of motion can also be derived and solved for ionization potentials and electron affinities, i.e., for operators which remove or add an electron to the ground state $|0\rangle$. In Section 5 we show how the discrete oscillator strength distribution of an RPA calculation can provide photoionization cross sections through numerical analytic continuation of the frequency-dependent polarizability for complex energy. Section 6 discusses the application of the equations of motion formalism and techniques to elastic and inelastic scattering of electrons by atoms and molecules.

2. Theory

In the equations of motion method we solve Rowe's equations of motion[9] for excitation operators $O_\lambda^\dagger$ defined by $O_\lambda^\dagger|0\rangle = |\lambda\rangle$. The solution of these equations of motion directly yields the excitation energies relative to the ground state and the corresponding transition densities. The equations are formally exact. The emphasis of the equations of motion is to calculate the relative quantities of direct physical interest. e.g., excitation energies, rather than absolute quantities such as total energies which are often of secondary importance.

Throughout this paper we will use the language of second quantization.[10] The creation and annihilation operators, $a_i^\dagger$ and a_j, respectively, satisfy the usual anticommutation relations

$$\{a_{i'}, a_j\} = \{a_i^\dagger, a_j^\dagger\} = 0$$
$$\{a_i^\dagger, a_j\} = \delta_{i'j'} \qquad (6)$$

Consider an operator $O_\lambda^\dagger$ which converts the ground state of a system $|0\rangle$ into an excited state $|\lambda\rangle$, i.e.,

$$O_\lambda^\dagger |0\rangle = |\lambda\rangle$$

Since $|0\rangle$ is the ground state, O_λ satisfies $O_\lambda |0\rangle = 0$. Clearly the operators $O_\lambda^\dagger$ and O_λ also satisfy the equations of motion

$$[H, O_\lambda^\dagger]|0\rangle = (E_\lambda - E_0)|0\rangle = \omega_\lambda |0\rangle \qquad (7)$$

and

$$[H, O_\lambda]|0\rangle = -\omega_\lambda O_\lambda |0\rangle \qquad (8)$$

Multiplying Eq. (7) by a variation in O_λ and Eq. (8) by a variation in $O_\lambda^\dagger$, closing with $\langle 0|$, and adding Eq. (7) to the Hermitian conjugate of Eq. (8), Rowe[1] obtains

$$\langle 0|[\delta O_\lambda, [H, O_\lambda^\dagger]]|0\rangle = \omega_\lambda \langle 0|[\delta O_\lambda, O_\lambda^\dagger]|0\rangle \qquad (9)$$

Eq. (9) is written in more symmetric form

$$\langle 0|[\delta O_\lambda, H, O_\lambda^\dagger]|0\rangle = \omega_\lambda \langle 0|[\delta O_\lambda, O_\lambda^\dagger]|0\rangle \qquad (10)$$

where the symmetric double commutator is defined,

$$2[A, B, C] = [A, [B, C]] + [[A, B], C] \qquad (11)$$

We can obtain the matrix element $\langle \lambda|W|0\rangle$ of the operator W from

$$\langle \lambda|W|0\rangle = \langle 0|O_\lambda W|0\rangle = \langle [O_\lambda, W]|0\rangle \qquad (12)$$

There are two approximations which must be made in solving Eq. (10). First, one must assume some approximate form for the ground state $|0\rangle$. A second approximation in solving Eq. (10) is the truncation of the expansion assumed for $O_\lambda^\dagger$. As an approximation the operator $O_\lambda^\dagger$ may be expanded as a sum over all single excitations (particle–hole pairs), i.e.,

$$O_\lambda^\dagger = \sum_{m'\nu'} Y_{m'\nu'} a_{m'}^\dagger a_{\nu'} \qquad (13)$$

where $m, n, \ldots$ are particle or virtual orbitals; $\gamma, \nu, \ldots$ are hole or orbitals occupied in the Hartree–Fock ground state; and primes designate spin-orbitals. Restricting $O_\lambda^\dagger$ to the form in Eq. (13) and evaluating the expectation value of the equations of motion with the Hartree–Fock wave function, one obtains the *single excitation* approximation of *configuration interaction* (SECI) or the *Tamm–Dancoff* approximation (TDA).[10]

We can write Eq. (13) more generally as

$$O_\lambda^\dagger = \sum_i (Y_i \eta_i^\dagger - Z_i \eta_i) \tag{14}$$

where i may be a particle–hole pair $(m'\nu')$ as in Eq. (13) or may include more complicated excitation operators. η_i is the Hermitian conjugate of $\eta_i^\dagger$. With the $O_\lambda^\dagger$ of Eq. (14), Eq. (10) yields[9]

$$\begin{pmatrix} \mathbf{A} & \mathbf{B} \\ -\mathbf{B}^* & -\mathbf{A}^* \end{pmatrix}\begin{pmatrix} Y \\ Z \end{pmatrix} = \omega_\lambda \begin{pmatrix} \mathbf{D} & O \\ O & \mathbf{D} \end{pmatrix}\begin{pmatrix} Y \\ Z \end{pmatrix} \tag{15}$$

where

$$A_{ij} = \langle 0 | [\eta_i, \hat{\mathcal{H}}, \eta_j^\dagger] | 0 \rangle$$
$$B_{ij} = -\langle 0 | [\eta_i, \hat{\mathcal{H}}, \eta_j] | 0 \rangle \tag{16}$$
$$D_{ij} = \langle 0 | [\eta_i, \eta_j^\dagger] | 0 \rangle$$

A is Hermitian, **B** is symmetric, and **D** is Hermitian.

A simple approximate solution to Eq. (10) can be obtained by restricting the summation in Eq. (14) to a particle–hole form, i.e.,

$$O_\lambda^\dagger = \sum_{m'\nu'} (Y_{m'\nu'} a_{m'}^\dagger a_{\nu'} - Z_{m'\nu'} a_{\nu'}^\dagger a_{m'}) \tag{17}$$

In this case **D** in Eq. (16) is the unit matrix if $|0\rangle$ is the Hartree–Fock wave function. The matrices in Eq. (15) are block diagonal by both space and spin symmetry. Eq. (17) can be written as

$$O_\lambda^\dagger(\Gamma SM) = \sum_{m\gamma(\Gamma)} [Y_{m\gamma}(\lambda S) C_{m\gamma}^\dagger(SM) - Z_{m\gamma}(\lambda S) C_{m\gamma}(\bar{S}\bar{M})] \tag{18}$$

where Γ is the irreducible representation of the molecular point group, SM specifies the spin state, and

$$C_{m\gamma}^\dagger(10) = \frac{1}{\sqrt{2}}(a_{m\alpha}^\dagger a_{\gamma\alpha} - a_{m\beta}^\dagger a_{\gamma\beta})$$

$$C_{m\gamma}^\dagger(00) = \frac{1}{\sqrt{2}}(a_{m\alpha}^\dagger a_{\gamma\alpha} + a_{m\beta}^\dagger a_{\gamma\beta}) \tag{19}$$

$$C_{m\gamma}(\bar{S}\bar{M}) = (-)^{S+M} C_{m\gamma}(S-M)$$

In Eq. (19), α and β refer to the spin states of the electron and unprimed indices are orbitals.

The Hamiltonian is written in second quantized form as

$$\mathcal{H} = \sum_{i'} \varepsilon_{i'} a_{i'}^\dagger a_{i'} + \tfrac{1}{2} \sum_{i'j'} \sum_{k'l'} V_{i'j'k'l'} a_{j'}^\dagger a_{i'}^\dagger a_{k'} a_{l'} - \sum_{i'j'} \sum_{\gamma'} (V_{i'\gamma'j'\gamma'} - V_{i'\gamma'\gamma'j'}) a_{i'}^\dagger a_{j'}$$

$$= \sum_i \varepsilon_i \sqrt{2}\, C_{ii}^\dagger(00) + \sum_{ij} [-\tfrac{1}{2} \sum_k V_{ikkj} + \sum_\gamma (V_{i\gamma\gamma j} - 2 V_{i\gamma j\gamma})] \sqrt{2}\, C_{ij}^\dagger(00) \qquad (20)$$

$$+ \sum_{ijkl} V_{ijkl} C_{ik}^\dagger(00) C_{jl}^\dagger(00) \qquad (21)$$

where ε_i is the orbital energy, i, j, k, and l are either particle or hole orbitals, and

$$V_{ijkl} = \int \varphi_i^*(1) \varphi_j^*(2) \frac{1}{r_{12}} \varphi_k(1) \varphi_l(2)\, d\tau_{12} \qquad (22)$$

Equations (21) and (23) along with the SCF approximation for the ground state yield

$$
\left.
\begin{aligned}
A_{m\gamma,n\delta} &= \delta_{mn}\, \delta_{\gamma\delta} (\varepsilon_m - \varepsilon_\gamma) + 2 V_{m\delta\gamma n} - V_{m\delta n\gamma} \\
B_{m\gamma,n\delta} &= 2 V_{mn\gamma\delta} - V_{mn\delta\gamma}
\end{aligned}
\right\} S = 0
$$

$$
\left.
\begin{aligned}
A_{m\gamma,n\delta} &= \delta_{mn}\, \delta_{\gamma\delta} (\varepsilon_m - \varepsilon_\gamma) - V_{m\delta n\gamma} \\
B_{m\gamma,n\delta} &= V_{mn\delta\gamma}
\end{aligned}
\right\} S = 1
$$

$$(23)$$

All matrix elements and orbital energies in Eq. (23) are obtained from the SCF ground state calculation. We solve Eq. (15) for the **Y** and **Z** particle–hole excitation amplitudes and energies. This approximation is called the closed-shell *random phase approximation* (RPA). Equation (12) for transition moments becomes, in the RPA,

$$M_{0\lambda} = \sum_{m\gamma} (Y_{m\gamma} + Z_{m\gamma}) \sqrt{2} M_{m\gamma} \qquad (24)$$

where $M_{m\gamma}$ is a particle–hole element of the transition moment matrix. The oscillator strength of a transition $|0\rangle \rightarrow |\lambda\rangle$ is given by

$$f_{0\lambda} = \tfrac{2}{3} G_\lambda \omega_{0\lambda} M_{0\lambda}^2 \qquad (25)$$

where G_λ is the degeneracy factor of the upper state $|\lambda\rangle$.

The **Z** amplitudes imply that ground state correlations are implicitly assumed in the ground state even though the expectation value of the double commutator is evaluated with an uncorrelated ground state. If the Z amplitudes are set equal to zero, and consequently reducing all **B** elements to zero, the equations are the Tamm–Dancoff approximation (TDA) or the single excitation configuration interaction (SECI).

In practice both the TDA and RPA give excitation energies that are usually too large. Table 1 gives the TDA and RPA excitation energies and transition moments for formaldehyde at the ground state experimental geometry.[6] The basis set is composed of a $[3s\,2p/1s]$ contracted Gaussian basis set to which several diffuse functions are added. The basis set is described in detail elsewhere.[5–7] No matrix over 30×30 is diagonalized in these calculations.

For a complete basis set the sum of the RPA oscillator strengths satisfies the Thomas–Reiche–Kuhn sum rule.[9] This property makes the discrete set of oscillator strengths obtained from an RPA calculation in a finite basis set particularly useful. For example, this oscillator strength distribution can be used to calculate second-order optical properties, e.g., the dynamic polarizability

$$\alpha(\omega) \cong \sum_n \frac{\tilde{f}_{on}}{\tilde{\omega}_{on}^2 - \omega^2} \tag{26}$$

and van der Waals coefficients, C_{ab} including anisotropies, for the long-range interactions between species A and B,

$$C_{ab} = \kappa(a, b) \sum_n \sum_m \frac{f_{on}^a f_{om}^b}{\omega_{on}^a \omega_{om}^b (\omega_{on}^a + \omega_{om}^b)} \tag{27}$$

where $\kappa(a, b)$ is a constant which depends on the species A and B. Application to He and H_2 gives accurate and encouraging results[11] These same spectra, i.e., $\{f_{on}\}$ and $\{\omega_{on}\}$, can be used to define $\alpha(z)$ for complex z which can then lead to photoionization cross sections through numerical analytic continuation. This application is discussed in Section 5.

A deficiency of the random phase approximation is that it can lead to instabilities for transitions to low-lying triplet excited state. These instabilities are characterized by complex eigenvalues for the excitation frequencies ω and have been discussed by Thouless.[12] They can usually be eliminated by going to larger and more accurate SCF calculations, although the excitation energies may still not be in good agreement with experiment. An example of such an instability is the $^3A_1$ $(\pi \to \pi^*)$ state of formaldehyde using a $[3s\,2p/1s]$ contracted Gaussian basis set.[5–7] The larger basis set of Table 1 gives an RPA excitation energy of 2.22 eV while higher-order calculations (see below) give 5.29 eV.

An obvious inconsistency in the RPA is that, whereas ground state correlations are implicitly assumed in the excitation process by including the

Table 1. Excitation Energies and Intensities in Formaldehyde[a]

State	Main transition	ΔE,[b] TDA	ΔE,[b] RPA	ΔE,[b] EOM	ΔE, observed	ΔE, CI	f, TDA	f, RPA	f, EOM	f, observed[h]
$^3A_2$	$n \to \pi^*$	3.67	3.54	3.46	3.6[c]	3.41[e]	—	—	—	—
$^1A_2$	$n \to \pi^*$	4.43	4.32	4.04	4.26[c]	3.81	—	—	—	—
$^3A_1$	$\pi \to \pi^*$	4.79	2.22	5.29	—	5.56	—	—	—	—
$^1A_1$	$\pi \to \pi^*$	10.66	10.31	10.10	—	9.90[f]	0.20	0.15	0.10	—
$^1B_1$	$\sigma \to \pi^*$	9.48	9.37	9.19	9.0[d]	9.03	0.003	0.002	0.002	—
$^3B_2$	$n \to 3s$	8.65	8.63	7.15	6.7–7.0[c]	—	—	—	—	—
$^1B_2$	$n \to 3s$	8.96	8.95	7.28	7.08	7.38	0.03	0.03	0.02	0.028
$^1B_2$	$n \to 3pa_1$	9.78	9.77	8.12	8.14	8.39	0.05	0.05	0.04	0.032
$^1A_1$	$n \to 3pb_2$	9.88	9.86	8.15	7.97	8.11	0.09	0.11	0.05	0.017
$^1A_2$	$n \to 3pb_1$	10.06	10.05	8.37	—	7.99[g]	—	—	—	—
$^1A_1$	$n \to 4pb_2$	11.23	11.23	9.40	9.58	—	0.01	0.01	0.004	i
$^1A_2$	$n \to 4pb_1$	11.26	11.26	9.47	—	—	—	—	—	—
$^1B_2$	$n \to 4pa_1$	11.31	11.31	9.55	9.63	—	0.07	0.06	0.04	0.032

[a] Vertical excitation energies.
[b] Results from the EOM method in the $(1p–1h)+(2p–2h)$ approximation. All energies in eV.
[c] Estimated vertical excitation energies. A. Chutjian, *J. Chem. Phys.* **61**, 4279 (1974).
[d] The next six experimental values, except for the $^3B_2$ state, are from J. E. Mentall, E. P. Gentieu, M. Krauss, and D. Neumann, *J. Chem. Phys.* **55**, 547 (1971).
[e] CI calculations of S. D. Peyerimhoff, R. J. Buenker, W. E. Kammer, and H. Hsu, *Chem. Phys. Lett.* **8**, 129 (1971).
[f] J. L. Whitten, *J. Chem. Phys.* **56**, 5458 (1972).
[g] J. L. Whitten and M. Hackmeyer, *J. Chem. Phys.* **51**, 5584 (1969).
[h] M. J. Weiss, C. E. Kuyatt, and S. Mielczarek, *J. Chem. Phys.* **54**, 4147 (1971).
[i] Observed to be much weaker than the transition to the $^1B_2(2b_2 4pa_1)$.

$Z_{m\nu}a_\nu^\dagger a_m$ term in $O_\lambda^\dagger$, the uncorrelated ground state is used to evaluate the expectation value in Eq. (9), i.e.,

$$O_\lambda|HF\rangle \neq 0 \tag{28}$$

To remove this deficiency, eliminate instabilities, and obtain better vertical excitation energies, the ground state wave function may be approximated by

$$|0\rangle \approx N_0(1+U)|HF\rangle \tag{29}$$

where

$$U = \tfrac{1}{2} \sum_{\substack{m\gamma \\ n\delta}} \sum_S C'_{m\gamma,n\delta}(S)C_{m\gamma}^\dagger(SO)C_{n\delta}^\dagger(\overline{SO}) \tag{30}$$

and $C'_{m\gamma,n\delta}(S)$ is a correlation coefficient. Equations (29) and (30) represent the explicit inclusion of a correlated ground state. Details of the solution of Eq. (10) and **A**, **B**, and **D** matrix elements are given elsewhere.[2,3] The correlation coefficients can be calculated from Rayleigh–Schrödinger perturbation theory or by iterating Eq. (10) and the condition $O_\lambda|0\rangle = 0$. The elements of the **A**, **B**, and **D** matrices now depend on the correlation coefficients. The resulting equations are called the *higher random phase approximation* (HRPA) or EOM $(1p-1h)$.

For many electronic transitions inclusion of only single particle–hole components in $O_\lambda^\dagger$ can be adequate. Inclusion of double excitations allows for a self-consistent readjustment of the core of basically ground state (hole) orbitals during the excitation process. Thus, to Eq. (18) must be added two-particle–two-hole creation and annihilation operators,

$$\sum_{\substack{(m\gamma) \\ (n\delta)}} \{Y_{m\gamma,n\delta}^{(2)}(\lambda S)\Gamma_{m\gamma,n\delta}^\dagger(SM) - Z_{m\gamma,n\delta}^{(2)}(\lambda S)\Gamma_{m\gamma,n\delta}(SM)\} \tag{31}$$

Again, details are given elsewhere.[3] To a good approximation

$$\omega = \omega^{(1p-1h)} - \Delta\omega \tag{32}$$

where $\omega^{(1p-1h)}$ is a solution of the HRPA equations and $\Delta\omega$ is obtained perturbatively.[3] These equations are the equations of motion including double-excitation mixing or the EOM $[(1p-1h)+(2p-2h)]$. In Table 1 EOM $[(1p-1h)+(2p-2h)]$ energies and oscillator strengths for the low-lying states in formaldehyde are also given. In general, agreement with experiment is good.

We have proposed a straightforward scheme for extending the equations of motion formalism to systems with simple open-shell ground states (OSRPA).[13] Although the method will not be discussed here in detail, we summarize the OSRPA procedure.

1. Perform an open-shell SCF calculation[14] to obtain an orthonormal restricted Hartree–Fock basis.

2. Rewrite the total Hamiltonian in terms of these SCF orbital energies, choosing the particle states to be eigenfunctions of the last open-shell Fock operator.

3. Use the restricted Hartree–Fock ground state $|HF\rangle$ as an approximation to $|0\rangle$ in Eq. (10).

4. Choose excitation operators $O_\lambda^\dagger$ such that the particle–hole excitation operators $\{\eta_i^\dagger\}$ operating on $|HF\rangle$ generate configurations which are eigenfunctions of $\hat{S}^2$ and $\hat{M}_S$. Furthermore all $\eta_i^\dagger$ are one-body operators except for those which change the spin of the open-shell electron or which move an electron between degenerate open-shell molecular orbitals.

5. The $\eta_i^\dagger$ are chosen so that $\eta_i^\dagger$ and η_i are tensor operators of the same rank and the Hermitian conjugate pairs transform in the same manner under rotation of the spin space.

This open-shell method is straightforward, although somewhat more involved than the closed-shell RPA. Other open-shell versions of the RPA have also been developed.[15,16]

3. Excitation Energies and Oscillator Strengths of Electronic Transitions in Molecules

We now discuss the results of the application of the equations of motion method to the electronic spectra of several diatomic and polyatomic molecules. These results include the excitation energies and oscillator strengths for the imporant transitions in N_2, CO, C_2H_4, CO_2, and C_6H_6. For N_2 and CO, potential energy curves for these excited states are also obtained by solving the equations of motion at several internuclear distances.

The first step of an equation of motion calculation is to carry out a Hartree–Fock calculation in order to generate a particle–hole basis. The occupied orbitals are hole states and the virtual orbitals are particle states. The SCF calculations are all done in a basis of Gaussian orbitals. The size of the basis determines the quality of the hole states and the number of particle states. The size of the contracted Gaussian basis used in these calculations varies from molecule to molecule, but in general consists of a $[4s\,3p]$ basis or a $[3s\,2p]$ basis, both augmented by some diffuse components. Details of the basis sets and the selection of the particle–hole pairs for each molecules can be found in the specific references.

3.1. States of N_2

The electron configuration of the ground state of N_2 is

$$(1\sigma_g)^2(1\sigma_u)^2(2\sigma_g)^2(2\sigma_u)^2(1\pi_u)^4(3\sigma_g)^2$$

Table 2 gives the excitation energies for transitions to the eleven low-lying states of N_2.[4] In these calculations we included excitations out of all hole levels except the $1\sigma_g$ and $1\sigma_u$ levels. All particle–hole excitations of the appropriate symmetry are included in the calculation on each state . With the particle–hole pairs specified the excitation energies of the excited states are first calculated in the $(1p-1h)$ approximation, i.e., only $1p-1h$ operators are included in $O_\lambda^\dagger$ and hence the excited state $|\lambda\rangle$ differs only by single particle–hole excitations relative to a correlated ground state. In the next stage of the calculation we introduce the effect of $2p-2h$ excitations on the excitation frequencies. From Eq. (15) it would seem that if N particle–hole pairs are included in $O_\lambda^\dagger$ then the resulting equations lead to an unsymmetric $2N \times 2N$ matrix. However, the special symmeteries of these matrices can be exploited in order to reduce the problem to eigenvalue equations for real symmetric matrices of order N.[4]

In the first column of Table 2 we list the symmetry and conventional spectroscopic designation of the various states. The next column shows the number of single particle–hole pairs used in setting up the equations of motion. The excitation frequencies in the $1p-1h$ approximation are listed in the third column. Comparison with the experimental vertical excitation energies shows that this approximation predicts all the states to lie about 1–3 eV above the experimental values. Inclusion of $2p-2h$ components lowers the $1p-1h$ excita-

Table 2. Equations of Motion Calculations: Excited States of N_2^a

State	N^b	\multicolumn{3}{c}{ΔE^c}		
		$(1p-1h)$	$(1p-1h)+(2p-2h)$	Experimental
$B\,^3\Pi_g(3\sigma \to \pi_g)^{\,d}$	15	9.6	7.5	8.1^e
$a\,^1\Pi_g^{\,f}$	15	11.5	8.8	9.3
$A\,^3\Sigma_u^+(\pi_u \to \pi_g)$	20	8.4	7.8	7.8
$B'\,^3\Sigma_u^{-\,g}$	8	11.3	10.2	9.7
$W\,^3\Delta_u$	—	10.1	9.4	8.9^h
$a'\,^1\Sigma_u^-$	8	11.3	10.6	9.9
$w\,^1\Delta_u$	—	12.0	11.0	10.3
$b'\,^1\Sigma_u^+$	20	16.8	15.0	14.4^i
$c'\,^1\Sigma_u^+(3\sigma_g \to \sigma_u)$	—	15.5	12.1	12.9
$C\,^3\Pi_u(2\sigma_u \to \pi_g)$	10	13.3	10.8	11.1
$b\,^1\Pi_u$	10	17.4	14.0	12.8

^{a} All calculations done at an equilibrium internuclear distance of 2.068 a.u.
^{b} Number of single particle–hole pairs used in the calculation.
^{c} In eV.
^{d} Indicates the main component of the excitation relative to the ground state.
^{e} The experimental results for this state and for the $a\,^1\Pi_g$, $A\,^3\Sigma_u^+$, $B\,^3\Sigma_u^-$, $a'\,^1\Sigma_u^-$, $w\,^1\Delta_u$, and $C\,^3\Pi_u$ states are those reported by W. Benesch, J. T. Vanderslice, S. G. Tilford, and P. G. Wilkinson, *Astrophys. J.* **142**, 1227 (1965). Their tabulations are based on high-resolution optical data.
^{f} Same designation as in the previous state.
^{g} The next five states have the same principal $1p-1h$ component type.
^{h} W. Benesch and K. A. Saum, *J. Phys. B* **4**, 732 (1971).
^{i} The experimental results for the $b'\,^1\Sigma_u^+$, $C\,^3\Pi_u$, and $b\,^1\Pi_u$ states are from the electron energy-loss spectrum of J. Geiger and B. Schroeder, *J. Chem. Phys.* **50**, 7 (1969).

Table 3. Oscillator Strengths for Transitions in N_2

Transition	f_{el}[a]	$q_{v'v''}$[b]	$f_{el}q_{v'v''}$	Experimental
$X\,^1\Sigma_g^+ \to c'\,^1\Sigma_u^+$	0.11	$q_{00} \sim 1$[c]	0.11	0.14 ± 0.04[d]
$X\,^1\Sigma_g^+ \to b\,^1\Pi_u$	0.64	—	—	—
$X\,^1\Sigma_g^+ \to b'\,^1\Sigma_u^+$	0.49	—	—	Large "measured" f_{el}[e]

[a]$f_{el} = \frac{2}{3}G \cdot \Delta E \cdot M^2$, where M is the dipole transition matrix element and G is the degeneracy of the upper state.
[b]Franck–Condon factors for the v' and v'' levels.
[c]G. M. Lawrence, D. L. Mickey, and K. Dressler, *J. Chem. Phys.* **48**, 1989 (1968).
[d]This is the measured f value for the 0–0 transition. See reference in footnote c.
[e]Weak due to intensity perturbations by $v' = 5$ and 6 of the $c'\,^1\Sigma_u^+$ and $v' = 0$ of the $e'\,^1\Sigma_u^+$ states. From shock-heated vibrationally excited N_2, $f_{el}(v' = 5 \to v' = 2) \approx 0.83$ and $f_{el}(v'' = 8 \to v' = 2) \approx 0.4$ [J. P. Appleton and M. Steinberg, *J. Chem. Phys.* **46**, 1521 (1967)].

tion energies by about 1–3 eV, resulting in excitation energies in good agreement with the experimental values. The calculations on the eleven states of N_2 listed in Table 2 required about 20 min on an IBM 370/155.

In Table 3 we compare the calculated oscillator strengths with available experimental data. The calculated f values for the $X\,^1\Sigma_g^+ \to c'^1\Sigma_u^+$ transition agree well with the measured value. Here a Franck–Condon (FC) factor of unity is assumed for the 0–0 transition. It is difficult to estimate FC factors for the $X^1\Sigma_g^+ \to b\,^1\Pi_u$ transition because of strong perturbation of the vibrational levels of the $b\,^1\Pi_u$ state by those of the $c\,^1\Pi_u$ state. Similar perturbations of the vibrational levels of the $b'\,^1\Sigma_u^+$ state by those of the $c'\,^1\Sigma_u^+$ state also make intensity measurements for the $X\,^1\Sigma_g^+ \to b'\,^1\Sigma_u^+$ band system quite complicated.

The transition energies in Table 2 are vertical excitation energies at the equilibrium geometry of the ground state. The equations of motion method can also be used to obtain the potential energy curves of the excited states and the corresponding transition moments. The approach is straightforward. The equations of motion are solved at several internuclear distances and hence we can obtain $\omega_{on}(R)$, i.e., the location of the excited state curve relative to the ground state at internuclear distance R. The solution of the equations of motion assumes that the ground state correlation coefficients are small compared with unity, and hence we can use this procedure to generate potential energy curves at internuclear distances for which this condition holds. If the correlation coefficients become large one must use an open-shell version of the equations of motion.[13] However, we can expect the closed-shell form of our theory to be applicable at geometries of practical spectroscopic interest, e.g., up to 30% change from equilibrium. Table 4 lists the excitation energies for the eleven states of N_2 at six internuclear distances.[17] Again, inclusion of both $1p$–$1h$ and $2p$–$2h$ operators in the excitation operator gives excitation energies in good agreement with experiment. Plots of these potential energy curves are shown in Ref. 17. The results for the $b'\,^1\Sigma_u^+$ and $c'\,^1\Sigma_u^+$ states are of current

Table 4.　Excitation Energies[a] : States of N_2

$X^1\Sigma_g^+$ $\rightarrow R$, Å	(1p–1h)	(1p–1h) +(2p–2h)	Observed[b]	(1p–1h)	(1p–1h) +(2p–2h)	Observed[b]
		$B\,^3\Pi_g$			$a\,^1\Pi_g$	
0.90	12.7	11.0	—	14.6	12.3	—
1.00	10.9	9.0	9.4	12.9	10.3	10.5
1.094[c]	9.6	7.5	8.1	11.5	8.8	9.3
1.20	8.2	5.9	6.8	10.0	7.2	8.0
1.30	7.1	4.8	5.8	8.9	6.1	7.0
1.40	6.3	4.0	4.9	8.0	5.3	6.0
		$A\,^3\Sigma_u^+$			$B'\,^3\Sigma_u^-$	
0.90	13.2	12.7	—	15.9	15.1	—
1.00	10.5	9.9	—	13.3	12.4	—
1.094	8.4	7.8	7.8	11.3	10.2	9.7
1.20	6.4	5.7	5.9	9.3	8.1	7.8
1.30	5.1	4.2	4.4	7.9	6.6	6.4
1.40	4.0	3.2	3.2	6.7	5.4	5.3
		$W\,^3\Delta_u$			$a'\,^1\Sigma_u^-$	
0.90	14.8	14.3	—	15.9	15.4	—
1.00	12.1	11.5	—	13.3	12.7	—
1.094	10.1	9.4	8.9	11.3	10.6	9.9
1.20	8.1	7.3	7.1	9.3	8.5	8.1
1.30	6.7	5.8	5.6	7.9	6.9	6.6
1.40	5.5	4.7	4.5	6.7	5.8	5.5
		$W\,^1\Delta_u$			$b'\,^1\Sigma_u^+$	
0.9	16.5	15.7	—	19.6	17.9	—
1.0	14.0	13.0	—	18.3	16.6	—
1.094	12.0	11.0	10.3	16.8	15.0	14.4
1.20	10.0	9.0	8.5	15.0	13.2	12.8
1.30	8.6	7.3	7.2	13.1	11.4	11.2
1.40	7.3	6.0	6.0	11.0	9.1	9.7
		$c'\,^1\Sigma_u^+$			$C\,^3\Pi_u$	
0.90	15.8	12.8	—	14.4	12.5	—
1.00	15.6	12.3	13.0	13.8	11.5	—
1.094	15.5	12.1	12.9	13.3	10.8	11.1
1.20	15.4	12.0	12.6	12.9	10.1	10.6
1.30	15.2	11.6	12.3	12.7	9.6	10.0
1.40	14.9	11.5	—	12.7	9.3	9.3
		$b\,^1\Pi_u$				
0.90	17.7	14.9	—			
1.00	17.6	14.5	—			
1.094	17.4	14.0	13.0			
1.20	17.2	13.4	12.0			
1.30	17.3	12.9	10.7			
1.40	17.5	12.7	9.4			

[a] In eV.
[b] The experimental results are from W. Benesch, J. T. Vanderslice, S. G. Tilford, and P. G. Wilkinson, *Astrophys. J.* **142**, 1227 (1965) and J. Geiger and B. Schroeder, *J. Chem. Phys.* **50**, 7 (1969).
[c] Experimental equilibrium internuclear distance.

Table 5. Oscillator Strengths for Transitions in N_2 [a]

$R, \text{Å}$	$X\,^1\Sigma_g^+ \to b'\,^1\Sigma_u^+$	$X\,^1\Sigma_g^+ \to c'\,^1\Sigma_u^+$	$X\,^1\Sigma_g^+ \to b\,^1\Pi_u$
0.90	0.13	0.07	0.58
1.00	0.13	0.09	0.62
1.094	0.49	0.11	0.64
1.20	0.59	0.001	0.58
1.30	0.31	0.12	0.58
1.40	0.13	0.22	0.60

[a] The electronic oscillator strength $f = \frac{2}{3} \cdot G \cdot \Delta E \cdot M^2$, where G is the degeneracy factor and the excitation energy ΔE and transition moment M are in the $1p$–$1h$ approximation.

experimental interest. Our results agree with the conclusions of Dressler[18] and Lefebvre-Brion,[19] i.e., the $b'\,^1\Sigma_u^+$ state is a $\pi_u \to \pi_g$ intravalence transition with $R_e \simeq 1.43$ Å and the $c'\,^1\Sigma_u^+$ a Rydberg $3\sigma_g \to 3\sigma_u$ transition with $R_e \simeq 1.11$ Å. The "deperturbed" curves[18,19] give $R_e = 1.45$ Å for the $b'\,^1\Sigma_u^+$ state and $R_e = 1.12$ Å for the $c'\,^1\Sigma_u^+$ state. These "deperturbed" curves correspond to hypothetical electronic states of the same symmetry that are permitted to cross each other.[18] For further details, including a similar discussion of the $b\,^1\Pi_u$ and $c\,^1\Pi_u$ states, we refer to Ref. 17.

Table 5 gives the electronic oscillator strengths for the $X\,^1\Sigma_g^+ \to b\,^1\Pi_u$, $b'\,^1\Sigma_u^+$, and $c'\,^1\Sigma_u^+$ transitions at several internuclear distances. These oscillator strengths do not contain any Franck–Condon factors. Only the oscillator strengths on the single particle–hole approximation are given since the transition moment changes negligibly due to the inclusion of $2p$–$2h$ components in $O_\lambda^\dagger$. The oscillator strengths in the $[(1p$–$1h)+(2p$–$2h)]$ approximations can then be derived from those in Table 5 by replacing the $(1p$–$1h)$ excitation energy by its value in the $[(1p$–$1h)+(2p-2h)]$ approximation and renormalization. The behavior of the oscillator strength of the $X\,^1\Sigma_g^+ \to b'\,^1\Sigma_u^+$ transition is very interesting. At smaller internuclear distances, i.e., $R_e = 0.90$ and 1.00 Å, the f value is about 0.1 but increases to 0.5–0.6 at $R = 1.094$ and 1.20 Å. This reflects the increasing valence character of the $b'\,^1\Sigma_u^+$ state at the larger internuclear distance. Note also the behavior of the f value of the $X\,^1\Sigma_g^+ \to c'\,^1\Sigma_u^+$ transition near $R = 1.20$ Å, close to the crossing point of $R = 1.22$ Å of the hypothetical "deperturbed" curves of the $b'\,^1\Sigma_u^+$ and $c'\,^1\Sigma_u^+$ states.[18]

3.2. States of CO

The electron configuration of the ground state of CO is

$$(1\sigma)^2\,(2\sigma)^2\,(3\sigma)^2\,(4\sigma)^2\,(1\pi)^4\,(5\sigma)^2$$

Table 6 lists the vertical excitation energies for nine low-lying states in CO at the ground state equilibrium internuclear distance of $R = 2.132$ a.u.[4] The

Table 6. Excited States of CO[a]

State	n^b	(1p–1h)	(1p–1h)+(2p–2h)	Experimental
			ΔE^c	
$a\,^3\Pi(5\sigma \to 2\pi)^{\,d}$	22	7.1	6.0	6.3[e]
$A\,^1\Pi$	22	10.3	8.5	8.4
$a'\,^3\Sigma^+(1\pi \to 2\pi)$	30	9.3	7.9	8.4
$e\,^3\Sigma^{-\,f}$	8	11.5	9.5	9.7
$d\,^3\Delta$	—	10.5	8.9	9.2
$I\,^1\Sigma^-$	8	11.5	9.8	9.9
$D\,^1\Delta$	—	12.0	10.0	10.5
$B\,^1\Sigma^+(\sigma \to \sigma^*)$	30	13.8	11.4	10.8
$C\,^1\Sigma^+(\sigma \to \sigma^*)$	—	13.4	11.4	11.4

[a] All calculations done at an equilibrium internuclear distance of 2.132 a.u.
[b] Number of single-particle hole pairs used in the calculation.
[c] In eV.
[d] Indicates the main component of the excitation relative to the ground state.
[e] The experimental results for the $A\,^1\Pi$, $B\,^1\Sigma^+$, and $C\,^1\Sigma^+$ states are from the electron energy-loss spectrum of V. Meyer, A. Skerbele, and E. Lassettre, *J. Chem. Phys.* **43**, 805 (1965). The experimental results for the other states are from G. Herzberg, T. Hugo, S. Tilford, and J. Simmons, *Can. J. Phys.* **48**, 3004 (1970).
[f] The next four states have the same principal component.

trends displayed in this table are very similar to those of N_2 discussed in Table 2. Again the calculated excitation energies agree well with the experimental values, e.g., a percentage error of 1%–6%. In this article we do not compare our calculated excitation energies with those of other methods since we are primarily interested in discussing and illustrating the equations of motion method.

In Table 7 we compare our calculated oscillator strengths with available experimental data. The $X\,^1\Sigma^+ \to A\,^1\Pi$ transition has been extensively studied by electron impact spectroscopy. Lassettre *et al.*[20] obtained a value of 0.043 for the transition to the $v' = 2$ level of the $A\,^1\Pi$ state form high-energy electron impact data. The calculated value of 0.052 for this transition agrees well with their result. The calculated f value of 0.12 for the $X\,^1\Sigma^+ \to C\,^1\Sigma^+$ transition agrees well with measured value of 0.16.[20] A recent reanalysis of the electron scattering data[21] for the $X\,^1\Sigma^+ \to B\,^1\Sigma^+$ transition indicates that the $B\,^1\Sigma^+$

Table 7. Oscillator Strengths for Transitions in CO

Transition	f_{el}	$q_{v'v''}{}^a$	$f_{el}q_{v'v''}$	Experimental
$X\,^1\Sigma^+ \to A\,^1\Pi$	0.22	$q_{20} \approx 0.24^b$	0.053	0.043[c]
$X\,^1\Sigma^+ \to C\,^1\Sigma^+$	0.12	$q_{00} \approx 1$	0.12	0.16
$X\,^1\Sigma^+ \to B\,^1\Sigma^+$	0.048	$q_{00} \approx 1$	0.048	0.016

[a] Franck–Condon factors for the v' and v'' levels.
[b] P. H. Krupenie, *Natl. Std. Ref. Data Ser., Natl. Bur. Std. (US)* **5** (1966).
[c] Measurements of Ref. 18.

state may be perturbed, suggesting that a more involved analysis of these data may be necessary.

Table 8 shows the excitation energies for these nine transitions at five internuclear distances.[22] These frequencies, together with the available ground state potential energy curves, provide the potential curves for their excited states. The agreement with experiment is quite good. Of special

Table 8. Excitation Energies[a] : States of CO

$X^1\Sigma^+$ $\to R$, Å	$(1p-1h)$	$(1p-1h)$ $+(2p-2h)$	Observed[b]	$(1p-1h)$	$(1p-1h)$ $+(2p-2h)$	Observed[b]
		$a\ ^3\Pi$			$A\ ^1\Pi$	
0.97	8.8	7.4	—	12.2	10.1	—
1.09	7.6	6.5	6.6	11.0	9.1	8.8
1.13	7.2	6.0	6.1	10.3	8.5	8.4
1.21	6.6	5.4	5.6	9.6	7.6	7.7
1.32	5.7	4.4	5.0	8.3	6.1	6.8
		$a'\ ^3\Sigma^+$			$e\ ^3\Sigma^-$	
0.97	13.1	11.2	—	15.6	13.1	—
1.09	10.5	9.1	9.3	12.9	11.0	10.8
1.13	9.5	8.1	8.5	11.9	10.0	9.8
1.21	8.3	6.6	7.0	10.6	8.3	8.2
1.32	6.6	4.5	5.5	8.8	6.0	6.6
		$d\ ^3\Delta$			$I\ ^1\Sigma^-$	
0.97	14.5	12.4	—	15.6	13.3	—
1.09	11.8	10.3	10.1	12.9	11.2	10.9
1.13	10.9	9.3	9.2	11.9	10.2	9.8
1.21	9.6	7.7	7.7	10.6	8.5	8.3
1.32	7.8	5.5	6.1	8.8	6.2	6.7
		$D\ ^1\Delta$			$B\ ^1\Sigma^{+\ c}$	
0.97	16.1	13.5	—	12.3	10.3	10.6[d]
1.09	13.4	11.5	—	12.9[e]	11.0	10.8
1.13	12.5	10.4	10.5	12.8	10.9	10.8
1.21	11.1	8.7	8.4	13.1[e]	11.0	10.8
1.32	9.3	6.3	6.8	12.9	10.5	10.8[d]
		$C\ ^1\Sigma^{+\ c}$				
0.97	13.0	10.9	—			
1.09	13.5[e]	11.4	—			
1.13	13.5	11.3	11.4			
1.21	13.8[e]	11.5	—			
1.32	14.9	12.4	—			

[a] In eV. The experimental R_e is 1.13 Å.
[b] The experimental results are from G. Herzberg, T. Hugo, S. Tilford, and J. Simmons, *Can. J. Phys.* **48**, 3004 (1970) and V. Meyer, A. Skerbele, and E. Lassettre, *J. Chem. Phys.* **43**, 805 (1965).
[c] The calculated excitation energies to the $b\ ^3\Sigma^+$ and $c\ ^3\Sigma^+$ states, the triplet states corresponding to the $B\ ^1\Sigma^+$ and $C\ ^1\Sigma^+$ states, are 10.5 and 11.2 eV at $R = 1.13$ Å compared to the observed values of 10.4 and 11.6 eV, respectively.
[d] Estimated from a plot of the measured values at $R = 1.09$, 1.13, *and* 1.21 Å.
[e] The excitation energies at this point were calculated without the diffuse $p\sigma$ function at the center of charge. This results in excitation frequencies too high by about 0.1 and 0.2 eV for the B and C states, respectively, but has a negligible effect on the frequencies of true valence states.

Table 9. Dependence of the Transition Moment
on Internuclear Distance in the Fourth Positive
Band of CO

R, Å	$M(R)^a$	$M_{obs}{}^b$
0.97	0.753	—
1.09	0.636	0.59
1.13	0.580	0.56
1.21	0.518	—
1.32	0.365	—

^{a} In a.u. See Ref. 4.
^{b} Ref. 25.

interest are the results for the $I\ {}^1\Sigma^-$ state which are all within 2%–7% of the observed values reported by Herzberg *et al.*[23] These results are quite different from those deduced by Krupenie and Weiss.[24] Our results, therefore, suggest that the experimental potential energy curve of Ref. 23 for the $I\ {}^1\Sigma^-$ state is the correct one for $R < 1.09$ Å.

In Table 9 we show the calculated values of the transition moment for the $X\ {}^1\Sigma^+ \to A\ {}^1\Pi$ transition at five internuclear distances. In the second column of this table we also list the experimental values[20] of $M(R)$ at $R \approx 1.09$ and 1.13 Å. This variation of the electronic transition moment with internuclear distance for this band system is important since it removes the discrepancy that originally existed between the total f values derived from lifetime data and electron energy loss spectra.[25,26]

3.3. The *T* and *V* States of Ethylene

Table 10 shows the excitation energies for the $N \to T$, $N \to V$, and $N \to R'''$ transitions of planar ethylene.[4] The T and V states are the triplet and singlet states arising primarily from the $\pi \to \pi^*$ transition, while the $N \to R'''$ transition is the first member of the $N \to nR'''$ Rydberg series according to Wilkinson's assignment.[27] Wilkinson[27] suggested that this R''' series arose from a $\pi \to n\,d\pi_x$ transition and the upper state is of the same symmetry as the V state. The excitation energies for the T and V states are 4.1 and 7.9 eV compared with the observed values of 4.6 and 7.6 eV, respectively. The calculated oscillator strength for the vertical transition is 0.40, compared to the experimental total f value of 0.34 for the $N \to V$ band. Our results yield a π^* orbital for the V state which, although more diffuse than the π^* orbital of the T state, would still be more appropriately characterized as a valence-like molecular orbital. Configuration interaction calculations[28] give a more diffuse π^* orbital, although the most recent and extensive CI calculations of Peyerimhoff lead to a π^* orbital of

Table 10. *The* $N \to T, N \to V,$ *and* $N \to R'''$ *Transitions of* $C_2H_4{}^a$

		ΔE ^{c}				
		(1p–1h)				
Transition	N^b	(1p–1h)	+(2p–2h)	Experimental	$\langle \pi^* \| z^2 \| \pi^* \rangle\ ^d$	$f_{calc}\ ^e$	f_{obs}
$N \to T$	22	4.8	4.1	4.6	2.7	—	—
$N \to V$	22	9.0	7.9	7.6^f	9.0	0.40	0.34^g
$N \to R'''$	22	10.4	8.9	9.05^h	83.3	0.02	—

^{a} Calculations are all done at approximately the ground state geometry (C–C bond length of 1.35 Å, C–H bond length of 1.07 Å, CH–C–H of 120°).
^{b} Number of $1p$–$1h$ pairs used in the calculation.
^{c} In eV.
^{d} The average value of z^2 (perpendicular to the molecular plane) for the π^* orbital (in a.u.^{2}).
^{e} Assuming a Franck–Condon factor of unity for the vertical excitation.
^{f} Maximum in the $N \to V$ absorption.
^{g} Total f value for the transition.
^{h} This is the $N \to R'''$ transition in Wilkinson's assignment. See text and Ref. 27 for discussion.

considerable valence character.[29] The calculated excitation energy of 8.9 eV for the R''' ($n\, d\pi_x$) state agrees well with the value of 9.05 eV observed by Wilkinson.[27]

3.4. States of CO_2

For CO_2 we have used the equations of motion method to calculate the excitation energies, optical oscillator strengths, and the generalized oscillator strengths.[5–7] The generalized oscillator strength, which is directly related to the differential cross sections in the Born approximation, is a very useful quantity in the comparison of experimental and theoretical results. If ΔE_n is the transition energy between the ground and excited state, and R and a_0 denote a Rydberg of energy and the Bohr radius, respectively, the generalized oscillator strength $f_n(K)$ is defined by

$$f_n(K) = (\Delta E_n/R) \frac{1}{(Ka_0)^2} |\varepsilon_n(K)|^2 \tag{33}$$

where K is the momentum transfer and $\varepsilon_n(K)$ is the matrix element-

$$\langle \psi_n | \sum_{j=1}^{N} e^{i\mathbf{K}\cdot\mathbf{r}_j} | \psi_0 \rangle$$

This quantity approaches the optical oscillator strength as the momentum transfer K goes to zero. For further details of the Born approximation in electron–molecule scattering we refer to the review article by Inokuti.[30]

Table 11 shows the assignments in the spectrum of CO_2 suggested by our calculations. The assignment of the $^1\Delta_u$ and $^1\Pi_g$ states in the optical and

Table 11. Assignments in the CO_2 Spectrum

	ΔE^a			
State	$(1p-1h)$ $+(2p-2h)$	Experimental	f	f_{exp}
$^3\Sigma_u^+$	7.35	7.5^b	—	—
$^3\Delta_u$	8.06	8.0^b	—	—
$^1\Delta_u$	8.56	8.41^c	—	—
$^1\Pi_g$	8.62	9.31^c	—	—
$^1\Sigma_u^+$	10.29	11.08^c	0.116	0.12^e
$^1\Pi_u$	10.97	11.4^d	0.168	—

^{a} See Ref. 33 for recent CI results on the spectra of this molecule.
^{b} M. J. Habin-Franskin and J. E. Collins, *Bull. Soc. R. Sci. Liege* **40**, 361 (1971).
^{c} Ref. 32.
^{d} Ref. 31.
^{e} E. C. Y. Inn, K. Watanabe, and M. Zeilikoff, *J. Chem. Phys.* **21**, 1648 (1953).

electron impact spectra is a problem of current interest. In the UV absorption spectrum the peaks observed at 9.31 and 8.41 eV have been assigned as two components of the $^1\Pi_g$ transition split by vibrational effects[31] and as the $^1\Pi_g$ and $^1\Delta_u$, respectively.[32] There is very little doubt that the $^1\Delta_u$ state of CO_2 does lie near 8.4–8.6 eV.[33] From Fig. 1 we see that the differential generalized oscillator strengths or the 8.61 and 9.16 eV energy loss features are qualitatively very much the same. Lassettre and Shiloff[31] assert, on the basis of this evidence and on vibronic arguments, that the two peaks observed are components of the $^1\Pi_g$ transition. The calculated generalized oscillator strengths of the $^1\Pi_g$ and $^1\Delta_u$ transitions are compared wtih the experimental data in Fig. 1. Since the b value normally used to obtain the generalized oscillator strength f (K) from the differential generalized oscillator through $f \approx bf'$ was not determined experimentally, we have used a b value found by a least-squares analysis to derive the generalized oscillator strength for purposes of a qualitative comparison. Clearly, the $f_n(K)$ curve of the $^1\Pi_g$ state is qualitatively much the same as the $f(K)$ for both the 9.16 and 8.61 eV peaks, whereas that of the $^1\Delta_u$ is not. Hence, it does seem that although the $^1\Delta_u$ state lies close to 8.6 eV, its generalized oscillator strength does not have the shape of the generalized oscillator strengths for the 8.61 and 9.16 eV energy loss features which Lassettre observes.[31] One could calculate the generalized oscillator strengths for the $^1\Pi_g$ state as a function of geometry. This would be useful in the assignment of these peaks. A resolution of these difficulties will require not only the potential energy curves for these states but also the generalized oscillator strengths.

Krauss *et al.*[34] and Lawrence[35] have suggested that there is an underlying continuum absorption in CO_2 in the range of 11–14 eV due to a transition to

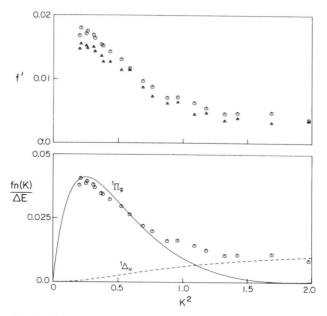

Fig. 1. Upper: Experimental differential generalized oscillator strengths for CO_2 measured at 8.61 eV (triangles) and 9.16 eV (circles) by Lassettre and Shiloff.[31] Lower: Comparison of theoretical results for the $^1\Pi_g$ (solid) and $^1\Delta_u$ states with the differential generalized oscillator strength of the 9.16 eV peak times 0.76.

a valence-like interloper state of $^1\Sigma_u^+$ symmetry. Our calculations yield a valence-like $^1\Sigma_u^+$ state occurring with a vertical transition energy of approximately 12–12.5 eV (from results of two independent calculations), which has an oscillator strength of $f = 0.64$. We cannot definitely assign this state to the continuum absorption without first computing its energy surface.

3.5. States of Benzene

As a final example we now present the results of calculations on the excited states of benzene as an application of the equations of motion method to a larger polyatomic molecule. These calculations were done with a relatively limited basis set and include results both with and without sigma–pi coupling.[5–7]

The SCF calculation in this $[3s2p/1s]$ basis provides 60 molecular orbitals. Table 12 lists the eigenvalues for the 21 molecular orbitals and the first 28 virtual (particle) states. In Table 13 we give the excitation energies for eight excited states of benzene in the π-electron approximation which therefore includes particle–hole pairs from the $1a_{2u}$ and $1e_{1g}$ hole states and $1b_{2g}$, $2a_{2u}$,

Table 12. SCF Orbital Eigenvalues for C_6H_6 [a]

MO	ε_i, a.u.	MO	ε_i, a.u.
$1a_{1g}$	-11.3218	$4e_{1u}$	0.4187
$1e_{1u}$	-11.3224	$4e_{2g}$	0.5149
$1e_{2g}$	-11.3215	$3b_{1u}$	0.5242
$1b_{1u}$	-11.3214	$5e_{2g}$	0.6441
$2a_{1g}$	-1.1945	$5e_{1u}$	0.6447
$2e_{1u}$	-1.0625	$1a_{2g}$	0.7572
$2e_{2g}$	-0.8704	$2b_{2u}$	0.9238
$3a_{1g}$	-0.7470	$2a_{2u}$	0.9897
$1b_{2u}$	-0.6768	$5a_{1g}$	0.9988
$2b_{2u}$	-0.6731	$4b_{1u}$	0.9996
$3e_{1u}$	-0.6276	$6e_{2g}$	1.0574
$1a_{2u}$	-0.5561	$2e_{1g}$	1.0859
$3e_{2g}$	-0.5336	$2e_{2u}$	1.1305
$1e_{1g}$	-0.3889	$6e_{1u}$	1.1735
$1e_{2u}$	0.0984	$2b_{2g}$	1.1995
$1b_{2g}$	0.3187	$2a_{2g}$	1.4051
$4a_{1g}$	0.3327		

[a] In a $[3s\ 2p/1s]$ basis. Total energy is -230.2184 a.u.

$2e_{1g}$, $2e_{2u}$, and $2b_{2g}$ particle states. In the $[(1p-1h)+(2p-2h)]$ approximation these excitation frequencies include the effect of relaxation of the π system in the excitation process. For the calculation including excitations out of the sigma system the hole states $3a_{1g}$, $2b_{1u}$, $1b_{1u}$, $3e_{1u}$, $1a_{2u}$, $3e_{2g}$, and $1e_{1g}$ and all the particle states in Table 12 are used to solve the equations of motion. These results are listed in Table 13. The ordering of the states is correct and the error range for the frequencies is reasonable considering the relatively small basis

Table 13. Excitation Energies of Benzene [a]

	π [b]		$\Sigma-\pi$ [d]		
State	$(1p-1h)$	$(1p-1h)$ $+(2p-2h)$	$(1p-1h)$	$(1p-1h)$ $+(2p-2h)$	Observed
$^1B_{2u}$	7.7	5.7	8.5	5.5	4.9
$^1B_{1u}$	8.4	7.8	8.7	7.4	6.1
$^1E_{2u}$	10.7	9.5^c	10.5	8.3^e	6.9
$^1E_{2g}$	12.9	11.3	12.8	10.9	—
$^3B_{1u}$	3.0	2.7	4.5	3.8	3.7
$^3E_{1u}$	6.1	5.2	7.1	5.2	4.6
$^3B_{2u}$	7.3	7.0	8.2	6.7	5.8
$^3E_{2g}$	8.5	7.9	9.4	8.1	—

[a] In eV.
[b] π approximation, see text.
[c] The transition moment for the $^1A_{1g} \rightarrow {}^1E_{1u}$ transition is 2.09 in this approximation, see text.
[d] See text.
[e] The transition moment for this transition is 1.74 in this approximation. The experimental moment is 1.61.

set. The error in the excitation frequencies becomes larger for the higher states where double and multiple excitations and a more flexible basis are expected to be more important. The results in the π approximation agree well with those including excitations out of the sigma and pi systems, except for the $^3B_{1u}$ and $^1E_{1u}$ states.

The only optically allowed transition of those listed in Table 13 is the $^1A_{1g} \rightarrow {}^1E_{1u}$ transition. The transition moment in the full calculation including $\sigma-\pi$ coupling is 1.74 a.u., which agrees well with the transition moment of 1.61 a.u. derived from the experimental oscillator strength of 0.88.[36] In the π approximation the transition moment is 2.09 a.u. This change in transition moment from 2.09 to 1.74 a.u. leads to a reduction of about 40% in the oscillator strength. Our results also give a $^1A_{2u}(\sigma \rightarrow \pi^*)$ state at 9.3 eV with a transition moment of 0.34 a.u.

4. Equations of Motion Method for Ionization Potentials and Electron Affinities

The determination of electron affinities and ionization potentials is an important problem of both theoretical and experimental interest. The locations and strengths of transitions are important in atmospheric, biological, and interstellar processes.

Theoretically, Koopmans' theorem has been the mainstay of many low-level calculations. From Koopmans' theorem we can say that ionization potentials and electron affinities can, in certain cases, be predicted by the canonical Hartree–Fock orbital energies. This result depends on correlation energy changes and relaxation effects upon electron removal or addition, being approximately equal but opposite in sign and hence cancelling. Although these effects are of opposite sign, there is no theoretical reason for the correlation energy changes and relaxation to be equal in magnitude. For example, in H_2CO using a $[4s3p/2s]$ contracted Gaussian basis set we obtain an orbital energy of -14.64 eV for an electron in the $1b_1$ orbital, while the experimental IP is 14.47 eV. However, the lowest ionization potential is, by Koopmans' theorem, 12.09 eV compared to 10.88 eV, experimental. An additional problem arises for electron affinities where basis sets in an SCF calculation must be large enough to allow for the SCF virtuals to converge to the more spatially diffuse negative ion orbitals.

To remedy the sporadic agreement of Koopmans' theorem with experiment, large-scale configuration interaction calculations can be performed on both the molecular and ionic states.[37] The resulting energy differences are ionization potentials and electron affinities. These are in excellent agreement with experiment but can involve tedious basis-set optimization and large amounts of computer time.

In view of the relative ease of these calculations and the results of the equations of motion method for excitation energies and oscillator strengths, we proposed[38] a similar way to calculate ionization potentials, electron affinities, and the positions of simple electron–molecule resonances. Independently, Simons[39] proposed a similar method. These methods are related to the Green's function method of Cederbaum *et al.*[40] and the propagator method of Purvis and Öhrn.[41]

Consider an operator $O_\lambda^\dagger$ which when operating on the exact initial state $|0\rangle$ with N electrons generates a state with one less or one more electron, i.e.,

$$O_\lambda^\dagger |0, N\rangle = |\lambda, N \pm 1\rangle \tag{34}$$

This is similar to the equations of motion operator $O_\lambda^\dagger$ for excited states,[2] except there $O_\lambda^\dagger$ is an operator which generates excited states of a system with the same number of particles as the ground state.

$O_\lambda^\dagger$ for ionization potentials and electron affinities can be written as a sum of operators with an odd number of electron creation and destruction operators, i.e., the net effect of $O_\lambda^\dagger$ must be the addition or removal of an electron. Rowe's equation of motion[9] can be used to determine the energy change associated with the operator $O_\lambda^\dagger$. Since $O_\lambda^\dagger$ has an odd number of creation and desctruction operators, the appropriate equation is Rowe's equation of motion for Fermi-like transfer operators,[9] i.e.,

$$\langle 0|\{\delta O_\lambda, H, O_\lambda^\dagger\}|0\rangle = \omega_\lambda \langle 0|\{\delta O_\lambda, O_\lambda^\dagger\}|0\rangle \tag{35}$$

where

$$\{A, B, C\} = \tfrac{1}{2}\{[A, B], C\} + \tfrac{1}{2}\{A, [B, C]\} \tag{36}$$

and ω_λ is the negative of the electron affinity or the ionization potential

$$\omega_\lambda = E_\lambda - E_0 \tag{37}$$

Equation (35) is exact. However, it cannot be solved exactly for most systems of chemical interest. There are two approximations which we can use to solve Eq. (35). $O_\lambda^\dagger$ can be written as a sum of odd numbers of creation and destruction operators. This sum may be truncated. For example, for electron affinities $O_\lambda^\dagger$ may be truncated after simple electron addition,

$$O_\lambda^\dagger = \sum_i Y_{i'} a_{i'}^\dagger \tag{38}$$

where the sum is over all spin-orbitals.

A second approximation is made in the choice for $|0\rangle$, e.g., we can choose the ground state to be the Hartree–Fock ground state. The use of the symmetric double anticommutator in Eq. (35) assures that the equations will be of low particle rank. That is, by writing Eq. (35) with as many commutators or anticommutators as we can, the resulting ionization potentials and electron affinities ω_λ will be relatively insensitive to the approximation used for the ground state.

So far all the equations have been completely general and apply to both electron affinities and ionization potentials. For the remainder of this section only ionization potentials will be considered. The theory for electron affinities is analogous. In fact, exactly the same equations result, so that one calculation may yield both ionization potentials and electron affinities.

Furthermore, we restrict either $|0\rangle$ or $|\lambda\rangle$ to closed-shell systems. This restriction is not severe since for many cases either the initial or final system is closed shell, e.g., to calculate the electron affinity of OH we can calculate the ionization potential of the closed shell OH⁻.

To determine a reasonable form for $O_\lambda^\dagger$, consider the initial state $|0\rangle$ to be the Hartree–Fock ground state and all possible double excitations, i.e.,

$$|0\rangle \approx N_0(|HF\rangle + |\chi\rangle) \qquad (39)$$

where N_0 is a normalization constant and $|\chi\rangle$ is a correlation function:

$$|\chi\rangle = \sum_{\substack{m \le n \\ \gamma \le \nu}} C_{\gamma\nu}^{mn} |_{\gamma\nu}^{mn}\rangle \qquad (40)$$

Equation (39) can be rewritten [2]

$$|0\rangle \approx N_0(|HF\rangle + \tfrac{1}{4} \sum_{mn\gamma\nu} [C_{m\gamma,n\delta}(O)(\tfrac{1}{2}a_{m\alpha}^\dagger a_{\gamma\alpha} a_{n\alpha}^\dagger a_{\delta\alpha}$$
$$+ a_{m\alpha}^\dagger a_{\gamma\alpha} a_{n\beta}^\dagger a_{\delta\beta} + a_{m\beta}^\dagger a_{\gamma\beta} a_{n\alpha}^\dagger a_{\delta\alpha} + \tfrac{1}{2}a_{m\beta}^\dagger a_{\gamma\beta} a_{n\beta}^\dagger a_{\delta\beta})$$
$$+ C_{m\gamma,n\delta}(1)(-\tfrac{1}{2}a_{m\alpha}^\dagger a_{\gamma\alpha} a_{n\alpha}^\dagger a_{\delta\alpha} + a_{m\alpha}^\dagger a_{\gamma\alpha} a_{n\beta}^\dagger a_{\delta\beta}$$
$$+ a_{m\beta}^\dagger a_{\gamma\beta} a_{n\alpha}^\dagger a_{\delta\alpha} - \tfrac{1}{2}a_{m\beta}^\dagger a_{\gamma\beta} a_{n\beta}^\dagger a_{\delta\beta})] \qquad (41)$$

All correlation coefficients are assumed to be small.

The important effects for ionization potentials are: (1) removal of an electron from a hole, (2) removal of an electron from a particle level, (3) removal of an electron from a hole and excitation of one of the remaining hole electrons; and (4) removal of an electron from a particle level and deexcitation of the remaining particle electron.

Effects (2)–(4) are higher-order processes for ionization potentials and (2) and (4) do not exist unless the initial state is correlated.

Hence we can write

$$O_\lambda^\dagger = \sum_i Y_i(-a_{i\beta}^\dagger) + \sum_{\substack{r \\ m\gamma\nu \\ \nu \ge \gamma}} Y_{(m\gamma\nu),r}^{(2)}\Gamma_{(m\gamma\nu),r}^\dagger - \sum_{\substack{r \\ pm\gamma \\ p \ge m}} Z_{(pm\gamma),r}^{(2)}\Theta_{(pm\gamma),r} \qquad (42)$$

In Eq. (42) the operators are spin adapted so the subscript r refers to the various possible spin couplings. $\Gamma^\dagger$ is an operator which has the effect of

Table 14. $\Gamma^\dagger$ and Θ Operators $(\nu > \gamma, p > m)$

$$\Gamma^\dagger_{m\gamma\gamma} = -a_{\gamma\beta}a^\dagger_{ma}a_{\gamma\alpha}$$

$$\Gamma^\dagger_{(m\nu\gamma)_1} = -(1/\sqrt{2})(a_{\nu\alpha}a^\dagger_{ma}a_{\gamma\beta} - a_{\nu\beta}a^\dagger_{ma}a_{\gamma\alpha})$$

$$\Gamma^\dagger_{(m\nu\gamma)_2} = \sqrt{2/3}(a_{\nu\beta}a^\dagger_{m\beta}a_{\gamma\beta} + \tfrac{1}{2}a_{\nu\alpha}a^\dagger_{ma}a_{\gamma\beta} + \tfrac{1}{2}a_{\nu\beta}a^\dagger_{ma}a_{\gamma\alpha})$$

$$\Theta_{mm\gamma} = -a_{m\beta}a^\dagger_{\gamma\alpha}a_{ma}$$

$$\Theta_{(pm\gamma)_1} = -(1/\sqrt{2})(a_{p\beta}a^\dagger_{\nu\alpha}a_{ma} - a_{p\alpha}a^\dagger_{\nu\alpha}a_{m\beta})$$

$$\Theta_{(pm\gamma)_2} = -\sqrt{2/3}(a_{m\beta}a^\dagger_{\nu\beta}a_{p\beta} + \tfrac{1}{2}a_{ma}a^\dagger_{\nu\alpha}a_{p\beta} + \tfrac{1}{2}a_{m\beta}a^\dagger_{\nu\alpha}a_{p\alpha})$$

removing an electron from a hole and exciting a different hole electron. Θ is an operator which removes an electron from a particle level and deexcites a remaining electron in a particle state. The $\Gamma^\dagger$ and Θ operators are given in Table 14.

If Eq. (42) is used in Eq. (35) we obtain the following matrix equations:

$$\mathbf{A}Y + (\mathbf{A}^{(1,2)} \quad \mathbf{B}^{(1,2)})\begin{pmatrix} Y^{(2)} \\ Z^{(2)} \end{pmatrix} = \omega_\lambda\left[\mathbf{D}Y + (\mathbf{D}^{(1,2)} \quad \mathbf{F}^{(1,2)})\begin{pmatrix} Y^{(2)} \\ Z^{(2)} \end{pmatrix}\right] \quad (43)$$

$$(\mathbf{A}^{(1,2)} \quad \mathbf{B}^{(1,2)})^\dagger Y + \begin{pmatrix} \mathbf{A}^{(2,2)} & \mathbf{C}^{(2,2)} \\ \mathbf{C}^{(2,2)} & -\mathbf{B}^{(2,2)} \end{pmatrix}\begin{pmatrix} Y^{(2)} \\ Z^{(2)} \end{pmatrix}$$

$$= \omega_\lambda\left[(\mathbf{D}^{(1,2)\dagger} \quad \mathbf{F}^{(1,2)\dagger})Y + \begin{pmatrix} \mathbf{D}^{(2,2)} & \mathbf{G}^{(2,2)} \\ \mathbf{G}^{(2,2)\dagger} & \mathbf{F}^{(2,2)} \end{pmatrix}\begin{pmatrix} Y^{(2)} \\ Z^{(2)} \end{pmatrix}\right] \quad (44)$$

where

$$A_{ij} = \langle 0|\{a^\dagger_{i\beta}, H, a_{j\beta}\}|0\rangle$$

$$D_{ij} = \langle 0|\{a^\dagger_{i\beta}, a_{j\beta}\}|0\rangle \equiv \delta_{ij}$$

$$A^{(1,2)}_{ij(m\gamma\nu)_r} = -\langle 0|\{a^\dagger_{i\beta}, H, \Gamma_{(m\gamma\nu)_r}\}|0\rangle$$

$$B^{(1,2)}_{i;(pm\gamma)_r} = \langle 0|\{a^\dagger_{i\beta}, H, \Theta_{(pm\gamma)_r}\}|0\rangle$$

$$D^{(1,2)}_{i;(m\gamma\nu)_r} = -\langle 0|\{a^\dagger_{i\beta}, \Gamma^\dagger_{(m\gamma\nu)_r}\}|0\rangle$$

$$F^{(1,2)}_{i;(pm\gamma)_r} = \langle 0|\{a^\dagger_{i\beta}, \Theta_{(pm\gamma)_r}\}|0\rangle \qquad (45)$$

$$A^{(2,2)}_{(m\gamma\nu)_r,(m\gamma\nu)_r} = \langle 0|\{\Gamma_{(m\gamma\nu)_r}, H, \Gamma^\dagger_{(m\gamma\nu)_r}\}|0\rangle$$

$$C^{(2,2)}_{(m\gamma\nu)_r,(pm\gamma)_r} = -\langle 0|\{\Gamma_{(m\gamma\nu)_r}, H, \Theta_{(pm\gamma)_r}\}|0\rangle$$

$$D^{(2,2)}_{(m\gamma\nu)_r,(m\gamma\nu)_r} = \langle 0|\{\Gamma_{(m\gamma\nu)_r}, \Gamma^\dagger_{(m\gamma\nu)_r}\}|0\rangle$$

$$G^{(2,2)}_{(m\gamma\nu)_r,(pm\gamma)_r} = \langle 0|\{\Gamma_{(m\gamma\nu)_r}, \Theta_{(pm\gamma)_r}\}|0\rangle$$

$$B^{(2,2)}_{(pm\gamma)_r,(pm\gamma)_r} = -\langle 0|\{\Theta^\dagger_{(pm\gamma)_r}, H, \Theta_{(pm\gamma)_r}\}|0\rangle$$

$$F^{(2,2)}_{(pm\gamma)_r,(pm\gamma)_r} = \langle 0|\{\Theta^\dagger_{(pm\gamma)_r}, \Theta_{(pm\gamma)_r}\}|0\rangle$$

The most important process for ionization potentials is single electron removal. All the matrices which involve an operator that is simply electron removal are in Eq. (43). Equation (44) is coupled to Eq. (43) through the ionization potential ω_λ and the vectors $Y^{(2)}$ and $Z^{(2)}$. Equation (44) can be solved for $\binom{Y^{(2)}}{Z^{(2)}}$, i.e.,

$$\binom{Y^{(2)}}{Z^{(2)}} = -\binom{\mathbf{A}^{(2,2)} - \omega_\lambda \, \mathbf{D}^{(2,2)} \quad \mathbf{C}^{(2,2)} - \omega_\lambda \, \mathbf{G}^{(2,2)}}{\mathbf{C}^{(2,2)\dagger} - \omega_\lambda \, \mathbf{G}^{(2,2)\dagger} \quad -\mathbf{B}^{(2,2)} - \omega_\lambda \, \mathbf{F}^{(2,2)}}^{-1}$$

$$\times (\mathbf{A}^{(1,2)} - \omega_\lambda \mathbf{D}^{(1,2)} \quad \mathbf{B}^{(1,2)} - \omega_\lambda \mathbf{F}^{(1,2)})^\dagger Y \tag{46}$$

$$= -\mathbf{A}^{-1}\mathbf{R}^\dagger Y \tag{47}$$

Combining (47) with (43)

$$\mathbf{A}Y - \mathbf{R}\mathbf{A}^{-1}\mathbf{R}^\dagger Y = \omega_\lambda Y \tag{48}$$

$$(\mathbf{A} - \Delta\mathbf{A}) Y = \omega_\lambda Y \tag{49}$$

Equation (49) must be solved iteratively for ω_λ since $\Delta\mathbf{A}$ depends on ω_λ through Eq. (46). For example, the first guess for ω_λ can be Koopmans' theorem value. It is used in constructing $\Delta\mathbf{A}$. A new ω_λ is chosen from the eigenvalues of Eq. (49) which is the closest to $\omega_\lambda^{\mathrm{Koop}}$ and this is used to form the new $\Delta\mathbf{A}$. This process continues until two successive iterations do not differ by more than a predetermined amount.

The formulas of Eq. (44) in terms of orbital energies and interaction matrix elements are derived using a formula-generating program.[39] The ground state is given in Eq. (41). If the correlation coefficient $C_{m\gamma,n\delta}$ is obtained from Rayleigh–Schrödinger perturbation theory, it is proportional to electron interaction matrix elements. All formulas used in Eq. (49) are derived so that Eq. (49) is third order in the electron interaction matrix elements, i.e., $\mathbf{A}$ formulas are generated to order εC^3 and VC^2, $\mathbf{R}$ formulas to order εC^2 and VC, and $\mathbf{A}$ formulas to order εC and V. This truncation by orders is reasonable since the terms most important for the ionization potentials are contained in $\mathbf{A}$.

$$A_{ij} = -\varepsilon_i \delta_{ij} - \sum_{\nu\gamma} \rho_{\nu\gamma}^{(2)} (2V_{i\nu\gamma j} - V_{i\nu\gamma j}) - \sum_{pq} \rho_{pq} (2V_{ipjq} - V_{ipqj}) \tag{50}$$

where

$$\rho_{\nu\gamma}^{(2)} = -\tfrac{1}{2} \sum_{pq\mu} \sum_{S} C'_{p\mu,q\nu}(S) C_{p\mu,q\gamma}(S)$$

$$\rho_{pq} = \tfrac{1}{2} \sum_{m\mu\nu} \sum_{S} C'_{m\mu,q\nu}(S) C_{m\mu,p\nu}(S) \tag{51}$$

and

$$C'_{p\nu,q\gamma}(0) = \tfrac{3}{4} C_{p\nu,q\gamma}(0) + \tfrac{1}{4} C_{p\nu,q\gamma}(1)$$

$$C'_{p\nu,q\gamma}(1) = \tfrac{1}{4} C_{p\nu,q\gamma}(0) + \tfrac{3}{4} C_{p\nu,q\gamma}(1) \tag{52}$$

Table 15. Summary of the Ionization Potential Calculations of Simons and Co-Workers

Initial molecule or ion	Final state	ΔE (eV), Koopmans' theorem	ΔE (eV), EOM	ΔE (eV), experiment
N_2 [a]	$^2\Sigma_g^+$	17.58	15.69	15.60^d
	$^2\Pi_u$	17.76	17.03	16.98
	$^2\Sigma_u^+$	21.75	18.63	18.78
HF^b	$^2\Pi$	17.79	15.87	16.01^e
	$^2\Sigma^+$	20.85	19.49	19.4
OH^- [c]	$^2\Pi$	3.06	1.76	1.825^f

[a] Ref. 43.
[b] Ref. 44.
[c] Ref. 45.
[d] D. W. Turner, C. Baker, A. D. Baker, and C. R. Brundle, *Molecular Photoelectron Spectroscopy*, Wiley, New York (1970).
[e] J. Berkowitz, W. A. Chupka, P. M. Guyon, J. H. Holloway, and R. Spohr, *J. Chem. Phys.* **54**, 5165 (1971).
[f] H. Hotop, T. A. Patterson, and W. C. Lineberger, *J. Chem. Phys.* **60**, 1806 (1974).

Note that through second order in interaction matrix elements, A_{ij} is purely on-diagonal and is given by Koopmans' theorem.

Matrices $\mathbf{D}^{(1,2)}$, $\mathbf{F}^{(1,2)}$, $\mathbf{C}^{(2,2)}$, and $\mathbf{G}^{(2,2)}$ are zero. Matrices $\mathbf{D}$, $\mathbf{D}^{(2,2)}$, and $\mathbf{F}^{(2,2)}$ are unit matrices. The formulas for matrices $\mathbf{A}^{(1,2)}$, $\mathbf{B}^{(1,2)}$, $\mathbf{A}^{(2,2)}$, and $\mathbf{B}^{(2,2)}$ are more complicated and are not given here.

In actual calculations, the matrix sizes are reduced by choosing $O_\lambda^\dagger$ operators which generate states of a specific spatial and spin symmetry. However, $\mathbf{A}$ is still very large, e.g., for a calculation of the $OH^- X\,^1\Sigma^+ \to {}^2\Pi_g$ state using a $\langle 4s3p2d_\pi/2s1p\rangle + Rs_0 + Rp_0 + Rs_H$ basis set the $\mathbf{A}$ matrix is 1050×1050. A reasonable approximation is to retain only on-diagonal terms in $\mathbf{A}$, similar to the inclusion of double excitation matrix elements in the EOM $[(1p-1h)+(2p-2h)]$ where only on-diagonal orbital energies are retained.

Simons et al.[43-45] have obtained excellent agreement using this method for the vertical ionization potentials of HF and N_2 and the vertical detachment energy of OH^-. Their method differs slightly from the one given here since they do not spin-symmetry adapt the electron removal operators, hence causing $\mathbf{A}$ to have a larger dimensionality. The results of Simons et al.[43-45] are given in Table 15. Using Slater basis sets Simons et al.[43] report agreement to ± 0.15 eV with experiment for diatomic molecules.

5. Photoionization

5.1. Method

Although experimental photoabsorption cross sections have been measured over a wide spectral range for a variety of atoms and molecules, accurate calculations of these cross sections present some difficulty. In most methods for computing photoionization cross sections the principal complication is in the calculation of continuum wave functions, which are particularly hard to construct in the molecular case. Alternatively, several procedures have been proposed for extracting photoionization cross sections from a discrete set of transition energies and oscillator strengths, calculated using bound-state methods.[46–51] The attractiveness of these schemes lies in their easy extension to molecular systems for which the computation of continuum wave functions is a formidable task.

In this section we describe one such L^2 method which we have applied using a discrete representation of the continuum generated by the equations of motion (EOM) method techniques. The method is based on a numerical analytic continuation of the frequency-dependent polarizability as originally suggested by Broad and Reinhardt.[46] It is important to point out that, unlike the RPA calculations of photoionization by Amus'ya *et al.*,[53] Jamieson[54] and Altick and Glassgold,[55] the technique we are describing employs square-integrable basis functions exclusively.

The frequency-dependent polarizability $\alpha(z)$ of an atom or molecule is given by the formal expression[56]

$$\alpha(z) = \sum_{i \neq 0} \frac{f_{oi}}{\omega_{oi}^2 - z^2} + \int_{\varepsilon_1}^{\infty} \frac{g(\varepsilon)\, d\varepsilon}{\varepsilon^2 - z^2} \tag{53}$$

where ω_{oi}, f_{oi}, and $g(\varepsilon)$ are the transition frequencies and the bound and continuum oscillator strengths, respectively, and ε_1 is the first ionization threshold of the system. Taking the limit $z \to \omega + i\eta$ in Eq. (53) yields

$$\alpha(\omega) = \sum_{i \neq 0} \frac{f_{oi}}{\omega_{oi}^2 - \omega^2} + P \int_{\varepsilon_1}^{\infty} \frac{g(\varepsilon)\, d\varepsilon}{\varepsilon^2 - \omega^2} + \frac{i\pi g(\omega)}{2\omega} \tag{54}$$

and thus the result[57]

$$\sigma(\omega) = \lim_{\eta \to 0} \frac{4\pi\omega}{c} I_m[\alpha(\omega + i\eta)], \quad \omega > \varepsilon_1 \tag{55}$$

for the photoionization cross section.

In this work, the polarizability is first approximated by a finite sum

$$\sigma(z) \approx \sum_{i=1}^{n} \frac{f_{oi}}{\omega_{oi}^2 - z^2} \tag{56}$$

Equation (56) can be viewed as a quadrature-like approximation[58] to the frequency-dependent polarizability and might thus be expected to provide an adequate representation of $\alpha(z)$ for complex values of z away from the poles ω_{oi}. This being the case, we construct a low-order rational-fraction representation[59,60] of the polarizability by fitting it to the approximate expression (56) at a number of points in the complex plane. The rational fraction can then be used to obtain smooth values of $\alpha(\omega)$, and hence $\sigma(\omega)$, for real energies where the original finite approximation is invalid. The rational-fraction representa-

$$R_{M,M} = \frac{\sum\limits_{l=0}^{M} P_l z^l}{1 + \sum\limits_{l=1}^{M} Q_l z^l} \tag{57}$$

tion of the polarizability is obtained by matching Eq. (57) to $\alpha(z_i)$ at $2M+1$ complex values of z_i. In actual computations, a continued fraction representation of $R_{M,M}$ is used, whose coefficients are calculated recursively.[61]

It is important to realize that $\alpha(z)$ in the form of Eq. (55) is itself a rational fraction $R_{2n-2,2n}$. Consequently, in fitting $\alpha(z)$ to a rational fraction in the complex plane it is essential to choose a low-order rational fraction so as not to reproduce the pole behavior of the finite sum approximation on the real axis. Since there is no *a priori* rule for choosing the order of the rational fraction or the fitting points, it is important to vary both of these in any numerical application to check that the results are stable.

Computation of photoionization cross sections by numerical analytic continuation requires a discrete set of oscillator strengths and excitation energies with which to construct a representation of the frequency-dependent polarizability as a finite sum as in Eq. (56). Solution of the RPA matrix equation (15) generally yields a set of states which correspond to bound excited states of the system as well as a number of states with transition energies above the ionization threshold of the ground state. The oscillator strengths for all these states may be computed with Eq. (24), and it may be shown that, for the RPA in the limit of a complete set of basis functions, the energy weighted sum rule $S(0) = N$ (where N is the number of electrons) and

$$S(k) = \sum_{i=1}^{N} \omega_{oi}^k f_{oi}$$

is exactly satisfied. The RPA (or HRPA) states are an exceptionally convenient choice for approximating $\alpha(z)$. Their ease of computation for molecular systems, and the small size of the matrices involved, make it feasible to consider photoionization calculations on systems of chemical interest. Furthermore, photoionization cross sections for excited states may be computed from the results of our EOM calculation by using Eq. (12) of Ref. 62 to construct oscillator strengths between the excited state of interest and all other states. Thus the frequency dependent polarizability for metastable states can be approximated and the photoionization cross section can be computed in the same way as that of the ground state, providing information on systems for which experimental measurement would be very difficult.

5.2. Examples

In this section we present some results of the application of the analytic continuation technique to the photoionization of atomic helium in its 1^1S, 2^1S, and 2^3S states as well as the photoionization of molecular hydrogen.[47,48]

The RPA equations for helium were solved using $[12s/8p]$ and $[10s/13p]$ sets of Gaussian basis functions. In each case the oscillator strengths of the two lowest bound solutions of P symmetry agree with experiment to better than 10%. We have also evaluated the energy weighted sums $S(k)$ for $-6 \leq k \leq 2$, and these are shown in Table 16 along with the accurate values of Pekeris and the calculated values of Dalgarno and Chan (4-term variational calculation).[63] A number of different numerical continuations showed that the calculated cross sections are relatively insensitive to the points used to construct the rational fraction. The study of the results in the 13-pole case led to cross sections which were indistinguishable on the scale of Fig. 2. Figure 2 compares the results of our 8- and 13-pole calculations with the experimental cross sections of Samson.[64]

Figures 3 and 4 show the calculated photoionization cross sections for the 2^1S and 2^3S metastable states of helium obtained from the $[12s/8p]$ basis. The fitting points for determining the rational fraction representation of $\alpha(z)$ were chosen with a real part between each pair of transition energies and the imaginary parts were varied over a region of the complex plane. For the different choices of the fitting points the calculated cross sections agree within 2%–8% of one another. In Figs. 3 and 4 we also plot the cross sections obtained by Norcross[65] and Jacobs[66] who used Hartree–Fock and correlated initial state wave functions, respectively, and close coupling final state wave functions. The agreement between these results and the present calculation is good. Within the experimental uncertainty of $\mp 14\%$ the various calculations agree well with the measured values.[67]

Table 16. *Energy-Weighted Oscillator Strength Sums of He*[a]

k	8-term RPA	13-term RPA	Dalgarno and Chan	"Accurate"
-6	1.730	1.731	1.9840	2.0672
-4	1.386	1.386	1.520	1.5616
-2	1.322	1.322	1.3788	1.3912
-1	1.478	1.479	1.5046	1.5050
0	2.001	2.000	2.0	2.0
1	3.883	4.121	3.91	4.35
2	13.306	30.345	13.32	30.32

[a]The 8- and 13-term RPA results are compared with the variational results of Dalgarno and Chan and the accurate results of Pekeris (see Ref. 63). Atomic units are used throughout. k refers to the power of ω_{oi} in the sum $\Sigma_i f_{oi} \omega_{oi}{}^k$.

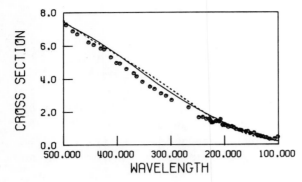

Fig. 2. Comparison of calculated photoionization cross-section of helium with the experimental results of Samson (○). The solid and dashed curves are the results of the 13- and 8-pole RPA calculations, respectively. In each case, the fitting points for the rational-fraction were chosen as $z_i = [(\omega_{oi+1} + \omega_{oi})/2, i2(\omega_{oi+1} - \omega_{oi})]$.

The RPA equations for H_2 were solved in a discrete Gaussian basis. The resulting spectrum consists of fourteen $X\,^1\Sigma_g^+ \to \,^1\Sigma_u^+$ transitions and seven $X\,^1\Sigma_g^+ \to \,^1\Pi_{ux}$ transitions. Hence the parallel and perpendicular components of $\alpha(z)$, Eq. (56), contain fourteen and seven poles, respectively. The calculations were done at the ground state equilibrium geometry of $R = 1.4$ a.u. The results of these calculations have been reported elsewhere,[11,68] where they were used to calculate the second-order optical properties and van der Waals coefficients in good agreement with experimental data as well as with other theoretical

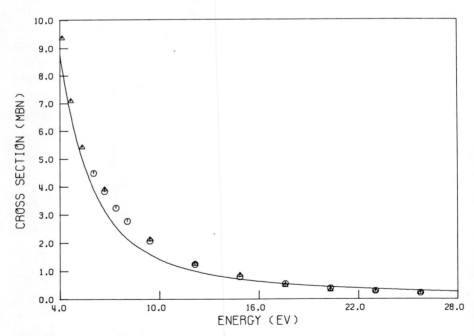

Fig. 3. Photoionization cross sections of the 2^1S state of helium in megabarns. The curve shows the present cross sections obtained by numerical analytic continuation. The triangles and octagons are the calculated results of Norcross (Ref. 65) and Jacobs (Ref. 66), respectively.

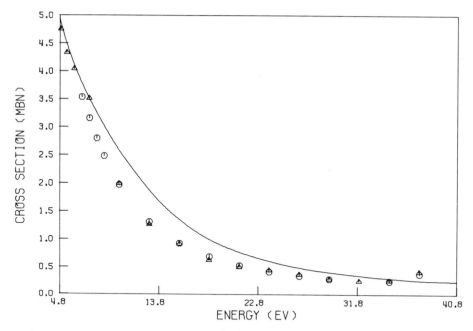

Fig. 4. Photoionization cross sections of the 2^3S state of helium in megabarns. The curve shows the present cross sections obtained by numerical analytic continuation. The triangles and octagons are the calculated results of Norcross (Ref. 65) and Jacobs (Ref. 66), respectively.

estimates. Of particular importance for the present purposes is that the resulting oscillator strength distribution satisfied the Thomas–Reiche–Kuhn sum rule exactly and several other energy-weighted sum rules approximately.[11,68] Moreover, the transition moments for the $X\,^1\Sigma_g^+ \leftarrow B\,^1\Sigma_u^+$ and $X\,^1\Sigma_g^+ \rightarrow C\,^1\Pi_u$ transitions agree very well with those of the extensive calculations of Wolniewicz.[70]

By analytic continuation of the parallel and perpendicular components of $\alpha(z)$ we can obtain the photoionization cross sections for the $^1\Sigma_u^+$ and $^1\Pi_u$ channels separately. The photoionization cross sections were obtained for several choices of fitting points. These points were chosen with a real part between each pair of frequencies and the imaginary parts were varied over a wide region of the complex plane. At photon frequencies near threshold, i.e., the vertical ionization potential of 16.4 eV for H_2, the calculated cross sections are within 5%–10% of one another for the different choices of the fitting points. At frequencies away from threshold the calculated cross sections are insensitive to the choice of fitting points for the analytic continuation.

In Fig. 5 we compare the total calculated photoionization cross sections for H_2 with the experimental results of Cook and Metzger,[71] Samson and Cairns,[72] and Rebbert and Ausloos.[73] The calculated cross sections are those for the vertical photoionization of H_2 and hence the threshold is at 16.4 eV.

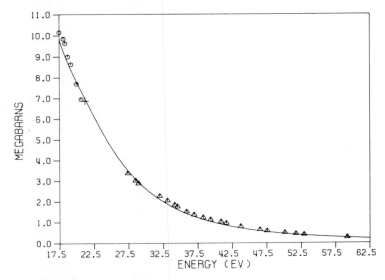

Fig. 5. Comparison of calculated total photoionization cross sections with experimental data. The points are taken from Refs. 71 and 72.

The $v' = 0$, $v'' = 0$ ionization threshold for H_2 is about 15.5 eV. Our calculated cross sections should be compared with the experimental ones only at frequencies above those for which the sum of all Franck–Condon factors for the open vibrational channels of H_2^+ are close to unity. From the Franck–Condon factors for the H_2–H_2^+ system, this begins at photon energies around 18 eV. From Fig. 5 we see that our calculated cross sections agree well with the measured cross sections. These cross sections are closer to the experimental cross sections than those of Kelly,[74] who used molecular continuum eigenfunctions and neglected electron correlation.

6. Electron Scattering Cross Sections

6.1. An Optical Potential and Elastic Scattering Cross Sections

Scattering of an electron by an N-electron atom or molecule is an $N + 1$ electron problem which has the difficulties of constructing continuum wave functions in addition to the usual complications of correlation effects in a many-body bound-state problem. The common approach of expanding the full wave function of the system in eigenstates of the target leads to an infinite set of coupled integrodifferential equations known as the close-coupling equations,[75] which, even in truncated form, are quite formidable (particularly for a molecular target). Alternatively, if one requires only information about elastic scattering cross sections, an effective one-electron potential may be found which reduces the $N+1$ electron problem to an equivalent single-particle potential scattering problem. Such a potential is called an "optical potential" and must contain the effects of energetically accessible and inaccessible inelastic processes.[76] Although optical potentials may, in principle, be constructed exactly, our inability to exactly describe a target containing more than one electron leads us to approximate formulations. The potential we derive here results from an RPA or TDA description of the target in conjunction with the standard Feshbach[77] formalism of the optical potential. Essentially the same results may be derived in a somewhat less straightforward manner by many-body Green's function techniques.[78–80] Here, we will first briefly review the basic ideas of the Feshbach optical model, and proceed to describe its implementation in the RPA. We will consider only closed-shell targets.

Feshbach theory[77] calls for the construction of an operator P which projects out the parts of the exact wave function that describe elastic scattering; that is, if ψ denotes the full wave function for the electron–atom (molecule) system, then P must satisfy the requirement that

$$P\psi \to \mathscr{A}F_0(r)\Phi_0(\{r_i\}) \tag{58}$$

where $\Phi_0(\{r_i\})$ is the ground state wave function of the target and $F_0(r)$, the optical wave function, gives the correct elastic scattering. In addition to the projection operator P, we also require its orthogonal complement Q such that $P+Q=1$. The full wave function ψ satisfies the Schrödinger equation

$$H\psi = E\psi \tag{59}$$

By using the property of idempotency of projection operators ($P^2 = P$, $Q^2 = Q$) and the fact that $P+Q=1$ we obtain

$$(E - PHP)P\psi = PHQ\, Q\psi \tag{60a}$$

$$(E - QHQ)Q\psi = QHP\, P\psi \tag{60b}$$

From Eqs. (60a) and (60b) we may derive an equation satisfied by $P\psi$ containing an effective Hamiltonian which describes elastic scattering:

$$(H_{\text{eff}} - E)P\psi = 0 \tag{61}$$

where

$$H_{\text{eff}} = PHP + PHQ \frac{1}{Q(E - H + i\eta)Q} QHP \tag{62}$$

To implement this theory for a many-electron system we must first construct approximations to the operators P and Q. Suppose, for example, that the ground state of the target is exactly described by the Hartree–Fock wave function $|HF\rangle$ and that its excited states are given by configuration interaction wave functions $|\xi\rangle$ computed with only single excitations from the Hartree–Fock ground state (TDA states). These states form an orthonormal set in which we can write an expansion of the full wave function

$$\psi = f_0(r)|HF\rangle + \sum_\xi f_\xi(r)|\xi\rangle \tag{63}$$

If we convert to the notation of second quantization the appropriate forms of P and Q are

$$P = \sum_p a_p^\dagger|HF\rangle\langle HF|a_p \tag{64a}$$

$$Q = \sum_{\substack{pp' \\ \xi\xi'}} a_p^\dagger|\xi\rangle S_{p\xi,p'\xi'}^{-1}\langle\xi'|a_{p'} \tag{64b}$$

where $S_{p\xi,p'\xi'}^{-1}$ is the inverse of the overlap matrix $\langle\xi|a_p a_p|\xi'\rangle$ and the sums over p and p' are over Hartree–Fock virtual orbitals. These operators satisfy the conditions necessary for them to be projection operators. But more importantly, we can see by expanding the functions f_0 and f_ξ in Eq. (63) in virtual HF orbitals and writing the result in second quantization

$$\psi = \sum_p C_p^0 a_p^\dagger|HF\rangle + \sum_{p\xi} C_p^\xi a_p^\dagger|\xi\rangle \tag{65}$$

that $P\psi$ has the form of Eq. (58). Thus we may proceed to evaluate the effective Hamiltonian of Eq. (62) and develop an expression for the optical potential. In a previous paper this was done and the result is called the TDA optical potential.[81]

We may instead use the RPA description of the target and obtain a more accurate form of the optical potential. If $|0\rangle$ denotes the ground state of the target, then excited states $|\lambda\rangle$ are generated by the RPA excitation operator $O_\lambda^\dagger$,

$$|\lambda\rangle = O_\lambda^\dagger|0\rangle$$

where

$$O_\lambda^\dagger = \sum_{m\alpha} [Y_{m\alpha}(\lambda)a_m^\dagger a_\alpha - Z_{m\alpha}(\lambda)a_\alpha^\dagger a_m]$$

and the sums on m and α run over particle–hole pairs.

Consider the projection operators

$$P = \sum_{pp'} a_p^\dagger|0\rangle M_{pp'}^{-1}\langle 0|a_{p'} \tag{66a}$$

$$Q = \sum_{\substack{pp' \\ \lambda\lambda'}} a_p^\dagger O_\lambda^\dagger|0\rangle S_{\lambda p,\lambda'p'}^{-1}\langle 0|O_{\lambda'}a_{p'} \tag{66b}$$

where

$$M_{pp'} = \langle 0|a_p a_p^\dagger|0\rangle \tag{67}$$

$$S_{\lambda p, \lambda' p'} = \langle 0|O_\lambda a_p a_{p'}^\dagger O_{\lambda'}^\dagger|0\rangle \tag{68}$$

To verify that $P \cdot Q$ is still zero, we must consider a matrix element of the form

$$\langle 0|a_p a_{p'}^\dagger O_\lambda^\dagger|0\rangle$$

Following Shibuya and McKoy,[2] we retain terms in $|0\rangle$ through first order in the correlation coefficients $C_{mn}^{\alpha\beta}$

$$|0\rangle \approx N_0 \left(1 + \sum_{\substack{m\alpha \\ n\beta}} C_{mn}^{\alpha\beta} a_m^\dagger a_n^\dagger a_\alpha a_\beta|HF\rangle\right) \tag{69}$$

With this choice for $|0\rangle$, it follows that the matrix element $\langle 0|a_p a_{p'}^\dagger O_\lambda^\dagger|0\rangle$ is zero and that the choices for P and Q are valid.

In discussing the RPA, we are faced with a problem that is not present when the TDA is used. The ground state $|0\rangle$ is assumed to be correlated, yet the RPA equations that are solved for the excitation operators $O_\lambda^\dagger$ do not specify the correlation coefficients. If the correlation coefficients in Eq. (69) are known (e.g., from Rayleigh–Schrödinger perturbation theory), the effective Hamiltonian can be constructed in a straightforward manner from the definitions of P and Q (although the matrix elements involved will be quite complicated). In this section, however, we will take an approach more consistent with the derivation of the RPA. We will replace matrix elements of the form $\langle 0|A|0\rangle$ by equivalent commutator expressions which lower the particle rank of the operator A, and then approximate $|0\rangle$ by the Hartree–Fock ground state.

We now discuss the construction of the operators in Eq. (62) with the above choice of projectors P and Q. The Hamiltonian may be written in second quantization as[82]

$$H = \sum_i \varepsilon_i : a_i^\dagger a_i : + \tfrac{1}{2} \sum_{ijkl} V_{ijkl} : a_i^\dagger a_j^\dagger a_l a_k : \tag{70}$$

where sums on i, j, k, and l go over hole and particle states (spin orbitals), V_{ijkl} is given by

$$V_{ijkl} = \int \phi_i^*(r_1)\phi_j^*(r_2)\frac{1}{|r_1 - r_2|}\phi_k(r_1)\phi_l(r_2)\,d^3r_1\,d^3r_2 \tag{71}$$

ε_i is a Hartree–Fock orbital energy, and the operators are normal ordered with respect to the HF ground state. The Hartree–Fock ground state energy has been chosen as the zero of energy and has thus been subtracted from the Hamiltonian.

Consider *PHQ* which requires matrix elements of the form

$$\langle 0|a_p H a_{p'}^\dagger O_\lambda^\dagger|0\rangle$$

If $O_\lambda^\dagger$ were an exact solution of the equations of motion for the excitation operator we would have the identity

$$O_\lambda|0\rangle = 0 \tag{1}$$

We apply this relation followed by the approximation $|0\rangle \approx |HF\rangle$ to obtain

$$\langle 0|a_p H a_{p'}^\dagger O_\lambda^\dagger|0\rangle = \langle 0|[a_p H a_{p'}^\dagger, O_\lambda^\dagger]|0\rangle$$
$$\approx \langle HF|[a_p H a_{p'}^\dagger, O_\lambda^\dagger]|HF\rangle \tag{72}$$

With this approximation *PHQ* may be easily evaluated,

$$PHQ = \sum_{\substack{pp'qq' \\ \lambda\lambda'm\alpha}} a_p^\dagger|0\rangle M_{pq}^{-1}\{Y_{m\alpha}(\lambda)\tilde{V}_{q\alpha p'm}$$

$$+ Z_{m\alpha}(\lambda)\tilde{V}_{qmp'\alpha}\}S_{p'\lambda,q'\lambda'}^{-1}\langle 0|O_{\lambda'}a_{q'} \tag{73}$$

where $\tilde{V}_{ijkl} = V_{ijkl} - V_{ijlk}$. To complete the construction of the effective Hamiltonian in this approximation we must evaluate the matrix elements of *QHQ* and then diagonalize. We may construct a diagonal approximation by following Dietrich and Hara[82] in assuming that states formed by adding an electron in an HF particle state to RPA excited states of the target are eigenfunctions of the full Hamiltonian,

$$H a_p^\dagger O_\lambda^\dagger|0\rangle = (\omega_\lambda + \varepsilon_p) a_p^\dagger O_\lambda^\dagger|0\rangle \tag{74}$$

where ω_λ is the RPA excitation energy. We also assume that $S_{\lambda p, \lambda'p'}$ is the unit matrix. To write an approximate effective Hamiltonian which can be constructed from the results of an RPA calculation on the target without knowledge of the correlation coefficients, we must replace $a_p^\dagger|0\rangle$ in Eqs. (66a) and (73) by $a_p^\dagger|HF\rangle$ which makes $M_{pp'}$ the unit matrix and PHP the same operator used in the TDA derivation. With these assumptions the effective Hamiltonian is

$$H_{\text{eff}} = \sum_p a_p^\dagger|HF\rangle\varepsilon_p\langle HF|a_p$$

$$+ \sum_{\substack{pp'q\lambda \\ m\gamma n\delta}} a_p^\dagger|HF\rangle\{Y_{m\gamma}(\lambda)\tilde{V}_{\gamma pmq} + Z_{m\gamma}(\lambda)\tilde{V}_{mp\gamma q}\}\frac{1}{E - \omega_\lambda - \varepsilon_q + i\eta}$$

$$\times \{Y_{n\delta}^*(\lambda)\tilde{V}_{nq\delta p'} + Z_{n\delta}^*(\lambda)\tilde{V}_{\delta qnp'}\}\langle HF|a_{p'} \tag{75}$$

Equation (75) for H_{eff} is an operator expression which is valid in the subspace of $N+1$ particle vectors spanned by the set $a_p^\dagger|HF\rangle$. It has the form

$$H_{\text{eff}} = \sum_{pp'} a_p^\dagger|HF\rangle H_{\text{eff}}^{pp'}\langle HF|a_{p'} \tag{76}$$

We can identify $H_{\text{eff}}^{pp'}$ as a matrix element of the effective one-body Hamiltonian H^{opt} that gives the optical model wave function $F_0(r)$ and satisfies the equation

$$(E + \tfrac{1}{2}\nabla_r^2)F_0(r) = H^{\text{opt}}(r, r')F(r') \tag{77}$$

$H^{\text{opt}}(r, r')$ is easily extracted from Eq. (75) as*

$$H^{\text{opt}}_{(r,r')} = V_{HF}(r, r')$$

$$+ \sum_{\substack{m\gamma \\ n\delta \\ q\lambda}} \{[Y_{m\gamma}(\lambda)V_{\gamma m}(r)\phi_q(r) + Z_{m\gamma}(\lambda)V_{m\gamma}(r)\phi_q(r)]$$

$$- [Y_{m\gamma}(\lambda)V_{\gamma q}(r)\phi_m(r) + Z_{m\gamma}(\lambda)V_{mq}(r)\phi_\gamma(r)]\}$$

$$\times \frac{1}{E - \omega_\lambda - \varepsilon_q + i\eta}\{[Y_{n\delta}^*(\lambda)V_{q\delta}(r')\phi_q^*(r') + Z_{n\delta}^*(\lambda)V_{\delta n}(r')\phi_q^*(r')]$$

$$- [Y_{n\delta}^*(\lambda)V_{q\delta}(r')\phi_n^*(r') + Z_{n\delta}^*(\lambda)V_{qn}(r')\phi_\delta^*(r')]\} \tag{78}$$

where the sums on m, n, and q run over particle states, the sums on γ and δ over hole states and

$$V_{ij}(r) \equiv \int \phi_i^*(r')\phi_j(r')\frac{1}{|r-r'|}d^3r' \tag{79}$$

The Hartree–Fock potential $v_{HF}(r, r')$ which came from PHP is

$$V_{HF}(r, r') = \sum \frac{Z_n}{R_{in}}\delta(r-r') + \sum_\gamma J_\gamma(r)\delta(r-r') - K_\gamma(r, r') \tag{80}$$

where the first term gives the nuclear attraction and J_γ and K_γ are the usual Coulomb and exchange operators.

Thus we have constructed an approximate optical potential which reduces the elastic scattering of electrons by an atom or molecule to a one-particle potential scattering problem. Cross sections for scattering from this potential can be computed in a variety of ways. For example, standard techniques of numerical integration of the differential equation may be applied or, in the case of a nonspherical target, recently developed basis-set methods for elastic electron–molecule scattering[83,84] can be used. The restriction to closed-shell systems is not a severe one, since most small molecules of interest have closed-shell ground states.

*See Note Added in Proof, p. 382.

6.2. Inelastic Cross Sections for Excitation of Electronically Excited States of Atoms and Molecules

In this section, we show how the EOM technique can be used to derive approximations to matrix elements of the transition operator that describe the inelastic scattering of an electron by a closed-shell atom or molecule.

A completely rigorous treatment of inelastic scattering would involve a coupling of the initial and final target electronic states to other energetically accessible (open channel) as well as virtual (closed channel) target states and the solution of the set of resulting coupled integrodifferential equations. From this point of view, elastic and inelastic electron scattering cannot be viewed as two separate problems, but rather as two different pieces of information that one extracts from the solution of coupled equations that describe the full many-electron problem.

For the purposes of this review, we will confine ourselves to the discussion of an approximation scheme which in collision theory is usually called the distorted-wave approximation. In this approximation, one only treats the initial and final electronic target states and regards the electron as being inelastically scattered by a static, one-body transition potential. It is in the construction of this potential that the EOM method can be applied.[85]

The equations of the distorted-wave approximation can be easily derived using the so-called two-potential formula of collision theory.[86] The full interaction potential V between an incident electron and an N-electron atom or molecule is given by

$$V = \sum_{\text{nuclei}} \left(-\frac{Z_n}{r_n}\right) + \sum_{i=1}^{N} \frac{1}{|r - r_i|} \tag{81}$$

The interaction potential is partitioned into two terms, $V = V_I + V_{II}$, in such a way that the amplitude for V_I can be easily found and that the effects of V_{II} are small. It is then easily shown that the exact inelastic transition matrix element T_{fi} is given by[86]

$$T_{fi} = T_{fi}^I + \langle k_f^- | V_{II} | \psi^+ \rangle \tag{82}$$

where T_{fi}^I is the scattering amplitude for V_I, and $|\psi^+\rangle$ is the exact $N+1$ particle wave function. $|k_i^\pm\rangle$ and $|k_f^\pm\rangle$ are the incoming (outgoing) wave functions of V_I at the appropriate initial and final energies. Energy conservation for this problem is given by the equation

$$E = E_{\text{initial}} + \tfrac{1}{2}k_i^2 = E_{\text{final}} + k_f^2 \tag{83}$$

where E_{initial} and E_{final} refer to the electronic target states and $k_{i(f)}$ is the momentum of the incident (scattered) electron. The inelastic cross section is given in terms of T_{fi} by the equation (atomic units are implied),

$$\sigma(k_f \leftarrow k_i) = \frac{k_f}{k_i} \frac{1}{4\pi^2} |T_{fi}|^2 \tag{84}$$

A reasonable choice for V_I is the ground state Hartree–Fock potential, so that

$$V_I = \sum_{\text{nuclei}} \left(-\frac{Z_n}{r_n}\right) + J - K \tag{85a}$$

$$V_{II} = \sum_{i=1}^{N} \frac{1}{|r - r_i|} - J + K \tag{85b}$$

where J and K are the usual Coulomb and exchange operators. Since V_I only contains the coordinates of the incident electron, $|k_i^{\pm}\rangle$ and $|k_f^{\pm}\rangle$, which are scattering solutions of the operator $-\frac{1}{2}\nabla_r^2 + H_{\text{target}} + V_I$, can be immediately written as

$$|k_i^{\pm}\rangle = \psi_0(r_1, \ldots, r_N)\phi_{k_i}^{\pm}(r) \tag{86a}$$

$$|k_f^{\pm}\rangle = \psi_f(r_1, \ldots, r_N)\phi_{k_f}^{\pm}(r) \tag{86b}$$

ψ_0 and ψ_f are the ground and final state target wave functions and $\phi_{k_i}^{\pm}$ and $\phi_{k_f}^{\pm}$ are Hartree–Fock continuum orbitals. Note that these solutions are antisymmetric under interchange of r with the internal target coordinates $\{r_1, \ldots, r_N\}$ since they come from a nonsymmetric Hamiltonian. Thus the first term of Eq. (82)

$$T_{fi}^I = \langle \psi_f(r_1, \ldots, r_N)e^{ik_f r} | V^I | \psi_0(r_1, \ldots, r_N)\phi_{k_i}^{+}(r)\rangle, \tag{87}$$

is rigorously zero due to the orthogonality of ψ_0 and ψ_f.

Equation (82) is exact; we arrive at the distorted-wave approximation by setting the exact $N+1$ electron scattering wave function ψ^+ equal to an antisymmetrized product of ψ_0 and $\phi_{k_i}^+$, giving

$$T_{fi} \sim \langle \psi_f(r_1, \ldots, r_N)\phi_{k_f}^{-}(r) | V_{II} | \mathcal{A}[\psi_0(r_1, \ldots, r_N)\phi_{k_i}^{+}(r)]\rangle \tag{88}$$

Without too much trouble, it can be shown that Eq. (88), which now contains direct and exchange contributions, can be conveniently grouped into two terms:

$$T_{fi} = N\left(\psi_f(r_1, \ldots, r_N)\phi_{k_f}^{-}(r) \left| \frac{1}{|r-r_1|} \right| \Psi_0(r_1, \ldots, r_N)\phi_{k_i}^{-}(r) - \Psi_0(r, \ldots, r_N)\phi_{k_i}^{+}(r_1)\right)$$

$$- N\left(\psi_f(r_1, \ldots, r_N)\phi_{k_f}^{-}(r) \left| \sum_{i \neq 1} \frac{1}{|r-r_i|} - J + K \right| \phi_{k_i}^{+}(r_1)\Psi_0(r, \ldots, r_N)\right) \tag{89}$$

We next show that, in the RPA, the first term above can be simplified and that the second term vanishes. In order to demonstrate this, the matrix elements in Eq. (89) must be rewritten in second-quantized form and hence permutation operators will be introduced so that the coordinates of the wave functions appear in identical order on both sides of the matrix elements. In second quantization, $\psi_0(r_1, \ldots, r_N)$ and $\psi_f(r_1, \ldots, r_N)$ become the state vectors $|0\rangle$ and $|f\rangle$.

For example, the first term in Eq. (89) can be expressed as

$$N\left(\psi_f(r_1, \ldots, r_N)\phi_{k_f}^-(r)\Big|\frac{1}{|r-r_1|}(1-P_{r,r_1})\Big|\Psi_0(r_1, \ldots, r_N)\phi_{k_i}^+(r)\right)$$

$$= \sum_{ij} \langle f|a_i^+ a_j|0\rangle u_{ij}$$

$$\equiv \langle f|\hat{U}|0\rangle \tag{90}$$

where P_{r,r_1} interchanges coordinates r and r_1,

$$u_{ij} = \left(\phi_i(r)\phi_{k_f}^-(r')\Big|\frac{1}{|r-r'|}\Big|\phi_j(r)\phi_{k_f}^+(r')\right)$$

$$- \left(\phi_i(r)\phi_{k_f}^-(r')\Big|\frac{1}{|r-r'|}\Big|\phi_{k_f}^+(r)\phi_j(r')\right)$$

$$\equiv (ik_f^-\|jk_f^+)_a \tag{91}$$

and the sums on i and j run over a complete set of (occupied and virtual) HF orbitals. Using the excitation operator $\hat{O}_f^+$, which generates the excited target state $|f\rangle$ from the ground state $|0\rangle$, this entire expression becomes

$$\langle 0|\hat{O}_f\hat{U}|0\rangle$$

We complete the derivation by making the same approximation, described in the previous section, of replacing the matrix element by the equivalent commutator expression, which lowers the particle rank, and then replacing $|0\rangle$ by the HF vacuum to get

$$\sum_{m,\alpha} u_{m\alpha}Y_{m\alpha}^{*f} + u_{\alpha m}Z_{m\alpha}^* \tag{92}$$

where the sums on m and α run over all particle–hole pairs.

Finally, we consider the second part of Eq. (89). The term involving $-J+K$ can again be written in second quantization as the matrix element of a one-electron operator,

$$-N(\psi_f(r_1, \ldots, r_N)\phi_{k_f}^-(r)|(-J+K)P_{r,r_1}|\phi_{k_i}^+(r)\Psi_0(r_1, \ldots, r_N)) = \sum_{ij} \langle f|a_i^+ a_j|0\rangle w_{ij} \tag{93}$$

where

$$w_{ij} = \delta_{i,k_i^+} \sum_u (k_f^- u\|ju)_a \tag{94}$$

In the RPA, this becomes

$$\sum_{\alpha,u} Y^{*(f)}_{k\,\bar{i},\alpha}(k_f^- u \| \alpha u)_a$$

In similar fashion, the term involving $\sum_{i \neq 1} 1/|r - r_1|$ can be written as the matrix element of a two-electron operator:

$$-\sum_{ijkl} \langle f | a_i^+ a_j^+ a_k a_l | 0 \rangle V_{ijkl}$$

with

$$V_{ijlk} = \delta_{i,\,k}^{\,+}(k_f^- j \| lk) \tag{95}$$

which, in the RPA, becomes

$$-\sum_{ijklm\alpha} \langle 0 | [Y^*_{m\alpha} a_\alpha^+ a_m - Z^{*(f)}_{m\alpha} a_m^+ a_\alpha,\; a_i^+ a_j^+ a_k a_l] | 0 \rangle V_{ijlk}$$

Straightforward evaluation of this expression gives

$$-\sum_{\alpha u} Y^{*(f)}_{k\,\bar{i},\alpha}(k_f^- u \| \alpha u)_a$$

which cancels the term in Eq. (93) identically, giving the simple expression for T_{fi}

$$T_{fi} = \sum_{m,\alpha} u_{m\alpha} Y^{*(f)}_{m\alpha} + u_{\alpha m} Z^{*(f)}_{m\alpha} \tag{96}$$

This is the final expression for the inelastic scattering amplitude in the distorted-wave random phase approximation.[85] This same expression has been derived in the context of many-body Green's functions by Csanak, Taylor, and Yaris[87] and applied by Taylor and co-workers[88] to the inelastic e^-–He problem with great success. Preliminary results on the application of this formalism to e^-–H_2 scattering indicate that it also provides a powerful method for describing the low-energy, inelastic e^-–molecule scattering problem.[89]

It is important to bear in mind that Eq. (88) is simply an expression for the scattering amplitude in the distorted-wave approximation. The RPA is only used to construct the static, transition potential and embodies some of the effects of target electron correlations. It *does not* describe correlations between the target electrons and the scattered electron. It should also be noted that some care is needed in evaluating the matrix elements $u_{m\alpha}$ of Eq. (96). The RPA transition potential can certainly be solved to arbitrary accuracy using a finite basis-set expansion, but the $u_{m\alpha}$ involve an integration over HF continuum orbitals which must reflect appropriate asymptotic boundary conditions. For atomic systems, these orbitals can be found by numerical integration, as was done by Taylor *et al.*[88] in their calculations on He, but this is considerably more difficult to do in molecular problems.[90] However, there are also promising basis-set expansion methods that can be readily applied in the molecular continuum problems.[84]

At high incident electron energies, it is customary to drop the exchange term in the definition of $u_{m\alpha}$ and to replace $\phi_k^{\pm}$ by a plane wave. The result is simply the Born approximation (coupled with the RPA) which has also been applied to high-energy electron scattering by various atoms and molecules.[5-7,91,92]

Finally, we should like to point out that, although this discussion has been confined to the RPA, one could easily go through the derivation using the HRPA. This would be especially useful in obtaining cross sections for the excitation of low-lying triplet states which are often poorly described by the RPA.[2,12]

Note Added in Proof

It has been called to our attention (Robert Yaris and Poul Jorgensen, private communication) that the second term in Eq. (78) is in error by a factor of 2. This results from the overcounting introduced by the definition of the Q operator in Eq. (66b). A suitable restriction of the summation over particle indices leads to the correct result with a factor of $\frac{1}{2}$.

References

1. D. J. Rowe, Equations-of-motion method and the extended shell model, *Rev. Mod. Phys.* **40**, 153–166 (1968).
2. T. Shibuya and V. McKoy, Higher random phase approximation as an approximation to the equations of motion, *Phys. Rev. A* **2**, 2208–2218 (1970).
3. T. Shibuya, J. Rose, and V. McKoy, Equations-of-motion method including renormalization and double-excitation mixing, *J. Chem. Phys.* **58**, 500–507 (1973).
4. J. Rose, T. Shibuya, and V. McKoy, Application of the equations-of-motion method to the excited states of N_2, CO, and C_2H_4, *J. Chem. Phys.* **58**, 74–83 (1973).
5. C. W. McCurdy, Jr. and V. McKoy, Equations of motion method: Inelastic electron scattering for helium and CO_2 in the Born approximation, *J. Chem. Phys.* **61**, 2820–2826 (1974).
6. D. L. Yeager and V. McKoy, Equations of motion method: Excitation energies and intensities of formaldehyde, *J. Chem. Phys.* **60**, 2714–2716 (1974).
7. J. Rose, T. Shibuya, and V. McKoy, Electronic excitations of benzene from the equations of motion method, *J. Chem. Phys.* **60**, 2700–2702 (1974).
8. D. J. Rowe, General variational equations for stationary and time-dependent states, *Nucl. Phys. A* **107**, 99–105 (1968).
9. D. J. Rowe, *Nuclear Collective Motion, Models, and Theory,* Methuen and Co. Ltd., London (1970).
10. See, for example, A. L. Fetter and J. D. Walecka, *Quantum Theory of Many-Particle Systems,* McGraw-Hill, New York (1971).
11. P. H. S. Martin, W. H. Henneker, and V. McKoy, Dipole properties of atoms and molecules in the random phase approximation, *J. Chem. Phys.* **62**, 69–79 (1975).

12. D. J. Thouless, Vibrational states of nuclei in the random phase approximation, *Nucl. Phys.* **22**, 78–95 (1961).

13. D. L. Yeager and V. McKoy, An equations of motion approach for open shell systems, *J. Chem. Phys.* **63**, 4861 (1975).

14. W. J. Hunt, T. H. Dunning, Jr., and W. A. Goddard, The orthogonality constrained basis set expansion method for treating off-diagonal Lagrange multipliers in calculations of electronic wave functions, *Chem. Phys. Lett.* **3**, 606–610 (1969).

15. P. Jorgensen, Electronic excitations of open-shell systems in the grand canonical and canonical time-dependent Hartree–Fock models. Applications on hydrocarbon radical ions, *J. Chem. Phys.* **57**, 4884–4892 (1972).

16. L. Armstrong, Jr., An open-shell random phase approximation, *J. Phys. B* **7**, 2320–2331 (1974).

17. W. Coughran, J. Rose, T. Shibuya, and V. McKoy, Equations of motion method: Potential energy curves for N_2, CO, and C_2H_4, *J. Chem. Phys.* **58**, 2699–2709 (1973).

18. K. Dressler, The lowest valence and Rydberg states in the dipole-allowed absorption spectrum of nitrogen. A survey of their interactions. *Can. J. Phys.* **47**, 547–561 (1969).

19. H. Lefebvre-Brion, Theoretical study of homogeneous perturbations. II. Least-squares fitting method to obtain "deperturbed" crossing Morse curves. Application to the perturbed $^1\Sigma_u^+$ states of N_2, *Can. J. Phys.* **47**, 541–546 (1969).

20. E. Lassettre and A. Skerbele, Absolute generalized oscillator strengths for four electronic transitions in carbon monoxide, *J. Chem. Phys.* **54**, 1597–1607 (1971).

21. K. N. Klump and E. N. Lassettre, Relative vibrational intensities for the $B\,^1\Sigma^+ \leftarrow X\,^1\Sigma^+$ transition in carbon monoxide, *J. Chem. Phys.* **60**, 4830–4832 (1974).

22. The basis set used in these calculations is different from that of Ref. 4. See Ref. 15 for details.

23. G. Herzberg, T. Hugo, S. Tilford, and J. Simmons, Rotational analysis of the forbidden $d^3\Delta_i \leftarrow X\,^1\Sigma^+$ absorption bands of carbon monoxide, *Can. J. Phys.* **48**, 3004–3015 (1970).

24. P. H. Krupenie and S. Weiss, Potential energy curves for CO and CO^+, *J. Chem. Phys.* **43**, 1529–1534 (1965).

25. V. D. Meyer, A. Skerbele, and E. N. Lassettre, Intensity distribution in the electron-impact spectrum of carbon monoxide at high-resolution and small scattering angles, *J. Chem. Phys.* **43**, 805–816 (1965).

26. M. J. Mumma, E. J. Stone, and E. C. Zipf, Excitation of the CO fourth positive band system by electron impact on carbon monoxide and carbon dioxide, *J. Chem. Phys.* **54**, 2627–2634 (1971).

27. P. G. Wilkinson, Absorption spectra of ethylene and ethylene-d_4 in the vacuum ultraviolet. II. *Can. J. Phys.* **34**, 643–652 (1956).

28. C. F. Bender, T. H. Dunning, Jr., H. F. Schaefer III, W. A. Goddard III, and W. J. Hunt, Multiconfiguration wavefunctions for the lowest $(\pi\pi^*)$ excited states of ethylene, *Chem. Phys. Lett.* **15**, 171–178 (1972).

29. R. J. Buenker and S. D. Peyerimhoff, All-valence-electron CM calculations for the characterization of the $^1(\pi, \pi^*)$ states of ethylene, in press.

30. M. Inokuti, Inelastic collisions of fast charged particles with atoms and molecules. The Bethe theory revisited, *Rev. Mod. Phys.* **43**, 297–347 (1971).

31. E. N. Lassettre and J. C. Shiloff, Collision cross-section study of CO_2, *J. Chem. Phys.* **43**, 560–571 (1965).

32. J. W. Rabalais, J. M. McDonald, V. Scherr, and S. P. McGlynn, Electronic spectroscopy of isoelectronic molecules. II. Linear triatomic groupings containing sixteen valence electrons, *Chem. Rev.* **71**, 73–108 (1971).

33. For the results of extensive CI calculations, see N. W. Winter, C. F. Bender, and W. A. Goddard III, Theoretical assignments of the low-lying electronic states of carbon dioxide, *Chem. Phys. Lett.* **20**, 489–492 (1973).

34. M. Krauss, S. R. Mielczarek, D. Neumann, and C. E. Kuyatt, Mechanism for production of the fourth positive band system of CO by electron impact on CO_2, *J. Geophys. Res.* **76**, 3733–3737 (1971).

35. G. M. Lawrence, Photodissociation of CO_2 to produce $CO(a^3\Pi)$, *J. Chem. Phys.* **56**, 3435–3442 (1972).

36. V. J. Hammond and W. C. Price, Oscillator strengths of the vacuum ultraviolet absorption bands of benzene and ethylene, *Trans. Faraday Soc.* **51**, 605–610 (1955).

37. See, for example, E. Clementi and A. D. McLean, Atomic negative ions, *Phys. Rev.* **133**, A419–A423 (1964).

38. D. L. Yeager, Ph.D. candidacy examination report, California Institute of Technology, March 1972.

39. J. Simons and W. D. Smith, Theory of electron affinities of small molecules, *J. Chem. Phys.* **58**, 4899–4907 (1973).

40. L. S. Cederbaum, G. Hohlneicher, and W. V. Niessen, Improved calculations of ionization potentials of closed-shell molecules, *Mol. Phys.* **26**, 1405–1424 (1973).

41. G. D. Purvis and Y. Öhrn, Atomic and molecular electronic spectra and properties from the electron propagator, *J. Chem. Phys.* **60**, 4063–4069 (1974).

42. D. L. Yeager, Ph.D. thesis, California Institute of Technology (February 1975).

43. T. Chen, W. Smith and J. Simons, Theoretical studies of molecular ions. Vertical ionization potentials of the nitrogen molecule, *Chem. Phys. Lett.* **26**, 296–300 (1974).

44. W. Smith, T. Chen, and J. Simons, Theoretical studies of molecular ions. Vertical ionization potentials of hydrogen fluoride, *J. Chem. Phys.* **61**, 2670–2674 (1974).

45. W. D. Smith, T. Chen, and J. Simons, Theoretical studies of molecular ions. Vertical detachment energy of OH^-, *Chem. Phys. Lett.* **27**, 499–502 (1974).

46. J. T. Broad and W. P. Reinhardt, Calculation of photoionization cross sections using L^2 basis sets, *J. Chem. Phys.* **60**, 2182–2183 (1974).

47. T. N. Rescigno, C. W. McCurdy, and V. McKoy, Calculation of helium photoionization in the random phase approximation using square-integrable basis functions, *Phys. Rev. A* **9**, 2409–2412 (1974).

48. P. H. S. Martin, T. N. Rescigno, V. McKoy, and W. H. Henneker, Photoionization cross sections for H_2 in the random phase approximation with a square-integrable basis, *Chem. Phys. Lett.* **29**, 496–501 (1974).

49. P. W. Langhoff, Stieltjes imaging of atomic and molecular photoabsorption profiles, *Chem. Phys. lett.* **22**, 60–64 (1973).

50. P. W. Langhoff and C. T. Corcoran, Stieltjes imaging of photoabsorption and dispersion profiles, *J. Chem. Phys.* **61**, 146–159 (1974).

51. A. Dalgarno, H. Doyle, and M. Oppenheimer, Calculation of photoabsorption processes in helium, *Phys. Rev. Lett.* **29**, 1051–1052 (1972).

52. H. Doyle, M. Oppenheimer, and A. Dalgarno, Bound-state expansion method for calculating resonance and nonresonance contributions to continuum processes: Theoretical development and application to the photoionization of helium, *Phys Rev. A* **11**, 909 (1975).

53. M. Ya. Amus'ya, N. A. Cherepkov, and L. V. Chernysheva, Cross sections for the photoionization of noble-gas atoms with allowance for multielectron correlations, *Sov. Phys.-JETP* **33**, 90–96 (1971).

54. M. J. Jamieson, Time-dependent Hartree–Fock theory for atoms, *Int. J. Quantum Chem.* **S4**, 103–115 (1971).

55. P. L. Altick and A. E. Glassgold, Correlation effects in atomic structure using the random phase approximation, *Phys. Rev.* **133**, A632–A646 (1964).

56. P. W. Langhoff and M. Karplus, Padé approximants to the normal dispersion expansion of dynamic polarizabilities, *J. Chem. Phys.* **52**, 1435–1449 (1970).

57. U. Fano and J. W. Cooper, Spectral distribution of atomic oscillator strengths, *Rev. Mod. Phys.* **40**, 441–507 (1968).

58. The quadrature-like approximation implicit in the use of an L^2-basis set has been examined in the context of Fredholm scattering calculations. See E. J. Heller, T. N. Rescigno, and W. P. Reinhardt, Extraction of accurate scattering information from Fredholm determinants calculated in an L^2 basis: A Chebyschev discretization of the continuum, *Phys. Rev. A* **8**, 2946–2951 (1973).

59. L. Schlessinger and C. Schwartz, Analyticity as a useful computational tool, *Phys. Rev. Lett.* **16**, 1173–1174 (1966).

60. L. Schlessinger, Use of analyticity in the calculation of nonrelativistic scattering amplitudes, *Phys. Rev.* **167**, 1411–1423 (1968).

61. H. S. Wall, *The Analytic Theory of Continued Fractions*, Van Nostrand, Princeton, New Jersey (1968).

62. D. L. Yeager, M. Nascimento, and V. McKoy, Some applications of excited state-excited state transition densities, *Phys. Rev. A* **11**, 1168 (1975).

63. Y. M. Chan and A. Dalgarno, The dipole spectrum and properties of helium, *Proc. Phys. Soc. London* **86**, 777–782 (1965).

64. J. A. R. Samson, The measurement of the photoionization cross sections of the atomic gases, in: *Advances in Atomic and Molecular Physics*, Vol. 2, pp. 177–261, Academic Press, New York (1966).

65. D. W. Norcross, Photoionization of the metastable states, *J. Phys. B* **4**, 652–657 (1971).

66. V. L. Jacobs, Photoionization from excited states of helium, *Phys. Rev. A* **9**, 1938–1946 (1974).

67. R. F. Stebbings, F. B. Dunning, F. K. Tittel, and R. D. Rundel, Photoionization of helium metastable atoms near threshold, *Phys. Rev. Lett.* **30**, 815–817 (1973).

68. P. H. S. Martin, W. H. Henneker, and V. McKoy, Second-order optical properties and Van der Waals coefficients of atoms and molecules in the random phase approximation, *Chem. Phys. Lett.* **27**, 52–56 (1974).

69. A. L. Ford and J. C. Broione, Direct-resolvent-operator computations on the hydrogen-molecule dynamic polarizability, Rayleigh, and Raman scattering, *Phys. Rev. A* **7**, 418–426 (1973).

70. L. Wolniewicz, Theoretical investigation of the transition probabilities in the hydrogen molecule, *J. Chem. Phys.* **51**, 5002–5008 (1969).

71. G. R. Cook and P. H. Metzger, Photoionization and absorption cross sections of H_2 and D_2 in the vacuum ultraviolet region, *J. Opt. Soc. Am.* **54**, 968–972 (1964).

72. J. A. R. Samson and R. B. Cairns, Total absorption cross sections of H_2, N_2, and O_2 in the region 550–200 Å, *J. Opt. Soc. Am.* **55**, 1035 (1965).

73. R. E. Rebbert and P. Ausloos, Ionization quantum yields and absorption coefficients of selected compounds at 58.4 and 73.6–74.4 nm, *J. Res. Nat. Bur. Stand. Sect A.* **75A**, 481–485 (1971).

74. H. P. Kelly, The photoionization cross section for H_2 from threshold to 30 eV, *Chem. Phys. Lett.* **20**, 547–550 (1973).

75. See P. G. Burke and M. J. Seaton, Numerical solutions of the integro-differential equations of electron–atom collision theory, in: *Methods of Computational Physics* (B. Alder, S. Fernbach, and M. Rotenberg, eds.), Vol. 10, pp. 1–80, Academic Press, New York (1971).

76. A. L. Fetter and K. M. Watson, The optical model, in: *Advances in Theoretical Physics* (K. Brueckner, ed.), Vol. 1, pp. 115–194, Academic Press, New York (1965).

77. H. Feshbach, A unified theory of nuclear reactions, *Ann. Phys. (N.Y.)* **5**, 357–390 (1958); A unified theory of nuclear reactions. II, *Ann. Phys. (N.Y.)* **14**, 287–313 (1962).

78. J. S. Bell and E. J. Squires, A formal optical model, *Phys. Rev. Lett.* **3**, 96–97 (1959).

79. B. Schneider, H. S. Taylor, and R. Yaris, Many-body theory of the elastic scattering of electrons from atoms and molecules, *Phys. Rev. A* **1**, 855–867 (1970).

80. B. S. Yarlagadda, Gy. Csanak, H. S. Taylor, B. Schneider, and R. Yaris, Application of many-body Green's functions to the scattering and bound-state properties of helium, *Phys. Rev. A* **7**, 146–154 (1973).

81. C. W. McCurdy, T. N. Rescigno, and V. McKoy, A many-body treatment of Feshbach theory applied to electron–atom and electron–molecule collisions, *Phys. Rev. A* **12**, 406 (1975).

82. K. Dietrich and K. Hara, On the many-body theory of nuclear reactions, *Nucl. Phys. A* **111**, 392–416 (1968).

83. T. N. Rescigno, C. W. McCurdy, and V. McKoy, Discrete basis set approach to nonspherical scattering, *Chem. Phys. Lett.* **27**, 401–404 (1974); Discrete basis set approach to nonspherical scattering. II, *Phys. Rev. A* **10**, 2240–2245 (1974).

84. T. N. Rescigno, C. W. McCurdy, and V. McKoy, Low-energy e^-–H_2 elastic scattering cross sections using discrete basis functions, *Phys. Rev. A* **11**, 825–829 (1975).

85. T. N. Rescigno, C. W. McCurdy, and V. McKoy, A relationship between the many-body theory of inelastic scattering and the distorted wave, *J. Phys. B* **7**, 2396–2402 (1974).

86. See for example, J. R. Taylor, *Scattering Theory*, p. 720, J. Wiley and Sons, New York (1972).

87. Gy. Csanak, H. S. Taylor, and R. Yaris, Many-body methods applied to electron scattering from atoms and molecules. II. Inelastic processes, *Phys. Rev. A* **3**, 1322–1328 (1971).
88. L. D. Thomas, B. S. Yarlagadda, Gy. Csanak, and H. S. Taylor, Analytical and numerical procedures in the application of many-body Green's function methods to electron–atom scattering problems, *Comput. Phys. Comm.* **6**, 316–330 (1973); The application of first order many-body theory to the calculation of differential and integral cross sections for the electron impact excitation of the $2^1S, 2^1P, 2^3S, 2^3P$ states of helium, *J. Phys. B* **7**, 1719–1733 (1974).
89. T. N. Rescigno, C. W. McCurdy, and V. McKoy, Excitation of the $b\,^3\Sigma_u^+$ state of H_2 by low energy electron impact in the distorted wave approximation (in preparation).
90. J. C. Tully and R. S. Berry, Elastic scattering of low energy electrons by the hydrogen molecule, *J. Chem. Phys.* **51**, 2056–2075 (1969).
91. B. Schneider, Inelastic scattering of high-energy electrons from atoms: The helium atom, *Phys. Rev. A* **2**, 1873–1877 (1970).
92. A. Szabo and N. S. Ostlund, Generalized oscillator strengths for the lowest $\Pi \leftarrow \Sigma$ transitions in CO and N_2, *Chem. Phys. Lett.* **17**, 163–166 (1972).

10

POLYATOM: *A General Computer Program for Ab Initio Calculations*

Jules W. Moskowitz
and
Lawrence C. Snyder

1. Introduction

The POLYATOM[1,2] system of computer programs was written to make quantitative wave mechanical descriptions of molecules. These programs employ a Gaussian basis set to compute determinantal electronic wave functions and corresponding derived properties. The computations are made in an *ab initio* style which includes all electrons and computes all integrals.

The philosophy of the programs has been guided by previous attempts to create such comprehensive systems. The main lessons learned from these are that the programs should be in as high level a language as possible, consistent with acceptable efficiency of object programs, and should also be written with the knowledge that they will later have to be altered, probably by someone other than the original author. There has been a major effort to document POLYATOM so that it can be understood and changed by subsequent users.

The significance of the POLYATOM system of programs is not so much that they represent the fastest or most efficient programs of their type, but rather that they have been a structure that many people have been able to understand, apply, and build upon.

Jules W. Moskowitz • Chemistry Department, New York University, New York, New York and *Lawrence C. Snyder* • Bell Laboratories, Murray Hill, New Jersey

In this chapter, we shall mainly discuss POLYATOM (Version 2),[2] with some reference to its predecessors and recent additions and improvements. For a more complete description of the programs, documentation and a listing can be obtained from the Quantum Chemistry Program Exchange.[1,2]

2. History and Evolution of POLYATOM

S. F. Boys first set forth, 25 years ago, a systematic scheme by which the wave functions of the stationary states of the electrons in the field of any arrangement of nuclei can be evaluated.[3,4] In the same papers he proposed the use of Gaussian orbitals as basis functions. Since all required integrals could be evaluated explicitly, the way was opened for the construction of general computer programs for molecular electronic structure.

The first implementation of these ideas was on the EDSAC computer at the Mathematical Laboratory of Cambridge University by Boys, Cook, Reeves, and Shavitt,[5] who automated both the mathematical analysis and the numerical computations in studies of BH, H_2O and H_3. The development of programs was continued by Reeves on an Elliot 402 computer of I.C.I., Ltd.[6] Reeves was joined by Harrison in program development and applications to ammonia on the Pegasus computer at Leeds University.[7]

Harrison moved to the Solid State and Molecular Theory Group at Massachusetts Institute of Technology, where he, together with Csizmadia, Moskowitz, Sutcliffe, and Barnett, authored the POLYATOM system[8,9] and implemented it on the IBM 709–90 computer. They documented POLYATOM and submitted it to the Quantum Chemistry Program Exchange (QCPE).[1] The major programs were: PA20 which listed integrals, PA30 which evaluated integrals over Guassian basis functions, PA40, which performed a closed-shell SCF calculation, and PA45 to compute the dipole moment.

Moskowitz and Harrison moved to New York University where the use and extension of POLYATOM continued. Neumann, Basch, Kornegay, and Snyder, at Bell Laboratories, joined Moskowitz, Hornback, and Liebmann of New York University in the continuation of improvements and additions to the programs. Open-shell SCF programs were added together with a simple CI program, a comprehensive properties package, and a program, to generate symmetry adapted basis vectors. They documented POLYATOM (Version 2) and submitted it to QCPE in 1970.[2] It was implemented on the GE-635 computer at Bell Laboratories and on the CDC-6600 computer at New York University. Metzgar and Bloor subsequently converted these programs for the IBM 360 series computers.[10]

A number of recent theoretical developments have been incorporated by Goutier, Macaulay and Duke into the PHANTOM programs, which constitute an extensively modified and somewhat expanded version of POLYATOM (Version 2).

The evolution of POLYATOM has been a notable event indeed. The work spans 25 years. It involves more than 20 persons, many of whom have never met. It was done in four countries at more than eight institutions and on a wide variety of computers.

3. General Description of POLYATOM

The POLYATOM system is oriented towards the MO–SCF approach, though it is intended that the system should be flexible enough to operate in other modes. Accordingly, the system consists of a number of independent main programs. The output of each program is on tape (disk), and the input is from tape (disk) and cards. All printed output is off-line. These main programs include PA20A and PA20B, which create a list of one- and two-electron integrals, respectively. The main programs, PA30A and PA30B, evaluate the one- and two-electron integrals, respectively; and PA40 performs a closed-shell SCF calculation. There are several supplementary main programs which employ a variety of open-shell methods and compute molecular properties.

With the exception of one or two basic subroutines the whole system is written in FORTRAN IV. There is much use of the subroutine facility, general rather than special-purpose subroutines, the use of labeled COMMON rather than blank COMMON, and a consistent method of using the intermediate files. With regard to the last point, a general set of subroutines has been written to manipulate files, without reducing the flexibility of the Fortran input–output statements. All intermediate files must be written and read with the aid of these routines. The routines are themselves simple and entirely written in FORTRAN IV.

In the entire POLYATOM system there are only two subroutines which use system-dependent functions. These are subroutines which obtain and print the time. Further, there are only two subroutines which are written in machine language. These pack and unpack integers into and from a data word.

The overall system has been designed to accept a maximum of 127 one-electron basis functions for a machine like the GE-635 with a 36-bit word length. The CDC-6600 with a 60-bit word length will accept 1024 basis functions. This limitation is due to the label generated by PA20. It is also designed for an unlimited number of centers, with unlimited symmetry properties. The present integral routines will do one- and two-electron integrals over s, p, d, and f Gaussian basis functions.

4. Basic Theory

In order to relate the POLYATOM system of programs to the larger body of theoretical chemical knowledge, we provide here a short discussion of nonempirical calculation of molecular electronic structure and properties. The language and notation used is that of our program descriptions.

A fundamental problem of theoretical chemistry is that of obtaining an approximate solution of the Schrödinger equation for electronic motion in the field of fixed nuclei:

$$\hat{H}\Psi_e = E_e\Psi_e \tag{1}$$

where H is the usual Hamiltonian operator, which is (in hartree a.u.)

$$\hat{H} = \sum_i -\tfrac{1}{2}\Delta^2(i) + \hat{V}(i) + \sum_{i>j}\frac{1}{r_{ij}} \equiv \sum_i \hat{h}(i) + \sum_{i>j}\frac{1}{r_{ij}} \tag{2}$$

the sums extending over all electrons. The Coulomb attraction of the ith electron to all nuclei is $\hat{V}(i)$. The electronic energy of the system is E_e and Ψ_e is the electronic wave function, which must be antisymmetric with respect to interchange of the coordinates of any two electrons. There is at present no method for obtaining an exact solution Ψ_e for many-electron systems, and some approximation is necessary.

The most powerful method for finding an approximate solution of (1) is by use of the variation principle, which states that any approximate solution Ψ is such that the quantity

$$E = \frac{\int \Psi^*\hat{H}\Psi \, d\tau}{\int \Psi^*\Psi \, d\tau} \tag{3}$$

where integration is over all space and spin coordinates of the electrons, will not be less than E_e and will only be equal to E_e if Ψ is the exact solution. Thus we attempt to approximate Ψ_e by choosing Ψ to minimize E.

The requirement that Ψ be antisymmetric makes it possible to expand Ψ without loss of generality as a linear combination of functions Φ_p which are themselves antisymmetric:

$$\Psi = \sum_p C_p\Phi_p \tag{4}$$

The state Ψ to be investigated belongs to an irreducible representation of the space and spin symmetry group of the molecule, and so we may restrict consideration of those Φ_p which belong to such a representation. Boys and Cook[4,5] refer to a Ψ of form (4) as a *polydetor* wave function and to the individual Φ_p as *codetors*. These Φ_p can be considered to be simple linear combinations of *detors*, i.e., the usual Slater determinants as shown for an N-electron problem in an abbreviated form

$$(N!)^{-1/2} \det|\varphi_{p1}(1)\omega_{p1}(1)\varphi_{p2}(2)\omega_{p2}(2)\cdots\varphi_{pN}(N)\omega_{pN}(N)|$$

The functions $\varphi_r(i)$ and $\omega_r(i)$ are functions of the spatial and spin coordinates, respectively, of a single electron i. It can be shown that if the functions φ_r form a mathematically complete set of one-electron functions, the set of all determinants which can be formed from them is also complete, so in principle, at least, a polydetor wave function can be made to approximate the exact solution to any required degree of accuracy.

By a careful choice of the φ_r, the expansion can usually be made in such a way that a single codetor predominates and that codetor alone provides a reasonable and simple approximation to the correct wave function. This may be achieved as follows. A single codetor wave function is constructed in which the φ_r are considered to be variable functions. In some cases, this codetor can be chosen to be a single determinant of doubly occupied orbitals,

$$\Phi_0 = \frac{1}{\sqrt{(2M)!}} \det|\varphi_1(1)\,\alpha(1)\varphi_1(2)\,\beta(2) \cdots$$
$$\times \varphi_M(2M-1)\alpha(2M-1)\varphi_M(2M)\beta(2M)|$$
$$= (|\varphi_1\bar{\varphi}_1 \cdots \varphi_M\bar{\varphi}_M|) \tag{5}$$

where the $2M$ by $2M$ determinant Φ_0, describes $2M$ electrons in M molecular orbitals. Here α and β stand for the two spin states of an electron. A state which can be represented by such a wave function will be referred to as a closed shell, a definition which includes the intuitive chemical concept of a closed-shell state. All other states will be referred to as open shells. As many molecules have closed-shell ground states, chemists are particularly interested in wave functions like Φ_0.

The energy of this wave function, for example Φ_0, is then minimized with respect to variations in the φ_r. In practice this minimization leads to the Hartree–Fock self-consistent field (SCF) equations and the resulting φ_r are called self-consistent field molecular orbitals (SCF–MOs).[12]

The total energy in the ground state, Φ_0, is given by the expression,

$$E = 2\sum_i H_i + \sum_{ij} (2J_{ij} - K_{ij}) \tag{6}$$

where i and j sum over all occupied molecular orbitals

$$H_i = \int \varphi_i^*(1)h(1)\varphi_i(1)\,dV_1 \tag{7}$$

and

$$J_{ij} = \int \varphi_i^*(1)\varphi_j^*(2)\frac{1}{r_{12}}\varphi_i(1)\varphi_j(2)\,dV_{12} \tag{8}$$

$$K_{ij} = \int \varphi_i^*(1)\varphi_j^*(2)\frac{1}{r_{12}}\varphi_i(2)\varphi_j(1)\,dV_{12} \tag{9}$$

are called the Coulomb and exchange integrals, respectively.

The treatment of the general open-shell case for both ground and excited states is far more complex than that for closed shells or those outlined above.

In the Roothaan expansion method[13] of calculating Hartree–Fock wave functions for open-shell electronic states it is usually necessary to introduce into the variational equations off-diagonal Langrangian multipliers connecting *molecular orbitals* (MOs) of the same spatial (point group) symmetry that appear in both the class of closed-shell and open-shell MOs. In the absence of the multipliers the closed- and open-shell MOs would be eigenfunctions of different one-electron SCF Hamiltonian operators and thus not constrained to be orthogonal. The method is generally called the *restricted Hartree–Fock* (RHF) approach.

In the Roothaan method the expectation value of the energy is given by

$$E = 2\sum_k H_k + \sum_{kl} + (2J_{kl} - K_{kl})$$
$$+ f\left[2\sum_m H_m + f\sum_{mn} (2aJ_{mn} - bK_{mn}) + 2\sum_{km} (2J_{km} - K_{km})\right] \qquad (10)$$

where, $a,$, b, and f are numerical constants depending on the particular case. In referring to the individual orbitals, we use k, l for the closed-shell orbitals and m, n for the open-shell orbitals. There are three important classes of states whose total energy can be represented by Eq. (10): (a) a half-closed shell consisting of singly occupied orbitals with all the spins parallel; (b) all the configurations arising from the configurations π^N, δ^N, $1 \le N \le 3$, of a linear molecule; and (c) all the states arising from the configuration p^N, $1 \le N \le 5$ of an atom.

We mention here an alternative procedure for treating open-shell excited electronic states. The idea is a considerable simplification, at the slight expense of certain restrictions on the basis set, of an iterative scheme proposed by Huzinaga which also rigorously eliminates the need for off-diagonal Lagrangian multipliers. The method has been implemented by Basch and Neumann,[14] and by Segal[15] and Goddard.[16] The latter named the approach the orthogonality constrained basis-set expansion method. A detailed explanation and critical analysis of the method has been given recently by Brinkley, Pople, and Dobash.[17]

In the unrestricted Hartree–Fock method, the molecular orbitals for electrons of spin α are permitted to differ from those of spin β. The energy of the single determinant is minimized with respect to the coefficients in both sets of orbitals. This leads to separate eigenvalue problems for the α and β spin-orbitals, as was detailed by Pople and Nesbet.[18]

In actual calculations on molecules it has not yet proven possible to determine the SCF–MOs exactly, in a manner analogous to that used in determining SCF atomic orbitals. Instead the φ_r are expressed as linear combinations of a set of basis functions η_i $(i = 1, N)$,

$$\varphi_r = \sum_{i=1}^{N} Y_{ri}\eta_i \qquad (11)$$

and the Y_{ri} are considered to be the variables in the SCF procedure. As the basis functions most often used are atomic orbitals, the method is called the *linear combination of atomic orbitals* (LCAO) MO–SCF method. The basis functions are considered fixed and the required molecular orbitals are found by determining the coefficients which minimize the energy of the single codetor wave function.

In the case of the closed-shell ground state and many open-shell states, Roothaan[12] has shown that the coefficients may be determined by a procedure involving the iterative solution of the eigenvalue problem

$$\mathbf{F}\tilde{\mathbf{Y}}_r = \mathbf{G}\tilde{\mathbf{Y}}_r \varepsilon_r \tag{12}$$

The scalar quantity ε_r is called the orbital energy of the orbital φ_r; $\mathbf{G}$ is the overlap matrix given by

$$G_{ij} = (\eta_i | \eta_j) = \int \eta_i(1)\eta_j(1)\, dv_1 \tag{13}$$

the integration being over the spatial coordinates of electron one. The Fock matrix $\mathbf{F}$ is given by

$$\mathbf{F} = \mathbf{H} + \mathbf{P} - \mathbf{Q} + \mathbf{R} \tag{14}$$

with

$$\mathbf{P} = 2\mathbf{J}^t - \mathbf{K}^t \tag{15}$$

$$\mathbf{Q} = 2\alpha\mathbf{J}^o - \beta\mathbf{K}^o \tag{16}$$

$$\mathbf{R} = \mathbf{G}\rho^t\mathbf{Q} + \mathbf{Q}\rho^t\mathbf{G} \tag{17}$$

The elements of the matrices are defined by

$$H_{ij} = (\eta_i | \hat{h} | \eta_j) = \int \eta_i(1)\hat{h}(1)\eta_j(1)\, dv_1 \tag{18}$$

where $\hat{h}$ is the one-electron operator, and by

$$J^{\sigma}_{ij} = \sum_{k=1}^{N} \sum_{l=1}^{N} \rho^{\sigma}_{kl}(\eta_i \eta_j | \eta_k \eta_l) \tag{19}$$

$$K^{\sigma}_{ij} = \sum_{k=1}^{N} \sum_{l=1}^{N} \rho^{\sigma}_{kl}(\eta_i \eta_k | \eta_j \eta_l) \tag{20}$$

with σ taking the symbols t (total) and o (open). The two-electron integral $(\eta_i \eta_j | \eta_k \eta_l)$ is defined by

$$(\eta_i \eta_j | \eta_k \eta_l) = \int \eta_i(1)\eta_k(2)\frac{1}{r_{12}}\eta_j(1)\eta_l(2)\, d\tau_1\, d\tau_2 \tag{21}$$

The "density matrices" ρ^t and ρ^o are given by

$$\rho^t_{kl} = \sum_{r}^{\substack{\text{occupied} \\ \text{closed}}} Y_{rk}Y_{rl} + f \sum_{r}^{\substack{\text{occupied} \\ \text{open}}} Y_{rk}Y_{rl} \tag{22}$$

$$\rho_{kl}^{o} = f \sum_{\substack{r \\ \text{open}}}^{\text{occupied}} Y_{rk} Y_{rl} \tag{23}$$

with the summations being over the open-shell or closed-shell orbitals only, and f being the fraction of the open shell which is occupied. The constants α and β given above depend on the state being investigated. The quantities α and β are simple functions of a and b. It is seen that when the state is a closed shell, then f is zero so $\boldsymbol{\rho}^{o}$, $\mathbf{Q}$, and $\mathbf{R}$ are all zero and the equation is greatly simplified. The total energy is given by

$$E = \mathrm{tr}\{(\mathbf{H}+\mathbf{F})\boldsymbol{\rho}^{t} - \mathbf{Q}[\boldsymbol{\rho}^{t} + (f-1)\boldsymbol{\rho}^{o}]\} \tag{24}$$

The solution of the equation is usually accomplished by an iterative process, since the matrix $\mathbf{F}$ involves the values of Y_{ri}. An estimate is made for $\mathbf{Y}$ initially, and this is used to compute the two density matrices, and hence the two $\mathbf{J}$ and two $\mathbf{K}$ matrices. These are assembled to form $\mathbf{P}$ and $\mathbf{Q}$, and the latter one is used with the overlap matrix $\mathbf{G}$ to form $\mathbf{R}$. The equation is solved to give a further estimate for $\mathbf{Y}$, and the process is repeated. For the simpler closed-shell case, this procedure often converges, but occasionally the solution is found to oscillate and some extrapolation procedure is necessary to force it to converge.

As can be seen, the solution of (12) will yield N new vectors $\mathbf{Y}_{r}$ each associated with an orbital energy ε_{r}. There are, however, only M molecular orbitals involved and these are called the *occupied orbitals* [the remaining $(N–M)$ being called *unoccupied* orbitals]. If it is desired to represent the ground state, the state of lowest energy, of a system, the occupied molecular orbitals are chosen as those which correspond to the lowest M orbital energies.

It should perhaps be noted at this point that if the basic orbitals (η) form a complete set, the molecular orbitals resulting from it would be the best possible orbitals within the single determinant restriction. The best possible molecular orbitals are referred to as the *Hartree–Fock* molecular orbitals.

The whole SCF procedure[12] can be made automatic after the basic orbitals (η) are chosen and the one-electron integrals $(\eta_i|\eta_j)$, $(\eta_i|h|\eta_j)$ and the two-electron integrals $(\eta_i\eta_j|\eta_k\eta_l)$ defined in (13), (18), and (21), are evaluated. Usually the basic orbitals are real, so there are at most $p = N(N+1)/2$ distinct integrals of each of the one-electron types and $q = p(p+1)/2$ of the two electron type. Since q varies roughly as the fourth power of N, the number of two-electron integrals rapidly becomes enormous as N increases. To illustrate this difficulty, the number of integrals to be evaluated for various values of N is tabulated in Table 1. It is this large number of integrals which is really the limiting factor in molecular orbital calculations.

Until recently the nodeless Slater functions were the most widely used atomic orbitals.[19] Their general form is

$$\eta = N r^{n-1} e^{-\alpha r} S_{l,m}(\theta, \Phi) \tag{25}$$

Table 1. The Number of One- and Two-Electron Integrals Associated with Various Size Basis Sets

Size of basis set	Number of molecular integrals	
	One-electron integrals	Two-electron integrals
10	55	4,540
20	210	22,155
30	465	108,345
40	820	336,610
50	1,275	814,725
100	5,050	12,751,250
200	20,100	202,015,050
300	45,150	1,019,261,250

where n, l, and m are quantum numbers ($n = 1, 2, 3, .. : l < n; -l \le m \le l$), N is a radial normalizing factor, and $S_{l,m}(\theta, \Phi)$ is a normalized real spherical harmonic. The radial distance from the center considered is r and α is the *orbital exponent*.

As can be seen the two-electron integrals $(\eta_i \eta_j | \eta_k \eta_l)$ may involve orbitals centered on one, two, three, and four different atoms. The one- and two-center integrals offer no particular difficulty and there are a number of good methods of evaluating them accurately and quickly. The three- and four-center integrals do, however, present considerable difficulties and few acceptable methods have been proposed for their evaluation. To obtain accurate values involves much complicated calculation.

To combat this difficulty it has been suggested[3] that the exponential portion $e^{-\alpha r}$ of the Slater orbital be replaced with a Gaussian function $e^{-\alpha r^2}$. This enables the many-center integrals to be done very easily and quickly. In principle, since the Gaussians form a complete set, the exact molecular orbitals may be expressed in terms of them.

In addition to the total electronic energy, chemists are also interested in many one-electron properties that may be computed from the one-electron wave function. Each property corresponds to the expectation value $\langle P \rangle$ of some operator P, where P can be written in the form of a sum of one particle operators for all electrons (i) and nuclei (μ) in the molecule,

$$P = \sum_i \hat{p}(i) + \sum_\mu \hat{p}(\mu) \tag{26}$$

For a closed-shell molecule, having the determinantal wave function defined above, the expectation value of a one-electron operator can be written in the simple form

$$\langle P \rangle = 2 \sum_j \langle \varphi_j | \hat{p} | \varphi_j \rangle + \sum_\mu \hat{p}(\mu) \tag{27}$$

Examples of such operators are the multipole moments of the charge distribution such as the dipole and quadrupole moments and the operator for the electric field and field gradient at an atomic nucleus.

Chemists have developed several ways to partition the electrons of a molecule among atoms or bonds. The most widely adopted definitions are those of Mulliken.[20] Let us refer to all basis functions on center μ by the index i, and all basis functions on other centers by j. For our closed-shell determinantal wave functions, the total gross population in basis function i is denoted by $N_{\mu i}$ and defined by

$$N_{\mu i} = 2 \sum_k Y_{ki}(Y_{ki} + \sum_{i \neq i'} Y_{ki'}S_{ii'}) + 2\sum_k Y_{ki}(\sum_j Y_{kj}S_{ij}) \tag{28}$$

where the index k runs over the filled molecular orbitals. S_{ij} is the overlap integral between basis functions i and j. The total gross population on atom μ, N_μ, is given by

$$N_\mu = \sum_i N_{\mu i} \tag{29}$$

The subtotal overlap population between basis functions i and j denoted by $n(i, j)$ is defined by

$$n(i, j) = 2 \sum_k 2 Y_{ki} Y_{kj} S_{ij} \tag{30}$$

The overlap population can be partitioned according to representation of the occupied molecular orbital. The overlap population between centers μ and ν denoted by $n_{\mu\nu}$ is then given by

$$n_{\mu\nu} = \sum_i \sum_j n(i, j) \tag{31}$$

Unfortunately, these definitions of gross atomic and overlap populations suffer from the fact that they are basis-set dependent.

5. POLYATOM *Programs*

5.1. General-Purpose Subroutines

A set of subroutines to facilitate the handling of data files on magnetic tape (disk) is included in the system. A data file is defined to be a block of information, composed of a number of records, and terminated by a special record to indicate the end of the data file.

Since information on magnetic tape must be read sequentially, it is necessary to simulate the random access of files so that minimum effort on the programmer's part is required to locate the desired data. This simulation allows the programmer to say in which order files will be written on a tape (disk) and to retrieve the files independent of that order. The file handling subroutines allow the programmer to manipulate files by recording the order of files generation on the tape (disk). The files are given names, and referred to subsequently by these names. The reading and writing of the information content of the files may be done in any way, including the standard Fortran binary and BCD tape operations, but without end-of-file marks. If a program is stopped for some reason, it can be restarted without any trouble at any point other than in the middle of a file, even if the tape (disk) position has been altered.

The only system dependent routines read the clock. There are only two machine language routines, the subroutines PACK and UNPACK. They pack and unpack six Fortran integers, which specify the basis functions, into a single word. Two words may be used on a machine having a shorter word length. Through this process the word length limits the maximum number of basis functions.

5.2. PA20A, PA20B: List One- and Two-Electron Integral Labels

The purpose of these programs is to eliminate those integrals which are identically zero, and to list the remainder to minimize as far as possible the number of integrals which must be evaluated independently. The program was originally written by M. C. Harrison at Massachusetts Institute of Technology and revised by C. Hornback at New York University.

The programs take specifications of the symmetry properties of a number of one-electron functions, and use them to create and write on a magnetic tape (disk) a number of files. These files contain lists of labels of the overlap, kinetic energy, nuclear attraction, and electronic repulsion integrals. The lists are organized so that all nonzero integrals are represented, and those integrals which are related by the symmetry properties of the basis functions are stored consecutively.

Two integrals involving one operator and various one-electron functions, $(i|\hat{O}|j)$ and $(i'|\hat{O}|j')$ or $(ij|kl)$ and $(i'j'|k'l')$, can be seen to be equal if a transformation of the dummy variables, $\vec{r}$ of Eq. (18), or $\vec{r}_1$ and $\vec{r}_2$ of Eq. (21), exists which leaves the operator unchanged and transforms the functions of one integral into those of the other. The present method is to apply to an integral a series of orthogonal transformations of space which leave the operator unaltered. By this method it is possible to generate a sublist of integrals which are equal in magnitude to the original. If it turns out that one of these is the negative of the original, then obviously the value of all these integrals must be zero.

The symmetry properties of the one-electron functions are read into the computer as a matrix, with each column corresponding to a transformation.

The entries in row i denote the effect of applying these transformations to function i. A nonzero entry, $\pm|m|$, indicates that the result of the transformation is to take function i into function $|m|$. Under the transformation there may or may not be a sign change. A zero entry indicates that the result of the transformation is not one of the functions under consideration. Symmetry operations which take a basis function into a hybrid, a linear combination of basis functions, are omitted.

The execution time of the present programs is by no means negligible compared with the time to evaluate the set of Gaussian integrals, and sometimes it has been necessary to use only a reduced number of transformations in order to minimize the combined time. In most cases, however, the label lists are used for many integral evaluations, and so the time to produce them is of little importance.

5.3. PA30A, PA30B: **Evaluate One- and Two-Electron Integrals**

These programs take specifications of a number of Gaussian one-electron functions (basis functions), from cards, and evaluate and store on magnetic tape (disk) the overlap (G), Kinetic energy (T), nuclear attraction (V), and electronic repulsion integrals (M), as specified by the lists of integral labels stored on the list file. The values of the integrals as well as their labels are stored on the integral file. The current program was written by H. Basch at Bell Laboratories.

The basic functions χ_p are restricted to the following form

$$\chi_p(\mathbf{r}_A) = N_p f_p(x_A, y_A, z_A) \sum C_{pu} N_u \exp(-\alpha_u r_A^2) \tag{32}$$

where the point of origin of the vector $\mathbf{r}_A$ is $\mathbf{A} = (A_x, A_y, A_z)$: $\mathbf{A}$ is known as the center of the function χ_p. The vector $\mathbf{A}$ is itself taken relative to some global origin, as is also the vector $\mathbf{r}$. Thus in (32), $z_A = x - A_x$, $y_A = y - A_y$, $z_A = z - A_z$, and $\mathbf{r}_A = \mathbf{r} - \mathbf{A}$. The function in (32) is known as a contracted Gaussian function and N_p is its normalizing factor.

The function

$$\eta(r_A) = N_u f_u(x_A, y_A, z_A) \exp(-\alpha_u r_A^2) \tag{33}$$

is known as a primitive Gaussian, and N_u is its normalizing factor.

The coefficients C_{pu} in (1) are the coefficients of the normalized primitives η_u. The prefactor f_p in (32) is restricted to be the same for each primitive in the sum. It can therefore be factored from the summation as has been done in (32).

The prefactor, $f(x, y, z)$, is further restricted to the following form

$$f(x, y, z) = x^l y^m z^n \tag{34}$$

Table 2. Function Types Allowed in PA30A and PA30B

Symbolic reference name	Form of $f(x, y, z)$	l	m	n
S	1	0	0	0
X	x	1	0	0
Y	y	0	1	0
Z	z	0	0	1
XX	x^2	2	0	0
YY	y^2	0	2	0
ZZ	z^2	0	0	2
XY	xy	1	1	0
XZ	xz	1	0	1
YZ	yz	0	1	1
XXX	x^3	3	0	0
YYY	y^3	0	3	0
ZZZ	z^3	0	0	3
XXY	x^2y	2	1	0
XXZ	x^2z	2	0	1
XYY	y^2x	1	2	0
YYZ	y^2z	0	2	1
XZZ	z^2x	1	0	2
YZZ	z^2y	0	1	2
XYZ	xyz	1	1	1

where $l+m+n$ must be less than or equal to 3. The form of the prefactor defines what will be referred to below as the "type" of a basis function. The allowed types and their symbolic reference names is given in Table 2.

The operators $\hat{G}$, $\hat{T}$, $\hat{V}$, and $\hat{M}$, are, respectively, the overlap, kinetic energy, nuclear attraction, and electronic repulsion operators specified as follows:

$$\hat{G} = 1$$

$$\hat{T} = -(1/2)\nabla^2$$

$$\hat{V} = \sum_m Z_m/r_m$$

$$\hat{M} = 1/r_{12}$$

where Z_m is the charge on nucleus m, and r_m is the distance of the general point from this nucleus. The required integrals are written as

$$(\chi_i|\hat{G}|\chi_j) \qquad \text{or} \qquad (i|j)$$

$$(\chi_i|\hat{T}|\chi_j) \qquad \text{or} \qquad (i|\hat{T}|j)$$

$$(\chi_i|\hat{V}|\chi_j) \qquad \text{or} \qquad (i|\hat{V}|j)$$

$$(\chi_i\chi_j|\hat{M}|\chi_k\chi_l) \qquad \text{or} \qquad (ij|kl)$$

where the notation for the last integral is defined to mean that i and j are functions of the coordinates of electron 1, and k and l are functions of the coordinates of electron 2.

The essential step in the reduction of the multicenter integrals is the application of the following theorem. The product of two Gaussians[21] having different centers A and B is itself a Gaussian (apart from a constant factor) with a center E somewhere on the line segment AB. Specifically

$$\exp[-\alpha(r_A^2 + r_B^2)] = \text{const} \exp(-\alpha r_E^2) \qquad (35)$$

The methods used for the evaluation of these integrals are variations of those suggested by Boys.[3] The formulas implemented in these programs are based on those given by Clementi and Davis[22] and those given by Taketa, Huzinaga, and O-Ohata.[23]

PA30A evaluates the G, T, and V integrals while PA30B evaluates the M integrals. In both programs, a list of integral labels is read in from a list file, and the integrals corresponding to those labels which are unique (i.e., have zero-tag) are evaluated while the other integrals are set equal to, or equal to minus, the last zero-tagged integral, and the "packed" labels and their values are written on an integral file.

5.4. PA40: Closed-Shell Self-Consistent Field Program

This program evaluates the molecular orbitals, orbital energies, and total energy for a single determinant LCAO–MO–SCF wave function. The input data consists of all the relevant integrals between basis functions on magnetic tape (disk), together with specifications on cards of the state to be investigated. The program was written by D. Neumann at New York University.

In the closed-shell case the solution of Roothaan's equations is implemented in the following way. The molecular orbitals φ_i are written as linear combinations of basis functions η_p, or contracted functions χ_p:

$$\Phi_i = \sum_p Y_{ip}\eta_p \qquad (36)$$

The problem to be solved is then

$$\mathbf{F}\tilde{\mathbf{Y}} = \mathbf{G}\tilde{\mathbf{Y}}\mathbf{E} \qquad (37)$$

where $\mathbf{G}$ is the matrix of overlap integrals

$$G_{pq} = (\eta_p|\eta_q) \qquad (38)$$

and where

$$\mathbf{F} = \mathbf{H} + 2\mathbf{J} - \mathbf{K} \qquad (39)$$

Here **H** is the "one electron part " of the Fock matrix, **F**, and **J** and **K** are the Coulomb and exchange matrices, respectively;

$$\mathbf{H} = \mathbf{T} + \mathbf{V} \tag{40}$$

where

$$T_{pq} = (\eta_p | -\tfrac{1}{2}\nabla^2 | \eta_q) \tag{41}$$

and

$$V_{pq} = (\eta_p | -\sum_{\alpha} Z_{\alpha} r_{\alpha}^{-1} | \eta_q) \tag{42}$$

$$J_{pq} = \sum_r \sum_s D_{rs} (pq|rs) \tag{43}$$

$$K_{pq} = \sum_r \sum_s D_{rs} (pr|qs) \tag{44}$$

Here the "one-electron density matrix" **D** is defined as

$$D_{rs} = 2 \sum_i f_i Y_{ir} Y_{is} \tag{45}$$

where f_i is the "fractional occupancy" of the ith molecular orbital. Thus if the ith MO contains n_i electrons, $f_i = 0.5 n_i$. Usually f_i has the values 0.0 or 1.0 for closed-shell calculations.

Equation (37) must be solved indirectly by forming, from the set $\{\eta_p\}$, a set of orthonormal symmetry functions $\{\sigma_r\}$

$$\sigma_r = \sum_p S_{rp} \eta_p \tag{46}$$

or

$$\{\sigma\} = \mathbf{S}\{\eta\}$$

where

$$(\sigma_r | \sigma_s) = \delta_{rs}$$

or

$$\mathbf{S}\mathbf{G}\tilde{\mathbf{S}} = \mathbf{1} \tag{47}$$

The solution to (37) is obtained by transforming (37) using **S** thus:

$$\mathbf{S}\mathbf{F}\tilde{\mathbf{S}}(\tilde{\mathbf{S}})^{-1}\tilde{\mathbf{Y}} = \mathbf{S}\mathbf{G}\tilde{\mathbf{S}}(\tilde{\mathbf{S}})^{-1}\tilde{\mathbf{Y}}\mathbf{E} \tag{48}$$

thus letting

$$\mathbf{F}' = \mathbf{S}\mathbf{F}\tilde{\mathbf{S}} \tag{49}$$

and

$$W = (\tilde{S})^{-1}\tilde{Y} \tag{50}$$

$$\tilde{Y} = \tilde{S}W \tag{51}$$

the actual problem to be solved becomes

$$F'W = WE \tag{52}$$

the matrix F' is blocked according to symmetry species since the transformation matrix S is ordered such that symmetry orbitals of the same irreducible representation are placed in adjacent rows of S.

As the matrix F in the pseudoeigenvalue equation (37) is itself dependent on the solutions of (37) through (43), (44), and (45), the problem of finding the molecular orbital coefficient matrix Y must be solved iteratively (i.e., until Y is self-consistent).

The matrix S is formed in the following manner. A matrix C is read from cards containing a set of nonorthonormal symmetry orbitals, symmetry orbitals of the same irreducible representation being in adjacent rows of C, thus,

$$\sigma'_r = \sum_p C_{rp}\eta_p \tag{53}$$

and

$$CG\tilde{C} = G' \tag{54}$$

where G' is now symmetry blocked. The matrix G' is then diagonalized by blocks

$$\tilde{U}G'U = g, \qquad g_{ij} = g_i\delta_{ij} \tag{55}$$

The above equation is then multiplied on the left and right by $g^{-1/2}$ to give

$$g^{-1/2}\tilde{U}G'Ug^{-1/2} = 1 \tag{56}$$

The matrix S is then

$$S = g^{-1/2}\tilde{U}C \tag{57}$$

and the rows of S represent functions which transform as irreducible representations and are orthonormal. It should be noted that these requirements are also met by any molecular orbital coefficient matrix which is a solution to (37).

The algorithm used in PA40 to determine the self-consistent solutions to (37) is the following:

A.1.   Form S [via Eqs. (54)–(57)]
A.2.   Form $H = T + V$ [via Eqs. (40)–(42)]
A.3.a. Symmetry block H $H' = SHS$
    b. Diagonalize H' by blocks. $H'W_o = W_o E_o$
    c. Form $\tilde{Y}_o = \tilde{S}W$

Thus, the initial guess at a set of molecular orbital coefficients is formed by diagonalizing the "one electron part" of the Fock matrix. At this point $\mathbf{Y} = \mathbf{Y}_o$.

B.1.   Take the matrix $\mathbf{Y}$ and form $\mathbf{D}$ [via Eq. (45)]

B.2.   Read the kinetic energy integrals into the matrix $\mathbf{F}$. $\mathbf{F} = \mathbf{T}$. Form the expectation value of the kinetic energy operator

$$\langle T \rangle = \mathrm{tr}\{\mathbf{DT}\} \tag{58}$$

B.3.   Add the nuclear attraction integrals into the matrix $\mathbf{F}$. $\mathbf{F} = \mathbf{T} + \mathbf{V} \equiv \mathbf{H}$. Form the expectation value of the nuclear attraction operator

$$\langle V1 \rangle = \mathrm{tr}\{\mathbf{DV}\} \tag{59}$$

B.4.   Add the coulomb, $\mathbf{J}$, and exchange, $\mathbf{K}$, matrices to $\mathbf{F}$, to form the Fock matrix of the present iteration. $\mathbf{F} = \mathbf{H} + 2\mathbf{J} - \mathbf{K}$.

B.5.   Form the expectation value of the Hartree–Fock operator

$$\langle h \rangle = \mathrm{tr}\{\mathbf{DF}\} = \mathrm{tr}\{\mathbf{D}(\mathbf{H} + 2\mathbf{J} - \mathbf{K})\} \tag{60}$$

B.6.   Form the expectation value of the electronic energy $\langle E \rangle$

$$\langle E \rangle = \{\langle T \rangle + \langle V1 \rangle\} + \tfrac{1}{2}\mathrm{tr}\{(2\mathbf{J} - \mathbf{K})\mathbf{D}\} \tag{61}$$

$$\langle E \rangle = \tfrac{1}{2}\{\langle T \rangle + \langle V1 \rangle + \langle h \rangle\} \tag{62}$$

and compute the virial $-V/T$,

$$-V/T = (\langle E \rangle - \langle T \rangle)/\langle T \rangle \tag{63}$$

B.7.   Compare the value of the electronic energy computed in step B.6 with the value computed the last time through that step. If they differ by less than some threshold value go to C.1.

B.8.a.   Symmetry block $\mathbf{F}$ using $\mathbf{S}$ if this is the first time through this step, (B.8.a). If this is not the first time through then symmetry block $\mathbf{F}$ using the last molecular orbital coefficient matrix (i.e., set $\mathbf{S} = \mathbf{Y}$). Then $\mathbf{F}' = \mathbf{SF\tilde{S}}$.

  b.   Diagonalize $\mathbf{F}$ by blocks. $\mathbf{F}'\mathbf{W}_1 = \mathbf{W}_1\mathbf{E}_1$

  c.   Transform $\mathbf{W}_1$. $\mathbf{\check{Y}}_1 = \mathbf{\tilde{S}W}_1$

B.9.   Set $\mathbf{Y} = \mathbf{Y}_1$ and go to step B.1.

C.1.   Self-consistency in the electronic energy is achieved. Print the results such as final orbital energies $\{E_i\}$, molecular orbital coefficients, $\mathbf{Y}$, last Fock matrix, $\mathbf{F}$, etc.

If an initial guess, $\mathbf{Y}_o$, of the molecular orbitals is given, then the algorithm begins with step B.1 with $\mathbf{Y} = \mathbf{Y}_o$.

If an initial density matrix, $\mathbf{D}_o$, is given, then the algorithm begins with step B.2 with $\mathbf{D} = \mathbf{D}_o$.

If desired extrapolations on the density matrix may be performed in step B.1, similarly extrapolations on the molecular orbital coefficients may be performed just prior to step B.1.

5.5. PA41, PA42, and PA43: Open-Shell SCF Programs

The PA41 program employs a single Hamiltonian for the closed and open shells developed by C. C. J. Roothaan[13] and outlined in Section 4. The double Hamiltonian method of Roothaan[13] is used in PA42. This method is often poorly convergent. The orthogonality constrained basis set methods appear to have improved convergence properties over both the above. The PA43 program is a spin-unrectricted Hartree–Fock program. It uses the method of Pople and Nesbet.[18] The implementation of these programs is analogous to that of PA40. These programs were written by D. Neumann at New York University.

5.6. PA50: Configuration Interaction of Selected Singly Excited Configurations

The purpose of the program is to transform selected two-electron integrals and to obtain configuration interactions of selected singly excited configurations (codetors). This program was written by H. Basch at Bell Telephone Laboratories.

The singly excited configuration resulting from the excitation of an electron from molecular orbital Φa to molecular orbital Φb of the closed-shell ground state having M doubly occupied orbitals is denoted by $S(a, b)$ and is defined as

$$S(a, b) = \frac{1}{\sqrt{2}} \{ (|\Phi_1 \bar{\Phi}_1 \cdots \Phi_a \bar{\Phi}_b \cdots \Phi_m \bar{\Phi}_m|) + (|\Phi_1 \bar{\Phi}_1 \cdots \Phi_b \bar{\Phi}_a \cdots \Phi_m \bar{\Phi}_m|) \} \tag{64}$$

for singlet configurations. The singly excited configuration for triplet configurations are denoted by $T(a, b)$ and are defined as

$$T(a, b) = (|\Phi_1 \bar{\Phi}_1 \cdots \Phi_a \Phi_b \cdots \Phi_m \bar{\Phi}_m|) \tag{65}$$

The matrix elements of the Hamiltonian are defined in the following way,

$$(T(a, b)|T(a, b)) = \varepsilon_b - \varepsilon_a - (bb|aa) \tag{66}$$

where ε_b and ε_a are energies of molecular orbitals a and b and $(bb|aa)$ represents the two electron repulsion integral defined over the molecular orbitals. The off-diagonal elements are defined as

$$(T(a, b)|T(c, d)) = -(bd|ca) \tag{67}$$

For singlet configurations the diagonal elements are defined as

$$(S(a, b)|S(a, b)) - \varepsilon_b - \varepsilon_a - (bb|aa) + 2(ab|ab) \tag{68}$$

and the off-diagonal elements are defined as

$$(S(a, b)|S(c, d)) = -(bd|ca) + 2(cd|ba) \tag{69}$$

5.7. PA59: **Population Analysis Package**

This program performs a Mulliken population analysis,[20] and an overlap population analysis. The quantities computed are the total atomic populations N_μ, and the overlap populations $n_{\mu\nu}$ between centers μ and ν separated by less than a maximum distance specified by the user. This program was written by D. Neumann at Bell Laboratories.

5.8. PA60: **One-Electron Properties Package**

This program computes expectation values of one-electron operators over the LCAO–GTO wave functions produced by PA40. The program was written by S. P. Liebmann at New York University.

Expectation values of the operators listed in Table 3 can be evaluated.

Table 3. *Operators for Which Expectation Values Can Be Computed by PA60*

1. Potential	$(x^2 + y^2 + z^2)^{-1/2} = 1/r$
2. Diamagnetic shielding	$\frac{3}{2}(r^2 - x^2)/r^3, \frac{3}{2}(r^2 - y^2)/r^3, \frac{3}{2}(r^2 - z^2)/r^3$
	$-\frac{3}{2}xy/r^3, -\frac{3}{2}xz/r^3, -\frac{3}{2}yz/r^3$
3. Electric field	$x/r^3, y/r^3, z/r^3$
4. Electric field gradient	$(3x^2 - r^2)/r^5, (3y^2 - r^2)/r^5, (3z^2 - r^2)/r^5$
	$3xy/r^5, 3xz/r^5, 3yz/r^5$
5. Dipole moment	x, y, z
6. Quadrupole moment	$\frac{1}{2}(3x^2 - r^2), \frac{1}{2}(3y^2 - r^2), \frac{1}{2}(3z^2 - r^2)$
	$\frac{3}{2}xy, \frac{3}{2}xz, \frac{3}{2}yz$
7. Diamagnetic susceptibilities	$\frac{3}{2}(r^2 - x^2), \frac{3}{2}(r^2 - y^2), \frac{3}{2}(r^2 - z^2), -\frac{3}{2}xy, -\frac{3}{2}xz, -\frac{3}{2}yz$
8. Second moment	$x^2, y^2, z^2, xy, xz, yz, r^2$
9. Octopole moment	$(5x^3 - 3xr^2)/2,$
	$(5xy^2 - xr^2)/2, (5xz^2 - xr^2)/2,$
	$(5x^2y - yr^2)/2, (5y^3 - 3yr^2)/2,$
	$(5yz^2 - yr^2)/2, (5x^2z - zr^2)/2,$
	$(5y^2z - zr^2)/2, (5z^3 - 3zr^2)/2, 5xyz/2$
10. Third moment	$x^3, xy^2, xz^2, x^2y, y^3, yz^2, x^2z, y^2z, z^3, xyz$
11. Third moment (combined)	xr^2, yr^2, zr^2
12. Hexadecapole moment	$(35x^4 - 30x^2r^2 + 3r^4)/8, (35y^4 - 30y^2r^2 + 3r^4)/8,$
	$(35z^4 - 30z^2r^2 + 3r^4)/8$
	$x^2r^2, y^2r^2, z^2r^2, r^4, xyr^2, xzr^2, yzr^2$
13. Fourth moment (even)	$x^4, y^4, z^4, x^2y^2, x^2z^2, y^2z^2, x^2r^2, y^2r^2, z^2r^2, r^4$
14. Fourth moment (odd)	$xyx^2, xy^3, xyz^2, xzx^2, xzy^2, xz^3, yzx^2, yzy^2, yz^3$
15. Overlap (charge or monopole)	1
16. Line charge density	$\delta(x), \delta(y), \delta(z)$
17. Planar charge density	$\delta(x)\delta(y), \delta(x)\delta(z), \delta(y)\delta(z)$
18. Charge density	$\delta(x)\delta(y)\delta(z)$

[a] All quantitites are given in atomic units.

5.9. PA25: **Generation of Symmetry-Adapted Basis Vectors**

The purpose of the program is to generate symmetry-adapted basis vectors (symmetry orbitals) to be used in the SCF programs. The program was written by Robert Kornegay at Bell Laboratories.

The program takes as input specifications of the symmetry properties of the one-electron functions that form a finite point group, determines all irreducible representations of the group and generates symmetry orbitals for each irreducible representation.

Let φ be a set of n functions $\varphi_1, \varphi_2, \ldots, \varphi_n$, associated in some way with a molecule. We assume that φ is sufficiently complete such that any symmetry operation R of the point group G of the molecule, transforms each of the functions into a linear combination of the other functions in the set

$$\varphi_i' = R\varphi_i = \sum_k \varphi_k A_{ki}(R) \qquad \text{for } i = 1, 2, \ldots, n \tag{70}$$

or in matrix notation,

$$\{\varphi'\} = \{\varphi\}\mathbf{A}(R) \tag{71}$$

The set of matrices $\mathbf{A}(R)$ forms a representation of the group G. This representation is, in general, an irreducible representation of G.

Projection operations are employed to construct the unitary matrix $\mathbf{U}$ which transforms the basis $\{\Phi\}$ to symmetry orbitals $\{\Psi\}$. The Ψ are a basis for the irreducible representations of G and are said to be symmetry-adapted to the group.

We use the character $\chi^\alpha(R)$ of the irreducible representation α and element R of the finite group G to generate the projection operator P^α,

$$\mathbf{P}^\alpha = \frac{n_\alpha}{g} \sum \chi^{(\alpha)*}(R)\mathbf{A}(R) \tag{72}$$

where g is the order of the group, n_α is the dimension of the irreducible representation, and the summation is over all elements in the group. The matrix $\mathbf{U}$ is constructed by repeated application of Eq. (72) for all irreducible representations of the group.[24] All zero vectors are deleted and all linear combination of vectors are moved by a Schmidt orthogonalization from each $\mathbf{P}^\alpha$. The vectors from all representations are then combined to form $\mathbf{U}$.

6. Recent Improvements and Additions

It is difficult to decide which of the many additions and improvements in the POLYATOM system to discuss here. We briefly list below those extensions with which we are personally familiar and which have been implemented at the authors' institutions. For the most part, these programs have not been submitted to QCPE.

6.1. Integral Evaluation

The integral evaluation program has been modified to compute integrals over all three components (x, y, z) of a p function and all six components $(x^2, y^2, z^2, xy, xz, yz)$ of the d function. All components of the basis function must have common contraction coefficients and exponents. This permits economies due to the simultaneous computation of blocks of integrals arising from the full sets of functions. The use of symmetry to reduce the number of integrals is limited to a single reflection plane for a molecular system. No label list file is generated.

The program for one-electron integrals is called PA31A. The program for two-electron integrals of s and p functions is called PA31B, and for s, p, and d functions PA31BD. These programs were written by H. Basch at Bell Laboratories and Bar-Ilan University.

6.2. Atomic Effective Potential

Chemists have long known that the essential features of molecular structure are determined by the valence electrons. As a result, a great deal of effort has gone into seeking ways of replacing the core electrons by a local effective-or pseudopotential. One such method has been developed by C. F. Melius, B. D. Olafson, L. Kahn, and W. A. Goddard III.[25,26] They found that the resulting effective potential depended on the angular momentum and could be expressed in the form

$$V^{EP}(r) = \sum_l V_l(r)|l\rangle\langle l| \tag{73}$$

where $V_l(r)$ is a radial function and $|l\rangle\langle l|$ is an angular momentum projection operator. In the early work $V_l(r)$ was obtained as a numerical function of r. However, it was found that a more efficient scheme resulted from expanding each radial function $V_l(r)$ in terms of Gaussians,

$$V_l(r) = \sum_k C_k r^{n_k} \exp(-\zeta_k r^2) \tag{74}$$

The general three-center integral $\langle i|V^{EP}|j\rangle$, for the atomic effective potential, can be reduced to a simple analytic form. As a result, the computation time is equivalent to that of a nuclear attraction integral. The effective potential program was written by C. F. Melius at the California Institute of Technology and has been integrated into PA30A.

6.3. SCF Programs

Recently Duke[33] has proposed an algorithm to store the elements of the **F** matrix in a linearized form. This speeds up the processing of two-electron integrals. This algorithm has been implemented by H. Basch at Bar-Ilan University and T. A Weber at Bell Laboratories for both open- and closed-shell programs.

In addition, an open-shell SCF program utilizing the orthogonality constrained basis set method has been implemented by H. Basch at Bar-Ilan University.

6.4. MC–SCF Programs

A set of multiconfigurations SCF programs has been written by H. Basch.

6.5. Generalized Valence Bond Method

The generalized valence-bond method has been implemented by Goddard and co-workers.[28]

6.6. Compton Profiles

A program has been written by T. A. Weber of Bell Laboratories to compute Compton profiles for gaseous and oriented molecules. The programs are referred to as PA70. Basis functions of s, p, d, and f types are accommodated. It has been applied to a large collection of molecular wave functions computed in a uniform basis set with POLYATOM.[29]

6.7. PHANTOM

Macaulay and Goutier at the University of Montreal have extensively modified POLYATOM to produce a set of programs they have named PHANTOM.[30] The PHANTOM system incorporates a number of recent theoretical developments.

It employs ideas of Dacre[31] and Elder[32] to reduce the integral file to the unique integrals. This has produced a very significant reduction of computing time.

The algorithm of Duke[33] is used to store integrals and construct Fock matrices. The programs can do restricted as well as unrestricted SCF calculations. The orthogonality contrained basis-set method is used for open-shell systems.

It is claimed that for a highly symmetrical molecule such as ethylene, PHANTOM is ten times as fast as POLYATOM (Version 2) on a CDC Cyber 74 computer.

7. Postscript

The creation of large computer program systems for chemistry such as POLYATOM requires many man-years of effort, often extending over a decade or more. The persons who are most productive originators of such programs, are frequently not those who will make the most significant applications.

For these reasons, the creation of POLYATOM and similar systems combined the contributions of a number of individuals, at least some of whom have a greater interest in programming and computer science. In order that the program contributions of these individuals of diverse interests fit together to form the larger productive system, there must be an agreed upon philosophy or format to guide their work. Moreover the programs must be documented to be built upon by persons other than the originators.

Even after a program system has been born, it must be maintained. There must be a way to report, record, and remove the inevitable program errors. And if the wider chemical community is to benefit through applications, the programs must be made accessible through physical distribution or other means.

The creation, maintenance, and distribution of chemical computation systems is an expensive activity with costs which often exceed the benefits to originating individuals and institutions. The benefits, however, are often large and distributed widely over the chemical community.

These circumstances retard the development of large and efficient program systems for chemical applications and provide one of the motivations behind proposals for a National Resource for Computation in Chemistry.[34,35]

References

1. I. G. Csizmadia, M. C. Harrison, J. W. Moskowitz, S. Seung, B. T. Sutcliffe, and M. P. Barnett, QCPE #47.1 POLYATOM-Program set for nonempirical molecular calculations, Quantum Chemistry Exchange Program, Indiana University, Bloomington, Indiana 47401.
2. D. B. Neumann, H. Basch, R. L. Kornegay, L. C. Snyder, J. W. Moskowitz, C. Hornback, and S. P. Liebmann, QCPE #199, The POLYATOM (Version 2) System of Programs for Quantitative Theoretical Chemistry.
3. S. F. Boys, Electronic wave functions. I. A general method of calculation for the stationary states of any molecular system, *Proc. R. Soc.* London, *Ser. A* **200**, 542 (1950).

4. S. F. Boys and G. B. Cook, Mathematical problems in the complete quantum predictions of chemical phenomena, *Rev. Mod. Phys.* **32**, 285 (1960).
5. S. F. Boys, G. B. Cook, C. M. Reeves, and I. Shavitt, Automatic fundamental calculations of molecular structure, *Nature* **178**, 1207 (1956).
6. C. M. Reeves, Use of Gaussian Functions in the calculation of wave functions for small molecules. I. Preliminary investigations, *J. Chem. Phys.* **39**, 1(1963).
7. C. M. Reeves and M. C. Harrison, Use of Gaussian functions in the calculation of wave functions for small molecules. II. The ammonia molecule, *J. Chem. Phys.* **39**, 11 (1963).
8. M. P. Barnett, Mechanized molecular calculation—The POLYATOM system, *Rev. Mod. Phys.* **35**, 571 (1963).
9. I. G. Csizmadia, M. C. Harrison, J. W. Moskowitz, and B. T. Sutcliffe, Nonempirical LCAO-MO-SCF-CI calculations on organic molecules with Gaussian-type functions. Part I. Introductory review and mathematical formalism, *Theor. Chim. Acta* **6**, 191 (1966).
10. T. D. Metzgar and J. E. Bloor, QCPE #238 POLYATOM: Version II (IBM 360).
11. D. Goutier, R. Macaulay, and A. J. Duke, QCPE #241, PHANTOM: *Ab initio* Quantum chemical programs for CDC 6000 and 7000 series computers.
12. C. C. J. Roothaan, New developments in molecular orbital theory, *Rev. Mod. Phys.* **23**, 69 (1951).
13. C. C. J. Roothaan, Self-consistent field theory for open shells of electronic systems, *Rev. Mod. Phys.* **32**, 179, (1960).
14. H. Basch and D. Neumann, 1969, unpublished.
15. G. A. Segal, Calculation of wavefunctions for the excited states of polyatomic molecules, *J. Chem. Phys.* **53**, 360 (1970).
16. W. J. Hunt, T. H. Dunning, and W. A. Goddard II, The orthogonality constrained basis set expansion method for treating off-diagonal Lagrange multipliers in calculations of electronic wave functions, *Chem. Phys. Lett.* **3**, 606 (1969).
17. J. S. Brinkley, J. A. Pople, and P. A. Dobash, The calculation of spin-restricted single-determinant wavefunctions, *Mol. Phys.* **28**, 1423 (1974).
18. J. A. Pople and R. K. Nesbet, Self-consistent orbitals for radicals, *J. Chem. Phys.* **22**, 571 (1954).
19. H. Eyring, J. Walter, and G. E. Kimball, *Quantum Chemistry*, J. Wiley and Sons, New York (1944).
20. R. S. Mulliken, Electronic populations analysis on LCAO-MO molecular wave functions. I., *J. Chem. Phys.* **23**, 1833 (1955).
21. I. Shavitt, The Gaussian function in calculations of statistical mechanics and quantum mechanics *Methods Comput. Phys.* **2**, (1963).
22. E. Clemente and D. R. Davis, Electronic structure of large molecular systems, *J. Comput. Phys.* **1**, 223 (1966).
23. H. Taketa, S. Huzinaga, and K. O-Ohata, Gaussian-expansion methods for molecular integrals, *J. Phys. Soc. Jpn.* **21**, 2313 (1966).
24. R. M. McWeeny, *Symmetry—An Introduction to Group Theory*, Pergamon Press, Oxford (1963).
25. L. R. Kahn and W. A. Goddard III, *Ab initio* effective potentials for use in molecular calculations, *J. Chem. Phys.* **56**, 2685 (1972).
26. C. F. Melius, B. D. Olafson, and W. A. Goddard III, Fe and Ni *ab initio* effective potentials for use in molecular calculations, *Chem. Phys. Lett.* **28**, 457 (1974).
27. H. Basch, Dimerization of methylenes by their least motion, coplanar approach: A multiconfiguration self-consistent field study, *J. Chem. Phys.* **55**, 1700 (1971).
28. W. A. Goddard III, T. H. Dunning, Jr., W. J. Hunt, and P. J. Hay, Generalized valence bond description of bonding in low-lying states of molecules, *Acc. Chem. Res.* **6**, 368 (1973).
29. L. C. Snyder and H. Basch, *Molecular Wave Functions and Properties*, J. Wiley and Sons, New York (1972).
30. R. Macaulay and D. Goutier, PHANTOM: A chemical software system for doing *ab initio* Gaussian-type calculations, unpublished.
31. P. D. Dacre, On the use of symmetry in SCF calculations, *Chem. Phys. Lett.* **7**, 47 (1970).

32. M. Elder, Use of molecular symmetry in SCF calculations, *Int. J. Quantum Chem.* **7**, 75 (1973).

33. A. J. Duke, An alternative procedure for setting up Fock matrices from randomly ordered lists of electron interaction integrals, *Chem. Phys. Lett.* **13**, 76 (1972).

34. *A Study of a National Center for Computation in Chemistry*, National Academy of Sciences, Washington, D.C. (1974).

35. *The Proposed National Resource for Computation in Chemistry*, A User-Oriented Facility, National Academy of Sciences, Washington, D.C. (1975).

11

Configuration Expansion by Means of Pseudonatural Orbitals

Wilfried Meyer

1. Introduction

The configuration interaction (CI) method as a general approach to solving the many-electron Schrödinger equation to—in principle—any desired accuracy, has been described in this volume by Shavitt. We refer to that chapter for all basic concepts of the CI method and an outline of its merits and its computational problems.

The most disturbing feature of the CI method is certainly the very slow convergence of the configuration expansion of the wave function. Much effort has therefore been devoted to finding out if those parts of the electron correlation can be separated which are relevant for describing a particular process or property and which may be cast into a few configurations only. This leads to the multiconfiguration SCF procedure that is dealt with in this volume by Wahl and Das. Apart from the difficulties in defining these relevant parts—success usually rests on a critical balance of competing effects—it seems quite obvious that many properties cannot be treated in this way, in particular those which involve electronic structure changes, e.g., ionization, excitation, or electron attachment. It is therefore not only of principal interest but actually needed to be able to account for high percentages of the electron correlation in the valence shell. Aiming at this, numerous attempts have been made to

Wilfried Meyer • Institut für Physikalische Chemie, Johannes Gutenberg Universität, Mainz, Germany improve the convergence behavior of the configuration expansion. Particularly worth mentioning is the expansion on the basis of the natural orbitals introduced by Löwdin[1] and extensively used by Davidson and co-workers.[2–5] The convergence is drastically improved at the beginning, but it stays very slow toward the end. There is a simple argument showing that the use of natural orbitals can be of limited help only: since a large fraction of the computational effort goes into the calculation of the basis integrals, one seeks to exhaust the capacity of a chosen basis set fully. In this case, no particular set of orthogonal orbitals is really superior to others. Indeed, many of today's most advanced calculations aiming at high percentages of correlation energies use thousands of configurations to construct the wave function. Modern techniques to handle such numbers of configurations are discussed in this volume, by Shavitt, and a particularly promising recent development is described in the chapter by Roos and Siegbahn.

The objective of this chapter is to describe a method that avoids such long expansions without loss of ability to recover large fractions of the correlation. The only way to achieve this seems to be to drop the basic requirement of the traditional CI: the orthogonality of all orbitals used to construct the configurations. The obvious argument against using nonorthogonal orbitals is the complicated structure of the resulting Hamiltonian matrix elements, e.g., for two Slater determinants of nonorthogonal spin-orbitals ψ_i and ψ_k' one gets for a two-electron operator $V(1, 2)$[1,6]

$$\langle A\psi_1\psi_2 \cdots \psi_N | V(1, 2) | A\psi_1'\psi_2' \cdots \psi_N' \rangle N(N-1)$$

$$= \sum_{i,j} \sum_{k,l} \langle \psi_i\psi_j | V(1, 2) | \psi_k\psi_l \rangle D_{ij,kl} \cdot \text{sign}(i-j) \cdot \text{sign}(k-l) \qquad (1)$$

with $D_{ij,kl}$ denoting that minor of the overlap matrix $S_{mn} = \langle \psi_n | \psi_m' \rangle$ from which the rows i and j and the columns k and l are deleted. Nonorthogonal orbitals have therefore been considered impractical for constructing many-electron wave functions, and they have only occasionally been used for very small systems.[7] From Eq. (1) it is just as obvious, however, that one does not necessarily have to go to a completely orthogonal set of orbitals in order to simplify the matrix elements significantly, but that only *most* of the elements of the overlap matrix should be zero. It is indeed the idea of exploiting a partial nonorthogonality of the orbitals that we wish to develop here.[8–10] The partially nonorthogonal orbitals, which play the crucial role, have been introduced by Edmiston and Krauss,[11] who called them pseudonatural orbitals. We shall therefore speak of a "pseudonatural orbital-configuration expansion" (PNO-CE). This expansion form has been used in conjunction with variational equations for the expansion coefficients (PNO-CI[9]) and with the coupled electron pair approximation (PNO-CEPA[10]). The usefulness of these methods rests on some features of atomic and molecular many-electron systems that are briefly summarized in Section 2. Section 3 deals with the optimal convergent configuration expansion, which becomes possible after dropping the orthogonality constraints, and Section 4 presents the computational procedure that is built upon this expansion. Finally, applications are presented in Section 5.

2. Features of the Electron Correlation in Molecular Systems

The following considerations about particular features of the electron correlation seem to apply to a large class of electronic states. It is beyond the scope of this article, however, to give a well-founded account of their ranges of validity. Instead, the reader may take these statements as simply indicating the area of applicability of the PNO-CE method.

2.1. The Reference Wave Function

The configuration expansion is usually dominated by only a few configurations if the orbitals and the coefficients of these configurations are optimized for minimal energy. We call this the reference wave function Ψ_0. Usually, it represents the independent particle model in the form appropriate to the system under consideration. For many electronic states, in particular, ground states of atoms and of molecules close to their equilibrium geometry, it reduces to a single configuration—the Hartree–Fock determinant. For valence excited states or a study of dissociation processes, one often needs a multiconfigurational reference function. The reference function is obtained by the self-consistent field methods that are the subject of other chapters of this volume. At this level the total energies are calculated to an accuracy in the order of 1% and the coefficients of further configurations are in the order of or smaller than 10^{-1}. Thus, the remainder of the configuration expansion can be considered a small correction to the reference wave function. The reference wave function is usually chosen to have all symmetry properties of the electronic state, in particular to be a spin eigenstate. Following Silverstone and Sinanoğlu,[12] we shall call the orbitals occupied in Ψ_0 "internal" orbitals. "External" orbitals are then defined as being orthogonal to all internal orbitals.

2.2. Doubly Substituted Configurations

Because of the dominance of the reference wave function, it is useful to classify the configurations according to the number of electrons that are outside the internal space, i.e., the number of internal orbitals being substituted by external orbitals. Since the Hamiltonian is built from one-electron and two-electron operators, configurations with more than two external orbitals have vanishing matrix elements with the reference function. The latter is usually determined as eliminating its coupling to singly substituted configurations (Brillouin theorem for self-consistent wave functions[13]). Therefore, the double substitutions play the crucial role in completing the configuration expansion. Suppressing at the moment all complications arising from a multiconfigurational reference function, the wave function including single and double substitutions can be written in the familiar form

$$\Psi = \Phi_0 + \sum_i \sum_a C_i^a \Phi_i^a + \sum_{i \leq j} \sum_{a \leq b} C_{ij}^{ab} \Phi_{ij}^{ab} \tag{2}$$

where Φ_0 is the reference determinant and Φ_{ij}^{ab} denotes a configuration obtained from Φ_0 by replacing the internal spin-orbitals ψ_i and ψ_j by external spin-orbitals ψ_a and ψ_b. If the ψ_a form an orthogonal set, then all configurations are orthogonal.

2.3. Multiple Substitutions

Since they are not coupled to Φ_0 they have very small coefficients. However, their number increases rapidly with the number of electron pairs and their consideration is essential if the norm of the wave function (2) deviates considerably from unity, i.e., a variational wave function restricted to the ansatz (2) deteriorates with increasing number of electrons.[15,16]

Because the perturbation due to the correlation is a rather short-range two-electron interaction,[17] the most important multiple substitutions are just products of double substitutions, i.e., they are of unlinked cluster type. Their coefficients and matrix elements are related to those of the double substitutions, and they can therefore be approximately incorporated into a configuration expansion that explicitly works with double substitutions only. Whereas the many-body perturbation theory[18] as well as cluster expansion formulations[19] require for this purpose the orthogonality of all orbitals, we have proposed a particularly simple coupled electron pair approach (CEPA)[10] that rests only on the orthogonality of the configurations. It yields modified equations for the optimal coefficients C_{ij}^{ab} of (2) and changes the way they enter formulas for expectation values,[20] but the structure of (2) is not changed. These methods are dealt with in detail in Kutzelnigg's chapter in this volume, and we will only draw the conclusion that there is good reason to restrict ourselves to an explicit treatment of double substitutions and try to find an optimal form for the expansion (2).

2.4. Perturbation Considerations

The smallness of the coefficients in (2) suggests using perturbation methods. As is well known,[21,22] the wave function is given to first-order accuracy by

$$\Psi^{(1)} = \Phi_0 - \sum_{a \leq b} \sum_{i \leq j} \Phi_{ij}^{ab} \langle \Phi_{ij}^{ab} | \hat{H} | \Phi_0 \rangle / E_{ij}^{ab} \tag{3}$$

where the energy denominators may be obtained from the eigenvalues of the Fock operator (Möller–Plesset perturbation theory) or from the diagonal elements of $\hat{H}$ (Epstein–Nesbet perturbation theory).[22] From $\Psi^{(1)}$ the energy is calculated to second order as

$$E^{(2)} = \langle \Phi_0 | \hat{H} | \Psi^{(1)} \rangle \tag{4}$$

and to third order as

$$E^{(3)} = \langle \Phi_0 | \hat{H} | \Phi_0 \rangle + \langle \Psi^{(1)} | \hat{H} - \langle \Phi_0 | \hat{H} | \Phi_0 \rangle | \Psi^{(1)} \rangle \tag{5}$$

The accuracy of the correlation energy to second order depends rather strongly on the particular molecular system and seems in general not better than 10%. It is therefore insufficient for many purposes. On the other hand, the third-order energy requires calculating the Hamiltonian matrix elements between all configurations of $\Psi^{(1)}$ and seems not to offer any advantage over the methods that obtain energy and wave function from solving variational or coupled pair equations. This is not quite true, however. From the computational point of view, it makes a considerable difference whether the matrix elements have to be obtained individually, as required for varying the coefficients C_{ij}^{ab}, or $E^{(3)}$ is calculated as a whole from given values of the coefficients.[23,24]

An obvious way to take advantage of the knowledge of the coefficients is to put expansion (2) into a more compact form before calculating the energy and other expectation values. In the simplest way this possibility follows from the fact that one may express a sum over determinants that differ in one particular column only, by a single determinant containing the sum of the differing columns:

$$\sum_a C^a |\psi_1 \cdots \psi_a \cdots \psi_N| = |\psi_1 \cdots \left(\sum_a C^a \psi_a \right) \cdots \psi_N| \tag{6}$$

This means that one sum of the two-electron clusters of (2) can be absorbed by defining new orbitals. For example, with

$$\psi_{a(ij)} = \sum_b \psi_b C_{ij}^{ab} / C_{ij}^a \tag{7}$$

one may write

$$\sum_{ab} C_{ij}^{ab} \Phi_{ij}^{ab} = \sum_a C_{ij}^a \Phi_{ij}^{aa_{ij}} \tag{8}$$

Since the $\Phi_{ij}^{aa_{ij}}$ are disjunct sums of orthogonal configurations, they are also mutually orthogonal. The number of configurations per pair (ij) is reduced from L^2 to L (L being the dimension of the external orbital space) at the cost of having one orbital in each configuration that is not orthogonal to all the others. This illustrates the basic idea of the PNO-CE approach: efficient approximate methods are used to determine particular orbitals, which allow for writing the configuration expansion in the most compact form. Individual matrix elements are then calculated for this reduced set of configurations, and their coefficients are determined from variational or coupled pair equations. The computational effort is larger than for just calculating $E^{(3)}$ but it is considered important to be able to refine the coefficients beyond their first-order value. Since the energy denominators in Eq. (3) may be rather small in molecular systems, this is certainly mandatory for calculating properties other than the energy. Of course, it is a matter of experience how well the approximations work that enter via the construction of the pseudonatural orbitals. This will be discussed in Section 5.

3. Optimal Configuration Expansion

For notational convenience we introduce creation and annihilation operators (see, e.g., March *et al.*[25]): Let η_i annihilate an electron from spin orbital ψ_i,

$$\eta_i \Psi(1, \ldots, N) = N^{1/2} \int d1 \psi_i^*(1)\Psi(1, \ldots, N) \tag{9}$$

and let $\eta_j^\dagger$ create an electron in spin orbital ψ_i, i.e.,

$$\eta_i^\dagger \Psi(2, \ldots, N) = N^{-1/2} \sum_{n=2}^{N} (1 - T_{1,n})\psi_i(1)\Psi(2, \ldots, N) \tag{10}$$

Antisymmetry requires the operators to satisfy the anticommutation relations:

$$\eta_i\eta_j + \eta_j\eta_i = 0, \qquad \eta_i^\dagger\eta_j^\dagger + \eta_j^\dagger\eta_i^\dagger = 0, \qquad \eta_i^\dagger\eta_j + \eta_j^\dagger\eta_i = \delta_{ij} \tag{11}$$

A Slater determinant may then be written as

$$|\psi_1\psi_2 \cdots \psi_N| = \eta_1^\dagger\eta_2^\dagger \cdots \eta_N^\dagger\Psi_{\text{vac}} \tag{12}$$

with the "vacuum state" satisfying $\eta_i\Psi_{\text{vac}} = 0$ for all i, and the first-order reduced density matrix[1,6] takes the form

$$\rho(\Psi_1\Psi_2|1, 1') = N \int d2 \cdots dN\Psi_1(1, 2, \ldots, N)\Psi_2^*(1', 2, \ldots, N)$$

$$= \sum_{ij} \psi_i(1)\psi_j^*(1')\langle\Psi_2|\eta_j^\dagger\eta_i|\Psi_1\rangle \tag{13}$$

3.1. The Natural Orbitals of a Two-Electron System

For a two-electron system a dramatic simplification of the configuration expansion can be achieved by choosing an appropriate set of spin-orbitals. This was first noted by Hurley *et al.*[26] for the singlet case and has been systematically investigated by Löwdin and Shull.[27] Since this possibility is at the heart of the PNO-CE method we reproduce the main results here.

An antisymmetric two-electron wave function with $s_z = 0$ may be expanded by means of an orthonormal set of spin orbitals $\varphi_i\alpha$, $\varphi_i\beta$ as

$$\Psi(1, 2) = \sum_{i,j=1}^{L} c_{ij}|\varphi_i\alpha\varphi_j\beta| \tag{14}$$

One of the summations can be absorbed by defining a set of "corresponding" orbitals

$$\varphi_i'\beta = \eta_{i\alpha}\Psi = \sum_j c_{ij}\varphi_j\beta \tag{15}$$

Their overlap matrix

$$\langle \varphi_k'|\varphi_i'\rangle = \sum_l c_{il}c_{kl}^* = (CC^\dagger)_{ik} \tag{16}$$

is Hermitian and may be transformed into a diagonal matrix by a unitary matrix U: $(UCC^\dagger U^\dagger)_{ij} = d_i^2 \delta_{ij}$. With the new orbitals

$$\tilde{\varphi}_i = \sum_j u_{ij}\varphi_j \tag{17}$$

$$d_i\tilde{\varphi}_i' = \sum_j (UCU^\dagger)_{ij}\tilde{\varphi}_j = \sum_j (UC)_{ij}\varphi_j \tag{18}$$

one obtains Ψ in a "diagonal representation"

$$\Psi = \sum_i d_i|\tilde{\varphi}_i\alpha\tilde{\varphi}_i'\beta| \tag{19}$$

constructed from an orthonormal set of spin orbitals $\tilde{\varphi}_i\alpha$, $\tilde{\varphi}_i'\beta$. Hence, the first-order density matrix also has diagonal form,

$$\rho(1, 1') = \sum_i d_i^2\{\tilde{\varphi}_i\alpha\tilde{\varphi}_i^*\alpha + \tilde{\varphi}_i'\beta\tilde{\varphi}_i'^*\beta\} \tag{20}$$

and the spin-orbitals $\tilde{\varphi}_i\alpha$, $\tilde{\varphi}_i'\beta$ are the natural spin-orbitals (NSOs) of Ψ as introduced by Löwdin.[1]

In general, the overlap matrix $\langle\tilde{\varphi}_i|\tilde{\varphi}_j'\rangle$ is nondiagonal. However, if we choose the orbitals φ_i and coefficients c_{ij} of (14) to be real, which can always be done for a real Hamiltonian operator, this overlap matrix becomes particularly simple for the two most important cases:

(1) For a pure singlet state $c_{ij} = c_{ji}$ and U can be chosen as real and diagonalizing C, which yields $\tilde{\varphi}_i = \tilde{\varphi}_i'$ and $\Psi = \sum_i d_i|\tilde{\varphi}_i\alpha\tilde{\varphi}_i\beta|$. In case the determinants of (14) are composed of orbitals from two different spatial symmetries γ and γ', the orbitals $\tilde{\varphi}_i$ appear in pairs $2^{1/2}(\tilde{\varphi}_{2i-1}\pm\tilde{\varphi}_{2i})$ with $\pm d_{2i}$, in which case one may write

$$\Psi = \sum_{i=1}^{L/2} d_{2i}\{|\tilde{\varphi}_{2i-1}^\gamma\alpha\tilde{\varphi}_{2i}^{\gamma'}\beta| + |\tilde{\varphi}_{2i}^{\gamma'}\alpha\tilde{\varphi}_{2i-1}^\gamma\beta|\} \tag{21}$$

(2) For a pure triplet state $c_{ij} = -c_{ji}$ and U can be chosen as real and yielding $UCU^{\dagger}$ in a form with two-by-two blocks $\left(\begin{smallmatrix} 0 & d \\ -d & 0 \end{smallmatrix}\right)$ in the diagonal. Hence, $\tilde{\varphi}'_{2i-1} = \tilde{\varphi}_{2i}$, $\tilde{\varphi}'_{2i} = -\tilde{\varphi}_{2i-1}$, and

$$\Psi = \sum_{i=1}^{L/2} d_{2i} \{ |\tilde{\varphi}_{2i-1} \alpha \tilde{\varphi}_{2i} \beta| - |\tilde{\varphi}_{2i} \alpha \tilde{\varphi}_{2i-1} \beta| \} \tag{22}$$

For the pure spin states, the natural spin orbitals correspond to a single orthonormal set of orbitals, which are referred to as the natural orbitals (NOs) of Ψ. The NOs not only yield a diagonal expansion of Ψ but the M orbitals with the largest "occupation numbers" $2d_i^2$ give the best approximation to Ψ that can be constructed from M orbitals, i.e., the NOs provide the basis for the most rapidly converging configuration expansion when ordered according to decreasing occupation numbers. The proof is quite elementary: the projection of Ψ into the subspace of all configurations that do not contain a particular spin orbital ψ_k is given by $(1 - \eta_k^{\dagger}\eta_k)\Psi$. The normalized projection is the best approximation to Ψ within the subspace. Its deviation from Ψ is measured by $1 - \langle \Psi | 1 - \eta_k^{\dagger}\eta_k | \Psi \rangle = \langle \Psi | \eta_k^{\dagger}\eta_k | \Psi \rangle$, i.e., by a diagonal element of the first-order reduced density matrix. As is well known from the variation principle, the smallest possible diagonal element is the smallest eigenvalue. Hence, the minimal error is introduced by deleting the natural orbital with smallest occupation number. This process may be repeated, since for a two-electron system the projection does not change the natural orbitals.

The rapid convergence of the configuration expansion of two-electron systems based on natural orbitals was first demonstrated by Löwdin and Shull[27] for helium and by Shull[28] for H_2. The natural orbital with the largest occupation was shown to be very similar to the Hartree–Fock orbital with an occupation number close to two.

3.2. Electron Pairs in Many-Electron Wave Functions

We now turn to the question of whether the configuration expansion of a many-electron wave function can be improved as well by using appropriate orbital sets.

As before one may define for any spin orbital ψ_i a corresponding $(N-1)$-electron function by $\Psi_i = \eta_i \Psi$. $\eta_i^{\dagger}\eta_i \Psi$ represents that part of Ψ in which one electron occupies ψ_i. Since $\sum_i \eta_i^{\dagger}\eta_i = N$, we obtain a "diagonal" expansion of Ψ in the form

$$\Psi = \frac{1}{N} \sum_i \eta_i^{\dagger}\eta_i \Psi = \frac{1}{N} \sum_i A(1/2, \ldots, N) \psi_i(1) \Psi_i(2, \ldots, N) \tag{23}$$

$[A(1/2, \ldots, N)$ is the antisymmetrizer given explicitly in Eq. (10).] The overlap matrix $\langle \Psi_i | \Psi_j \rangle = \langle \Psi | \eta_i^{\dagger}\eta_i | \Psi \rangle$ gives the coefficients of the first-order reduced-density matrix. Thus, for ψ_i being the set of natural spin orbitals, the functions Ψ_i are mutually orthogonal. Moreover, Coleman[29] has shown that of any such expansion truncated to M terms, the best one is obtained from the set of natural spin orbitals ordered according to decreasing occupation numbers. This does not imply rigorously that the first M NSOs represent the best basis of M spin orbitals for a configuration expansion because an expansion of the functions $\Psi_i(2, \ldots, N)$ for $i \leq M$ does require spin orbitals ψ_j with $j > M$. (In contrast to the argument given for two-electron systems, the projection of Ψ into a configurational subspace changes its NSOs in the case of a many-electron wave function.) Still, the NSOs are good approximations to the optimal orbitals as long as M is sufficiently large for the Ψ_i being fairly well expanded in configurations of this basis. The natural orbitals are therefore very useful if one wants to reduce the size of a given orbital set to be employed for a configuration expansion.[2-5,30,31] However, in contrast to the two-electron case, the NSOs do not simplify the structure of the configuration expansion of a many-electron wave function. Expression (23) does not even imply the partitioning of Ψ into mutually orthogonal parts, as was the case for two-electron systems.

As an alternative, let us therefore consider the possibility of exploiting the advantages offered by the diagonal expansion of a two-electron wave function. In a formal way, subsystems of electron pairs in a many-electron system can be formed by defining two-electron functions

$$u_{Pm}(1, 2) = \binom{N}{2}^{1/2} \int d3 \cdots dN \Phi_{Pm}^*(3, \ldots, N)\Psi(1, 2, 3, \ldots, N) \quad (24)$$

on the basis of a set of orthogonal $(N-2)$-electron configuration functions Φ_{Pm}. (m is used to denote different configuration functions that are fully related by symmetry requirements; see below.) In analogy to (23), Ψ can be expanded as

$$\Psi = \binom{N}{2}^{-1} \sum_{Pm} A(1, 2/3, \ldots, N)u_{Pm}(1, 2)\Phi_{Pm}(3, \ldots, N) \quad (25)$$

and one may now look for diagonal expansions of the pair functions u_{Pm}. As before, the most rapidly converging expansion of the form (25) is obtained by a unitary transformation of the Φ_{Pm} to yield the natural pairs $\tilde{u}_{Pm}$ and the natural $(N-2)$-electron functions $\tilde{\Phi}_{Pm}$, which have the disadvantage, however, of being of a complicated configurational structure and of being defined with respect to an unknown function Ψ.

From the practical point of constructing a configuration expansion for Ψ, the following considerations have to be taken into account:

(a) The Φ_{Pm} retained in a truncated expansion (25) should approximately span the space of the leading natural functions $\tilde{\Phi}_{Pm}$, but be of a simple configurational structure. That means in effect that they should be configuration functions that are constructed from a limited but optimized set of orbitals. Of course, this task is most directly solved by setting up a self-consistent reference function Ψ_0, which dominates the expansion of Ψ, and subsequently deriving the Φ_{Pm} by deleting two of the electrons. The expansion of Ψ is then limited to double substitutions with respect to Ψ_0, but we have already indicated why this restriction should be acceptable.

(b) In order to take full advantage of the symmetry of the system, the Φ_{Pm} should be symmetry adapted (i.e., elements of a basis that is irreducible with respect to symmetry operations). This implies that the functions u_{Pm} describe symmetry-irreducible pairs as introduced by Sinanoğlu.[32] In particular, these pairs should be pure singlet or triplet functions to allow for the most convenient natural expansion.

(c) For the sake of tractable expressions of the Hamiltonian matrix elements, the orbitals appearing in Ψ_0 and Φ_{Pm} must be mutually orthogonal and orthogonal to all other orbitals. This constitutes two classes of orbitals as already mentioned: a small set of "internal" orbitals (henceforth denoted by indices i, j, etc.) and "external" orbitals (denoted by indices a, b, etc.), which satisfy the orthogonality relations

$$\langle \varphi_i | \varphi_j \rangle = \delta_{ij}, \qquad \langle \varphi_i | \varphi_a \rangle = 0 \tag{26}$$

Based on the internal and external orbital subspaces, the total wave function can be divided up as follows:

$$\Psi = \psi_0 + \psi_{int} + \psi_{semi\text{-}int} + \psi_{ext} \tag{27}$$

The internal and semi-internal parts are defined by having no electrons and one electron in the external subspace, respectively. They appear only in the open-shell case and are composed of relatively few configurations. Thus our main concern is with Ψ_{ext}. Substituting Ψ_{ext} for Ψ in Eq. (24), one obtains external pair functions by means of which one may write

$$\Psi_{ext} = \sum_{Pm} A(1, 2/3, \ldots, N) u_{Pm}^{ext}(1, 2) \Phi_{Pm}(3, \ldots, N) \tag{28}$$

[The factor $\binom{N}{2}^{-1}$ of (25) disappears here since the Φ_{Pm} are now confined to the internal space. In the following we shall use u_{Pm} for external pairs only and therefore drop the superscript.] Now, the orthogonality of the Φ_{Pm} is sufficient for the different parts of Ψ_{ext} to be mutually orthogonal. Thus, each pair function can be transformed independently into its natural form without creating nonorthogonal configurations. As demonstrated in Section 3.1, the natural orbitals of the pairs (NOPs) constitute orthogonal sets of external orbitals if the pairs are spin adapted. But quite obviously the natural orbitals of different pairs are not orthogonal in the general case. As will be shown later on, this does not significantly complicate the calculation of matrix elements, however.

A diagonal expansion of electron pairs in many-electron wave functions was first proposed by Hurley *et al.*[26] in connection with the separated electron pair approximation. In this case, the core function Φ_{Pm} of a pair u_{Pm} is the product of all other $(N-2)/2$ pair functions and thus of rather complicated structure. A tractable energy expression can be derived only if all pair functions are strongly orthogonal. The strong orthogonality is equivalent to expanding the pair functions by means of distinct subsets of orthogonal orbitals. In this case the diagonal form can be obtained by unitary transformations within each subset, retaining the orthogonality of the total orbital set. However, the optimal partitioning of the orbital space has turned out to be rather difficult,[33,34] and the quality of the pair functions is substantially limited due to the considerable penetration of the pairs in a molecule. Moreover, in this method one completely neglects the interpair correlation, which has been shown to be as important as intrapair correlation even in systems with well-localized electrons.[10,35,36] In a system with appreciable interpair correlation, strong orthogonality cannot be imposed on all important pair functions— their diagonal expansion will therefore unavoidably lead to nonorthogonal orbitals. This is without any consequences if one follows the independent electron pair treatment of the electron correlation as proposed by Sinanoğlu[17,32] and Nesbet.[15] In this case, one calculates separately variational wave functions of the form $\Psi_0 + \sum_m A u_{Pm} \Phi_{Pm}$ and approximates the total correlation energy by the sum of the pair correlation energies. The natural orbitals of such independent pair functions are the "pseudonatural orbitals" (PNOs), introduced by Edmiston and Krauss.[11] Following their suggestion, the PNOs of a particular "representative" pair have often been used as an optimal orbital set for a traditional CI since these orbitals are properly located in space and are therefore superior to the virtual SCF orbitals.[11,37,38] A more genuine use of the PNOs has been made by Kutzelnigg and co-workers[36,39,40] in correlation calculations on small polyatomic hydrides: employing a scheme for a direct determination of approximate PNOs (without setting up a pair CI matrix) they circumvented the integral transformation and obtained independent electron pair correlation energies very efficiently. The independent pair treatment has serious drawbacks, though. The sum of the pair correlation energies often depends strongly on the definition of the pairs, i.e., it varies considerably with unitary transformations of the internal orbitals that leave Ψ_0 unchanged.[9,41] For localized orbitals this sum usually overshoots the correct correlation energy by 5 to 15%,[10,42,43] whereas for very delocalized electrons this approximation may completely fail (e.g., see Section 3.2). For a wave function restricted to intraorbital pairs, which may be considered a simplified separated pair ansatz,[44] Ahlrichs and Kutzelnigg[8] considered calculating variational energies while expanding the pairs in terms of PNOs. In this case the Hamiltonian matrix elements between different pair functions require just overlap integrals of the PNOs. They dismissed this scheme in favor of the independent electron pair approximation because of the unlinked cluster problem, which, as pointed out in Section 2.3, makes variational wave functions unsuitable for larger molecules. It was then realized by this author that the unlinked cluster problem could be solved in a fair approximation by the coupled electron pair approach without requiring orthogonal orbitals, and that the coupling matrix elements between any type of pair functions are rather simple irrespective of their expansion in orthogonal and nonorthogonal orbital sets. The considerable computational advantages of the PNO expansion led to the proposal of the variational PNO-CI and the nonvariational PNO-CEPA schemes for calculating correlated wave functions.[10,11]

3.3. Construction of Core and Pair Functions

It has been noted that the core functions should be derived from the reference wave function in such a way that the external pairs turn out to be symmetry adapted, and that they constitute mutually orthogonal parts of the total wave function. This is most directly achieved by starting from the complete orthogonal set of symmetry-adapted core functions that can be constructed from the internal orbitals, and coupling external pairs to them following standard group-theoretical techniques. But the set of configuration functions generated in this way does not comply with a further request, which is of considerable practical importance: that the configuration function basis should exactly span the "interaction space" of Ψ_0,[45,46] i.e., it should not be possible to find a linear combination of configuration functions of the same configuration (i.e., the same orbital occupancy) that has a vanishing Hamiltonian matrix element with Ψ_0. Such a linear combination would not contribute to the wave function in first order of perturbation theory and should be projected out. In the open-shell case, with the above-defined core functions, some external pairs will be equal to first order and a certain linear combination of the corresponding core functions would be sufficient.

A rather simple but powerful procedure for deriving the appropriate core functions and external configuration functions can be derived from a brief analysis of the Hamiltonian operator. As is well known, $\hat{H}$ may be written by means of creation and annihilation operators [see Eqs. (9)–(13)] as follows[25]:

$$\hat{H} = \sum_{r,s} \langle \varphi_r | \hat{H}^0 | \varphi_s \rangle (\eta_r^\dagger \eta_s + \bar{\eta}_r^\dagger \bar{\eta}_s)$$

$$+ \frac{1}{2} \sum_{r,s} \sum_{t,u} \langle \varphi_r \varphi_s | r_{12}^{-1} | \varphi_t \varphi_u \rangle (\eta_r^\dagger \eta_s^\dagger \eta_t \eta_u + \bar{\eta}_r^\dagger \eta_s^\dagger \bar{\eta}_t \eta_u + \eta_r^\dagger \bar{\eta}_s^\dagger \eta_t \bar{\eta}_u + \bar{\eta}_r^\dagger \bar{\eta}_s^\dagger \bar{\eta}_t \bar{\eta}_u)$$

$$(29)$$

(r, s, t, u run over internal and external orbitals, and the overbar denotes creation/annihilation of a spin-down electron). Considering the permutational invariances of the two-electron integrals, it is easily verified by interchanging pairs of indices that the second term of (29) may be rearranged into

$$
\sum_{r \leq s} \sum_{t \leq u} \langle \varphi_r \varphi_s | r_{12}^{-1} | \varphi_t \varphi_u + \varphi_u \varphi_t \rangle (\eta_r^\dagger \bar{\eta}_s^\dagger + \eta_s^\dagger \bar{\eta}_r^\dagger)(\eta_t \bar{\eta}_u + \eta_u \bar{\eta}_t)/2(1 + \delta_{rs})(1 + \delta_{tu})
$$
$$
+ \langle \varphi_r \varphi_s | r_{12}^{-1} | \varphi_t \varphi_u - \varphi_u \varphi_t \rangle \{ \eta_r^\dagger \eta_s^\dagger \eta_t \eta_u + \bar{\eta}_r^\dagger \bar{\eta}_s^\dagger \bar{\eta}_t \bar{\eta}_u
$$
$$
+ (\eta_r^\dagger \bar{\eta}_s^\dagger - \eta_s^\dagger \bar{\eta}_r^\dagger)(\eta_t \bar{\eta}_u - \eta_u \bar{\eta}_t)/2 \}
\tag{30}
$$

We introduce the spin-orbital pair annihilation operators

$$
(\eta_r \eta_s | pm) = \begin{cases} \eta_r \eta_s & m = 1, p = 1 \\ (\eta_r \bar{\eta}_s + p \eta_s \bar{\eta}_r)(2 + 2\delta_{rs})^{-1/2} & m = 0, p = \pm 1 \\ \eta_r \eta_s & m = -1, p = 1 \end{cases}
\tag{31}
$$

and the corresponding pair creation operators $(\eta_r^\dagger \eta_s^\dagger | pm)$. For convenience in later expressions, we use $p = 1 - 2s$ for differentiating between singlet $(p = 1)$ and triplet $(p = -1)$ pairs. $\hat{H}$ may now be written as

$$
\hat{H} = \sum_{r,s} \langle \varphi_r | h | \varphi_s \rangle (\eta_r^\dagger \eta_s | 10)
$$
$$
+ \sum_{r \leq s} \sum_{t \leq u} \sum_{p=1,-1} \langle \varphi_r \varphi_s | r_{12}^{-1} | \varphi_t \varphi_u + p \varphi_u \varphi_t \rangle (1 + \delta_{rs})^{-1/2}(1 + \delta_{tu})^{-1/2}
$$
$$
\times \sum_m (\eta_r^\dagger \eta_s^\dagger | pm)(\eta_t \eta_u | pm)
\tag{32}
$$

Since Ψ_{ext} has two electrons in external orbitals, we obtain finally

$$
\langle \Psi_{\text{ext}} | \hat{H} | \Psi_0 \rangle = \sum_{a \leq b} \sum_{i \leq j} \sum_{p=1,-1} \langle \varphi_a \varphi_b | r_{12}^{-1} | \varphi_i \varphi_j + p \varphi_j \varphi_i \rangle (1 + \delta_{ab})^{-1/2}(1 + \delta_{ij})^{-1/2}
$$
$$
\times \left\langle \Psi_{\text{ext}} \left| \sum_m (\eta_a^\dagger \eta_b^\dagger | pm)(\eta_i \eta_j | pm) \right| \Psi_0 \right\rangle
\tag{33}
$$

If spatial symmetries are disregarded, all terms of the right-hand side are linearly independent. Thus, the configuration functions

$$
\Phi_{ij p}^{ab} = C_{ijp} \sum_m (\eta_a^\dagger \eta_b^\dagger | pm)(\eta_i \eta_j | pm) \Psi_0, \qquad a \leq b, i \leq j
\tag{34}
$$

span exactly the external space interacting with Ψ_0. The external pairs obtained in the form

$$
u_{ijpm} = \sum_{a \leq b} C_{ijp}^{ab} (\eta_a^\dagger \eta_b^\dagger | pm) \Psi_{\text{vac}}
\tag{35}
$$

are spin adapted as desired, but the core functions

$$
\Phi_{ijpm} = (\eta_i \eta_j | pm) \Psi_0
\tag{36}
$$

are neither spin adapted nor normalized if i or j refers to an open-shell orbital. This is not at all a handicap due to the simple coupling explicitly given in Eq. (34). Since $\hat{H}$ is a spin-free operator it is evident that the Φ_{ijp}^{ab} have the same spin properties as Ψ_0. In order to guarantee normalized configuration functions, the constant C_{ijp} in (34) is defined as

$$C_{ijp} = \left(\sum_{m=-s}^{s} \langle \Phi_{ijpm} | \Phi_{ijpm} \rangle \right)^{-1/2} \tag{37}$$

If applied to a single-determinant Ψ_0, the above procedure is identical with the electron pair treatment of McWeeny and Steiner,[47] which is based on a group-function formalism. It can be used without any difficulty as long as Ψ_0 is a single configuration function. If Ψ_0 has a multiconfigurational structure some Φ_{ijpm} of the same spin type p may not be orthogonal or even become linearly dependent, however. In order to retain the orthogonality of the configuration functions one has then to orthogonalize the core functions before coupling them to external pairs. The annihilation operators in Eq. (34) turn into linear combinations of pair annihilation operators and the external pairs can no longer be associated with a particular pair of internal orbitals. This is why we prefer to replace the subscripts ijp by P, indicating a more general structure of the core function. It should be noted that by insisting on orthogonal sets of core functions (for each spin type p) one avoids the complications with "cross-correlation terms" that appear in Sinanoğlu's pair theory of the open-shell case.[12,32]

The expression (32) for $\hat{H}$ may also be used for constructing the semi-internal interacting space. The spin coupling used in the second term, which in the semi-internal case becomes $\sum_m (\eta_a^\dagger \eta_k^\dagger | pm)(\eta_i \eta_j | pm)$, is adequate as long as both i and j refer to doubly occupied orbitals. Otherwise the two "substitution operators," which differ just in p, may yield nonorthogonal configurations. [This did not appear in the external case since there orthogonality is maintained due to the different spin types of the external pairs; but $(\eta_a^\dagger \eta_k^\dagger | pm)$ may not be spin adapted for k being an open-shell electron.] It is then more appropriate to spin-couple the creation and annihilation operators referring to open-shell orbitals, i.e., if k and j denote open-shell orbitals, the substitution operators

$$\sum_m (\eta_a^\dagger \eta_i | pm)(\eta_k^\dagger \eta_j | pm)$$

will generally lead to vanishing or orthogonal configurations. In the special case of $k = j$, this operator results in single substitutions for $p = 1$ since $(\eta_k^\dagger \eta_k | 10) = 1/\sqrt{2}$ and in so-called spin-polarizing configurations[12] for $p = -1$. If there is more than one open shell the polarization type configurations for equal i but different j are not mutually orthogonal. Although orthogonalization is simple, in the context of electron pair theories this asks again for a generalization of the definition of a pair. We shall not go into further detail since our main interest concerns the external parts of the wave function.

It is straightforward to extend the procedure described above to include spatial symmetries, if the orbitals are symmetry adapted. Then symmetry causes further interrelations between two-electron integrals and linearly independent terms of $\hat{H}$ consist of totally symmetric combinations of the substitution operators in Eqs. (33) and (34).

4. The PNO-CE Method

The particular feature of the configuration expansion we want to work with is the diagonal expansion of the external pairs in terms of their natural orbitals. This leads to several technical problems specific to the PNO-CE method: the derivation of matrix elements for configurations with nonorthogonal orbitals, the determination of suitable approximations for the NOPs, and the calculation of a particular set of two-electron orbital integrals.

4.1. The Structure of the Hamiltonian Matrix Elements

The complications due to the partial nonorthogonality of the external orbitals remain relatively minor because of the strong orthogonality between external pair functions u_{Pm} and core function Φ_{On}, i.e.,

$$\int d3 u_{Pm}(1, 3)\Phi_{On}(3, \ldots, N) = 0 \tag{38}$$

which follows from u and Φ being confined to orthogonal complements of the total spin-orbital space. In his group function formalism, McWeeny[6,48] has demonstrated how the strong orthogonality reduces Hamiltonian matrix elements. Following his approach we shall first analyze the general structure of the Hamiltonian matrix elements between doubly substituted functions of the type

$$\Psi_P(1, \ldots, N) = C_P A(1, 2/3, \ldots, N) \sum_m u_{pm}(1, 2)\Phi_{Pm}(3, \ldots, N) \tag{39}$$

Inserting then particular expansions for u_{Pm} yields the matrix elements for individual configuration functions Φ_P^{ab}. For two functions Ψ_P and Ψ_Q, one has

$$\langle \Psi_Q | \hat{H} - E_0 | \Psi_P \rangle = C_Q C_P \sum_{mn} \langle u_{Qn}(1, 2)\Phi_{Qn}(3, \ldots, N)$$

$$\times | \hat{H} - E_0 | A^2 u_{Pm}(1, 2)\Phi_{Pm}(3, \ldots, N)\rangle \tag{40}$$

where

$$A^2 = 1 - \sum_{\mu=3}^{N} (T_{1\mu} + T_{2\mu}) + \sum_{\mu=3}^{N} \sum_{\nu=\mu+1}^{N} T_{1\mu}T_{2\nu}, \qquad E_0 = \langle \Psi_0 | \hat{H} | \Psi_0 \rangle \tag{41}$$

Because of the strong orthogonality (38) the transpositions of A^2 can give nonvanishing contributions only in connection with those parts of $\hat{H}$ that involve the same electron coordinates, i.e., for each $r_{1\mu}^{-1}$ only the term $1 - T_{1\mu}$ has to be considered. Sorting the items of $\hat{H}$ according to the electron indices, one obtains

$$\langle \Psi_Q | \hat{H} - E_0 | \Psi_P \rangle = C_Q C_P \sum_{mn} \langle u_{Qn}(1, 2) | \hat{H}^{(1,2)} | u_{Pm}(1, 2) \rangle \langle \Phi_{Qn} | \Phi_{Pm} \rangle$$

$$+ \langle \Phi_{Qn}(3, \ldots, N) | \hat{H}^{(3,\ldots,N)} - E_0 | \Phi_{Pm}(3, \ldots, N) \rangle \langle u_{Qn} | u_{Pm} \rangle$$

$$+ \langle u_{Qn}(1, 2) | R(\rho(\Phi_{Pm}\Phi_{Qn})|1) + R(\rho(\Phi_{Pm}\Phi_{Qn})|2) | u_{Pm}(1, 2) \rangle \qquad (42)$$

Here $\rho(\Phi_{Pm}\Phi_{Qn})$ is the first-order reduced density matrix

$$\rho(\Phi_{Pm}\Phi_{Qn}|3, 3') = (N-2) \int d4 \cdots dN \Phi_{Pm}(3, \ldots, N) \Phi_{Qn}^*(3', 4, \ldots, N) \qquad (43)$$

and the one-electron potential $R(\rho|1)$ is defined via

$$\langle \psi_i(1) | R(\psi_k \psi_l|1) | \psi_j(1) \rangle = \int d1\, d2\, \psi_i^*(1) \psi_l^*(2) r_{12}^{-1} (1 - T_{12}) \psi_j(1) \psi_k(2) \qquad (44)$$

(note that ρ is a linear form in spin-orbital products $\psi_k \psi_l^*$). Since the external pairs represent pure spin states, their space and spin parts may be factored according to

$$u_{Pm}(1, 2) = u_P(1, 2) \chi_{pm}(1, 2)$$

i.e.,

$$\chi_{pm}(1, 2) = \begin{cases} \alpha(1)\alpha(2) & m = 1 \\ 2^{-1/2}(\alpha(1)\beta(2) - p\beta(1)\alpha(2)) & m = 0 \\ \beta(1)\beta(2) & m = -1 \end{cases} \qquad (45)$$

$$u_P(1, 2) = \sum_{a \leq b} C_P^{ab} [\varphi_a(1)\varphi_b(2) + p\varphi_b(1)\varphi_a(2)](2 + 2\delta_{ab})^{-1/2}$$

$$= \sum_{a \leq b} C_P^{ab} A(ab, p) \varphi_a(1)\varphi_b(2) \qquad (46)$$

In the last line we have defined $A(ab, p) = (2 + 2\delta_{ab})^{-1/2}(1 + pT_{ab})$.

The density matrices $\rho(u_{Pm}u_{Qn})$ may consequently also be factored into $\rho(u_P u_Q)\chi_{pmqn}$ with

$$\chi_{pmqn}(1, 1') = \int d2\, \chi_{pm}(1, 2)\chi_{qn}(1', 2)$$

If we define spin-free Coulomb and exchange operators by

$$\langle \varphi_i(1) | J(\varphi_k\varphi_l|1) | \varphi_j(1) \rangle = \langle \varphi_i(1)\varphi_l(2) | r_{12}^{-1} | \varphi_j(1)\varphi_k(2) \rangle$$

$$\langle \varphi_i(1) | K(\varphi_k\varphi_l|1) | \varphi_j(1) \rangle = \langle \varphi_i(1)\varphi_l(2) | r_{12}^{-1} | \varphi_k(1)\varphi_j(2) \rangle \qquad (47)$$

and construct by spin integration the spin-free density matrices

$$\rho_{PQ}^{J}(1, 1) = C_P C_Q \sum_m \int ds_1 \, \rho(\Phi_{Pm}\Phi_{Qn}|1, 1)$$

$$\rho_{PQ}^{K}(1, 1') = C_P C_Q \sum_{mn} \int ds_1 \, ds_{1'} \, \rho(\Phi_{Pm}\Phi_{Qn}|1, 1')\chi_{pmqn}(1', 1) \quad (48)$$

then after spin integration (42) becomes

$$\langle \Psi_Q | \hat{H} - E_0 | \Psi_P \rangle = \langle u_Q(1, 2) | \delta_{PQ} r_{12}^{-1} + F_{QP}(1) + F_{QP}(2) | u_P(1, 2) \rangle \quad (49)$$

where the Fock-type potential is given by

$$F_{QP}(1) = \delta_{QP}\hat{H}^0(1) + \delta_{pq} J(\rho_{PQ}^{J}|1) - K(\rho_{PQ}^{K}|1)$$

$$+ \delta_{pq} C_P C_Q \sum_m \langle \Phi_{Qn} | \hat{H}^{(3,...,N)} - E_0 | \Phi_{Pm} \rangle / 2 \quad (50)$$

From (49) it is quite obvious that any nonorthogonality of the external orbitals in which u_Q and u_P may be expanded introduces only a trivial complication. For particular Φ_P^{ab} and Φ_Q^{cd}, (49) implies

$$\langle \Phi_Q^{cd} | \hat{H} - E_0 | \Phi_P^{ab} \rangle = A(ab, p)A(cd, q)\{\delta_{QP}\langle \varphi_c \varphi_d | r_{12}^{-1} | \varphi_a \varphi_b \rangle$$

$$+ \langle \varphi_c | F_{QP} | \varphi_a \rangle \langle \varphi_d | \varphi_b \rangle + \langle \varphi_d | F_{QP} | \varphi_b \rangle \langle \varphi_c | \varphi_a \rangle \quad (51)$$

If we introduce the "hole" pair function

$$U_{Pm}(1, 2) = \binom{N}{2}^{1/2} \int d3 \cdots dN \, \Phi_{Pm}^*(3, \ldots, N)\Psi_0(1, 2, 3, \ldots, N) \quad (52)$$

then strong orthogonality yields immediately

$$\langle \Psi_P | \hat{H} | \Psi_0 \rangle = C_P \sum_m \langle u_{Pm} | r_{12}^{-1} | U_{Pm} \rangle \quad (53)$$

Carrying out the spin integration yields the pure space function

$$U_P = C_P^2 \sum_m \int ds_1 \, ds_2 \, U_{Pm}(1, 2)\chi_{pm}^*(1, 2) \quad (54)$$

and

$$\langle \Psi_P | \hat{H} | \Psi_0 \rangle = C_P^{-1}\langle u_P | r_{12}^{-1} | U_P \rangle \quad (55)$$

U_P, and the potentials F_{QP} depend on the configurational structure of Ψ_0, and in the general case it can only be stated that they are made up from orthogonal internal orbitals and can be evaluated by standard techniques.

Let us further consider the important case of Ψ_0 being a single determinant. Then $P \equiv ijp$ and (52) and (54) yield $U_P = A(ij,p)\varphi_i\varphi_j$. Therefore

$$\langle \Phi_P^{ab} | \hat{H} | \Psi_0 \rangle = C_P^{-1}A(ab, p)A(ij, p)(\varphi_a\varphi_b | r_{12}^{-1} | \varphi_i\varphi_j) \quad (56)$$

Using the Fock operator $F_0 = \hat{H}^0 + R(\rho(\Psi_0\Psi_0))$, F_{QP} can be simplified due to the relations

$$\langle\Phi_{Qn}(3,\ldots,N)|\hat{H}^{(3,\ldots,N)} - E_0|\Phi_{Pm}(3,\ldots,N)\rangle$$

$$= \langle U_{Pm}(1,2)|-F_0(1)-F_0(2)+r_{12}^{-1}|U_{Qn}(1,2)\rangle \tag{57}$$

$$\rho(\Phi_{Pm}\Phi_{Qn}) = \rho(\Psi_0\Psi_0)\langle U_{Qn}|U_{Pm}\rangle - \rho(U_{Qn}U_{Pm}) \tag{58}$$

These relations show the well-known particle–hole symmetry with respect to a single determinant Ψ_0 and are easily verified by means of the familiar Slater–Condon rules. Thus F_{QP} is expressed in simple internal pair functions. It may be worked out quite easily although it is somewhat cumbersome to list all the various cases that arise from different occupations of the internal orbitals which appear in P and Q.

In the most simple case, a closed-shell Ψ_0, all U_{Pm} are spin adapted and their spin part can be factored out. Since in this case $C_P = (2-p)^{-1/2}$ and

$$\int d1\, d2\, \chi_{pmqn}(1,1)\chi_{qnpm}(2,2) = \delta_{pq}(2-p) \tag{59}$$

$$\int d1\, d2\, \chi_{pmqn}(1,2)\chi_{qnpm}(2,1) = (2-p)(2-q)/2$$

one gets

$$F_{QP} = \delta_{QP}F_0 - \delta_{pq}J(\rho(U_QU_P)) + \tfrac{1}{2}[(2-p)(2-q)]^{1/2}K(\rho(U_QU_P))$$

$$+ \delta_{pq}\langle U_P|-F_0(1)-F_0(2)+r_{12}^{-1}|U_Q\rangle/2 \tag{60}$$

where

$$\rho(U_{klq}U_{ijp}) = 2A(kl,q)A(ij,p)\varphi_k\varphi_i^*\delta_{lj} \tag{61}$$

With respect to matrix elements involving semi-internal parts of the wave function, we note that semi-internal pairs are still strongly orthogonal to all cores but they are not spin adapted. Thus (42) together with (52) and (53) are still valid, but spin integration is somewhat more complicated. As a final example demonstrating the minor complications caused by the natural expansion of the external pairs, we give the matrix element between a doubly substituted function Ψ_P and a singly substituted function Φ_k^a for the case of a single-determinant reference function:

$$\langle\Phi_k^a|\hat{H}|\Psi_P\rangle = C_P \sum_m \langle\psi_a(1)U_{Pm}(2,3)|F_0(2)+r_{12}^{-1}-r_{23}^{-1}|u_{Pm}(1,2)\psi_k(3)\rangle \tag{62}$$

In the closed-shell case, spin integration yields

$$\langle\Phi_k^a|\hat{H}|\Phi_{ijp}^{bc}\rangle = A(ij,p)A(bc,p)$$

$$\times \{\langle\varphi_a|\varphi_b\rangle\langle\varphi_c|F_0\delta_{jk} - J(\varphi_j\varphi_k)|\varphi_i\rangle + \langle\varphi_a|K(\varphi_b\varphi_c)|\varphi_i\rangle\delta_{jk}\} \tag{63}$$

4.2. Calculation of Approximate Pseudonatural Orbitals

The exact natural orbitals of the pairs u_p (the NOPs) are not available prior to the knowledge of the u_p themselves. But if the expansion coefficients C_P^{ab} have first to be obtained in the traditional way from the overall secular equation, the transformation to the compact natural form is merely a cosmetic for the final wave function. Thus the usefulness of our approach rests on the assumption that sufficiently accurate approximations to the NOPs can be obtained in an efficient way. In particular, we have to look for a method that does not require the complete transformation of the two-electron basis integrals to orbital integrals, since this is one of the bottlenecks of the traditional CI that we want to circumvent.

In Section 2.4, we have pointed to the fact that in case Ψ_0 dominates the wave function, an energy expectation value close to optimal third-order accuracy can be obtained by using natural orbitals, which are derived from a wave function correct to first order only. As a first step we may disregard all matrix elements between configurations of different pairs P. This leads to the independent electron-pair approximation (IEPA), mentioned in the previous section, and the related sets of pseudonatural orbitals. Methods for a determination of approximate PNOs directly from the basis integrals have first been developed by Kutzelnigg and co-workers.[49-51] They have been based on successively solving a hierarchy of variational equations with projection operators guaranteeing the orthogonality of the orbitals. A simpler and faster procedure was later proposed by the present author.[10] It uses a particular approximation for the two-electron orbital integrals and yields very good approximations to the true PNOs. We shall describe this scheme for the closed-shell case.

For the independent electron pair wave function $\Psi_0 + \Psi_P$, the Schrödinger equation $\hat{H} - E_0 - e_P|\Psi_0 + \Psi_P\rangle = 0$ yields by integrating with Ψ_0 and Φ_P^{ab} and using the matrix elements (49) and (55), the following equations for the spin-free external pair function u_P defined in Eq. (46):

$$C_P^{-1}\langle u_P|r_{12}^{-1}|U_P\rangle - e_P = 0 \tag{64}$$

$$C_P^{-1}\langle\varphi_a\varphi_b|r_{12}^{-1}|U_P\rangle + \langle\varphi_a\varphi_b|F_{PP}(1) + F_{PP}(2) + r_{12}^{-1} - e_P|u_P\rangle = 0 \tag{65}$$

The first term of (65) may be written $A(ij, p)\langle\varphi_a|K(\varphi_i\varphi_j)|\varphi_b\rangle$ so that all terms can be derived from Coulomb and exchange operators of internal orbitals except for

$$\langle\varphi_a\varphi_b|r_{12}^{-1}|u_P\rangle = \sum_{cd} C_P^{cd} A(cd, p)\langle\varphi_a\varphi_b|r_{12}^{-1}|\varphi_c\varphi_d\rangle \tag{66}$$

Neglecting true exchange-type integrals we may approximate the two-electron integrals in this expression by

$$\langle\varphi_a\varphi_b|r_{12}^{-1}|\varphi_c\varphi_d\rangle \approx \langle\varphi_a|J(\varphi_d\varphi_d)|\varphi_c\rangle\delta_{bd} + \langle\varphi_b|J(\varphi_c\varphi_c)|\varphi_d\rangle\delta_{ac}(1 - \delta_{bd}\delta_{ac}/2) \tag{67}$$

Since J is a long-range potential and the orbitals of weight in the expansion of u_P are localized in the region of U_P, we may assume

$$J(\varphi_b\varphi_b), J(\varphi_c\varphi_c) \approx J_P \equiv \tfrac{1}{2}J(\rho(U_PU_P))$$

Then

$$\langle\varphi_a\varphi_b|r_{12}^{-1}|u_P\rangle = \langle\varphi_a\varphi_b|J_P(1)+J_P(2)|u_P\rangle - \tfrac{1}{2}(\langle\varphi_a|J_P|\varphi_a\rangle+\langle\varphi_b|J_P|\varphi_b\rangle)\langle\varphi_a\varphi_b|u_P\rangle \tag{68}$$

If we now choose the set of external orbitals φ_a to obey

$$\langle\varphi_a|F_{PP}+J_P|\varphi_b\rangle = e_a\delta_{ab} \tag{69}$$

i.e., as being the eigenfunctions of $\Omega^+(F_{PP}+J_P)\Omega$, where Ω projects into the external space, Eq. (66) is immediately rewritten as

$$C_P^{ab} = -A(ab,p)C_P^{-1}\langle\varphi_a\varphi_b|r_{12}^{-1}\|U_P\rangle/(e_a+e_b-\tfrac{1}{2}[\langle\varphi_a|J_P|\varphi_a\rangle+\langle\varphi_b|J_P|\varphi_b\rangle]-e_P) \tag{70}$$

Equations (64) and (70) are easily solved iteratively, but one may well neglect the usually small pair energy e_P in (70) since the energy denominator is approximate anyway. Then Eq. (70) is simply a first-order perturbation expression for the expansion coefficients, based on the eigenfunctions of the potential $F_{PP}+J_P$. This is an $(N-1)$-particle potential similar to those widely used in atomic many-body perturbation calculations[18] but here adequately adjusted to each pair P. Its significance lies in the fact that the off-diagonal matrix elements of a pair CI based on its eigenfunctions are reduced to the order of exchange-type integrals.

After the approximate expansion coefficients C_P^{ab} are determined, it is straightforward to derive the corresponding natural orbitals following the procedure of Section 3.1. If the so obtained approximate PNOs are used to construct diagonal configurations, the variation of their expansion coefficients yields an independent pair wave function the correlation energy of which has been found for a variety of closed- and open-shell systems to differ by less than 1% from the result of a full pair CI. Some examples are given in Table 1. Ahlrichs and Driessler[52] have recently shown how the above procedure can be extended to yield the exact PNOs. To this end, Eq. (70) is generalized into

$$C_P^{ab(n+1)} - C_P^{ab(n)} = A(ab,p)R_P^{ab}/(e_a+e_b-\tfrac{1}{2}[\langle\varphi_a|J_P|\varphi_a\rangle+\langle\varphi_b|J_P|\varphi_b\rangle]-e_P) \tag{71}$$

where R_P^{ab} stands for the left-hand side of Eq. (65) with u_P replaced by $u_P^{(n)}$. In the first step $u_P = 0$ and (71) reduces to (70). Convergence is usually fast, but

Table 1. Comparison of Pair Correlation Energies from Diagonal Expansions by Means
of Approximate PNOs with Those from Full Pair CI Calculations

System		Internal pair		Diagonal expansion approximate PNOs	Full pair CI
$H_2{}^a$	$1\sigma_g-1\sigma_g$	S		0.038608	0.038651
$H_3{}^b$	$1\sigma_g-1\sigma_g$	S	external	0.026377	0.026397
			semi-internal	0.019279	0.019318
	$1\sigma_u-1\sigma_g$	S		0.010575	0.010611
		T		0.005003	0.005011
Li^c	$1s-1s$	S	external	0.038178	0.038212
			semi-internal	0.000065	0.000065
	$1s-2s$	S		0.001270	0.001274
		T		0.000949	0.000950
$CH_4{}^d$	$2a_1-2a_1$	S		0.00947	0.00951
	$1f_2-1f_2$	S		0.01572	0.01576
	$2a_1-1f_2'$	S		0.01343	0.01349
		T		0.00568	0.00575
	$1f_2-1f_2$	S		0.01540	0.01548
		T		0.01496	0.01508

[a] Basis 5s,3p; r_{HH} = 1.4 bohr.
[b] Basis 5s,3p; r_{HH} = 1.7 bohr.
[c] Basis 10s,4p,1d.
[d] Basis A' of Meyer[10]; r_{CH} = 2.05 bohr.

per iteration one has to calculate the exchange operator $K(u_P^{(n)})$ from the two-electron integrals. It makes little sense to insist on exact PNOs, however, since the error introduced by decoupling the pairs—that is, by taking the PNOs for the NOPs—is certainly of the same order as the above-mentioned 1%. In fact, it must even be expected to be somewhat larger because some of the coupling matrix elements contain Coulomb-type integrals.

Before discussing these errors further let us point to the important fact that the rapid convergence of the natural expansion offers the possibility of effectively reducing the size of the expansion by truncation according to a certain energy threshold. An estimate for the energy contribution of a particular configuration can simply be obtained from the second-order expression $C_P^{ab}\langle\varphi_P^{ab}|\hat{H}|\Psi_0\rangle$. A threshold of about 10^{-4} hartree has been found to lead in typical cases to truncation errors of 1 to 2% of the correlation energy, which is in the order of other errors of the method. For intraorbital pairs and next–neighbor interorbital pairs, such a threshold results in expansions of 10 to 15 terms per pair.

In Table 2 we show for BH the size of the errors introduced successively by going from the full CI in double substitutions (full-diagonal expansion in NOPs) to a full-diagonal expansion using exact PNOs, to a full-diagonal expansion using the approximate PNOs, and finally to a truncated expansion with approximate PNOs. The errors are 0.9, 0.6, and 1.4%, respectively.

According to our experience, this is typical for ground states of small molecules and applies also to open-shell cases. Since basis set defects must be expected to amount to 10–20% even for large-scale calculations, the approximations inherent in the PNO-CE scheme seem tolerable. Of course, the errors must be expected to increase when low-lying external orbitals are available and the dominance of Ψ_0 is reduced. An example in that direction is the ground state of BH examined in Table 2. At equilibrium distance, the substitution $3\sigma^2 \to 1\pi^2$ has a coefficient of 0.2 and has a drastic effect on the dipole moment. Although the errors in the nergy are in the range given above, the correlation contribution to the dipole moment is not satisfactorily reproduced by the diagonal PNO expansions. This defect can easily be removed by adding just a few off-diagonal configurations that involve the heavily occupied 1π-orbital, but it indicates that care has to be exercised by calculating properties and that in this case the option for off-diagonal configurations is important.

One may wonder if the PNOs are equally useful in the case of delocalized electrons for which the independent electron pair approximation is known to break down. By comparing PNO-CI calculations based on canonical and localized orbitals, we have never observed differences between the calculated correlation energies larger than 2%. As an example, let us consider a "molecule" of four well-separated H_2 in tetrahedral arrangement. It is seen from Table 3 that the independent pair energies sum to less than 60% of the correct correlation energy in the delocalized case, but the PNOs are still good for 98%. It looks like the neglect of the pair coupling term affects mainly the coefficients of the natural configurations rather than the natural orbitals themselves, but we cannot give a general argument why these orbitals should be particularly insensitive to approximations.

Table 2. *Comparison of Various Steps in Approximating the Configuration Expansion for BH*[a]

Pair		All double substitutions	All diagonal doubles		Diagonal doubles with $\Delta E \geq 0.0001$, approximate PNOs
			exact PNOs	approximate PNOs	
2σ–2σ	S	0.028548	0.028499	0.028330	0.028157
3σ–3σ	S	0.035356	0.035060	0.033865	0.033793
2σ–3σ	S	0.021767	0.021522	0.021586	0.021074
2σ–3σ	T	0.007437	0.007592	0.007748	0.007515
Total		0.093354	0.092674	0.091528	0.090539
$\Delta\mu^{corr}$		0.10685	0.010917	0.13760	0.13726

[a] Basis $11s, 6p, 2d, 1f/6s, 2p, 1d_\sigma$; all singles included; atomic units.

Table 3. Correlation Energies from PNO Configuration Expansions for Localized and Delocalized Internal Orbitals of $(H_2)_4$ in Tetrahedral Configuration [a]

Internal orbitals	Localized	Delocalized
Variational CI	0.11143	0.10913
Coupled electron pair approximation	0.11864[b]	0.11741
Independent electron pair approximation	0.11864	0.06801

[a] $r_{HH} = 1.7$ bohr, $r_{H_2-H_2} = 10$ bohr; energy in hartrees.
[b] Exact for the given basis of $5s$ per H.

We note finally that one may define natural orbitals for semi-internal substitutions simply by summing orbitals according to (7). The coefficients of the orbitals may be estimated in the same perturbational fashion as described above for the external configurations. Likewise, single substitutions could in principle be contracted into one configuration per orbital, but because they do not couple to Ψ_0 we have no estimate for their coefficients. For this reason and considering their importance for properties, they should be included in the expansion uncontracted. For their construction we usually choose the PNOs of the corresponding intraorbital pair. The ten most heavily occupied orbitals have been found to be sufficient for all purposes.

4.3. The Calculation of Two-Electron Orbital Integrals

Our situation is determined by the fact that we may have a large number of orbitals—up to full external sets per pair—but that we need only a small subset of all possible integrals. Therefore we cannot proceed along straightforward integral transformation.

From the matrix elements (49) and (55) one sees that electron repulsion terms appear in two different ways:

(a) The interaction between external and internal electrons is given in the form of matrix elements of the Fock-type operators F_{PO}. They can all be calculated from the Coulomb and exchange operators $J(\varphi_i\varphi_j)$ and $K(\varphi_i\varphi_j)$. These are $N(N+1)$ operators for N internal orbitals and they are required in any case for the calculation of the PNOs, so that only little extra work goes into calculating this part of the electron repulsion. Likewise, the interaction between two internal electrons can also be obtained from the above operators.

(b) Repulsion terms between two external orbitals appear only in matrix elements between configurations of the same pair P. Due to the structure of the natural expansion they have the form $\langle \varphi_a\varphi_{a'}|r_{12}^{-1}|\varphi_b\varphi_{b'} + p\varphi_{b'}\varphi_b \rangle$, where a, a' denote pairs of natural orbitals as in (19). For an average of M configurations per pair P we need $M(M+1)/2$ integrals of this type. Their calculation is the time-determining step of the method.

The computational procedure is considerably facilitated by the fact that the number of this type of integrals is usually so small that all of them—or at least all of a particular symmetry type—can be held in the high-speed memory. Due to the form of the integrals is it most useful to start from a file of two-electron basis integrals

$$I^p_{\mu\nu,\rho\sigma} = \langle \varphi_\mu\varphi_\nu | r_{12}^{-1} | \varphi_\rho\varphi_\sigma + p\varphi_\sigma\varphi_\rho \rangle [(1+\delta_{\mu\nu})(1+\delta_{\rho\sigma})(1+\delta_{\mu\rho}\delta_{\nu\sigma})]^{-1} \quad (72)$$

in double-triangular order, i.e.,

$$\mu \geq \nu, \qquad \mu \geq \rho, \qquad \bar\sigma \geq \sigma; \qquad \bar\sigma = \rho + \delta_{\mu\rho}(\nu - \rho) \quad (73)$$

with the later indices changing first. One verifies easily that

$$\langle \varphi_a\varphi_{a'} | r_{12}^{-1} | \varphi_b\varphi_{b'} + p\varphi_{b'}\varphi_b \rangle$$
$$= 2 \sum_{\mu=1}^{L} \sum_{\nu=1}^{\mu} \sum_{\rho=1}^{\mu} \sum_{\sigma=1}^{\bar\sigma} I^p_{\mu\nu,\rho\sigma}(1+T_{ab}T_{a'b'})(1+pT_{aa'})(1+pT_{bb'})c^\mu_a c^\nu_{a'} c^\rho_b c^\sigma_{b'} \quad (74)$$

where T_{ab} permutes the indices a and b, and the c^μ_a are the expansion coefficients of the orbital φ_a. The computation then proceeds as follows:

(1) For a particular index pair $\mu\nu$ one derives for all PNO pairs φ_a, $\varphi_{a'}$:

$$I^p_{\mu\nu,aa'} = \sum_{\rho=1}^{\mu} \left\{ c^\rho_{a'} \sum_{\sigma=1}^{\bar\sigma} I^p_{\mu\nu,\rho\sigma} c^\sigma_a + c^\rho_a \sum_{\sigma=1}^{\bar\sigma} I^p_{\mu\nu,\rho\sigma} c^\sigma_{a'} \right\} \quad (75)$$

(2) One accumulates at the storage places for the orbital integrals

$$I^p_{\mu\nu,aa'}(c^\mu_a c^\nu_{a'} + pc^\mu_{a'} c^\nu_a) + I^p_{\mu\nu,bb'}(c^\mu_b c^\nu_{b'} + pc^\mu_{b'} c^\nu_b) \quad (76)$$

Thus as an intermediate storage one needs only $N^{\text{pairs}} \times M$ places for the quantities $I^p_{\mu\nu,aa'}$. If one starts from basis integrals in a different order this storage must be larger, e.g., for integrals in the usual canonical order $N^{\text{pairs}} \times M^2$ places would be required for the quantities $I_{\mu a,\rho b}$. Additional pairs appear in step (1) for matrix elements between single and double substitutions. But the computing time is not changed significantly since in all cases it is dominated by the inner sum in Eq. (75), which requires about $(L^4/8) \times N^{\text{pairs}} \times M$ operations. The second step involves only $L^2 \times M^2 \times N^{\text{pairs}}$ operations. It should be noted that M does not necessarily increase with L but is determined by the desired accuracy and is usually in the order of 10 only. Thus for a particular molecule with N^{pairs} fixed, the above process goes with a power of L between 4 and 5. By considering increasing size of the molecule with roughly constant ratio between L and the number of electrons, one should take into account that by using localized orbitals, M becomes very small or drops to zero for nonneighbor pairs. Therefore $N^{\text{pairs}} \times M$ goes basically like L for larger molecules, so that we end up with an L^5 process like traditional integral transformation. However, due to the small number of the integrals finally needed, we circumvent the heavy I/O problems connected with the traditional transformation. Furthermore, setting up and diagonalizing the CI matrix becomes a trivial matter because of its small dimension.

5. Applications of the PNO-CI and PNO-CEPA Methods

5.1. General Remarks

In this section we shall briefly discuss some studies of electron correlation effects with the intention of demonstrating the capability of our method and indicating its limits. PNO-CI programs were coded by the present author in 1970 for closed- and open-shell states[9,10] and by Kutzelnigg and co-workers for closed-shell states.[53] So far, both programs are restricted to single-determinant reference wave functions.

The first PNO-CI/CEPA calculations were performed for the ground state and several ionized states of the water molecule.[9] In this calculation about 85% of the empirically estimated total correlation energy was recovered by the PNO-CI, providing an upper bound to the total energy of 0.063 au (0.1%) above the experimental nonrelativistic energy. The PNO-CEPA gave 89% of the total correlation, implying an estimate of 4% for the contribution of higher-order substitutions. These fractions of recovered correlation energies were significantly larger than had been obtained for comparable molecules at that time.[54]

The investigation of ionization energies as observed in photoelectron spectroscopy was later extended to CH_4,[10] NH_3, HF, N_2,[55] and new molecules like H_2CCS.[56] Potential curves and related spectroscopic constants have been calculated for the diatomic hydrides from LiH to HCl,[57] their positive ions,[58] and the molecules N_2 and F_2.[59,60] Near-equilibrium potential surfaces have been investigated for CH_4 and H_2O,[59,61] and the correlation contributions to the inversion barriers of H_3O^+, CH_3^-, and NH_3,[52] and to the rotation barrier of C_2H_6[60] have been determined. Dipole moment curves have been calculated for the diatomic hydrides[57] and polarizabilities have been investigated for Ne, HF, H_2O, NH_3, CH_4, CO,[63] and the atoms up to Ca.[64,65]

In most of these applications, the PNO-CI yielded the lowest variational energies obtained so far. Before presenting some of the applications in more detail let us mention some technical points:

(a) *Basis sets.* Gaussian type basis functions have been used throughout. Sets of the size $11s$, $6p$, $2d$, $1f$ for first row atoms have been found to be capable of accounting for 80 to 85% of the valence-shell correlation energy. The results are rather insensitive with respect to the values of the d and f exponents. The latter may therefore be related to atom-optimized p exponents by simple rules of thumb.[10,57] Recently published molecule-optimized d and f exponents[60,62] turned out to follow these rules closely. For the hydrogen atom, sets of 5s, 1p or 5s, 2p are used. Smaller sets of the size 9s, 5p, 1d/5s, 1p suffice to get about 70% of the valence shell correlation energy but usually they do not well reproduce properties that depend on the charge distribution.

   (*b*) *Thresholds.* The optimal threshold for configuration selection depends somewhat on the size of the basis set, but due to the rapid convergence of the PNO expansion its choice is not at all a critical point. A value of 10^{-4} hartree has been found appropriate for the larger basis sets. In this case, the truncation error amounts to about 1 to 3% and is considerably smaller than the basis set deficiency.

   (*c*) *CEPA versus CI.* For all applications the CEPA results are in better agreement with experiment than the variational CI results. This indicates that the CEPA method provides a reasonable approximation for the effects of higher-order substitutions. These substitutions should clearly contribute to quantities like ionization energies and dissociation energies since their number increases rapidly with the number of electrons. Apparently they also affect the calculation of properties like dipole moments and polarizabilities. The correlation contributions of triple and quadruple substitutions as recently calculated by Sazaki and Yoshimine[66] for the first row atoms amount to 70 to 80% of our CEPA values, in reasonable accord with their estimate for the fraction recovered.[66] All results quoted below refer to PNO-CEPA calculations.

5.2. Correlation Contributions to Ionization Energies

   The correlation energy of an atomic or molecular system is likely to change quite drastically by removing an electron. For the first ionization potentials (IPs) this change is generally a decrease so that the Hartree–Fock IPs (the differences between the H–F energies of the neutral and the ionized states) are too low. Often they are hardly better than the IPs obtained from Koopmans' theorem, i.e., by simply identifying them with the orbital energies, since in this case there is some compensation of the correlation error by the neglect of charge rearrangement.[67] The situation is different for ionized states produced by removing an electron from an inner orbital. As first demonstrated by Bagus' SCF calculations for "hole states" of atoms,[68] the rearrangement energy becomes quite large and the correlation energy may behave anomalously, i.e., the ionized state may have more correlation energy than the neutral state. This effect is due to strong semi-internal correlation, which becomes possible after the removal of an inner electron.[69] Such hole states are highly excited states of the ion and it is not possible to impose explicitly orthogonality constraints. Instead, the proper wave function is determined by the condition that the leading term in the configuration expansion, the Hartree–Fock determinant, be of a particular orbital occupancy. We have not found it difficult to obtain the corresponding solution of the CI or CEPA eigenvalue equations.

Table 4. Calculated and Experimental Vertical Ionization Energies (in eV)

System	Ejected electron	Koopmans'	SCF	PNO-CEPA	Experiment
H_2O^a	$1b_1$	13.86	11.10	12.48	12.61
	$3a_1$	15.87	13.32	14.68	14.73
	$1b_2$	19.50	17.59	18.85	18.7
	$2a_1$	36.77	34.22	32.35	32.2
	$1a_1$	559.51	539.14	539.63	539.9
NH_3^b	$3a_1$	11.65	9.47	10.83	10.9
	$1e$	17.07	15.44	16.41	16.3
	$2a_1$	31.04	29.06	27.40	27.0
	$1a_1$	422.81	405.32	405.40	405.6
CH_4^c	$1f_2$	14.85	13.67	14.29	14.4
	$2a_1$	25.70	24.31	23.38	22.9
	$1a_1$	304.87	290.78	290.70	290.91
N_2^d	$3\sigma_g$	17.26	15.92	15.66	15.59
	$1\pi_u$	16.71	15.31	16.58	16.98
	$2\sigma_u$	21.20	20.22	19.07	18.75
	1σ	426.73	410.23	409.89	409.92

[a] See Meyer[9] for theoretical values and references therein for experimental values.
[b] Theoretical values unpublished (Rosmus and Meyer[55]), experimental values from W. H. E. Schwartz, *J. Electron. Spectroscop.* **6**, 377 (1975); A. W. Potts and W. Price, *Proc. Roy. Soc. (London)* **A326**, 181 (1972); T. D. Thomas and R. W. Shaw, *J. Electron Spectroscop.* **5**, 1081 (1974).
[c] See Rosmus and Meyer.[55]
[d] Theoretical values unpublished (Rosmus and Meyer[55]), experimental values from RKR curves; see P. E. Cade, K. W. Sales, and A. C. Wahl, *J. Chem. Phys.* **44**, 1973 (1966).

The IPs calculated for some small molecules are listed in Table 4. The results show that the CEPA correlation contributions greatly improve the agreement between theoretical and experimental values, the remaining errors amounting to only a few tenths of an electron volt. It appears that these remaining discrepancies can be attributed largely to the deficiency in the calculated correlation energies of 10 to 15%: by scaling the correlation to 100% (estimates for the total correlation energies are available for the smaller molecules) the agreement can be improved significantly with only a few exceptions. One should note, however, that for polyatomic molecules comparison with experiment is complicated due to vibrational effects, which may also amount to a few tenths of an electron volt.

For a detailed discussion of the *ab initio* work on ionization potentials see the chapter by Schwartz in Volume 4.

5.3. Correlation Contributions to Potential Surfaces and Related Spectroscopic Constants

The quality of the Hartree–Fock (H–F) approximation is dependent on the geometrical configuration of a molecular system. This affects the various parameters of a potential surface to very different degrees: Whereas H–F bond distances and angles of the equilibrium geometry are usually correct to within 1–2%,[54] the errors of harmonic force constants are about 10–15% for hydrides but may go up to 30% for molecules like N_2 or CO (see the chapter by Pulay in Volume 4). The dissociation energies derived from H–F calculations are known to be particularly dissatisfying when a bond is broken.[54]

Several methods have been devised to obtain reasonable potential surfaces while treating the electron correlation only very restrictively. In particular, the multiconfiguration SCF method has been used very successfully, although it suffers somewhat from the difficulty connected with defining the "relevant" configurations. Since there seems to be no unambiguous way of separating the "extra" molecular correlation, one would like to treat fully the correlation in the valence shell of that part of the molecule that is involved in the deformation process. The main difficulty arises from the fact that there are usually several configurations of rapidly changing importance that should be treated on equal footing with the leading H–F configuration, i.e., as part of the reference wave function Ψ_0. As long as the H–F configuration is sufficiently separated energetically from the other configurations, it should be possible to stick to a single-configuration reference wave function when using the CEPA method, which may be looked at as transferring appropriate parts of the correlation, which are explicitly calculated for the H–F configuration, to the other configurations.[20,57] Experience has shown that significant deviations from the experimental potential curves do not occur for bond distances shorter than twice the equilibrium distance.[20,60]

In order to establish the degree of reliability that can be attributed to spectroscopic constants obtained from PNO-CEPA potential surfaces, a systematic investigation has been performed for the ground state potential curves of the diatomic hydrides from LiH to HCl[57] as well as their ions.[58] The comparison of the spectroscopic constants derived from the CEPA potential curves with experiment (Table 5) reveals a high reliability of the theoretical values: the mean deviations over both rows are as follows:

$$r_e \sim 0.003 \text{ Å}, \qquad \omega_e \sim 16 \text{ cm}^{-1}, \qquad \alpha_e \sim 0.005 \text{ cm}^{-1}, \qquad \omega_e x_e \sim 1.5 \text{ cm}^{-1}$$

As shown by the calculated harmonic force constants for H_2O and CH_4 (Table 6), the near equilibrium energy surface of polyatomic hydrides can be reproduced with about the same accuracy. The deviations from experiment are below 2%, with the exception of the symmetric stretching constant of CH_4, F_{11}, for which the experimental harmonic value has been questioned, however.[10,61]

The dissociation energies calculated for the diatomic hydrides show a systematic error of the calculated value that goes up to 0.3 eV for HF. Here again, it can be traced to the deficiency of 5 to 15% in the recovered total correlation. Taking this systematic defect into account, the theoretical values

Table 5. Calculated and Observed Spectroscopic Constants for the First Row Diatomic Hydrides[a]

Hydride	r_e, Å	α_e, cm^{-1}	ω_e, cm^{-1}	$\omega_e x_e$, cm^{-1}	D_e, eV
LiH	1.606	0.189	1428	20.1	1.48
	1.599	0.212	1402	22.5	2.48
	1.595	0.213	1406	23.2	2.52
BeH	1.338	0.259	2145	30.1	2.18
	1.344	0.299	2064	36.4	2.15
	1.343	0.303	2060	36.2	2.11
BH	1.221	0.375	2485	42.5	2.78
	1.238	0.406	2352	46.6	3.49
	1.232	0.412	2367	49.4	3.57
CH	1.104	0.474	3044	55.5	2.47
	1.122	0.532	2842	64.4	3.47
	1.120	0.534	2858	63.0	3.63
NH	1.018	0.570	3546	66.0	2.10
	1.039	0.648	3269	78.8	3.38
	1.037	0.648	3282	78.3	3.40
OH	0.950	0.656	4054	74.5	2.99
	0.971	0.724	3744	84.9	4.34
	0.971	0.714	3740	86.4	4.63
FH	0.898	0.744	4476	83.8	4.32
	0.917	0.787	4169	90.4	5.83
	0.917	0.795	4139	90.1	6.12

[a] Results taken from Meyer and Rosmus,[57] Tables IV–X. Upper numbers, SCF; middle numbers, PNO-CEPA; lower numbers, experimental values.

Table 6. Parameters of the Near-Equilibrium Energy Surface of H_2O and CH_4[a]

System	Parameter	SCF[b]	PNO–CEPA[b]	Experiment[c]
H_2O	r_e	0.9405	0.9550	0.9572
	α_e	106.41	105.07	104.52
	f_{rr}	9.684	8.642	8.454
	$f_{\alpha\alpha}$	0.752	0.720	0.698
	$f_{rr'}$	−0.061	−0.096	−0.101
	$f_{r\alpha}$	0.233	0.243	0.219
	$f_{\alpha\alpha\alpha}$	−0.127	−0.112	−0.127
	f_{rrr}	−10.92	−9.81	−9.98
	f_{rrrr}	15.4	15.7	16.8
	D_e	6.88	9.45	10.1
CH_4	r_e	1.038	1.091	1.085[d]
	F_{11}	5.881	5.472	5.842[d]
	F_{22}	0.642	0.589	0.580
	F_{33}	5.667	5.377	5.383
	F_{34}	0.221	0.222	0.225
	F_{44}	0.612	0.560	0.545
	D_e	14.33	17.62	18.25

[a] Bond distances in angstroms, angles in degrees, force constants in aJÅ^{-n} (n = number of stretching coordinates), dissociation energies in electron volts.
[b] Theoretical values from Meyer,[9,59] (H_2O), and Meyer[10,59] and Pulay and Meyer,[61] (CH_4).
[c] H_2O: K. Kuchitsu and Y. Morino, Bull. Chem. Soc. Japan 38, 814 (1965); A. R. Hoy, I. M. Mills, and G. Strey, Mol. Phys. 24, 1265 (1972); CH_4: J. L. Duncan and I. M. Mills, Spectrochem. Acta 20, 523 (1964); D_e from B. deB. Darwent, Nat. Nat. Ref. Data Ser., Nat. Bur. Std. (U.S.) 31 (1970).
[d] These values are probably in error due to overestimated anharmonicity corrections; see Meyer[10] and Pulay and Meyer[61] for further discussion.

are believed to allow for predictions with an accuracy of about 0.05 eV for the experimentally uncertain dissociation energies in the second row.

The difficulties of calculating accurate dissociation energies are underlined by the results for H_2O and CH_4 given in Table 6. Only 80–85% of the empirical correlation contributions are calculated, although 85–90% of the total valence shell correlation is considered in the molecular calculations. Since it seems very difficult to improve the latter fraction significantly, attempts have been made to develop a semiempirical extrapolation scheme for estimating complete basis limits by comparing the convergence behavior of PNO-CE wave functions for correlated systems.[70] For ionization potentials, electron affinity of F, and the dissociation energy of HF, agreement with experiment to better than 0.1 eV has been obtained, but experience so far is insufficient to assess the usefulness of this scheme for larger systems.

5.4. Dipole Moments and Polarizabilities

Due to the variation principle, the energy is the least sensitive quantity with respect to errors in the wave function. Thus, the approximation inherent in using PNOs in place of the exact natural orbitals of the pairs is more critical for calculating other properties.

Dipole moment curves have been calculated for the diatomic hydrides, and this provides a sensitive test on the possible errors: for several hydrides there are energetically low-lying orbitals that are unoccupied in the SCF wave function but gain considerable weight in the correlated wave function. The molecule BH has already been discussed in Section 4.2. Table 7 shows for the hydrides LiH to CH a considerable effect on the dipole moment when the

Table 7. *Calculated Dipole Moments μ_0 for Small Hydrides* [a]

System	SCF[b]	PNO-CEPA[c]		Experiment
LiH	−6.091	−5.954	−5.896	−5.882
BeH	−0.302	−0.310	−0.205	
BH	1.722	1.241	1.330	1.27±0.21
CH	1.688	1.424	1.392	1.46±0.06
NH	1.640	1.551	1.556	
OH	1.749	1.693		1.668
HF	1.901	1.855		1.827
H_2O	1.993	1.845		1.840
NH_3	1.576	1.474	1.488	1.472

[a] Results taken from Meyer and Rosmus[57] (diatomic hydrides) and Werner and Meyer[63] (polyatomic hydrides), respectively. See these references also for the experimental values. All values in Debye.
[b] μ_0 refers to the calculated potential curves for the diatomic hydrides but to experimental equilibrium and mean geometries for the polyatomic hydrides.
[c] Left column, standard PNO expansion with diagonal substitutions only. Right column, off-diagonal configurations added for the most strongly occupied PNOs.

Table 8. Comparison of Calculated and Experimental Isotropic
Polarizabilities [a]

System	SCF	PNO-CEPA	Experiment
Li	170.3	164.5	164.0 ± 3.4
Na	192.8	165.0	159.2 ± 3.4
K	418.0	287.6	292.8 ± 6.1
Ne	2.368	2.676	2.669
Ar	10.69	11.10	11.08
HF	4.98	5.67	5.60
H_2O	8.68	9.86	9.82
NH_3	13.61	14.96	14.82
CH_4	16.69	17.22	17.3
CO	12.40	13.13	13.1

[a] Theoretical values taken from Werner and Meyer[64] and Reinsch and Meyer[65] for the atoms and Werner and Meyer[63] for the molecules. See these references also for the experimental values quoted here. Werner and Meyer[64] and Reinsch and Meyer[65] give calculated polarizabilities for all atoms up to calcium, but listed here are only those for which accurate experimental values are available. The molecular results include an approximate treatment of the zero-point vibration. All values in a_0^3.

standard PNO expansion of the wave function is extended by adding off-diagonal configurations for the most heavily occupied PNOs. The number of these configurations is small though and does not increase the computational effort significantly. The effect of the additional configurations decreases rapidly toward the right-hand side of the periodic table and has been found to be below 1% of the total dipole moment for saturated polyatomic hydrides. For the diatomic hydrides, agreement with the available experimental data is seen to be quite satisfactory. The remaining errors may well be due to basic set defects. For the hydrides H_2O and NH_3 we have obtained dipole moments from basis sets that have been designed for the calculation of polarizabilities and have been tested to be saturated with respect to the dipole moment. Table 8 shows that in these cases agreement with experiment is excellent after vibrational averaging has been taken into account.

Static dipole polarizabilities have been calculated for the atoms up to $Ca^{(64,65)}$ and the molecules HF, H_2O, NH_3, CH_4, and $CO^{(63)}$ by means of the finite perturbation method[71,72]: The dipole moment is calculated under the presence of a finite perturbing electric field and the polarizability tensor $d\mu_i/df_j$ is obtained by numerical differentiation. In this way, the second-order property can be obtained just by applying the standard procedures for constructing ground state wave functions. As an example, Table 9 shows for HF the convergence behavior of correlation energy, dipole moment, and polarizability as a function of the threshold value for configuration selection. It is interesting to note that the first few configurations overcorrect the SCF error. For the two dipole properties convergence appears to have been reached at a threshold of 10^{-3} hartree or about 65% of the valence shell correlation, which is obtained

Table 9. *Convergence Behavior of Correlation Energy, Dipole Moment, and Polarizability with Decreasing Energy Threshold for Configuration Selection for HF*[a]

Energy threshold	$-E_{corr}$	μ_e	$\alpha_{e\parallel}$	Number of double substitutions
SCF	0.0	0.7572	5.751	0
10^{-2}	0.07902	0.6691	6.388	6
$10^{-2.5}$	0.16966	0.7046	6.372	23
10^{-3}	0.20225	0.7105	6.289	43
$10^{-3.6}$	0.22350	0.7132	6.276	79
10^{-4}	0.22802	0.7134	6.271	107

[a] PNO-CI; all single substitutions included; see Werner and Meyer[64] for basis set.

with 45 spin-adapted configurations in this case. The calculated polarizabilities given in Table 8 show deviations from experiment of less than 2% after inclusion of vibrational effects, which amount to 1 to 3%. The correlation effects are found to contribute between 4 and 13%.

6. Conclusions

We have described a method for deriving many-electron wave functions in their most compact form. Perturbation methods are used to obtain sets of approximate natural orbitals for each external pair. These orbitals are used to describe the external electron pairs by their natural expansion, the coefficients of which are varied to make the energy stationary. In this way, the number of Hamiltonian matrix elements to be calculated individually is reduced drastically, but sufficient flexibility is left for improving the expansion coefficients beyond their first-order perturbation estimate. Computationally, the advantages of this method consist in avoiding a complete integral transformation and in reducing the size of the large data sets usually required for the two-electron orbital integrals and the CI matrix elements. Thus, setting up the numerical Hamiltonian matrix becomes very simple. The nonorthogonality of the pseudonatural orbitals of different pairs has been shown to introduce only minor complications independent of the complexity of the reference function.

The approximations inherent in using the PNOs in place of the true natural orbitals of the external pairs imply errors in the calculated correlation energies of only 0.5 to 2%. They may be much more severe for calculating correlation effects on properties, however. A further development of the PNO-CE method therefore asks for an improvement of the PNOs. This can be done only iteratively by calculating the natural orbitals in the effective field of the other external pairs. The method of self-consistent electron pairs recently developed by the author[24] achieves this goal by simultaneously improving orbital and configuration expansion coefficients. The appropriate formulation of this method is in terms of external pair expansion coefficients referring directly to the basis functions, so that the natural orbitals lose their significance. We have therefore omitted discussing this new method here.

References

1. P.-O. Löwdin, *Phys. Rev.* **97**, 1474 (1955).
2. C. F. Bender and E. R. Davidson, *Phys. Rev.* **138**, 23 (1969).
3. E. R. Davidson, *Reduced Density Matrices in Quantum Chemistry*, Academic Press, New York (1975).
4. C. F. Bender and E. R. Davidson, *J. Chem. Phys.* **47**, 4972 (1967).
5. A. K. Q. Siu and E. R. Davidson, *Int. J. Quant. Chem.* **4**, 223 (1974).
6. R. McWeeny and B. T. Sutcliffe, *Methods of Molecular Quantum Mechanics*, Academic Press, London (1969).
7. V. V. Kibartas, V. I. Karetskis, and A. P. Yutsis, *Zhur. Eksp. Teor. Fis.* **29**, 623 (1955) [*Transl.: JETP* **2**, 476 (1956)].
8. R. Ahlrichs and W. Kutzelnigg, *J. Chem. Phys.* **49**, 1819 (1968).
9. W. Meyer, *Int. J. Quant. Chem.* **S5**, 341 (1971).
10. W. Meyer, *J. Chem. Phys.* **58**, 1017 (1973).
11. C. Edmiston and M. Krauss, *J. Chem. Phys.* **42**, 1119 (1965); **45**, 1833 (1966).
12. H. J. Silverstone and O. Sinanoğlu, *J. Chem. Phys.* **44**, 1899, 3608 (1966).
13. L. Brillouin, *Actualités Sci. Ind.* **159** (1934).
14. B. Levy, *Int. J. Quant. Chem.* **IV**, 297 (1970).
15. R. K. Nesbet, *Phys. Rev.* **109**, 1632 (1958); *Adv. Chem. Phys.* **9**, 311 (1965); **14**, 1 (1969).
16. E. R. Davidson, *in*: *The World of Quantum Chemistry* (L. R. Daudel and P. Pullman, eds.), p. 17, Reidel, Dordrecht, Holland (1974).
17. O. Sinanoğlu, *J. Chem. Phys.* **36**, 706 (1962).
18. H. P. Kelly, *Phys. Rev.* **131**, 684 (1963); **144**, 39 (1966); *Adv. Chem. Phys.* **XIV**, 129 (1969) and references therein.
19. J. Čižek, *J. Chem. Phys.* **45**, 4256 (1966); *Adv. Chem. Phys.* **14**, 35 (1969).
20. W. Meyer, *Theor. Chim. Acta* **35**, 277 (1974).
21. C. Møller and M. S. Plesset, *Phys. Rev.* **46**, 618 (1934).
22. R. K. Nesbet, *Proc. Roy. Soc.* (*London*) **A230**, 312 (1955).
23. B. Roos, *Chem. Phys. Lett.* **15**, 153 (1972).
24. W. Meyer, *J. Chem. Phys.* **64**, 2901 (1976).
25. N. H. March, W. H. Young, and S. Shampanthar, *The Many-Body Problem in Quantum Mechanics*, Cambridge University Press, Cambridge (1967).
26. A. C. Hurley, J. Lennard-Jones, and J. A. Pople, *Proc. Roy. Soc.* (*London*) **A220**, 446 (1953).
27. P.-O. Löwdin and H. Shull, *Phys. Rev.* **101**, 1730 (1956).
28. H. Shull, *J. Chem. Phys.* **30**, 1405 (1959).
29. A. J. Coleman, *Rev. Mod. Phys.* **35**, 668 (1963).
30. P. J. Hay, *J. Chem. Phys.* **59**, 2468 (1973).
31. A. K. Q. Siu and E. F. Hayes, *J. Chem. Phys.* **61**, 37 (1974).
32. O. Sinanoğlu, *Adv. Chem. Phys.* **14**, 237 (1969).
33. K. J. Miller and K. Ruedenberg, *J. Chem. Phys.* **48**, 3414 (1968).
34. D. M. Silver, K. Ruedenberg, and E. L. Mehler, *J. Chem. Phys.* **52**, 1206 (1970).
35. O. Sinanoğlu and B. Skutnik, *Chem. Phys. Lett.* **1**, 699 (1968).
36. M. Jungen and R. Ahlrichs, *Theor. Chim. Acta* (*Berlin*) **17**, 339 (1970).
37. A. W. Weiss, *Phys. Rev.* **162**, 71 (1967); **188**, 119 (1968); **A3**, 126 (1971).
38. B. Liu, *Int. J. Quant. Chem.* **S5**, 123 (1971).
39. R. Ahlrichs and W. Kutzelnigg, *Theor. Chim. Acta* **10**, 377 (1968).
40. M. Gelus, R. Ahlrichs, V. Staemmler, and W. Kutzelnigg, *Chem. Phys. Lett.* **7**, 503 (1970); *Theor. Chim. Acta* **21**, 63 (1971).

41. E. R. Davidson and C. F. Bender, *J. Chem. Phys.* **49**, 465 (1968).
42. R. K. Nesbet, T. L. Barr, and E. R. Davidson, *Chem. Phys. Lett.* **4**, 203 (1969).
43. R. Ahlrichs, F. Driessler, H. Lischka, V. Staemmler, and W. Kutzelnigg, *J. Chem. Phys.* **62**, 1235 (1975).
44. M. Krauss and A. W. Weiss, *J. Chem. Phys.* **40**, 80 (1964).
45. A. Bunge, *J. Chem. Phys.* **53**, 20 (1970).
46. A. D. McLean and B. Liu, *J. Chem. Phys.* **58**, 1066 (1973).
47. R. McWeeny and E. Steiner, *Adv. Quant. Chem.* **2**, 93 (1975).
48. R. McWeeny, *Rev. Mod. Phys.* **32**, 335 (1960).
49. W. Kutzelnigg, *Theor. Chim. Acta* **1**, 327 (1963).
50. W. Kutzelnigg, *in: Selected Topics in Molecular Physics* (E. Clementi, ed.), pp. 91–102, Verlag Chemie, Weinheim (1972).
51. R. Ahlrichs, W. Kutzelnigg, and W. A. Bingel, *Theor. Chim. Acta* **5**, 289, 305 (1966).
52. R. Ahlrichs and F. Driessler, *Theoret. Chim. Acta* **36**, 275 (1975).
53. R. Ahlrichs, H. Lischka, V. Staemmler, and W. Kutzelnigg, *J. Chem. Phys.* **62**, 1225 (1975).
54. H. F. Shaefer III, *The Electronic Structure of Atoms and Molecules: A Survey of Rigorous Quantum Mechanical Results*, Addison-Wesley, Reading, Massachusetts (1972).
55. P. Rosmus and W. Meyer, to be published.
56. P. Rosmus, B. Solouki, and H. Bock, *Chem. Phys.* (in press).
57. W. Meyer and P. Rosmus, *J. Chem. Phys.* **63**, 2356 (1975).
58. P. Rosmus and W. Meyer, *J. Chem. Phys.* (in press).
59. W. Meyer, *Proc. SRC Atlas Symp. No. 4, Quantum Chemistry—The State of the Art*, Chilton, Berkshire, England (1974).
60. R. Ahlrichs, H. Lischka, B. Zurawski, and W. Kutzelnigg, *J. Chem. Phys.* **63**, 4685 (1975).
61. P. Pulay and W. Meyer, to be published.
62. R. Ahlrichs, F. Keil, H. Lischka, W. Kutzelnigg, and V. Staemmler, *J. Chem. Phys.* **63**, 455 (1975).
63. H.-J. Werner and W. Meyer, *Mol. Phys.* **31**, 855 (1976).
64. H.-J. Werner and W. Meyer, *Phys. Rev. A* **13**, 13 (1976).
65. E.-A. Reinsch and W. Meyer, *Phys. Rev. A* **14**, 915 (1976).
66. F. Sasaki and M. Yoshimine, *Phys. Rev. A* **9**, 17 (1974).
67. R. S. Mulliken, *J. Chem. Phys.* **46**, 497 (1949).
68. P. S. Bagus, *Phys. Rev.* **139**, A619 (1965).
69. I. Öksüz and O. Sinanoğlu, *Phys. Rev.* **181**, 42, 54 (1969).
70. E. L. Mehler and W. Meyer, *Chem. Phys. Lett.* **38**, 144 (1976).
71. H. D. Cohen and C. C. J. Roothaan, *J. Chem. Phys.* **44**, 505 (1966).
72. J. A. Pople, J. W. McIver Jr., and N. S. Ostlund, *J. Chem. Phys.* **49**, 2960 (1968).

Author Index

Boldface page numbers indicate a chapter in this volume.

447

Subject Index

455

Date Due

BJJH
